MATLAB® for Engineers

MATLAB® for Engineers

Third Edition

HOLLY MOORE
Salt Lake Community College
Salt Lake City, Utah

PEARSON

Boston • Columbus • Indianapolis • New York
San Francisco • Upper Saddle River • Amsterdam
Cape Town • Dubai • London • Madrid • Milan
Munich • Paris • Montreal • Toronto • Delhi
Mexico City • Sao Paulo • Sydney • Hong Kong
Seoul • Singapore • Taipei • Tokyo

Vice President and Editorial Director, Engineering/Computer Science: *Marcia J. Horton*
Executive Editor: *Holly Stark*
Editorial Assistant: *William Opaluch*
Marketing Manager: *Tim Galligan*
Production Manager: *Pat Brown*
Art Director: *Jayne Conte*
Cover Designer: *Bruce Kenselaar*
Media Editor: *Daniel Sandin*
Full-Service Project Management: *Pavithra Jayapaul, TexTech International*
Composition: *TexTech International*
Printer/Binder: *Edwards Brothers*
Cover Printer: *Lehigh-Phoenix Color/Hagerstown*

Credits and acknowledgments borrowed from other sources and reproduced, with permission, in this textbook appear on appropriate page within text.

MATLAB® and Simulink® are registered trademarks of The Mathworks, Inc., 3 Apple Hill Drive, Natick MA 01760-2098.

Many of the designations by manufacturers and seller to distinguish their products are claimed as trademarks. Where those designations appear in this book, and the publisher was aware of a trademark claim, the designations have been printed in initial caps or all caps.

Library of Congress Cataloging–in–Publication Data

Moore, Holly.
  MATLAB® for engineers / Holly Moore. — 3rd ed.
      p. cm.
  Includes index.
  ISBN-13: 978-0-13-210325-1
  ISBN-10: 0-13-210325-7
    1. Engineering mathematics—Data processing.   2. MATLAB®.   I. Title.
TA345.M585 2011
620.001'51—dc23

                                     2011022739

10 9 8 7 6 5 4

ISBN 10: 0-13-210325-7
ISBN 13: 978-0-13-210325-1

Contents

4 • MANIPULATING MATLAB® MATRICES   121

5 • PLOTTING   149

6 • USER-DEFINED FUNCTIONS   205

7 • USER-CONTROLLED INPUT AND OUTPUT   240

About This Book

This book grew out of my experience teaching MATLAB® and other computing languages to freshmen engineering students at Salt Lake Community College. I was frustrated by the lack of a text that "started at the beginning." Although there were many comprehensive reference books, they assumed a level of both mathematical and computer sophistication that my students did not possess. Also, because MATLAB® was originally adopted by practitioners in the fields of signal processing and electrical engineering, most of these texts provided examples primarily from those areas, an approach that didn't fit with a general engineering curriculum. This text starts with basic algebra and shows how MATLAB® can be used to solve engineering problems from a wide range of disciplines. The examples are drawn from concepts introduced in early chemistry and physics classes and freshman and sophomore engineering classes. A standard problem-solving methodology is used consistently.

The text assumes that the student has a basic understanding of college algebra and has been introduced to trigonometric concepts; students who are mathematically more advanced generally progress through the material more rapidly. Although the text is not intended to teach subjects such as statistics or matrix algebra, when the MATLAB® techniques related to these subjects are introduced, a brief background is included. In addition, sections describing MATLAB® techniques for solving problems by means of calculus and differential equations are introduced near the end of appropriate chapters. These sections can be assigned for additional study to students with a more advanced mathematics background, or they may be useful as reference material as students progress through an engineering curriculum.

The book is intended to be a "hands-on" manual. My students have been most successful when they read the book while sitting beside a computer and typing in the examples as they go. Numerous examples are embedded in the text, with more complicated numbered examples included in each chapter to reinforce the concepts introduced. Practice exercises are included in each chapter to give students an immediate opportunity to use their new skills, and complete solutions are available online at: www.pearsonhighered.com/moore.

The material is grouped into three sections. The first, *An Introduction to Basic MATLAB® Skills*, gets the student started and contains the following chapters:

- Chapter 1 shows how MATLAB® is used in engineering and introduces a standard problem-solving methodology.
- Chapter 2 introduces the MATLAB® environment and the skills required to perform basic computations. This chapter also introduces M-files, and the concept of organizing code into cells. Doing so early in the text makes it easier for students to save their work and develop a consistent programming strategy.
- Chapter 3 details the wide variety of problems that can be solved with built-in MATLAB® functions. Background material on many of the functions is provided to help the student understand how they might be used. For example, the difference between Gaussian random numbers and uniform random numbers is described, and examples of each are presented.

- Chapter 4 demonstrates the power of formulating problems by using matrices in MATLAB® and expanding on the techniques employed to define those matrices. The **meshgrid** function is introduced in this chapter and is used to solve problems with two variables. The difficult concept of meshing variables is revisited in Chapter 5 when surface plots are introduced.
- Chapter 5 describes the wide variety of both two-dimensional and three-dimensional plotting techniques available in MATLAB®. Creating plots via MATLAB® commands, either from the command window or from within an M-file, is emphasized. However, the extremely valuable techniques of interactively editing plots and creating plots directly from the workspace window are also introduced.

  MATLAB® is a powerful programming language that includes the basic constructs common to most programming languages. Because it is a scripting language, creating programs and debugging them in MATLAB® is often easier than in traditional programming languages such as C++. This makes MATLAB® a valuable tool for introductory programming classes. The second section of the text, *Programming in MATLAB®,* introduces students to programming and consists of the following chapters:

- Chapter 6 describes how to create and use user-defined functions. This chapter also teaches students how to create a "toolbox" of functions to use in their own programming projects.
- Chapter 7 introduces functions that interact with the program user, including user-defined input, formatted output, and graphical input techniques. The use of MATLAB®'s debugging tools is also introduced.
- Chapter 8 describes logical functions such as **find** and demonstrates how they vary from the **if** and **if/else** structures. The **switch case** structure is also introduced. The use of logical functions over control structures is emphasized, partly because students (and teachers) who have previous programming experience often overlook the advantages of using MATLAB®'s built-in matrix functionality.
- Chapter 9 introduces repetition structures, including **for** loops, **while** loops, and midpoint break loops which utilize the **break** command. Numerous examples are included because students find these concepts particularly challenging.

Chapters 1 through 9 should be taught sequentially, but the chapters in Section 3, *Advanced MATLAB® Concepts,* do not depend upon each other. Any or all of these chapters could be used in an introductory course or could serve as reference material for self-study. Most of the material is appropriate for freshmen. A two-credit course might include Chapters 1 through 9 plus Chapter 10, while a three-credit course might include Chapters 1 through 14, but eliminate Sections 12.4, 12.5, 13.4, 13.5, and 13.6, which describe differentiation techniques, integration techniques, and solution techniques for differential equations. Chapters 15 and 16 will be interesting to more advanced students, and might be included in a course delivered to sophomore or junior students instead of to freshmen. The skills developed in these will be especially useful as students become more involved in solving engineering problems:

- Chapter 10 discusses problem solving with matrix algebra, including dot products, cross products, and the solution of linear systems of equations. Although matrix algebra is widely used in all engineering fields, it finds early application in the statics and dynamics classes taken by most engineering majors.

- Chapter 11 is an introduction to the wide variety of data types available in MATLAB®. This chapter is especially useful for electrical engineering and computer engineering students.
- Chapter 12 introduces MATLAB®'s symbolic mathematics package, built on the MuPad engine. Students will find this material especially valuable in mathematics classes. My students tell me that the package is one of the most valuable sets of techniques introduced in the course. It is something they start using immediately.
- Chapter 13 presents numerical techniques used in a wide variety of applications, especially curve fitting and statistics. Students value these techniques when they take laboratory classes such as chemistry or physics or when they take the labs associated with engineering classes such as heat transfer, fluid dynamics, or strengths of materials.
- Chapter 14 examines graphical techniques used to visualize data. These techniques are especially useful for analyzing the results of numerical analysis calculations, including results from structural analysis, fluid dynamics, and heat transfer codes.
- Chapter 15 introduces MATLAB®'s graphical user interface capability, using the GUIDE application. Creating their own GUI's gives students insight into how the graphical user interfaces they use daily on other computer platforms are created.
- Chapter 16 introduces Simulink®, which is a simulation package built on top of the MATLAB® platform. Simulink® uses a graphical user interface that allows programmers to build models of dynamic systems. Simulink® has found significant acceptance in the field of Electrical Engineering but has wide application across the engineering spectrum.

Appendix A lists all of the functions and special symbols (or characters) introduced in the text. Appendix B describes strategies for scaling data, so that the resulting plots are linear. Appendix C includes the complete MATLAB® code to create the Ready_Aim_Fire graphical user interface described in Chapter 15. An instructor web-site includes the following material:

- M-files containing solutions to practice exercises
- M-files containing solutions to example problems
- M-files containing solutions to homework problems
- PowerPoint slides for each chapter
- All of the figures used in the text, suitable for inclusion in your own PowerPoint presentations
- A series of lectures (including narration) suitable for use with online classes or as reviews

ABOUT THE THIRD EDITION

New versions of MATLAB® are rolled out every 6 months, which makes keeping any text up-to-date a challenge. The major changes included in this edition are as follows:

- All of the screen shots throughout the book were updated to reflect the 2011a release.
- The introduction to cell mode was moved to Chapter 2 from Chapter 7. The description of the cell mode publishing features was expanded and updated in Chapter 7.

- Information on debugging features was added to Chapters 7 and 8.
- Based on student and instructor feedback, Chapter 8 was significantly revised and split into two chapters.
  - The new Chapter 8 introduces MATLAB®'s logical functions such as **find**, and the more traditional selection structures **if**, **if/else**, and **switch/case**.
  - The new Chapter 9 deals exclusively with repetition structures.
- The symbolic toolbox was changed significantly in the 2007b edition, which required changes to the symbolic algebra materials in Chapter 12.
- Two additional chapters were added in an attempt to make the text useful to a wider audience.
  - Chapter 15 describes graphical user interfaces.
  - Chapter 16 is an introduction to Simulink®.
- Problems were added at the end of each chapter.
- Additional example problems were added.
- A number of new functions are introduced throughout the book, suggested to us by adopters of the text.

Dedication and Acknowledgments

This project would not have been possible without the support of my family, which endured reading multiple drafts of the text and ate a lot of frozen pizza while I concentrated on writing. Thanks to Mike, Heidi, Meagan, and David, and to my husband, Dr. Steven Purcell. I also benefited greatly from the suggestions for problems related to electricity from Lee Brinton and Gene Riggs of the SLCC Electrical Engineering Department. Their cheerful efforts to educate me on the mysteries of electricity are much appreciated. I'd also like to thank Dr. Ghassan Hamarneh for his careful review of the second edition, which helped tremendously as I prepared this latest manuscript.

This book is dedicated to my father, Professor George Moore, who taught in the Department of Electrical Engineering at the South Dakota School of Mines and Technology for almost 20 years. Professor Moore earned his college degree at the age of 54 after a successful career as a pilot in the United States Air Force and was a living reminder that you are never too old to learn. My mother, Jean Moore, encouraged both him and her two daughters to explore outside the box. Her loving support made it possible for both my sister and I to enjoy careers in engineering—something few women attempted in the early 1970s. I hope that readers of this text will take a minute to thank those people in their lives who've helped them make their dreams come true. Thanks Mom and Dad.

1

About MATLAB®

Objectives

After reading this chapter, you should be able to:

- Understand what MATLAB® is and why it is widely used in engineering and science

- Understand the advantages and limitations of the student edition of MATLAB®
- Formulate problems by using a structured problem-solving approach

1.1 WHAT IS MATLAB®?

MATLAB® is one of a number of commercially available, sophisticated mathematical computation tools, which also include Maple, Mathematica, and MathCad. Despite what proponents may claim, no single one of these tools is "the best." Each has strengths and weaknesses. Each allows you to perform basic mathematical computations. They differ in the way they handle symbolic calculations and more complicated mathematical processes, such as matrix manipulation. For example, MATLAB® (short for **Mat**rix **Lab**oratory) excels at computations involving matrices, whereas Maple excels at symbolic calculations. At a fundamental level, you can think of these programs as sophisticated computer-based calculators. They can perform the same functions as your scientific calculator—and **many more**. If you have a computer on your desk, you may find yourself using MATLAB® instead of your calculator for even the simplest mathematical applications—for example, balancing your checkbook. In many engineering classes, the use of programs such as MATLAB® to perform computations is replacing more traditional computer programming. Although programs such as MATLAB® have become a standard tool for engineers and scientists, this doesn't mean that you shouldn't learn a high-level language such as C++, JAVA, or FORTRAN.

Because MATLAB® is so easy to use, you can perform many programming tasks with it, but it isn't always the best tool for a programming task. It excels at numerical calculations—especially matrix calculations—and graphics, but you wouldn't want to use it to write a word-processing program. For large applications, such as operating systems or design software, C++, JAVA, or FORTRAN would be the programs of choice. (In fact, MATLAB®, which *is* a large application program, was originally written in FORTRAN and later rewritten in C, a precursor of C++.) Usually, high-level programs do not offer easy access to graphing—an application at which MATLAB® excels. The primary area of overlap between MATLAB® and high-level programs is "number crunching"—repetitive calculations or the processing of large quantities of data. Both MATLAB® and high-level programs are good at processing numbers. A "number-crunching" program is generally easier to write in MATLAB®, but usually it will execute faster in C++ or FORTRAN. The one exception to this rule is calculations involving matrices. MATLAB® is optimized for matrices. Thus, if a problem can be formulated with a matrix solution, MATLAB® executes substantially faster than a similar program in a high-level language.

KEY IDEA
MATLAB® is optimized for matrix calculations

MATLAB® is available in both a professional and a student version. The professional version is probably installed in your college or university computer laboratory, but you may enjoy having the student version at home. MATLAB® is updated regularly; this textbook is based on MATLAB® 7.12. If you are using earlier versions such as MATLAB® 6, you may notice some minor differences between it and MATLAB® 7.12. There are substantial differences in versions that predate MATLAB® 5.5.

The standard installation of the professional version of MATLAB® is capable of solving a wide variety of technical problems. Additional capability is available in the form of function toolboxes. These toolboxes are purchased separately, and they may or may not be available to you. You can find a complete list of the MATLAB® product family at The MathWorks web site, www.mathworks.com.

1.2 STUDENT EDITION OF MATLAB®

The professional and student editions of MATLAB® are very similar. Beginning students probably won't be able to tell the difference. Student editions are available for Microsoft Windows, Mac OSX, and Linux operating systems and can be purchased from college bookstores or online from The MathWorks at www.mathworks.com.

KEY IDEA
MATLAB® is regularly updated

The MathWorks packages its software in groups called *releases*, and MATLAB® 7.12 is featured, along with other products, such as Simulink® 7.7, in Release R2011a. New versions are released every 6 months. The release number is the same for both the student and professional edition, but the student version may lag the professional version by several months. The student edition of R2011a includes the following features:

- Full MATLAB®
- Simulink®, with the ability to build models with up to 1000 blocks (the professional version allows an unlimited number of blocks)
- Symbolic Math Toolbox
- Control System Toolbox
- Signal Processing Toolbox
- DSP System Toolbox
- Statistics Toolbox
- Optimization Toolbox
- Image Processing Toolbox
- Software manuals for both MATLAB® 7 and Simulink®
- A CD containing the full electronic documentation
- A single-user license, limited to students for use in their classwork (the professional version is licensed either singly or to a group)

Toolboxes other than those included with the student edition may be purchased separately. You should be aware that if you are using a professional installation of MATLAB®, all of the toolboxes available in the student edition may not be available to you.

The biggest difference you should notice between the professional and student editions is the command prompt, which is

```
>>
```

in the professional version and

```
EDU>>
```

in the student edition.

1.3 HOW IS MATLAB® USED IN INDUSTRY?

The ability to use tools such as MATLAB® is quickly becoming a requirement for many engineering positions. A recent job search on Monster.com found the following advertisement:

> . . . is looking for a System Test Engineer with Avionics experience. . . . Responsibilities include modification of MATLAB® scripts, execution of Simulink® simulations, and analysis of the results data. Candidate MUST be very familiar with MATLAB®, Simulink®, and C++. . .

This ad isn't unusual. The same search turned up 660 different companies that specifically required MATLAB® skills for entry-level engineers. Widely used in all engineering and science fields, MATLAB® is particularly popular for electrical engineering applications. The sections that follow outline a few of the many applications currently using MATLAB®.

KEY IDEA
MATLAB® is widely used in engineering

1.3.1 Electrical Engineering

MATLAB® is used extensively in electrical engineering for signal-processing applications. For example, Figure 1.1 includes several images created during a research program at the University of Utah to simulate collision-detection algorithms used by the housefly (and adapted to silicon sensors in the laboratory). The research resulted in the design and manufacture of a computer chip that detects imminent collisions. This has potential use in the design of autonomous robots using vision for navigation and especially in automobile safety applications.

1.3.2 Biomedical Engineering

Medical images are usually saved as dicom files (the Digital Imaging and Communications in Medicine standard). Dicom files use the file extension .dcm.

Figure 1.1
Image processing using a fisheye lens camera to simulate the visual system of a housefly's brain. (Used by permission of Dr. Reid Harrison, University of Utah.)

Figure 1.2
Horizontal slices through the brain, based on the sample data file included with MATLAB®.

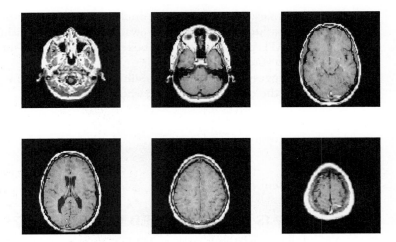

The MathWorks offers an Image Processing Toolbox that can read these files, making their data available to MATLAB®. (The Image Processing Toolbox is included with the student edition and is optional with the professional edition.) The Image Processing Toolbox also includes a wide range of functions, many of them especially appropriate for medical imaging. A limited MRI data set that has already been converted to a format compatible with MATLAB® ships with the standard MATLAB® program. This data set allows you to try out some of the imaging functions available both with the standard MATLAB® installation and with the expanded imaging toolbox, if you have it installed on your computer. Figure 1.2 shows six images of horizontal slices through the brain based on the MRI data set.

The same data set can be used to construct a three-dimensional image, such as either of those shown in Figure 1.3. Detailed instructions on how to create these images are included in the MATLAB® tutorial, accessed from the help button on the MATLAB® toolbar.

1.3.3 Fluid Dynamics

Calculations describing fluid velocities (speeds and directions) are important in a number of different fields. Aerospace engineers in particular are interested in the behavior of gases, both outside an aircraft or space vehicle and inside the combustion chambers. Visualizing the three-dimensional behavior of fluids is tricky, but MATLAB®

Figure 1.3
Three-dimensional visualization of MRI data, based on the sample data set included with MATLAB®.

Figure 1.4
Quiver plot of gas behavior in a thrust-vector control device.

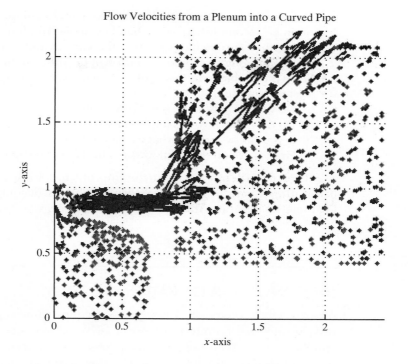

Flow Velocities from a Plenum into a Curved Pipe

offers a number of tools that make it easier. In Figure 1.4, the flow-field calculation results for a thrust-vector control device are represented as a quiver plot. Thrust-vector control is the process of changing the direction in which a nozzle points (and hence the direction a rocket travels) by pushing on an actuator (a piston-cylinder device). The model in the figure represents a high-pressure reservoir of gas (a plenum) that eventually feeds into the piston and thus controls the length of the actuator.

KEY IDEA

Always use a systematic problem-solving strategy

1.4 PROBLEM SOLVING IN ENGINEERING AND SCIENCE

A consistent approach to solving technical problems is important throughout engineering, science, and computer programming disciplines. The approach we outline here is useful in courses as diverse as chemistry, physics, thermodynamics, and engineering design. It also applies to the social sciences, such as economics and sociology. Different authors may formulate their problem-solving schemes differently, but they all have the same basic format:

- **State the problem**.
  - Drawing a picture is often helpful in this step.
  - If you do not have a clear understanding of the problem, you are not likely to be able to solve it.
- **Describe the input** values (knowns) **and** the required **outputs** (unknowns).
  - Be careful to include units as you describe the input and output values. Sloppy handling of units often leads to wrong answers.
  - Identify constants you may need in the calculation, such as the ideal-gas constant and the acceleration due to gravity.
  - If appropriate, label a sketch with the values you have identified, or group them into a table.

- Develop an algorithm to solve the problem. In computer applications, this can often be accomplished with a **hand example**. You'll need to
  - Identify any equations relating the knowns and unknowns.
  - Work through a simplified version of the problem by hand or with a calculator.
- **Solve** the problem. In this book, this step involves creating a **MATLAB® solution**.
- **Test the solution**.
  - Do your results make sense physically?
  - Do they match your sample calculations?
  - Is your answer really what was asked for?
  - Graphs are often useful ways to check your calculations for reasonableness.

If you consistently use a structured problem-solving approach, such as the one just outlined, you'll find that "story" problems become much easier to solve. Example 1.1 illustrates this problem-solving strategy.

EXAMPLE 1.1

THE CONVERSION OF MATTER TO ENERGY

Albert Einstein (Figure 1.5) is arguably the most famous physicist of the 20th century. Einstein was born in Germany in 1879 and attended school in both Germany and Switzerland. While working as a patent clerk in Bern, he developed his famous theory of relativity. Perhaps the best-known physics equation today is his

$$E = mc^2$$

This astonishingly simple equation links the previously separate worlds of matter and energy and can be used to find the amount of energy released as matter is changed in form in both natural and human-made nuclear reactions.

Figure 1.5
Albert Einstein.
(Courtesy of the Library
of Congress, LC-
USZ62-60242.)

The sun radiates 385×10^{24} J/s of energy, all of which is generated by nuclear reactions converting matter to energy. Use MATLAB® and Einstein's equation to determine how much matter must be converted to energy to produce this much radiation in one day.

1. **State the Problem**
   Find the amount of matter necessary to produce the amount of energy radiated by the sun every day.

2. **Describe the Input and Output**

   Input

   Energy: $E = 385 \times 10^{24}$ J/s which must be converted into the total energy radiated during one day

   Speed of light: $c = 3.0 \times 10^{8}$ m/s

   Output

   Mass m in kg

3. **Develop a Hand Example**
   The energy radiated in one day is

   $$385 \times 10^{24} \text{J/s} \times 3600 \text{ s/h} \times 24 \text{ h/day} \times 1 \text{ day} = 3.33 \times 10^{31} \text{ J}$$

   The equation $E = mc^2$ must be solved for m and the values for E and c substituted. We have

   $$m = \frac{E}{c^2}$$

   $$m = \frac{3.33 \times 10^{31} \text{ J}}{(3.0 \times 10^{8} \text{m/s})^2}$$

   $$= 3.7 \times 10^{14} \tfrac{\text{J}}{\text{m}^2\text{s}^2}$$

   We can see from the output criteria that we want the mass in kg, so what went wrong? We need to do one more unit conversion:

   $$1 \text{J} = 1 \text{ kg m}^2/\text{s}^2$$

   $$= 3.7 \times 10^{14} \frac{\text{kg m}^2/\text{s}^2}{\text{m}^2/\text{s}^2} = 3.7 \times 10^{14} \text{ kg}$$

4. **Develop a MATLAB® Solution**
   At this point, you have not learned how to create MATLAB® code. However, you should be able to see from the following sample code that MATLAB® syntax is similar to that used in most algebraic scientific calculators. MATLAB® commands are entered at the prompt (>>), and the results are reported on the next line. The code is as follows:

```
>> E=385e24    The user types in this information
E =
   3.8500e+026    This is the computer's response
>> E=E*3600*24
E =
   3.3264e+031
>> c=3e8
c =
   300000000
```

```
>> m=E/c^2
m =
    3.6960e+014
```

From this point on, we will not show the prompt when describing interactions in the command window.

5. Test the Solution

The MATLAB® solution matches the hand calculation, but do the numbers make sense? Anything times 10^{14} is a really large number. Consider, however, that the mass of the sun is 2×10^{30} kg. We can calculate how long it would take to consume the mass of the sun completely at a rate of 3.7×10^{14} kg/day. We have

$$\text{Time} = \frac{\text{Mass of the sun}}{\text{Rate of consumption}}$$

$$\text{Time} = \frac{2 \times 10^{30} \text{ kg}}{3.7 \times 10^{14} \text{ kg/day}} \times \frac{\text{year}}{365 \text{ days}} = 1.5 \times 10^{13} \text{ years}$$

That's 15 trillion years! We don't need to worry about the sun running out of matter to convert to energy in our lifetimes.

2

MATLAB® Environment

Objectives

After reading this chapter, you should be able to:

- Start the MATLAB® program and solve simple problems in the command window
- Understand MATLAB®'s use of matrices
- Identify and use the various MATLAB® windows
- Define and use simple matrices
- Name and use variables
- Understand the order of operations in MATLAB®

- Understand the difference between scalar, array, and matrix calculations in MATLAB®
- Express numbers in either floating-point or scientific notation
- Adjust the format used to display numbers in the command window
- Save the value of variables used in a MATLAB® session
- Save a series of commands in an M-file

2.1 GETTING STARTED

Using MATLAB® for the first time is easy; mastering it can take years. In this chapter, we will introduce you to the MATLAB® environment and show you how to perform basic mathematical computations. After reading this chapter, you should be able to start using MATLAB® for homework assignments or on the job. Of course, you will be able to do more things as you complete the rest of the chapters.

Because the procedure for installing MATLAB® depends upon your operating system and your computing environment, we will assume that you have already installed MATLAB® on your computer or that you are working in a computing laboratory with MATLAB® already installed. To start MATLAB® in either the Windows or Apple environment, click on the icon on the desktop, or use the start menu to find the program. In the UNIX environment, type Matlab at the shell prompt. No matter how you start it, once MATLAB® opens, you should see the MATLAB® prompt (>> or EDU>>), which tells you that MATLAB® is ready for you to enter a command. When you have finished

Figure 2.1
MATLAB® opening window. The MATLAB® environment consists of a number of windows, four of which open in the default view. Others open as needed during a MATLAB® session.

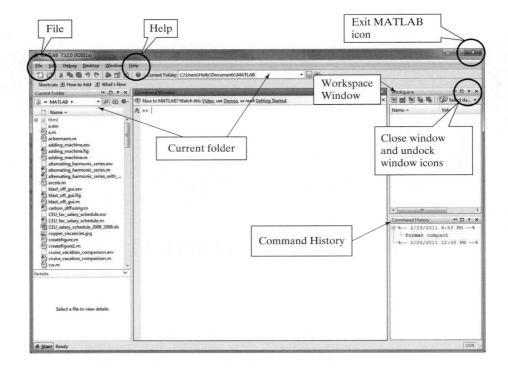

your MATLAB® session, you can exit MATLAB® by typing `quit` or `exit` at the MATLAB® prompt. MATLAB® also uses the standard Windows menu bar, so you can exit the program by choosing EXIT MATLAB from the File menu or by selecting the close icon (**x**) at the upper right-hand corner of the screen. The default MATLAB® screen, which opens each time you start the program, is shown in Figure 2.1.

To start using MATLAB®, you need be concerned only with the command window (in the center of the screen). You can perform calculations in the command window in a manner similar to the way you perform calculations on a scientific calculator. Even most of the syntax is the same. For example, to compute the value of 5 squared, type the command

> **5^2**

The following output will be displayed:

> **ans =**
>
> **25**

Or, to find the value of cos (π), type

> **cos(pi)**

which results in the output

> **ans =**
>
> **-1**

KEY IDEA
MATLAB® uses the standard algebraic rules for order of operation

MATLAB® uses the standard algebraic rules for order of operation, which becomes important when you chain calculations together. These rules are discussed in Section 2.3.2. Notice that the value of pi is built into MATLAB®, so you do not have to enter it yourself.

HINT

You may think some of the examples are too simple to type in yourself—that just reading the material is sufficient. However, you will remember the material better if you both read it and type it!

Before going any further, try Practice Exercise 2.1.

PRACTICE EXERCISE 2.1

Type the following expressions into MATLAB® at the command prompt, and observe the results:

1. 5 + 2
2. 5 * 2
3. 5/2
4. 3 + 2 * (4 + 3)
5. 2.54 * 8/2.6
6. 6.3 − 2.1045
7. 3.6^2
8. 1 + 2^2
9. sqrt(5)
10. cos(pi)

HINT

You may find it frustrating to learn that when you make a mistake, you cannot just overwrite your command after you have executed it. This occurs because the command window is creating a list of all the commands you have entered. You cannot "un-execute" a command, or "un-create" it. What you can do is enter the command correctly and then execute your new version. **MATLAB**® offers several ways to make this easier for you. One way is to use the arrow keys, usually located on the right-hand side of your keyboard. The up arrow, ↑, allows you to move through the list of commands you have executed. Once you find the appropriate command, you can edit it and then execute your new version.

2.2 MATLAB® WINDOWS

MATLAB® uses several display windows. The default view, shown in Figure 2.1, includes in the middle a large *command window*, located on the right, the *command history window* and *workspace* windows, and located on the left the *current folder window*. Older versions of MATLAB® also included a *launch pad* window, which has been replaced by the *start* button in the lower left-hand corner. In addition, *document windows*, *graphics windows*, and *editing windows* will automatically open when needed. Each is described in the sections that follow. MATLAB® also includes a built-in help tutorial that can be accessed from the menu bar, as shown in Figure 2.1. To personalize your desktop, you can resize any of these windows, stack them on top of each other, close the ones you are not using with the close icon (the **x** in the upper right-hand corner of each window), or "undock" them with the undock icon, ⌐, also located in the upper right-hand corner of each window. You can restore the default configuration by selecting Desktop on the menu bar, then navigating to Desktop Layout, and then to Default.

2.2.1 Command Window

KEY IDEA

The command window is similar to a scratch pad

The command window is located in the center pane of the default view of the MATLAB® screen, as shown in Figure 2.1. The command window offers an environment similar to a scratch pad. Using it allows you to save the values you calculate, but not the **commands** used to generate those values. If you want to save the command sequence, you will need to use the editing window to create an **M-file**. M-files are described in Section 2.4.2. Both approaches are valuable. Before we introduce M-files, we will concentrate on using the command window.

2.2.2 Command History

The **command history** window records the commands you issued in the command window. When you exit MATLAB®, or when you issue the `clc` command, the command window is cleared. However, the command history window retains a list of all your commands. You may clear the command history with the edit menu. If you work on a public computer, as a security precaution, MATLAB®'s defaults may be set to clear the history when you exit MATLAB®. If you entered the earlier sample commands listed in this book, notice that they are repeated in the command history window. This window is valuable for a number of reasons, among them that it allows you to review previous MATLAB® sessions and that it can be used to transfer commands to the command window. For example, first clear the contents of the command window by typing

KEY IDEA

The command history records all of the commands issued in the command window

```
clc
```

This action clears the command window but leaves the data in the command history window intact. You can transfer any command from the command history window to the command window by double-clicking (which also executes the command) or by clicking and dragging the line of code into the command window. Try double-clicking

```
cos(pi)
```

in the command history window. The command is copied into the command window and executed. It should return

```
ans =
     -1
```

Now click and drag

```
5^2
```

from the command history window into the command window. The command will not execute until you hit Enter, and then you will get the result:

```
ans =
    25
```

You will find the command history useful as you perform more and more complicated calculations in the command window.

2.2.3 Workspace Window

The *workspace window* keeps track of the *variables* you have defined as you execute commands in the command window. These variables represent values stored in the computer memory, which are available for you to use. If you have been doing the examples, the workspace window should show just one variable, ans, and indicate that it has a value of 25 and is a double array:

Name	Value	Class
⊞ ans	25	double

(Your view of the workspace window may be slightly different, depending on how your installation of MATLAB® is configured.)

Set the workspace window to show more about the displayed variables by right-clicking on the bar with the column labels. (This feature is new to MATLAB® 7 and will not work if you have an older version.) Check size and bytes, in addition to name, value, and class. Your workspace window should now display the following information, although you may need to resize the window to see all the columns:

Name	Value	Size	Bytes	Class
⊞ ans	25	1×1	8	double

The yellow grid-like symbol indicates that the variable **ans** is an array. The size, 1×1, tells us that it is a single value (one row by one column) and therefore a scalar. The array uses 8 bytes of memory. MATLAB® was written in C, and the class designation tells us that in the C language, ans is a double-precision floating-point array. For our needs, it is enough to know that the variable ans can store a floating-point number (a number with a decimal point). Actually, MATLAB® considers every number you enter to be a floating-point number, whether you insert a decimal point or not.

In addition to information about the size of the arrays and type of data stored in them, you can also choose to display statistical information about the data. Once again right click the bar in the workspace window that displays the column headings. Notice that you can select from a number of different statistical measures, such as the max, min, and standard deviation.

You can define additional variables in the command window, and they will be listed in the workspace window. For example, typing

```
A = 5
```

returns

```
A =
     5
```

Notice that the variable **A** has been added to the workspace window, which lists variables in alphabetical order. Variables beginning with capital letters are listed first, followed by variables starting with lowercase letters.

Name	Value	Size	Bytes	Class
⊞ A	5	1 × 1	8	double
⊞ ans	25	1 × 1	8	double

In Section 2.3.2 we will discuss in detail how to enter matrices into MATLAB®. For now, you can enter a simple one-dimensional matrix by typing

```
B = [1, 2, 3, 4]
```

This command returns

```
B =
    1    2    3    4
```

The commas are optional; you would get the same result with

```
B = [1 2 3 4]
B =
    1    2    3    4
```

Notice that the variable **B** has been added to the workspace window and that it is a 1 × 4 array:

Name	Value	Size	Bytes	Class
⊞ A	5	1 × 1	8	double
⊞ B	[1 2 3 4]	1 × 4	32	double
⊞ ans	25	1 × 1	8	double

You can define two-dimensional matrices in a similar fashion. Semicolons are used to separate rows. For example,

```
C = [1 2 3 4; 10 20 30 40; 5 10 15 20]
```

returns

```
C =
    1     2     3     4
   10    20    30    40
    5    10    15    20
```

Name	Value	Size	Bytes	Class
⊞ A	5	1 × 1	8	double
⊞ B	[1 2 3 4]	1 × 4	32	double
⊞ C	<3 × 4 double>	3 × 4	96	double
⊞ ans	25	1 × 1	8	double

Notice that **C** appears in the workspace window as a 3 × 4 matrix. To conserve space, the values stored in the matrix are not listed.

You can recall the values for any variable by typing in the variable name. For example, entering

    A

returns

    A =
         5

Although the only variables we have introduced are matrices containing numbers, other types of variables are possible.

In describing the command window, we introduced the `clc` command. This command clears the command window, leaving a blank page for you to work on. However, it does not delete from memory the actual variables you have created. The `clear` command deletes all of the saved variables. The action of the `clear` command is reflected in the workspace window. Try it out by typing

    clear

in the command window. The workspace window is now empty:

Name	Value	Size	Bytes	Class

If you suppress the workspace window (closing it either from the file menu or with the close icon in the upper right-hand corner of the window), you can still find out which variables have been defined by using the `whos` command:

    whos

If executed before we entered the `clear` command, `whos` would have returned

Name	Size	Bytes	Class
A	1×1	8	double
B	1×4	32	double
C	3×4	96	double
ans	1×1	8	double

2.2.4 Current Folder Window

The current folder window lists all the files in the active directory. When MATLAB® either accesses files or saves information, it uses the current folder unless told differently. The default for the location of the current folder varies with your version of the software and the way it was installed. However, the current folder is listed at the top of the main window. The current folder can be changed by selecting another directory from the drop-down list located next to the directory listing or by browsing through your computer files. Browsing is performed with the browse button, located next to the drop-down list (see Figure 2.2).

Figure 2.2
The *Current Folder Window* lists all the files in the active directory. You can change the current folder by using the drop-down menu or the browse button.

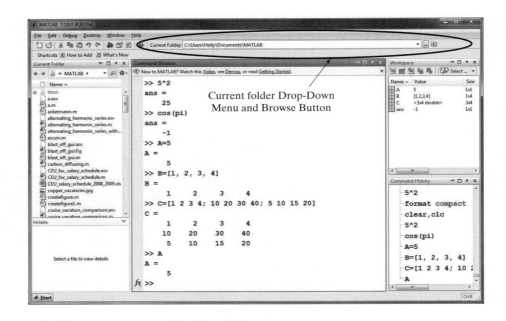

Current folder Drop-Down Menu and Browse Button

KEY IDEA

A semicolon suppresses the output from commands issued in the command window

2.2.5 Document Window

Double-clicking on any variable listed in the workspace window automatically launches a document window, containing the **variable editor**. Values stored in the variable are displayed in a spreadsheet format. You can change values in the array editor, or you can add new values. For example, if you have not already entered the two-dimensional matrix C, enter the following command in the command window:

```
C = [1 2 3 4; 10 20 30 40; 5 10 15 20];
```

Placing a semicolon at the end of the command suppresses the output so that it is not repeated in the command window. However, **C** should now be listed in the workspace window. If you double-click on it, a document window will open above the command window, as shown in Figure 2.3. You can now add more values to the **C** matrix or change existing values.

The document window/variable editor can also be used in conjunction with the workspace window to create entirely new arrays. Run your mouse slowly over the icons in the shortcut bar at the top of the workspace window. If you are patient, you should see the function of each icon appear. The new variable icon looks like a grid with a large asterisk behind it. Select the new variable icon, and a new variable called unnamed should appear on the variable list. You can change its name by right-clicking and selecting **rename** from the pop-up menu. To add values to this new variable, double-click on it and add your data from the array editor window. The new variable button is a new feature in MATLAB® 7; if you are using an older version, you will not be able to create variables this way.

When you are finished creating new variables, close the array editor by selecting the close window icon in the upper right-hand corner of the window.

2.2.6 Graphics Window

The graphics window launches automatically when you request a graph. To demonstrate this feature, first create an array of x values:

```
x = [1 2 3 4 5];
```

Figure 2.3
The *Document Window* displays the *Variable Editor.*

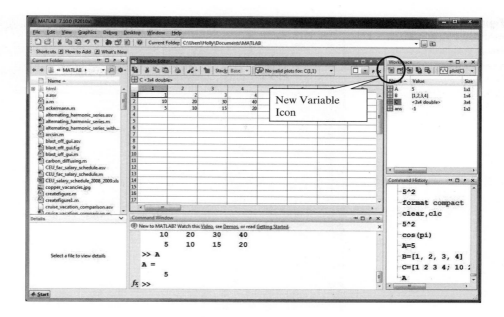

(Remember, the semicolon suppresses the output from this command; however, a new variable, x, appears in the workspace window.)

Now create a list of y values:

```
y = [10 20 30 40 50];
```

To create a graph, use the plot command:

```
plot(x,y)
```

KEY IDEA

Always add a title and axis labels to graphs

The graphics window opens automatically (see Figure 2.4). Notice that a new window label appears on the task bar at the bottom of the windows screen. It will be titled either **<Student Version> Figure...** or simply **Figure 1**, depending on whether you are using the student or professional version, respectively, of the software. Any additional graphs you create will overwrite Figure 1, unless you specifically command MATLAB® to open a new graphics window.

MATLAB® makes it easy to modify graphs by adding titles, *x* and *y* labels, multiple lines, etc. Annotating graphs is covered in a separate chapter on plotting. Engineers and scientists **never** present a graph without labels!

2.2.7 Edit Window

To open the edit window, choose **File** from the menu bar, then **New**, and, finally **Script** (**File** → **New** → **Script**). This window allows you to type and save a series of commands without executing them. You may also open the edit window by typing **edit** at the command prompt or by selecting the **New Script** button on the toolbar.

2.2.8 Start Button

The start button is located in the lower left-hand corner of the MATLAB® window. It offers alternative access to the various MATLAB® windows, as well as to the help function, Internet products, demos and MATLAB® toolboxes. Toolboxes provide additional MATLAB® functionality for specific content areas. The symbolic toolbox in particular is highly useful to scientists and engineers. The start button is new to MATLAB® 7 and replaces the launchpad window used in MATLAB® 6.

Figure 2.4
MATLAB® makes it easy to create graphs.

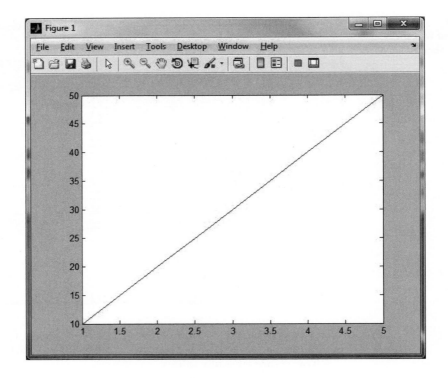

2.3 SOLVING PROBLEMS WITH MATLAB®

The command window environment is a powerful tool for solving engineering problems. To use it effectively, you will need to understand more about how MATLAB® works.

2.3.1 Using Variables

Although you can solve many problems by using MATLAB® like a calculator, it is usually more convenient to give names to the values you are using. MATLAB® uses the naming conventions that are common to most computer programs:

• All names must start with a letter. The names can be of any length, but only the first 63 characters are used in MATLAB® 7. (Use the `namelengthmax` command to confirm this.) Although MATLAB® will let you create long variable names, excessive length creates a significant opportunity for error. A common guideline is to use lowercase letters and numbers in variable names and to use capital letters for the names of constants. However, if a constant is traditionally expressed as a lowercase letter, feel free to follow that convention. For example, in physics textbooks the speed of light is always lowercase c. Names should be short enough to remember and should be descriptive.

• The only allowable characters are letters, numbers, and the underscore. You can check to see if a variable name is allowed by using the `isvarname` command. As is standard in computer languages, the number 1 means that something is true and the number 0 means false. Hence,

```
isvarname time
ans =
    1
```

indicates that `time` is a legitimate variable name, and

```
isvarname cool-beans
ans =
       0
```

tells us that `cool-beans` is not a legitimate variable name. (Recall that the dash is not an allowed character.)

- Names are case sensitive. The variable **x** is different from the variable **X**.

- MATLAB® reserves a list of keywords for use by the program, which you cannot assign as variable names. The `iskeyword` command causes MATLAB® to list these reserved names:

```
iskeyword
ans =
  'break'
  'case'
  'catch'
  'classdef'
  'continue'
  'else'
  'elseif'
  'end'
  'for'
  'function'
  'global'
  'if'
  'otherwise'
  'parfor'
  'persistent'
  'return'
  'spmd'
  'switch'
  'try'
  'while'
```

- MATLAB® allows you to reassign built-in function names as variable names. For example, you could create a new variable called `sin` with the command

```
sin = 4
```

which returns

```
sin =
     4
```

This is clearly a dangerous practice, since the `sin` (i.e., sine) function is no longer available. If you try to use the overwritten function, you'll get an error statement:

```
sin(3)
??? Index exceeds matrix dimensions.
```

You can check to see if a variable is a built-in MATLAB® function by using the `which` command:

```
which sin
sin is a variable.
```

You can reset `sin` back to a function by typing

```
clear sin
```

Now when you ask

```
which sin
```

the response is

```
built-in (C:\ProgramFiles\MATLAB\R2011a\toolbox\matlab\elfun\
  @double\sin)
% double method
```

which tells us the location of the built-in function.

PRACTICE EXERCISE 2.2

Which of the following names are allowed in MATLAB®? Make your predictions, then test them with the `isvarname`, `iskeyword`, and `which` commands.

1. test
2. Test
3. if
4. my-book
5. my_book
6. Thisisoneverylongnamebutisitstillallowed?
7. 1stgroup
8. group_one
9. zzaAbc
10. z34wAwy?12#
11. sin
12. log

KEY IDEA
The matrix is the primary data type in MATLAB® and can hold numeric as well as other types of information

VECTOR
A matrix composed of a single row or a single column

2.3.2 Matrices in MATLAB®

The basic data type used in MATLAB® is the *matrix*. A single value, called a *scalar*, is represented as a 1 × 1 matrix. A list of values, arranged in either a column or a row, is a one-dimensional matrix called a *vector*. A table of values is represented as a two-dimensional matrix. Although we'll limit ourselves to scalars, vectors, and two-dimensional matrices in this chapter, MATLAB® can handle higher order arrays. (The terms matrix and array are used interchangeably by MATLAB® users, even though they are technically different in a mathematical context.)

In mathematical nomenclature, matrices are represented as rows and columns inside square brackets:

$$A = [5] \quad B = [2 \ \ 5] \quad C = \begin{bmatrix} 1 & 2 \\ 5 & 7 \end{bmatrix}$$

In this example, *A* is a 1 × 1 matrix, *B* is a 1 × 2 matrix, and *C* is a 2 × 2 matrix. The advantage in using matrix representation is that whole groups of information can be represented with a single name. Most people feel more comfortable assigning a name to a single value, so we'll start by explaining how MATLAB® handles scalars and then move on to more complicated matrices.

Table 2.1 Arithmetic Operations Between Two Scalars (Binary Operations)

Operation	Algebraic Syntax	MATLAB® Syntax
Addition	$a + b$	**a + b**
Subtraction	$a - b$	**a − b**
Multiplication	$a \times b$	**a * b**
Division	$\dfrac{a}{b}$ or $a \div b$	**a / b**
Exponentiation	a^b	**a^b**

SCALAR

A single-valued matrix

Scalar Operations

MATLAB® handles arithmetic operations between two scalars much as do other computer programs and even your calculator. The syntax for addition, subtraction, multiplication, division, and exponentiation is shown in Table 2.1. The command

```
a = 1 + 2
```

should be read as "a is assigned a value of 1 plus 2," which is the addition of two scalar quantities. Arithmetic operations between two scalar variables use the same syntax. Suppose, for example that you have defined **a** in the previous statement and that b has a value of 5:

```
b = 5
```

Then

```
x = a + b
```

returns the following result:

```
x =
     8
```

A single equals sign (=) is called an assignment operator in MATLAB®. The assignment operator causes the result of your calculations to be stored in a computer memory location. In the preceding example, **x** is assigned a value of 8. If you enter the variable name

```
x
```

into MATLAB®, you get the following result:

```
x =
     8
```

KEY IDEA

The assignment operator is different from an equality

The assignment operator is significantly different from an equality. Consider the statement

```
x = x + 1
```

This is not a valid algebraic statement, since **x** is clearly not equal to **x + 1**. However, when interpreted as an assignment statement, it tells us to replace the current value of **x** stored in memory with a new value that is equal to the old **x** plus **1**.

Since the value stored in **x** was originally 8, the statement returns

```
x =
     9
```

indicating that the value stored in the memory location named **x** has been changed to 9. The assignment statement is similar to the familiar process of saving a file. When you first save a word-processing document, you assign it a name. Subsequently, after you've made changes, you resave your file, but still assign it the same name. The first and second versions are not equal: You've just assigned a new version of your document to an existing memory location.

Order of Operations

In all mathematical calculations, it is important to understand the order in which operations are performed. MATLAB® follows the standard algebraic rules for the order of operation:

- First perform calculations inside parentheses, working from the innermost set to the outermost.
- Next, perform exponentiation operations.
- Then perform multiplication and division operations, working from left to right.
- Finally, perform addition and subtraction operations, working from left to right.

To better understand the importance of the order of operations, consider the calculations involved in finding the surface area of a right circular cylinder.

The surface area is the sum of the areas of the two circular bases and the area of the curved surface between them, as shown in Figure 2.5. If we let the height of the cylinder be 10 cm and the radius 5 cm, the following MATLAB® code can be used to find the surface area:

```
radius = 5;
height = 10;
surface_area = 2*pi*radius^2 + 2*pi*radius*height
```

The code returns

```
surface_area =
            471.2389
```

In this case, MATLAB® first performs the exponentiation, raising the radius to the second power. It then works from left to right, calculating the first product and then the second product. Finally, it adds the two products together. You could instead formulate the expression as

```
surface_area = 2*pi*radius*(radius + height)
```

Figure 2.5
Finding the surface area of a right circular cylinder involves addition, multiplication, and exponentiation.

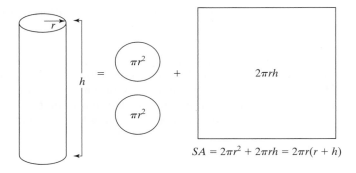

$$SA = 2\pi r^2 + 2\pi rh = 2\pi r(r + h)$$

which also returns

```
surface_area =
        471.2389
```

In this case, MATLAB® first finds the sum of the radius and height and then performs the multiplications, working from left to right. If you forgot to include the parentheses, you would have

```
surface_area = 2*pi*radius*radius + height
```

in which case the program would have first calculated the product of `2*pi*radius*radius` and then added `height`—obviously resulting in the wrong answer. Note that it was necessary to include the multiplication operator before the parentheses, because MATLAB® does not assume any operators and would misinterpret the expression

```
radius(radius + height)
```

as follows. The value of radius plus height is 15 (radius = 10 and height = 5), so MATLAB® would have looked for the 15th value in an array called radius. This interpretation would have resulted in the following error statement.

```
??? Index exceeds matrix dimensions.
```

It is important to be extra careful in converting equations into MATLAB® statements. There is no penalty for adding extra parentheses, and they often make the code easier to interpret, both for the programmer and for others who may use the code in the future. Here's another common error that could be avoided by liberally using parentheses. Consider the following mathematical expression

$$e^{\frac{Q}{RT}}$$

In MATLAB® the mathematical constant e is evaluated as the function, `exp`, so the appropriate syntax is

```
exp(-Q/(R*T))
```

Unfortunately, leaving out the parentheses as in

```
exp(-Q/R*T)
```

gives a very different result. Since the expression is evaluated from left to right, first Q is divided by R, then the result is multiplied by T—not at all what was intended.

Another way to make computer code more readable is to break long expressions into multiple statements. For example, consider the equation

$$f = \frac{\log(ax^2 + bx + c) - \sin(ax^2 + bx + c)}{4\pi x^2 + \cos(x - 2) * (ax^2 + bx + c)}$$

It would be very easy to make an error keying in this equation. To minimize the chance of that happening, break the equation into several pieces. For example, first assign values for x, a, b, and c:

```
x = 9;
a = 1;
b = 3;
c = 5;
```

Then define a polynomial and the denominator:

```
poly = a*x^2 + b*x + c;
denom = 4*pi*x^2 + cos(x - 2)*poly;
```

Combine these components into a final equation:

```
f = (log(poly) - sin(poly))/denom
```

The result is

```
f =
    0.0044
```

As mentioned, this approach minimizes your opportunity for error. Instead of keying in the polynomial three times (and risking an error each time), you need key it in only once. Your MATLAB® code is more likely to be accurate, and it's easier for others to understand.

KEY IDEA

Try to minimize your opportunity for error

HINT

MATLAB® does not read "white space," so you may add spaces to your commands without changing their meaning. A long expression is easier to read if you add a space before and after plus (+) signs and minus (−) signs but not before and after multiplication (*) and division (/) signs.

PRACTICE EXERCISES 2.3

Predict the results of the following MATLAB® expressions, then check your predictions by keying the expressions into the command window:

1. 6/6 + 5
2. 2*6^2
3. (3 + 5) * 2
4. 3 + 5 * 2
5. 4*3/2*8
6. 3 − 2/4 + 6^2
7. 2^3^4
8. 2^(3^4)
9. 3^5 + 2
10. 3^(5 + 2)

Create and test MATLAB® syntax to evaluate the following expressions, then check your answers with a handheld calculator.

11. $\dfrac{5 + 3}{9 - 1}$

12. $2^3 - \dfrac{4}{5 + 3}$

13. $\dfrac{5^{2+1}}{4 - 1}$

14. $4\dfrac{1}{2} * 5\dfrac{2}{3}$

15. $\dfrac{5 + 6 * \dfrac{7}{3} - 2^2}{\dfrac{2}{3} * \dfrac{3}{3 * 6}}$

EXAMPLE 2.1

SCALAR OPERATIONS

Wind tunnels (see Figure 2.6) play an important role in our study of the behavior of high-performance aircraft. In order to interpret wind tunnel data, engineers need to understand how gases behave. The basic equation describing the properties of gases is the ideal gas law, a relationship studied in detail in freshman chemistry classes. The law states that

$$PV = nRT$$

where P = pressure in kPa,
V = volume in m³,
n = number of kmoles of gas in the sample,
R = ideal gas constant, 8.314 kPa m³/kmol K, and
T = temperature, expressed in kelvins (K).

In addition, we know that the number of kmoles of gas is equal to the mass of the gas divided by the molar mass (also known as the molecular weight) or

$$n = m/\text{MW}$$

where
m = mass in kg and
MW = molar mass in kg/kmol.

Different units can be used in the equations if the value of R is changed accordingly.

Figure 2.6
Wind tunnels are used to test aircraft designs. (Louis Bencze/Getty Images Inc., Stone Allstock.)

(*continued*)

Now suppose you know that the volume of air in the wind tunnel is 1000 m³. Before the wind tunnel is turned on, the temperature of the air is 300 K, and the pressure is 100 kPa. The average molar mass (molecular weight) of air is approximately 29 kg/kmol. Find the mass of the air in the wind tunnel.

To solve this problem, use the following problem-solving methodology:

1. **State the Problem**
   When you solve a problem, it is a good idea to restate it in your own words: Find the mass of air in a wind tunnel.

2. **Describe the Input and Output**

   Input

Volume	$V = 1000$ m³
Temperature	$T = 300$ K
Pressure	$P = 100$ kPa
Molecular weight	MW $= 29$ kg/kmol
Gas constant	$R = 8.314$ kPa m³/kmol K

   Output

Mass	$m = ?$ kg

3. **Develop a Hand Example**
   Working the problem by hand (or with a calculator) allows you to outline an algorithm, which you can translate to MATLAB® code later. You should choose simple data that make it easy to check your work. In this problem, we know two equations relating the data:

   $$PV = nRT \quad \text{ideal gas law}$$
   $$n = m/\text{MW} \quad \text{relationship between mass and moles}$$

   Solve the ideal gas law for n, and plug in the given values:

   $$n = PV/RT$$
   $$= \frac{100 \text{ kPa} \times 1000 \text{ m}^3}{8.314 \text{ kPa m}^3/\text{kmol K} \times 300\text{K}}$$
   $$= 40.0930 \text{ kmol}$$

   Convert moles to mass by solving the conversion equation for the mass m and plugging in the values:

   $$m = n \times \text{MW} = 40.0930 \text{ kmol} \times 29 \text{ kg/mol}$$
   $$m = 1162.70 \text{ kg}$$

4. **Develop a MATLAB® Solution**
   First, clear the screen and memory:

   ```
   clear, clc
   ```

   Now perform the following calculations in the command window:

   ```
   P = 100
   P =
           100
   T = 300
   ```

```
T =
      300
V = 1000
V =
        1000
MW = 29
MW =
        29
R = 8.314
R =
       8.3140
n = (P*V)/(R*T)
n =
       40.0930
m = n*MW
m =
       1.1627e+003
```

There are several things you should notice about this MATLAB® solution. First, because no semicolons were used to suppress the output, the values of the variables are repeated after each assignment statement. Notice also the use of parentheses in the calculation of *n*. They are necessary in the denominator, but not in the numerator. However, using parentheses in both makes the code easier to read.

5. Test the Solution

In this case, comparing the result with that obtained by hand is sufficient. More complicated problems solved in MATLAB® should use a variety of input data, to confirm that your solution works in a variety of cases. The MATLAB® screen used to solve this problem is shown in Figure 2.7.

Figure 2.7
MATLAB® screen used to solve the ideal gas problem.

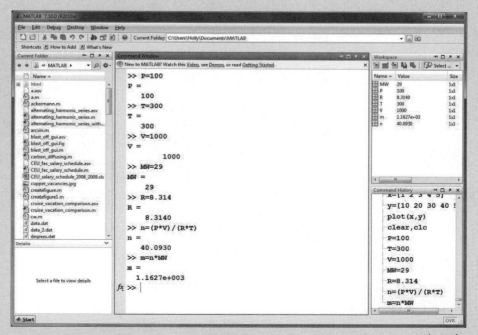

(continued)

Notice that the variables defined in the command window are listed in the workspace window. Notice also that the command history lists the commands executed in the command window. If you were to scroll up in the command history window, you would see commands from previous MATLAB® sessions. All of these commands are available for you to move to the command window.

EXPLICIT LIST
A list identifying each member of a matrix

Array Operations
Using MATLAB® as a glorified calculator is fine, but its real strength is in matrix manipulations. As described previously, the simplest way to define a matrix is to use a list of numbers, called an *explicit list*. The command

```
x = [1 2 3 4]
```

returns the row vector

```
x =
     1 2 3 4
```

Recall that, in defining this vector, you may list the values either with or without commas. A new row is indicated by a semicolon, so a column vector is specified as

```
y = [1; 2; 3; 4]
```

and a matrix that contains both rows and columns is created with the statement

```
a = [1 2 3 4; 2 3 4 5 ; 3 4 5 6]
```

and will return

```
a =
     1 2 3 4
     2 3 4 5
     3 4 5 6
```

HINT

It's easier to keep track of how many values you've entered into a matrix if you enter each row on a separate line—the semicolon is optional.

```
a = [1 2 3 4;
 2 3 4 5;
 3 4 5 6]
```

While a complicated matrix might have to be entered by hand, evenly spaced matrices can be entered much more readily. The command

```
b = 1:5
```

and the command

```
b = [1:5]
```

are equivalent statements. Both return a row matrix

```
b =
     1 2 3 4 5
```

(The square brackets are optional.) The default increment is 1, but if you want to use a different increment, put it between the first and final values on the right side of the command. For example,

```
c = 1:2:5
```

indicates that the increment between values will be 2 and returns

```
c =
      1     3     5
```

If you want MATLAB® to calculate the spacing between elements, you may use the linspace command. Specify the initial value, the final value, and how many total values you want. For example,

```
d = linspace(1, 10, 3)
```

returns a vector with three values, evenly spaced between 1 and 10:

```
d =
      1     5.5     10
```

You can create logarithmically spaced vectors with the logspace command, which also requires three inputs. The first two values are powers of 10 representing the initial and final values in the array. The final value is the number of elements in the array. Thus,

```
e = logspace(1, 3, 3)
```

returns three values:

```
e =
    10 100 1000
```

Notice that the first element in the vector is 10^1 and the last element in the array is 10^3.

HINT

New MATLAB® users often err when using the **logspace** command by entering the actual first and last values requested, instead of the corresponding power of 10. For example,

```
logspace(10,100,3)
```

is interpreted by MATLAB® as: Create a vector from 10^{10} to 10^{100} with three values. The result is

```
ans =
      1.0e+100 *
      0.0000 0.0000 1.0000
```

A common multiplier (1×10^{100}) is specified for each result, but the first two values are so small in comparison to the third, that they are effectively 0.

HINT
■
■ You can include mathematical operations inside a matrix definition state-
■ ment. For example, you might have a = [0 : pi/10 : pi].
■

Matrices can be used in many calculations with scalars. If a = [1 2 3], we can add 5 to each value in the matrix with the syntax

    b = a + 5

which returns

    b =
           6   7   8

KEY IDEA

Matrix multiplication is different from element-by-element multiplication

This approach works well for addition and subtraction; however, multiplication and division are a little different. In matrix mathematics, the multiplication operator (*) has a specific meaning. Because all MATLAB® operations can involve matrices, we need a different operator to indicate element-by-element multiplication. That operator is .* (called *dot multiplication or array multiplication*). For example,

    a.*b

results in element 1 of matrix a being multiplied by element 1 of matrix b,
          element 2 of matrix a being multiplied by element 2 of matrix b,
          element *n* of matrix a being multiplied by element *n* of matrix b.

For the particular case of our a (which is **[1 2 3]**) and our b (which is **[6 7 8]**),

    a.*b

returns

    ans =
           6   14   24

(Do the math to convince yourself that these are the correct answers.)

When you multiply a scalar times an array you may use either operator (* or .*), but when you multiply two arrays together they mean something quite different. Just using * implies a matrix multiplication, which in this case would return an error message, because a and b here do not meet the rules for multiplication in matrix algebra. The moral is, be careful to use the correct operator when you mean element-by-element multiplication.

Similar syntax holds for exponentiation (.^) and element-by-element division (./) of individual elements:

KEY IDEA

Unless you are specifically performing matrix algebra calculations, use the dot operators

    a.^2
    a./b

Unfortunately, when you divide a scalar by an array you still need to use the ./ syntax, because the / means taking the matrix inverse to MATLAB®. As a general rule, unless you specifically are doing problems involving linear algebra (matrix mathematics), you should use the dot operators.

As an exercise, predict the values resulting from the preceding two expressions, and then test your predictions by executing the commands in MATLAB®.

PRACTICE EXERCISES 2.4

As you perform the following calculations, recall the difference between the * and .* operators, as well as the / and ./ and the ^ and .^ operators:

1. Define the matrix a = [2.3 5.8 9] as a MATLAB® variable.
2. Find the sine of **a**.
3. Add 3 to every element in **a**.
4. Define the matrix b = [5.2 3.14 2] as a MATLAB® variable.
5. Add together each element in matrix **a** and in matrix **b**.
6. Multiply each element in **a** by the corresponding element in **b**.
7. Square each element in matrix **a**.
8. Create a matrix named **c** of evenly spaced values from 0 to 10, with an increment of 1.
9. Create a matrix named **d** of evenly spaced values from 0 to 10, with an increment of 2.
10. Use the `linspace` function to create a matrix of six evenly spaced values from 10 to 20.
11. Use the `logspace` function to create a matrix of five logarithmically spaced values between 10 and 100.

KEY IDEA

The matrix capability of MATLAB® makes it easy to do repetitive calculations

The matrix capability of MATLAB® makes it easy to do repetitive calculations. For example, suppose you have a list of angles in degrees that you would like to convert to radians. First put the values into a matrix. For angles of 10, 15, 70, and 90, enter

```
degrees = [10 15 70 90];
```

To change the values to radians, you must multiply by $\pi/180$:

```
radians = degrees*pi/180
```

This command returns a matrix called `radians`, with the values in radians. (Try it!) In this case, you could use either the * or the .* operator, because the multiplication involves a single matrix (`degrees`) and two scalars (pi and 180). Thus, you could have written

```
radians = degrees.*pi/180
```

HINT

The value of π is built into MATLAB® as a floating-point number called `pi`.

Because π is an irrational number, it cannot be expressed *exactly* with a floating-point representation, so the MATLAB® constant `pi` is really an approximation. You can see this when you find `sin(pi)`. From trigonometry, the answer should be 0. However, MATLAB® returns a very small number, 1.2246e–016. In most calculations, this won't make a difference in the final result.

Another useful matrix operator is transposition. The transpose operator changes rows to columns and vice versa. For example,

```
degrees'
```

returns

```
ans =
        10
        15
        70
        90
```

This makes it easy to create tables. For example, to create a table that converts degrees to radians, enter

```
table = [degrees', radians']
```

which tells MATLAB® to create a matrix named `table`, in which column 1 is degrees and column 2 is radians:

```
table =
        10.0000    0.1745
        15.0000    0.2618
        70.0000    1.2217
        90.0000    1.5708
```

If you transpose a two-dimensional matrix, all the rows become columns and all the columns become rows. For example, the command

```
table'
```

results in

```
        10.0000    15.0000    70.0000    90.0000
         0.1745     0.2618     1.2217     1.5708
```

Note that `table` is not a MATLAB® command but merely a convenient variable name. We could have used any meaningful name, say, **conversions** or **degrees_to_radians.**

EXAMPLE 2.2

MATRIX CALCULATIONS WITH SCALARS

Scientific data, such as data collected from wind tunnels, is usually in SI (Système International) units. However, much of the manufacturing infrastructure in the United States has been tooled in English (sometimes called American Engineering or American Standard) units. Engineers need to be fluent in both systems and should be especially careful when sharing data with other engineers. Perhaps the most notorious example of unit confusion problems is the Mars Climate Orbiter (Figure 2.8), which was the second flight of the Mars Surveyor Program. The spacecraft burned up in the orbit of Mars in September of 1999 because of a lookup table embedded in the craft's software. The table, probably generated from wind-tunnel testing, used pounds force (lbf) when the program expected values in newtons (N).

Figure 2.8
Mars Climate Orbiter.
(Courtesy of NASA/Jet
Propulsion Laboratory.)

In this example, we'll use MATLAB® to create a conversion table of pounds force to newtons. The table will start at 0 and go to 1000 lbf, at 100-lbf intervals. The conversion factor is

$$1 \text{ lbf} = 4.4482216 \text{ N}$$

1. **State the Problem**
   Create a table converting pounds force (lbf) to newtons (N).
2. **Describe the Input and Output**

   Input

The starting value in the table is	0 lbf
The final value in the table is	1000 lbf
The increment between values is	100 lbf
The conversion from lbf to N is	1 lbf = 4.4482216 N

   Output

   Table listing pounds force (lbf) and newtons (N)

3. **Develop a Hand Example**
   Since we are creating a table, it makes sense to check a number of different values. Choosing numbers for which the math is easy makes the hand example simple to complete, but still valuable as a check:

0	*	4.4482216 =	0
100	*	4.4482216 =	444.82216
1000	*	4.4482216 =	4448.2216

4. **Develop a MATLAB® Solution**

```
clear, clc
lbf = [0:100:1000];
N = lbf * 4.44822;
[lbf',N']
ans =
  1.0e+003 *
       0        0
  0.1000   0.4448
  0.2000   0.8896
  0.3000   1.3345
```

(*continued*)

0.4000	1.7793
0.5000	2.2241
0.6000	2.6689
0.7000	3.1138
0.8000	3.5586
0.9000	4.0034
1.0000	4.4482

It is always a good idea to clear both the workspace and the command window before starting a new problem. Notice in the workspace window (Figure 2.9) that lbf and **N** are 1×11 matrices and that ans (which is where the table we created is stored) is an 11×2 matrix. The output from the first two commands was suppressed by adding a semicolon at the end of each line. It would be very easy to create a table with more entries by changing the increment to 10 or even to 1. Notice also that you'll need to multiply the results shown in the table by 1000 to get the correct answers. MATLAB® tells you that this is necessary directly above the table, where the common scale factor is shown.

5. Test the Solution

Comparing the results of the MATLAB® solution with the hand solution shows that they are the same. Once we've verified that our solution works, it's easy to use the same algorithm to create other conversion tables. For instance, modify this example to create a table that converts newtons (N) to pounds force (lbf), with an increment of 10 N, from 0 N to 1000 N.

Figure 2.9
The MATLAB® workspace window shows the variables as they are created.

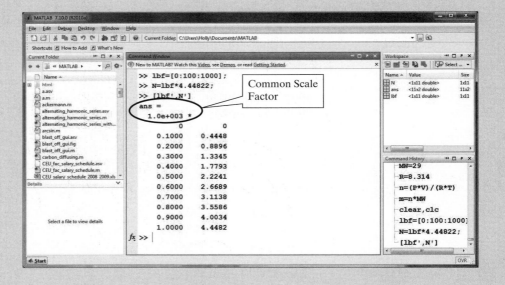

EXAMPLE 2.3

CALCULATING DRAG

One performance characteristic that can be determined in a wind tunnel is drag. The friction related to drag on the Mars Climate Observer (caused by the atmosphere of Mars) resulted in the spacecraft's burning up during course corrections. Drag is extremely important in the design of terrestrial aircraft as well (see Figure 2.10).

Drag is the force generated as an object, such as an airplane, moves through a fluid. Of course, in the case of a wind tunnel, air moves past a stationary model, but the equations are the same. Drag is a complicated force that depends on many factors. One factor is skin friction, which is a function of the surface properties of the aircraft, the properties of the moving fluid (air in this case), and the flow patterns caused by the shape of the aircraft (or, in the case of the Mars Climate Observer, by the shape of the spacecraft). Drag can be calculated with the drag equation

$$\text{drag} = C_\text{d}\frac{\rho V^2 A}{2}$$

where C_d = drag coefficient, which is determined experimentally, usually in a wind tunnel,
ρ = air density,
V = velocity of the aircraft,
A = reference area (the surface area over which the air flows).

Although the drag coefficient is not a constant, it can be taken to be constant at low speeds (less than 200 mph). Suppose the following data were measured in a wind tunnel:

drag	20,000 N
ρ	$1 \times 10^{-6}\,\text{kg/m}^3$
V	100 mph (you'll need to convert this to meters per second)
A	1 m^2

Calculate the drag coefficient. Finally, use this experimentally determined drag coefficient to predict how much drag will be exerted on the aircraft at velocities from 0 mph to 200 mph.

Figure 2.10
Drag is a mechanical force generated by a solid object moving through a fluid.

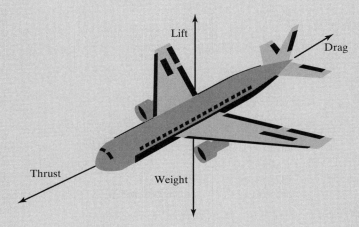

(continued)

1. State the Problem

   Calculate the drag coefficient on the basis of the data collected in a wind tunnel. Use the drag coefficient to determine the drag at a variety of velocities.

2. Describe the Input and Output

   Input

Drag	20,000 N
Air density ρ	$1 \times 10^{-6} \, kg/m^3$
Velocity V	100 mph
Surface area A	$1 \, m^2$

   Output

   Drag coefficient
   Drag at velocities from 0 to 200 mph

3. Develop a Hand Example

   First find the drag coefficient from the experimental data. Notice that the velocity is in miles/h and must be changed to units consistent with the rest of the data (m/s). The importance of carrying units in engineering calculations cannot be overemphasized!

   $$C_d = \frac{drag \times 2}{\rho \times V^2 \times A}$$

   $$= \frac{(20{,}000 \text{ N} \times 2)}{1 \times 10^{-6} kg/m^3 \times \left(100 \text{ miles/h} \times 0.4470 \dfrac{m/s}{miles/h} \right)^2 \times 1 m^2}$$

   $$= 2.0019 \times 10^7$$

   Since a newton is equal to a kg m/s², the drag coefficient is dimensionless. Now use the drag coefficient to find the drag at different velocities:

   $$drag = C_d \times \rho \times V^2 \times A/2$$

   Using a calculator, find the value of the drag with $V = 200$ mph:

   $$drag = \frac{2.0019 \times 10^7 \times 1 \times 10^{-6} kg/m^3 \times \left(200 \text{ miles/h} \times 0.4470 \dfrac{m/s}{miles/h} \right)^2 \times 1 \text{ m}^2}{2}$$

   $$drag = 80{,}000 \text{ N}$$

4. Develop a MATLAB® Solution

   ```
   drag = 20000;
   density = 0.000001;
   velocity = 100*0.4470;
   area = 1;
   cd = drag*2/(density*velocity^2*area)
   cd =
    2.0019e+007
   velocity = 0:20:200;
   velocity = velocity*0.4470;
   ```

   Define the variables, and change *V* to SI units.

   Calculate the coefficient of drag.

   Redefine *V* as a matrix. Change it to SI units and calculate the drag.

```
drag = cd*density*velocity.^2*area/2;
table = [velocity', drag']
table =
 1.0e+004 *
      0          0
  0.0009     0.0800
  0.0018     0.3200
  0.0027     0.7200
  0.0036     1.2800
  0.0045     2.0000
  0.0054     2.8800
  0.0063     3.9200
  0.0072     5.1200
  0.0080     6.4800
  0.0089     8.0000
```

Notice that the equation for drag, or

```
drag = cd * density * velocity.^2 * area/2;
```

uses the .^ operator, because we intend that each value in the matrix velocity be squared, not that the entire matrix velocity be multiplied by itself.
Using just the exponentiation operator (^) would result in an error message.
We could have used the .* operator as well in places where * was used, but since
all the other quantities are scalars it doesn't matter. Unfortunately, it is possible
to compose problems in which using the wrong operator does not give us an
error message but does give us a wrong answer. This makes step 5 in our
problem-solving methodology especially important.

Figure 2.11
The command history
window creates a record of
previous commands.

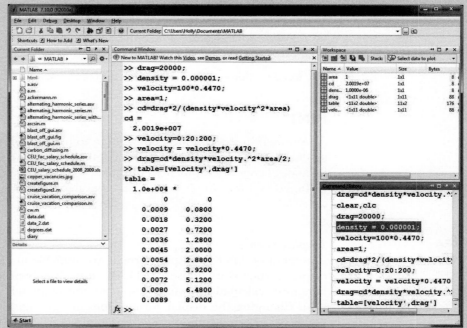

(*continued*)

5. Test the Solution

Comparing the hand solution with the MATLAB® solution (Figure 2.11), we see that they give the same results. Once we have confirmed that our algorithm works with sample data, we can substitute new data and be confident that the results will be correct. Ideally, the results should also be compared with experimental data, to confirm that the equations we are using accurately model the real physical process.

SCIENTIFIC NOTATION
A number represented as a value between one and ten times ten to an appropriate power

2.3.3 Number Display

Scientific Notation

Although you can enter any number in decimal notation, that isn't always the best way to represent very large or very small numbers. For example, a number that is used frequently in chemistry is Avogadro's constant, whose value, to four significant digits, is 602,200,000,000,000,000,000,000. Similarly, the diameter of an iron atom is approximately 140 picometers, which is 0.000000000140 m. Scientific notation expresses a value as a number between 1 and 10, multiplied by a power of 10 (the exponent). Thus, Avogadro's number becomes 6.022×10^{23}, and the diameter of an iron atom, 1.4×10^{-10} m. In MATLAB®, values in scientific notation are designated with an **e** between the decimal number and the exponent. (Your calculator probably uses similar notation.) For example, you might have

```
Avogadro's_constant = 6.022e23;
Iron_diameter = 140e-12; or
Iron_diameter = 1.4e-10;
```

It is important to omit blanks between the decimal number and the exponent. For instance, MATLAB® will interpret

```
6.022 e23
```

as two values (6.022 and 10^{23}). Since putting two values in an assignment statement is an error, MATLAB® will generate the message:

```
Error: Unexpected MATLAB® expression.
```

HINT

Although it is a common convention to use e to identify a power of 10, students (and teachers) sometimes confuse this nomenclature with the mathematical constant e, which is equal to 2.7183. To raise e to a power, use the exp function, for example exp(3) is equivalent to e^3.

KEY IDEA
MATLAB® does not differentiate between integers and floating-point numbers, unless special functions are invoked

Display Format

A number of different display formats are available in MATLAB®. No matter which display format you choose, MATLAB® uses double-precision floating-point numbers in its calculations, which results in approximately 16 decimal digits of precision. Changing the display format does not change the accuracy of the results. Unlike some other computer programs, MATLAB® handles both integers and decimal numbers as floating-point numbers.

KEY IDEA
No matter what display format is selected, calculations are performed using double-precision floating-point numbers

When elements of a matrix are displayed in MATLAB®, integers are always printed without a decimal point. However, values with decimal fractions are printed in the default short format that shows four digits after the decimal point. Thus,

```
A = 5
```

returns

```
A =
     5
```

but

```
A = 5.1
```

returns

```
A =
     5.1000
```

and

```
A = 51.1
```

returns

```
A =
     51.1000
```

MATLAB® allows you to specify other formats that show additional digits. For example, to specify that you want values to be displayed in a decimal format with 15 digits after the decimal point, use the command

```
format long
```

which changes all subsequent displays. Thus, with `format long` specified,

```
A
```

now returns

```
A =
     51.100000000000001
```

Notice that the final digit in this case is 1, which represents a round-off error. Two decimal digits are displayed when the format is specified as `format bank`:

```
A =
     51.10
```

The bank format displays only real numbers, so it's not appropriate when complex numbers need to be represented. Thus the command

```
A = 5+3i
```

returns the following using bank format

```
A =
     5.00
```

Using `format long` the same command returns

```
A =
     5.000000000000000 + 3.000000000000000i
```

You can return the format to four decimal digits with the command

```
format short
```

To check the results, recall the value of **A**:

```
A
A =
    5.0000 + 3.0000i
```

When numbers become too large or too small for MATLAB® to display in the default format, it automatically expresses them in scientific notation. For example, if you enter Avogadro's constant into MATLAB® in decimal notation as

```
a = 602000000000000000000000
```

the program returns

```
a =
    6.0200e+023
```

You can force MATLAB® to display *all* numbers in scientific notation with `format short e` (with four decimal digits) or `format long E` (with 15 decimal digits). For instance,

```
format short e
x = 10.356789
```

returns

```
x =
    1.0357e+001
```

Another pair of formats that are often useful to engineers and scientists, `format short eng` and `format long eng`, are similar to scientific notation but require the power of 10 to be a multiple of three. This corresponds to common naming conventions. For example,

$$1 \text{ millimeter} = 1 \times 10^{-3} \text{meters}$$
$$1 \text{ micrometer} = 1 \times 10^{-6} \text{ meters}$$
$$1 \text{ nanometer} = 1 \times 10^{-9} \text{ meters}$$
$$1 \text{ picometer} = 1 \times 10^{-12} \text{ meters}$$

Consider the following example. First change to engineering format and then enter a value for **y**.

```
format short eng
y = 12000
```

which gives the result

```
y =
    12.0000e+003
```

When a matrix of values is sent to the screen, and if the elements become very large or very small, a common scale factor is often applied to the entire matrix. This scale factor is printed along with the scaled values. For example, when the command window is returned to

```
format short
```

the results from Example 2.3 are displayed as

```
table =
1.0e+005 *
                 0          0
            0.0002     0.0400
            0.0004     0.1602
            0.0006     0.3603
            0.0008     0.6406      etc . . .
```

Two other formats that you may occasionally find useful are `format +` and `format rat`. When a matrix is displayed in `format +`, the only characters printed are plus and minus signs. If a value is positive, a plus sign will be displayed; if a value is negative, a minus sign will be displayed. If a value is zero, nothing will be displayed. This format allows us to view a large matrix in terms of its signs:

```
format +
B = [1, -5, 0, 12; 10005, 24, -10,4]
B =
        +- +
        ++-+
```

RATIONAL NUMBER
A number that can be represented as a fraction

The `format rat` command displays numbers as rational numbers (i.e., as fractions). Thus,

```
format rat
x = 0:0.1:0.5
```

returns

```
x =
     0    1/10    1/5    3/10    2/5    1/2
```

If you're not sure which format is the best for your application, you may select `format short g` or `format long g`. This format selects the best of fixed-point or floating-point representations.

The `format` command also allows you to control how tightly information is spaced in the command window. The default (`format loose`) inserts a line feed between user-supplied expressions and the results returned by the computer. The `format compact` command removes those line feeds. The examples in this text use the compact format to save space. Table 2.2 shows how the value of π is displayed in each format.

Table 2.2 Numeric Display Formats

MATLAB® Command	Display	Example
format short	4 decimal digits	**3.1416** **123.4568**
format long	14 decimal digits	**3.14159265358979** **1.234567890000000e+002**
format short e	4 decimal digits scientific notation	**3.1416e+000** **1.2346e+002**
format long e	14 decimal digits scientific notation	**3.141592653589793e+000** **1.234567890000000e+002**

(Continued)

Table 2.2 (Continued)

MATLAB® Command	Display	Example
format bank	2 decimal digits only real values are displayed	**3.14**
format short eng	4 decimal digits engineering notation	**3.1416e+000** **123.4568e+000**
format long eng	14 decimal digits engineering notation	**3.141592653589793e+000** **123.456789000000e+000**
format +	+, −, blank	**+**
format rat	fractional form	**355/113**
format short g	MATLAB® selects the best format	**3.1416** **123.46**
format long g	MATLAB® selects the best format	**3.14159265358979** **123.456789**

If none of these predefined numeric display formats is right for you, you can control individual lines of output with the `fprintf` function, described in a later chapter.

2.4 SAVING YOUR WORK

Working in the command window is similar to performing calculations on your scientific calculator. When you turn off the calculator or when you exit the program, your work is gone. It *is* possible to save the *values* of the variables you defined in the command window and that are listed in the workspace window, but while doing so is useful, it is more likely that you will want to save the list of commands that generated your results. The `diary` command allows you to do just that. Also we will show you how to save and retrieve variables (the results of the assignments you made and the calculations you performed) to MAT-files or to DAT-files. Finally we'll introduce script M-files, which are created in the edit window. Script M-files allow you to save a list of commands and to execute them later. You will find script M-files especially useful for solving homework problems. When you create a program in MATLAB®, it is stored in an M-file.

2.4.1 Diary

The diary function allows you to record a MATLAB® session in a file and retrieve it for later review. Both the MATLAB® commands and the results are stored—including all your mistakes. To activate the diary function simply type

```
diary
```

or

```
diary on
```

at the command prompt. To end a recording session type `diary` again, or `diary off`. A file named diary should appear in the current folder. You can retrieve the file by double-clicking on the file name in the current folder window. An editor window will open with the recorded commands and results. You can also open the file in any text editor, such as Notepad. Subsequent sessions are added to the end of the file. If you prefer to store the diary session in a different file, specify the filename

```
diary' <filename>
```

or

```
diary('filename')
```

In this text we'll use angle brackets (<>) to indicate user-defined names. Thus, to save a diary session in a file named My_diary_file type

```
diary My_diary_file
```

or

```
diary('My_diary_file')
```

2.4.2 Saving Variables

To preserve the variables you created in the **command window** (check the **work-space window** on the left-hand side of the MATLAB® screen for the list of variables), you must save the contents of the **workspace window** to a file. The default format is a binary file called a MAT-file. To save the workspace (remember, this is just the variables, not the list of commands in the command window) to a file, type

```
save <file_name>
```

at the prompt. Recall that, although `save` is a MATLAB® command, **file_name** is a user-defined file name. It can be any name you choose, as long as it conforms to the naming conventions for variables in MATLAB®. Actually, you don't even need to supply a file name. If you don't, MATLAB® names the file **matlab.mat**. You could also choose

```
File  →  Save Workspace As
```

from the menu bar, which will then prompt you to enter a file name for your data. To restore a workspace, type

```
load <file_name>
```

Again, `load` is a MATLAB® command, but **file_name** is the user-defined file name. If you just type `load`, MATLAB® will look for the default **matlab.mat** file.

The file you save will be stored in the current folder.

For example, type

```
clear, clc
```

This command will clear both the workspace and the command window. Verify that the workspace is empty by checking the workspace window or by typing

```
whos
```

Now define several variables—for example,

```
a = 5;
b = [1,2,3];
c = [1, 2; 3,4];
```

Check the workspace window once again to confirm that the variables have been stored. Now, save the workspace to a file called my_example_file:

```
save my_example_file
```

Confirm that a new file has been stored in the current folder. If you prefer to save the file to another directory (for instance, onto a flash drive), use the browse button (see Figure 2.2) to navigate to the directory of your choice. Remember that in a public computer lab the current folder is probably purged after each user logs off the system.

Now, clear the workspace and command window by typing

```
clear, clc
```

The workspace window should be empty. You can recover the missing variables and their values by loading the file (my_example_file.mat) back into the workspace:

```
load my_example_file
```

The file you want to load must be in the current folder, or MATLAB® won't be able to find it. In the command window, type

```
a
```

which returns

```
a =
     5
```

Similarly,

```
b
```

returns

```
b =
     1   2   3
```

and typing

```
c
```

returns

```
c =
     1   2
     3   4
```

MATLAB® can also store individual matrices or lists of matrices into a file in the current folder with the command

```
save <file_name> <variable_list>
```

where **file_name** is the user-defined file name designating the location in memory at which you wish to store the information, and **variable_list** is the list of variables to be stored in the file. For example,

```
save my_new_file a b
```

would save just the variables a and b into **my_new_file.mat**.

If your saved data will be used by a program other than MATLAB® (such as C or C++), the .mat format is not appropriate, because .mat files are unique to MATLAB®. The ASCII format is standard between computer platforms and is more appropriate if you need to share files. MATLAB® allows you to save files as ASCII files by modifying the save command to

```
save <file_name> <variable_list> -ascii
```

Figure 2.12
Double-clicking the file name in the command directory launches the Import Wizard.

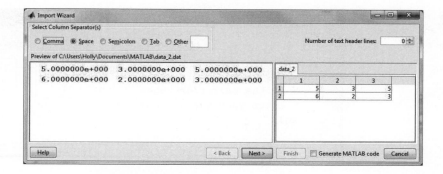

ASCII
Binary data storage format

The command `-ascii` tells MATLAB® to store the data in a standard eight-digit text format. ASCII files should be saved into a .dat file or .txt file instead of a .mat file; be sure to add .the extension to your file name:

```
save my_new_file.dat a b -ascii
```

KEY IDEA
When you save the workspace, you save only the variables and their values; you do not save the commands you've executed

If you don't add .dat, MATLAB® will default to .mat.
If more precision is needed, the data can be stored in a 16-digit text format:

```
save file_name variable_list -ascii -double
```

You can retrieve the data from the current folder with the load command:

```
load <file_name>
```

For example, to create the matrix z and save it to the file **data_2.dat** in eight-digit text format, use the following commands:

```
z = [5 3 5; 6 2 3];
save data_2.dat z -ascii
```

Together, these commands cause each row of the matrix **z** to be written to a separate line in the data file. You can view the data_2.dat file by double-clicking the file name in the current folder window (see Figure 2.12). Perhaps the easiest way to retrieve data from an ASCII .dat file is to enter the **load** command followed by the file name. This causes the information to be read into a matrix with the same name as the data file. However, it is also quite easy to use MATLAB®'s interactive Import Wizard to load the data. When you double-click a data file name in the current folder to view the contents of the file, the Import Wizard will automatically launch. Just follow the directions to load the data into the workspace, with the same name as the data file. You can use this same technique to import data from other programs, including Excel spreadsheets, or you can select **File → Import Data . . .** from the menu bar.

2.4.3 Script M-Files

Using the command window for calculations is an easy and powerful tool. However, once you close the MATLAB® program, all of your calculations are gone. Fortunately, MATLAB® contains a powerful programming language. As a programmer, you can create and save code in files called M-files. These files can be reused anytime you wish to repeat your calculations. An M-file is an ASCII text file similar to a C or FORTRAN source-code file. It can be created and edited with the MATLAB® M-file

Figure 2.13
The MATLAB® edit window, also called the editor/debugger.

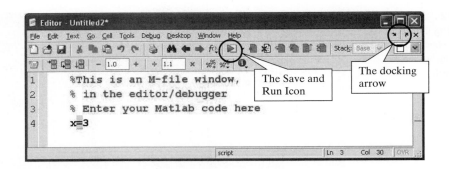

editor/debugger (the edit window discussed in Section 2.2.7), or you can use another text editor of your choice. To open the editing window, select

```
File  →  New  →  Script
```

from the MATLAB® menu bar, or select the New Script icon, located directly below the file menu. The MATLAB® edit window is shown in Figure 2.13. Many programmers prefer to dock the editing window onto the MATLAB® desktop, using the docking arrow in the upper right-hand corner of the window. This allows you to see both the contents of the M-file and the results displayed when the program is executed. The results from an M-file program are displayed in the command window.

If you choose a different text editor, make sure that the files you save are ASCII files. Notepad is an example of a text editor that defaults to an ASCII file structure. Other word processors, such as WordPerfect or Word, will require you to specify the ASCII structure when you save the file. These programs default to proprietary file structures that are not ASCII compliant and may yield some unexpected results if you try to use code written in them without specifying that the files be saved in ASCII format.

M-FILE
A list of MATLAB® commands stored in a separate file

When you save an M-file, it is stored in the current folder. You'll need to name your file with a valid MATLAB® variable name—that is, a name starting with a letter and containing only letters, numbers, and the underscore (_). Spaces are not allowed (see Section 2.3.1).

KEY IDEA
The two types of M-files are scripts and functions

There are two types of M-files, called scripts and functions. A script M-file is simply a list of MATLAB® statements that are saved in a file with a .m file extension. The script can use any variables that have been defined in the workspace, and any variables created in the script are added to the workspace when the script executes. You can execute a script created in the MATLAB® edit window by selecting the Save and Run icon from the menu bar, as shown in Figure 2.13. (The Save and Run icon changed appearance with MATLAB® 7.5. Previous versions of the program used an icon similar to an exclamation point.) You can also execute a script by typing a file name or by using the run command from the command window as shown in Table 2.3. No matter how you do it, you can only run an M-file if it is in the current folder.

You can find out what M-files and MAT files are in the current folder by typing

```
what
```

into the command window. You can also browse through the current folder by looking in the current folder window.

Using script M-files allows you to work on a project and to save the list of commands for future use. Because you will be using these files in the future, it is a good

Table 2.3 Approaches to Executing a Script M-File from the Command Window

MATLAB® Command	Comments
myscript	Type the file name, for example **myscript**. The .m file extension is assumed.
run myscript	Use the run command with the file name.
run('myscript')	Use the functional form of the run command.

idea to sprinkle them liberally with comments. The comment operator in MATLAB® is the percentage sign, as in

```
% This is a comment
```

MATLAB® will not execute any code on a commented line.

You can also add comments after a command, but on the same line:

```
a = 5      %The variable a is defined as 5
```

Here is an example of MATLAB® code that could be entered into an M-file and used to solve Example 2.3 :

```
clear, clc
% A Script M-file to find Drag
% First define the variables
drag = 20000;                    %Define drag in Newtons
density= 0.000001;               %Define air density in kg/m^3
velocity = 100*0.4470;          %Define velocity in m/s
area = 1;                        %Define area in m^2
% Calculate coefficient of drag
cd = drag *2/(density*velocity^2*area)
% Find the drag for a variety of velocities
velocity = 0:20:200;             %Redefine velocity
velocity = velocity*.4470        %Change velocity to m/s
drag = cd*density*velocity.^2*area/2;    %Calculate drag
table = [velocity',drag']        %Create a table of results
```

KEY IDEA

Liberally comment
MATLAB® code

This code can be run either from the M-file or from the command window. The results will appear in the command window in either case, and the variables will be stored in the workspace. The advantage of an M-file is that you can save your program to run again later.

HINT

You can execute a portion of an M-file by highlighting a section and then right-clicking and selecting **Evaluate Section**. You can also comment or "uncomment" whole sections of code from this menu; doing so is useful when you are creating programs while you are still debugging your work.

Example 2.4 uses a script M-file to find the velocity and acceleration that a spacecraft might reach in leaving the solar system.

EXAMPLE 2.4

CREATING AN M-FILE TO CALCULATE THE ACCELERATION OF A SPACECRAFT

In the absence of drag, the propulsion power requirements for a spacecraft are determined fairly simply. Recall from basic physical science that

$$F = ma$$

In other words, force (F) is equal to mass (m) times acceleration (a). Work (W) is force times distance (d), and since power (P) is work per unit time, power becomes force times velocity (v):

$$W = Fd$$

$$P = \frac{W}{t} = F \times \frac{d}{t} = F \times v = m \times a \times v$$

This means that the power requirements for the spacecraft depend on its mass, how fast it's going, and how quickly it needs to speed up or slow down. If no power is applied, the spacecraft just keeps traveling at its current velocity. As long as we don't want to do anything quickly, course corrections can be made with very little power. Of course, most of the power requirements for spacecraft are not related to navigation. Power is required for communication, for housekeeping, and for science experiments and observations.

The *Voyager 1* and *2* spacecraft explored the outer solar system during the last quarter of the 20th century (see Figure 2.14). *Voyager 1* encountered both Jupiter and Saturn; *Voyager 2* not only encountered Jupiter and Saturn but continued on to Uranus and Neptune. The *Voyager* program was enormously successful, and the *Voyager* spacecraft continue to gather information as they leave the solar system. The power generators (low-level nuclear reactors) on each spacecraft are expected to function until at least 2020. The power source is a sample of plutonium-238, which, as it decays, generates heat that is used to produce electricity. At the launch of each spacecraft, its generator produced about 470 watts of power. Because the plutonium is decaying, the power production had decreased to about 335 watts in 1997, almost 20 years after launch. This power is used to operate the science

Figure 2.14
The *Voyager 1* and *Voyager 2* spacecraft were launched in 1977 and have since left the solar system. (Courtesy of NASA/Jet Propulsion Laboratory.)

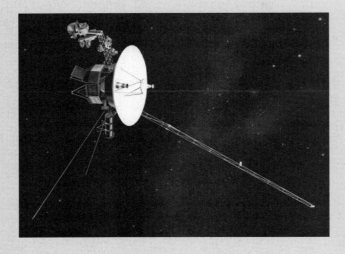

package, but if it were diverted to propulsion, how much acceleration would it produce in the spacecraft? *Voyager 1* is currently traveling at a velocity of 3.50 AU/year (an AU is an astronomical unit), and *Voyager 2* is traveling at 3.15 AU/year. Each spacecraft weighs 721.9 kg.

1. State the Problem

Find the acceleration that is possible with the power output from the spacecraft power generators.

2. Describe the Input and Output

Input

Mass = 721.9 kg
Power = 335 watts = 335 J/s
Velocity = 3.50 AU/year (*Voyager 1*)
Velocity = 3.15 AU/year (*Voyager 2*)

Output

Acceleration of each spacecraft, in m/s/s

3. Develop a Hand Example

We know that

$$P = m \times a \times v$$

which can be rearranged to give

$$a = \frac{P}{m \times v}$$

The hardest part of this calculation will be keeping the units straight. First let's change the velocity to m/s. For *Voyager 1*,

$$v = 3.50 \frac{\text{AU}}{\text{year}} \times \frac{150 \times 10^9 \text{m}}{\text{AU}} \times \frac{\text{year}}{365 \text{ days}} \times \frac{\text{day}}{24 \text{ h}} \times \frac{\text{h}}{3600 \text{ s}} = 16{,}650 \text{ m/s}$$

Then we calculate the acceleration:

$$a = \frac{335 \text{ J/s} \times 1 \text{ kg} \times \text{m}^2/\text{s}^2 \text{J}}{721.9 \text{ kg} \times 16{,}650 \text{ m/s}} = 2.7 \times 10^{-5} \text{ m/s}^2$$

4. Develop a MATLAB® Solution

```
clear, clc
%Example 2.4
%Find the possible acceleration of the Voyager 1
%and Voyager 2 Spacecraft using the on board power
%generator
format short
mass=721.9;              %mass in kg
power=335;               %power in watts
velocity=[3.5 3.15];     %velocity in AU/year
%Change the velocity to m/sec
velocity=velocity*150e9/365/24/3600
%Calculate the acceleration
acceleration=power./(mass.*velocity)
```

(*continued*)

Figure 2.15
The results of an M-file execution print into the command window. The variables created are reflected in the workspace and the M-file is listed in the current folder window. The commands issued in the M-file are not mirrored in the command history.

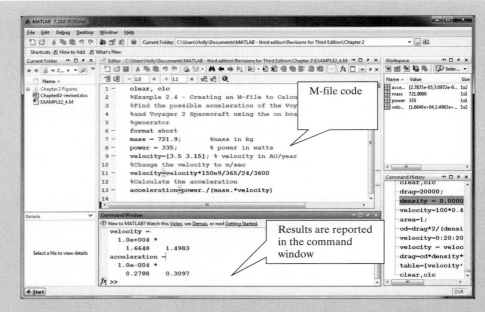

To evaluate the program, select the Save and Run icon. The results are printed in the command window, as shown in Figure 2.15.

5. Test the Solution

   Compare the MATLAB® results with the hand example results. Notice that the velocity and acceleration calculated from the hand example and the MATLAB® solution for *Voyager 1* match. The acceleration seems quite small, but applied over periods of weeks or months such an acceleration can achieve significant velocity changes. For example, a constant acceleration of $2.8 \times 10^{-5}\,\mathrm{m/s^2}$ results in a velocity change of about 72 m/s over the space of a month:

   $$2.8 \times 10^{-5}\,\mathrm{m/s^2} \times 3600\,\mathrm{s/h}$$
   $$\times\ 24\,\mathrm{h/day} \times 30\,\mathrm{days/month} = 72.3\,\mathrm{m/s}$$

   Now that you have a MATLAB® program that works, you can use it as the starting point for other, more complicated calculations.

2.4.4 Cell Mode

KEY IDEA
Cell mode is new to MATLAB® 7

KEY IDEA
Cell mode allows you to execute portions of the code incrementally

CELL
A section of MATLAB® code located between cell dividers (%%)

New to MATLAB® 7 is a utility that allows the user to divide M-files into sections, or cells, that can be executed one at a time. This feature is particularly useful as you develop MATLAB® programs. To activate the cell mode, select

```
Cell → Enable Cell Mode
```

from the menu bar in the edit window, as shown in Figure 2.16. Once the cell mode has been enabled, the cell toolbar appears, as shown in Figure 2.17.

To divide your M-file program into cells, you can create cell dividers by using a double percentage sign followed by a space. If you want to name the cell, just add a name on the same line as the cell divider:

```
%% Cell Name
```

Figure 2.16
You can access the cell mode from the menu bar in the edit window.

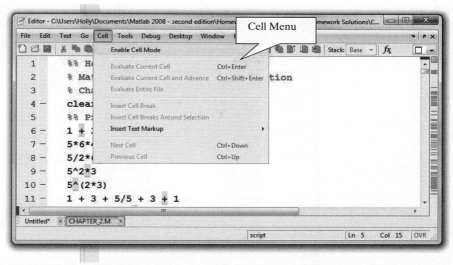

Figure 2.17
The cell toolbar allows the user to execute one cell, or section, at a time.

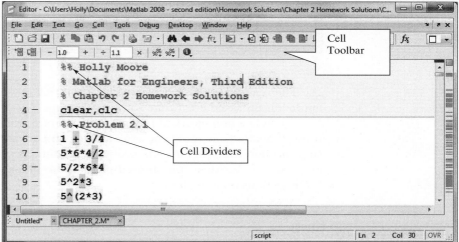

It's important to include the space after the double percentage sign (%%). If you don't, the line is recognized as a comment, not a cell divider.

Once the cell dividers are in place, if you position the cursor anywhere inside the cell, the entire cell turns pale yellow. For example, in Figure 2.17, the first four lines of the M-file program make up the first cell. Now we can use the evaluation icons on the cell toolbar to evaluate a single section, evaluate the current section and move on to the next section, or evaluate the entire file. Also on the cell toolbar is an icon that lists all the cell titles in the M-file, as shown in Figure 2.18.

Figure 2.18 shows the first 14 lines of an M-file written to solve some homework problems. By dividing the program into cells, it was possible to work on each problem separately. Be sure to save any M-files you've developed this way by selecting **Save** or **Save As** from the file menu:

> **File** → **Save**

or

> **File** → **Save As**

Figure 2.18
The show cell titles icon lists all the cells in the M-file.

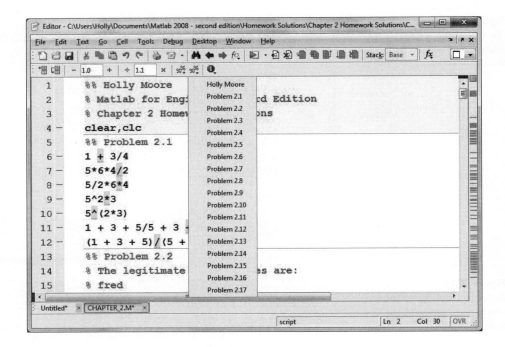

The reason for using these commands is that in cell mode, the program is not automatically saved every time you run it.

Dividing a homework M-file into cells offers a big advantage to the person who must evaluate it. By using the **evaluate cell and advance** function, the grader can step through the program one problem at a time. Even more important, the programmer can divide a complicated project into manageable sections and evaluate these sections independently.

SUMMARY

In this chapter, we introduced the basic MATLAB® structure. The MATLAB® environment includes multiple windows, four of which are open in the default view:

- Command window
- Command history window
- Workspace window
- Current folder window

In addition, the

- Document window
- Graphics window
- Edit window

open as needed during a MATLAB® session.

Variables defined in MATLAB® follow common computer naming conventions:

- Names must start with a letter.
- Letters, numbers, and the underscore are the only characters allowed.

- Names are case sensitive.
- Names may be of any length, although only the first 63 characters are used by MATLAB®.
- Some keywords are reserved by MATLAB® and cannot be used as variable names.
- MATLAB® allows the user to reassign function names as variable names, although doing so is not good practice.

The basic computational unit in MATLAB® is the matrix. Matrices may be

- Scalars (1×1 matrix)
- Vectors ($1 \times n$ or $n \times 1$ matrix, either a row or a column)
- Two-dimensional arrays ($m \times n$ or $n \times m$)
- Multidimensional arrays

Matrices often store numeric information, although they can store other kinds of information as well. Data can be entered into a matrix manually or can be retrieved from stored data files. When entered manually, a matrix is enclosed in square brackets, elements in a row are separated by either commas or spaces, and a new row is indicated by a semicolon:

```
a = [1 2 3 4; 5 6 7 8]
```

Evenly spaced matrices can be generated with the colon operator. Thus, the command

```
b = 0:2:10
```

creates a matrix starting at 0, ending at 10, and with an increment of 2. The `linspace` and `logspace` functions can be used to generate a matrix of specified length from given starting and ending values, spaced either linearly or logarithmically. The `help` function or the MATLAB® Help menu can be used to determine the appropriate syntax for these and other functions.

MATLAB® follows the standard algebraic order of operations. The operators supported by MATLAB® are listed in the "MATLAB® Summary" section of this chapter.

MATLAB® supports both standard (decimal) and scientific notation. It also supports a number of different display options, described in the "MATLAB® Summary" section. No matter how values are displayed, they are stored as double-precision floating-point numbers.

MATLAB® variables can be saved or imported from either .MAT or .DAT files. The .MAT format is proprietary to MATLAB® and is used because it stores data more efficiently than other file formats. The .DAT format employs the standard ASCII format and is used when data created in MATLAB® will be shared with other programs.

Collections of MATLAB® commands can be saved in script M-files. This is the best way to save the list of commands used to solve a problem so that they can be reused at a later time. Cell mode allows the programmer to group M-file code into sections and to run each section individually. It is especially convenient when one M-file is used to solve multiple problems.

MATLAB® SUMMARY

The following MATLAB® summary lists all the special characters, commands, and functions that were defined in this chapter:

Special Characters	
[ ]	forms matrices
( )	used in statements to group operations
	used with a matrix name to identify specific elements
,	separates subscripts or matrix elements
;	separates rows in a matrix definition
	suppresses output when used in commands
:	used to generate matrices
	indicates all rows or all columns
=	assignment operator assigns a value to a memory location; not the same as an equality
%	indicates a comment in an M-file
%%	cell divider
+	scalar and array addition
-	scalar and array subtraction
*	scalar multiplication and multiplication in matrix algebra
.*	array multiplication (dot multiply or dot star)
/	scalar division and division in matrix algebra
./	array division (dot divide or dot slash)
^	scalar exponentiation and matrix exponentiation in matrix algebra
.^	array exponentiation (dot power or dot caret)

Commands and Functions	
ans	default variable name for results of MATLAB® calculations
ascii	indicates that data should be saved in standard ASCII format
Clc	clears command window
Clear	clears workspace
Diary	creates a copy of all the commands issued in the workspace window, and most of the results
exit	terminates MATLAB®
format +	sets format to plus and minus signs only
format compact	sets format to compact form
format long	sets format to 14 decimal places
format long e	sets format to scientific notation with 14 decimal places
format long eng	sets format to engineering notation with 14 decimal places
format long g	allows MATLAB® to select the best format (either fixed point or floating point), using 14 decimal digits
format loose	sets format to the default, noncompact form
format short	sets format to the default, 4 decimal places
format short e	sets format to scientific notation with 4 decimal places
format short eng	sets format to engineering notation with 4 decimal places

Commands and Functions	
format short g	allows MATLAB® to select the best format (either fixed point or floating point), using 4 decimal digits
format rat	sets format to rational (fractional) display
help	invokes help utility
linspace	linearly spaced vector function
load	loads matrices from a file
logspace	logarithmically spaced vector function
namelengthmax	finds the maximum variable name length
pi	numeric approximation of the value of π
quit	terminates MATLAB®
save	saves variables in a file
who	lists variables in memory
whos	lists variables and their sizes

KEY TERMS

arguments	current folder	prompt
array	document window	scalar
array editor	dot operators	scientific notation
array operators	edit window	script
ASCII	function	start button
assignment	graphics window	transpose
cell mode	M-file	vector
command history	matrix	workspace
command window	operator	

PROBLEMS

You can either solve these problems in the command window, using MATLAB® as an electronic calculator, or you can create an M-file of the solutions. If you are solving these problems as a homework assignment, or if you want to keep a record of your work, the best strategy is to use an M-file, divided into cells with the cell divider %%.

Getting Started

2.1   Predict the outcome of the following MATLAB® calculations:

$$1 + 3/4$$
$$5*6*4/2$$
$$5/2*6*4$$
$$5\wedge2*3$$
$$5\wedge(2*3)$$
$$1 + 3 + 5/5 + 3 + 1$$
$$(1 + 3 + 5)(5 + 3 + 1)$$

Check your results by entering the calculations into the command window.

Using Variables

2.2 Identify which name in each of the following pairs is a legitimate MATLAB® variable name:

fred	fred!
book_1	book-1
2ndplace	Second_Place
#1	No_1
vel_5	vel.5
tan	while

Test your answers by using `isvarname`—for example,

```
isvarname fred
```

Remember, `isvarname` returns a 1 if the name is valid and a 0 if it is not. Although it is possible to reassign a function name as a variable name, doing so is not a good idea. Use `which` to check whether the preceding names are function names—for example,

```
which sin
```

In what case would MATLAB® tell you that `sin` is a variable name, not a function name?

Figure P2.4(a)

Scalar Operations and Order of Operations

2.3 Create MATLAB® code to perform the following calculations:

$$5^2$$

$$\frac{5+3}{5 \cdot 6}$$

$$\sqrt{4+6^3} \qquad (\textit{Hint: A square root is the same thing as a 1/2 power.})$$

$$9\frac{6}{12} + 7 \cdot 5^{3+2}$$

$$1 + 5 \cdot 3/6^2 + 2^{2-4} \cdot 1/5.5$$

Check your code by entering it into MATLAB® and performing the calculations on your scientific calculator.

2.4 As you answer the following questions, consider the shapes shown in Figure P2.4.

(a) The area of a circle is πr^2. Define r as 5, then find the area of a circle, using MATLAB®.

(b) The surface area of a sphere is $4\pi r^2$. Find the surface area of a sphere with a radius of 10 ft.

(c) The volume of a sphere is $4/3\pi r^3$. Find the volume of a sphere with a radius of 2 ft.

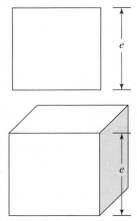

Figure P2.5 (a–c)

2.5 As you answer the following questions, consider the shape shown in Figure P2.5.

(a) The area of a square is the edge length squared ($A = \text{edge}^2$). Define the edge length as 5, then find the area of a square, using MATLAB®.

(b) The surface area of a cube is 6 times the edge length squared ($\text{SA} = 6 \times \text{edge}^2$). Find the surface area of a cube with edge length 10.

(c) The volume of a cube is the edge length cubed ($V = \text{edge}^3$). Find the volume of a cube with edge length 12.

Figure P2.6
The geometry of a barbell can be modeled as two spheres and a cylindrical rod.

2.6 Consider the barbell shown in Figure P2.6.

(a) Find the volume of the figure, if the radius of each sphere is 10 cm, the length of the bar connecting them is 15 cm, and the diameter of the bar is 1 cm. Assume that the bar is a simple cylinder.

(b) Find the surface area of the figure.

2.7. The ideal gas law was introduced in Example 2.1. It describes the relationship between pressure (P), temperature (T), volume (V), and the number of moles of gas (n).

$$PV = nRT$$

The additional symbol, R, represents the ideal gas constant. The ideal gas law is a good approximation of the behavior of gases when the pressure is low and the temperature is high. (What constitutes low pressure and high temperature varies with different gases.) In 1873, Johannes Diderik van der Waals (Figure P2.7) proposed a modified version of the ideal gas law that better models the behavior of real gases over a wider range of temperature and pressure.

$$\left(P + \frac{n^2 a}{V^2}\right)(V - nb) = nRT$$

In this equation the additional variables a and b represent values characteristic of individual gases.

Use both the ideal gas law and van der Waals' equation to calculate the temperature of water vapor (steam), given the following data.

Pressure, P	220 bar	
Moles, n	2 mol	
Volume, V	1 L	
a	5.536 L^{2}bar/mol^2	*
B	0.03049 L/mol	*
Ideal gas constant, R	0.08314472 L bar/K mol	

*Source: Weast, R. C. (Ed.), *Handbook of Chemistry and Physics (53rd Edn.)*, Cleveland: Chemical Rubber Co., 1972.

Figure P2.8(a)

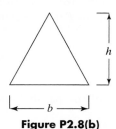

Figure P2.8(b)

Array Operations

2.8 (a) The volume of a cylinder is $\pi r^2 h$. Define r as 3 and h as the matrix

    h = [1, 5, 12]

Find the volume of the cylinders (see Figure P2.8a).

(b) The area of a triangle is 1/2 the length of the base of the triangle, times the height of the triangle. Define the base as the matrix

    b = [2, 4, 6]

and the height h as 12, and find the area of the triangles (see Figure P2.8b).

(c) The volume of any right prism is the area of the base of the prism, times the vertical dimension of the prism. The base of the prism can be any shape—for example, a circle, a rectangle, or a triangle.

Find the volume of the prisms created from the triangles of part (b). Assume that the vertical dimension of these prisms is 6 (see Figure P2.8c).

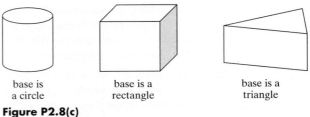

| base is a circle | base is a rectangle | base is a triangle |

Figure P2.8(c)

2.9 The response of circuits containing resistors, inductors, and capacitors depends upon the relative values of the resistors and the way they are connected. An important intermediate quantity used in describing the response of such circuits is *s*. Depending on the values of *R*, *L*, and *C*, the values of *s* will be either both real values, a pair of complex values, or a duplicated value.

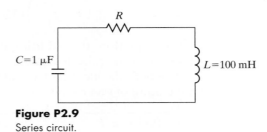

Figure P2.9
Series circuit.

The equation that identifies the response of a particular series circuit (Figure P2.9) is

$$S = -\frac{R}{2L} \pm \sqrt{\left(\frac{R}{2L}\right)^2 - \frac{1}{LC}}$$

(a) Determine the values of *s* for a resistance of 800 Ω.
(b) Create a vector of values for *R* ranging from 100 to 1000Ω and evaluate *s*. Refine your values of *R* until you find the approximate size of resistor that yields a pure real value of *s*. Describe the effect on *s* as *R* increases in value.

Hint:
1 μF = 1e-6F
1 mH = 1e-3H

2.10 The equation that identifies the response parameter, *s*, of the parallel circuit shown in Figure P2.10 is

$$S = -\frac{1}{2RC} \pm \sqrt{\left(\frac{1}{2RC}\right)^2 - \frac{1}{LS}}$$

(a) Determine the values of *s* for a resistance of 200 Ω.
(b) Create a vector of values for *R* ranging from 100 to 1000 Ω and evaluate *s*. Refine your values of *R* until you find the size of resistor that yields a pure real value of *s*. Describe the effect on *s* as *R* decreases.

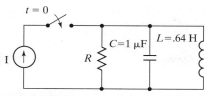

Figure P2.10
Parallel circuit.

2.11   Burning one gallon of gasoline in your car produces 19.4 pounds of CO_2. Calculate the amount of CO_2 emitted during a year for the following vehicles, assuming they all travel 12,000 miles per year. The reported fuel-efficiency numbers were extracted from the manufacturers' websites based on the EPA 2010 criteria; they are an average of the city and highway estimates.

2010	Smart Car Fortwo	37 mpg
2010	Civic Coupe	29 mpg
2010	Civic Hybrid	43 mpg
2010	Chevrolet Cobalt	31 mpg
2010	Toyota Prius (Hybrid)	48 mpg
2010	Toyota Yaris	32 mpg

2.12   **(a)** Create an evenly spaced vector of values from 1 to 20 in increments of 1.
   (b) Create a vector of values from zero to 2π in increments of $\pi/10$.
   (c) Create a vector containing 15 values, evenly spaced between 4 and 20. (*Hint:* Use the `linspace` command. If you can't remember the syntax, type `help linspace`.)
   (d) Create a vector containing 10 values, spaced logarithmically between 10 and 1000. (*Hint:* Use the `logspace` command.)

2.13   **(a)** Create a table of conversions from feet to meters. Start the feet column at 0, increment it by 1, and end it at 10 feet. (Look up the conversion factor in a textbook or online.)
   (b) Create a table of conversions from radians to degrees. Start the radians column at 0 and increment by 0.1π radian, up to π radians. (Look up the conversion factor in a textbook or online.)
   (c) Create a table of conversions from mi/h to ft/s. Start the mi/h column at 0 and end it at 100 mi/h. Print 15 values in your table. (Look up the conversion factor in a textbook or online.)
   (d) The acidity of solutions is generally measured in terms of pH. The pH of a solution is defined as $-\log_{10}$ of the concentration of hydronium ions. Create a table of conversions from concentration of hydronium ion to pH, spaced logarithmically from .001 to .1 mol/liter with 10 values. Assuming that you have named the concentration of hydronium ions H_conc, the syntax for calculating the negative of the logarithm of the concentration (and thus the pH) is

```
pH = -log10(H_conc)
```

2.14 The general equation for the distance that a freely falling body has traveled (neglecting air friction) is

$$d = \frac{1}{2} gt^2$$

Assume that $g = 9.8 \text{ m/s}^2$. Generate a table of time versus distance traveled for values of time from 0 to 100 seconds. Choose a suitable increment for your time vector. (*Hint*: Be careful to use the correct operators; t^2 is an array operation!)

2.15 In direct current applications, electrical power is calculated using Joule's law as

$$P = VI$$

where P is power in watts
  V is the potential difference, measured in volts
  I is the electrical current, measured in amperes
Joule's law can be combined with Ohm's law

$$V = IR$$

to give

$$P = I^2R$$

where R is resistance measured in ohms.

The resistance of a conductor of uniform cross section (a wire or rod for example) is

$$R = \rho \frac{l}{A}$$

where
  ρ is the electrical resistivity measured in ohm-meters
  l is the length of the wire
  A is the cross-sectional area of the wire
This results in the equation for power

$$P = I^2 \rho \frac{l}{A}$$

Electrical resistivity is a material property that has been tabulated for many materials. For example

Material	Resistivity, ohm-meters (measured at 20°C)
Silver	1.59×10^{-8}
Copper	1.68×10^{-8}
Gold	2.44×10^{-8}
Aluminum	2.82×10^{-8}
Iron	1.0×10^{-7}

Calculate the power that is dissipated through a wire with the following dimensions for each of the materials listed.

diameter	0.001 m
length	2.00 m

Assume the wire carries a current of 120 amps.

2.16 Repeat the previous problem for 10 wire lengths, from 1 m to 1 km. Use logarithmic spacing.

2.17 Newton's law of universal gravitation tells us that the force exerted by one particle on another is

$$F = G\frac{m_1 m_2}{r^2}$$

where the universal gravitational constant G is found experimentally to be

$$G = 6.673 \times 10^{-11} \, \text{N m}^2/\text{kg}^2$$

The mass of each particle is m_1 and m_2, respectively, and r is the distance between the two particles. Use Newton's law of universal gravitation to find the force exerted by the earth on the moon, assuming that

the mass of the earth is approximately 6×10^{24} kg,
the mass of the moon is approximately 7.4×10^{22} kg, and
the earth and the moon are an average of 3.9×10^8 m apart.

2.18 We know that the earth and the moon are not always the same distance apart. Based on the equation in the previous problem, find the force the moon exerts on the earth for 10 distances between 3.8×10^8 m and 4.0×10^8 m. Be careful when you do the division to use the correct operator.

2.19 Recall from Problem 2.7 that the ideal gas law is:

$$PV = nRT$$

and that the van der Waals modification of the ideal gas law is

$$\left(P + \frac{n^2 a}{V^2}\right)(V - nb) = nRT$$

Using the data from Problem 2.7, find the value of temperature (T), for

(a) 10 values of pressure from 0 bar to 400 bar for volume of 1 L
(b) 10 values of volume from 0.1 L to 10 L for a pressure of 220 bar

Number Display

2.20 Create a matrix **a** equal to $[-1/3, 0, 1/3, 2/3]$, and use each of the built-in format options to display the results:

```
format short (which is the default)
format long
format bank
format short e
format long e
```

```
format short eng
format long eng
format short g
format long g
format +
format rat
```

Saving Your Work in Files

2.21 • Create a matrix called D_to_R composed of two columns, one representing degrees and the other representing the corresponding value in radians. Any value set will do for this exercise.
   • Save the matrix to a file called degrees.dat.
   • Once the file is saved, clear your workspace and then load the data from the file back into MATLAB®.

2.22 Create a script M-file and use it to do the homework problems you've been assigned from this chapter. Your file should include appropriate comments to identify each problem and to describe your calculation process. Don't forget to include your name, the date, and any other information your instructor requests. Divide the script up into convenient sections, using cell mode.

3

Built-In MATLAB® Functions

Objectives

After reading this chapter, you should be able to:

- Use a variety of common mathematical functions
- Understand and use trigonometric functions in MATLAB®
- Compute and use statistical and data analysis functions
- Generate uniform and Gaussian random-number matrices
- Understand the computational limits of MATLAB®
- Recognize and be able to use the special values and functions built into MATLAB®

INTRODUCTION

The vast majority of engineering computations require quite complicated mathematical functions, including logarithms, trigonometric functions, and statistical analysis functions. MATLAB® has an extensive library of built-in functions to allow you to perform these calculations.

3.1 USING BUILT-IN FUNCTIONS

Many of the names for MATLAB®'s built-in functions are the same as those defined not only in the C programming language, but in Fortran and Java as well. For example, to take the square root of the variable **x**, we type

```
b = sqrt(x)
```

A big advantage of MATLAB® is that function arguments can generally be either scalars or matrices. In our example, if **x** is a scalar, a scalar result is returned. Thus, the statement

```
x = 9;
b = sqrt(x)
```

returns a scalar:

```
b =
        3
```

However, the square-root function, sqrt, can also accept matrices as input. In this case, the square root of each element is calculated, so

```
x = [4, 9, 16];
b = sqrt(x)
```

returns

```
b =
    2    3    4
```

KEY IDEA
Most of the MATLAB®
function names are the
same as those used in other
computer programs

All functions can be thought of as having three components: a name, input, and output. In the preceding example, the name of the function is sqrt, the required input (also called the *argument*) goes inside the parentheses and can be a scalar or a matrix, and the output is a calculated value or values. In this example, the output was assigned the variable name b.

ARGUMENT
Input to a function

Some functions require multiple inputs. For example, the remainder function, rem, requires two inputs: a dividend and a divisor. We represent this as rem(x,y), so

```
rem(10,3)
```

calculates the remainder of 10 divided by 3:

```
ans =
        1
```

The size function is an example of a function that returns two outputs, which are stored in a single array. It determines the number of rows and columns in a matrix. Thus,

```
d = [1, 2, 3; 4, 5, 6];
f = size(d)
```

returns the 1×2 result matrix

```
f =
    2    3
```

You can also assign variable names to each of the answers by representing the left-hand side of the assignment statement as a matrix. For example,

```
[rows,cols] = size(d)
```

gives

```
rows =
        2
cols =
        3
```

A useful feature of the more recent versions of MATLAB® is the adaptive help capability. As you type a function name, a screen tip appears showing the correct function format. It also includes a link to the function's help page.

NESTING
Using one function as the
input to another

You can create more complicated expressions by nesting functions. For instance,

```
g = sqrt(sin(x))
```

finds the square root of the sine of whatever values are stored in the matrix named **x**. If **x** is assigned a value of 2,

```
x = 2;
```

the result is

```
g =
    0.9536
```

Nesting functions can result in some complicated MATLAB® code. Be sure to include the arguments for each function inside their own set of parentheses. Often, your code will be easier to read if you break nested expressions into two separate statements. Thus,

```
a = sin(x);
g = sqrt(a)
```

gives the same result as g = sqrt(sin(x)) and is easier to follow.

HINT

■ You can probably *guess* the name and syntax for many MATLAB® functions. However, check to make sure that the function of interest is working the way you assume it is, before you do any important calculations.

3.2 USING THE HELP FEATURE

MATLAB® includes extensive help tools, which are especially useful in understanding how to use functions. There are two ways to get help from within MATLAB®: a command-line help function (help) and an HTML-based set of documentation available by selecting Help from the menu bar, selecting the help icon (a question mark) or by using the *F1* function key, usually located at the top of your keyboard (or found by typing helpwin in the command window). There is also an online help set of documentation, available through the Start button or the Help icon on the menu bar. However, the online help usually just reflects the HTML-based documentation. You should use both help options, since they provide different information and insights into how to use a specific function.

To use the command-line help function, type help in the command window:

```
help
```

KEY IDEA

Use the help function to help you use MATLAB®'s built-in functions

A list of help topics will appear:

```
HELP topics:

MATLAB\general       - General-purpose commands
MATLAB\ops           - Operators and special characters
MATLAB\lang          - Programming language constructs
MATLAB\elmat         - Elementary matrices and matrix
                       manipulation
MATLAB\elfun         - Elementary math functions
MATLAB\specfun       - Specialized math functions

and so on
```

To get help on a particular topic, type help <topic>. (Recall that the angle brackets, < >, identify where you should type your input; they are not included in your actual MATLAB® statement.)

For example, to get help on the `tangent` function, type

```
help tan
```

The following should be displayed:

```
TAN        Tangent of argument in radians.
  TAN(X) is the tangent of the elements of X.
See also atan, tand, atan2.
```

To use the windowed help screen, select **Help** → **Product Help** from the menu bar. A windowed version of the help list will appear (see Figure 3.1). You can then navigate to the appropriate topic. To access this version of the help utility directly from the command window, type `doc <topic>`. Thus, to access the windowed help for tangent, type

```
doc tan
```

The contents of the two methods for getting help on a function are different. If your question isn't immediately answered by whichever method you try first, it's often useful to try the other technique. The windowed help utility includes a MATLAB® tutorial that you will find extremely useful. The list in the left-hand window is a table of contents. Notice that it includes a link to a list of functions, organized both by category and alphabetically by name. You can use this link to find out what MATLAB® functions are available to solve many problems. For example, you might want to round a number you've calculated. Use the MATLAB® help window to determine whether an appropriate MATLAB® function is available.

Select the **MATLAB® Functions-By Category** link (see Figure 3.1) and then the **Mathematics** link (see Figure 3.2).

Figure 3.1
The MATLAB® help environment.

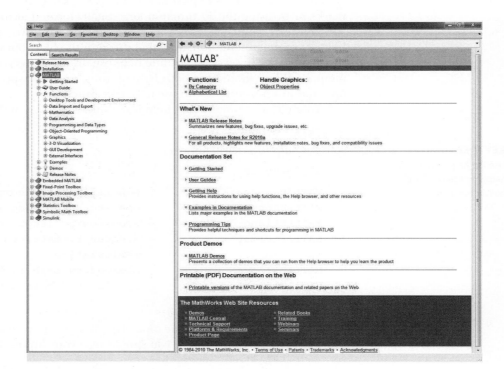

Figure 3.2
Functions-By Category help window. Notice the link to Mathematics functions in the right-hand pane.

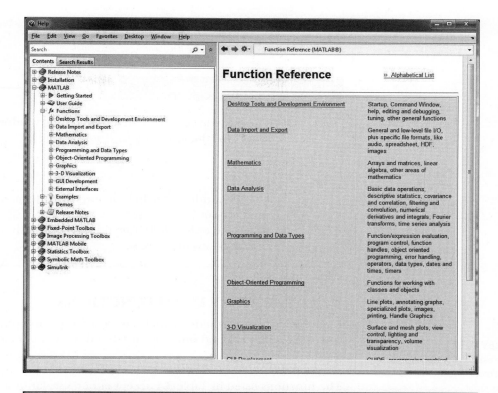

Figure 3.3
Mathematics help window.

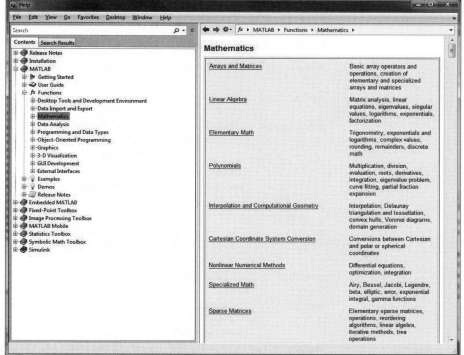

Near the middle of the page is the category Elementary Math (Figure 3.3), which lists rounding as a topic. Follow the links and you will find a whole category devoted to rounding functions. For example, round rounds to the nearest integer.

You could have also found the syntax for the round function by selecting **Functions—Alphabetical List.**

PRACTICE EERCISES 3.1

1. Use the help command in the command window to find the appropriate syntax for the following functions:
   a. cos
   b. sqrt
   c. exp
2. Use the windowed help function from the menu bar to learn about the functions in Exercise 1.
3. Go to the online help function at www.mathworks.com to learn about the functions in Exercise 1.

3.3 ELEMENTARY MATH FUNCTIONS

Elementary math functions include logarithms, exponentials, absolute value, rounding functions, and functions used in discrete mathematics.

KEY IDEA

Most functions accept scalars, vectors, or matrices as input

3.3.1 Common Computations

The functions listed in Table 3.1 accept either a scalar or a matrix of x values.

Table 3.1 Common Math Functions

abs(x)	Finds the absolute value of **x**.	**abs(−3)** **ans = 3**
sqrt(x)	Finds the square root of **x**.	**sqrt(85)** **ans = 9.2195**

nthroot(x,n)	Finds the real *n*th root of **x**. This function will not return complex results. Thus,
	
	$$(-2)\wedge(1/3)$$
	
	does not return the same result, yet both answers are legitimate third roots of −2.
		nthroot(−2, 3)
		ans =
		  −1.2599
		
		(−2)^(1/3)
		
		ans =
		  0.6300 + 1.0911i

sign(x)	Returns a value of −1 if **x** is less than zero, a value of 0 if **x** equals zero, and a value of +1 if **x** is greater than zero.	**sign(−8)** **ans = −1**
rem(x,y)	Computes the remainder of **x/y**.	**rem(25,4)** **ans = 1**
exp(x)	Computes the value of e^x, where *e* is the base for natural logarithms, or approximately 2.7183.	**exp(10)** **ans = 2.2026e +** **004**
log(x)	Computes ln(x), the natural logarithm of **x** (to the base *e*).	**log(10)** **ans = 2.3026**
log10(x)	Computes $\log_{10}(\mathbf{x})$, the common logarithm of **x** (to the base 10).	**log10(10)** **ans = 1**

HINT

As a rule, the function `log` in all computer languages means the `natural logarithm`. Although not the standard in mathematics textbooks, it is the standard in computer programming. Not knowing this distinction is a common source of errors, especially for new users. If you want logarithms to the base 10, you'll need to use the `log10` function. A `log2` function is also included in MATLAB®, but logarithms to any other base will need to be computed; there is no general logarithm function that allows the user to input the base.

PRACTICE EXERCISES 3.2

1. Create a vector **x** from −2 to +2 with an increment of 1. Your vector should be

$$\mathbf{x} = \begin{bmatrix} -2, & -1, & 0, & 1, & 2 \end{bmatrix}$$

   a. Find the absolute value of each member of the vector.
   b. Find the square root of each member of the vector.
2. Find the square root of both −3 and +3.
   a. Use the `sqrt` function.
   b. Use the `nthroot` function. (You should get an error statement for −3.)
   c. Raise −3 and +3 to the ½ power.
   How do the results vary?
3. Create a vector **x** from −9 to 12 with an increment of 3.
   a. Find the result of **x** divided by 2.
   b. Find the remainder of **x** divided by 2.
4. Using the vector from Exercise 3, find $\mathbf{e^x}$.
5. Using the vector from Exercise 3:
   a. Find $\ln(\mathbf{x})$ (the natural logarithm of **x**).
   b. Find $\log_{10}(\mathbf{x})$ (the common logarithm of **x**). Explain your results.
6. Use the `sign` function to determine which of the elements in vector **x** are positive.
7. Change the `format` to `rat`, and display the value of the **x** vector divided by 2.
   (Don't forget to change the format back to `format short` when you are done with this exercise set.)

HINT

The mathematical notation and MATLAB® syntax for raising *e* to a power are not the same. To raise *e* to the third power, the mathematical notation would be e^3. However, the MATLAB® syntax is `exp(3)`. Students also sometimes confuse the syntax for scientific notation with exponentials. The number `5e3` should be interpreted as 5×10^3.

EXAMPLE 3.1

USING THE CLAUSIUS–CLAPEYRON EQUATION

Meteorologists study the atmosphere in an attempt to understand and ultimately predict the weather (see Figure 3.4). Weather prediction is a complicated process, even with the best data. Meteorologists study chemistry, physics, thermodynamics, and geography, in addition to specialized courses about the atmosphere.

One equation used by meteorologists is the Clausius–Clapeyron equation, which is usually introduced in chemistry classes and examined in more detail in advanced thermodynamics classes. Rudolf Clausius and Emile Clapeyron were physicists responsible for the early development of thermodynamic principles during the mid-1800s (see Figures 3.5a and Figure 3.5b).

In meteorology, the Clausius–Clapeyron equation is employed to determine the relationship between saturation water-vapor pressure and the atmospheric temperature. The saturation water-vapor pressure can be used to calculate relative humidity, an important component of weather prediction, when the actual partial pressure of water in the air is known.

The Clausius–Clapeyron equation is

$$\ln\left(\frac{P^0}{6.11}\right) = \left(\frac{\Delta H_v}{R_{\text{air}}}\right) * \left(\frac{1}{273} - \frac{1}{T}\right)$$

Figure 3.4
View of the earth's weather from space. (Courtesy of NASA/Jet Propulsion Laboratory.)

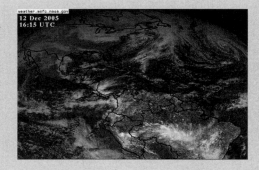

Figure 3.5
Portraits of (a) Rudolf Clausius and (b) Emile Clapeyron.

(a)

(b)

where

$$P^0 \quad = \quad \text{saturation vapor pressure for water, in mbar, at temperature } T$$
$$\Delta H_v \quad = \quad \text{latent heat of vaporization for water, } 2.453 \times 10^6 \text{ J/kg}$$
$$R_{\text{air}} \quad = \quad \text{gas constant for moist air, } 461 \text{ J/kg}$$
$$T \quad = \quad \text{temperature in kelvins (K)}.$$

It is rare that temperatures on the surface of the earth are lower than $-60°$F or higher than $120°$F. Use the Clausius–Clapeyron equation to find the saturation vapor pressure for temperatures in this range. Present your results as a table of Fahrenheit temperatures and saturation vapor pressures.

1. State the Problem

   Find the saturation vapor pressure at temperatures from $-60°$F to $120°$F, using the Clausius–Clapeyron equation.

2. Describe the Input and Output

   Input

   $$\Delta H_v = 2.453 \times 10^6 \text{ J/kg}$$
   $$R_{\text{air}} = 461 \text{ J/kg}$$
   $$T = -60°\text{F to } 120°\text{F}$$

   Since the number of temperature values was not specified, we'll choose to recalculate every $10°$F.

   Output

   Saturation vapor pressures

3. Develop a Hand Example

   The Clausius–Clapeyron equation requires that all the variables have consistent units. This means that temperature (T) needs to be in kelvins. To change degree Fahrenheit to kelvin, we use the conversion equation

   $$T_k = \frac{(T_f + 459.6)}{1.8}$$

   (There are lots of places to find units conversions. The Internet is one source, as are science and engineering textbooks.)

   Now we need to solve the Clausius–Clapeyron equation for the saturation vapor pressure P^0. We have

   $$\ln\left(\frac{P^0}{6.11}\right) = \left(\frac{\Delta H_v}{R_{\text{air}}}\right) \times \left(\frac{1}{273} - \frac{1}{T}\right)$$

   $$P^0 = 6.11 \times e^{\left(\left(\frac{\Delta H_v}{R_{\text{air}}}\right) \times \left(\frac{1}{273} - \frac{1}{T}\right)\right)}$$

   Next, we solve for one temperature—for example, $T = 0°$F. Since the equation requires temperature in kelvins we must perform the unit conversion to obtain

   $$T = \frac{(0 + 459.6)}{1.8} = 255.3333 \text{ K}$$

   Finally, we substitute values to get

   $$P^0 = 6.11 \times e^{\left(\left(\frac{2.453 \times 10^6}{461}\right) \times \left(\frac{1}{273} - \frac{1}{255.3333}\right)\right)} = 1.5836 \text{ mbar}$$

4. Develop a MATLAB® Solution

Create the MATLAB® solution in an M-file, and then run it in the command environment:

```
%Example 3.1
%Using the Clausius-Clapeyron Equation, find the
%saturation vapor pressure for water at different
%temperatures

TempF=[-60:10:120];              %Define temp matrix in F
TempK=(TempF + 459.6)/1.8;       %Convert temp to K
Delta_H=2.45e6;                  %Define latent heat of
                                 %vaporization

R_air = 461;                     %Define ideal gas constant
                                 %for air

%
%Calculate the vapor pressures
Vapor_Pressure=6.11*exp((Delta_H/R_air)*(1/273 - 1./TempK));
%Display the results in a table
my_results = [TempF',Vapor_Pressure']
```

When you create a MATLAB® program, it is a good idea to comment liberally (lines beginning with %). This makes your program easier for others to understand and may make it easier for you to "debug." Notice that most of the lines of code end with a semicolon, which suppresses the output. Therefore, the only information that displays in the command window is the table my_results:

```
my_results =
   -60.0000       0.0698
   -50.0000       0.1252
   -40.0000       0.2184
   ...
   120.0000     118.1931
```

5. Test the Solution

Compare the MATLAB® solution when $T = 0°F$ with the hand solution:

Hand solution: $P^0 = 1.5888$ mbar
MATLAB® solution: $P^0 = 1.5888$ mbar

The Clausius–Clapeyron equation can be used for more than just humidity problems. By changing the values of ΔH and R, you could generalize the program to deal with any condensing vapor.

3.3.2 Rounding Functions

MATLAB® contains functions for a number of different rounding techniques (Table 3.2). You are probably most familiar with rounding to the closest integer; however, you may want to round either up or down, depending on the situation.

For example, suppose you want to buy apples at the grocery store. The apples cost $0.52 a piece. You have $5.00. How many apples can you buy? Mathematically,

$$\frac{\$5.00}{\$0.52/\text{apple}} = 9.6154 \text{ apples}$$

Table 3.2 Rounding Functions

round(x)	Rounds **x** to the nearest integer.	**round(8.6)** **ans = 9**
fix(x)	Rounds (or truncates) **x** to the nearest integer toward zero. Notice that 8.6 truncates to 8, not 9, with this function.	**fix(8.6)** **ans = 8** **fix(−8.6)** **ans = −8**
floor(x)	Rounds **x** to the nearest integer toward negative infinity.	**floor(−8.6)** **ans = −9**
ceil(x)	Rounds **x** to the nearest integer toward positive infinity.	**ceil(−8.6)** **ans = −8**

But clearly, you can't buy part of an apple, and the grocery store won't let you round to the nearest number of apples. Instead, you need to round it down. The MATLAB® function to accomplish this is `fix`. Thus,

```
fix(5/0.52)
```

returns the maximum number of apples you can buy:

```
ans =
     9
```

3.3.3 Discrete Mathematics

MATLAB® includes functions to factor numbers, find common denominators and multiples, calculate factorials, and explore prime numbers (Table 3.3). All of these functions require integer scalars as input. In addition, MATLAB® includes the `rats` function, which expresses a floating-point number as a rational number—that is, a fraction. Discrete mathematics is the mathematics of whole numbers. Factoring, calculating common denominators, and finding least common multiples are procedures usually covered in intermediate algebra courses. Factorials are usually covered in statistics or probability courses and may not be familiar to beginning engineering students.

A factorial is the product of all the positive integers from 1 to a given value. Thus 3 factorial (indicated as 3!) is $3 \times 2 \times 1 = 6$. Many problems involving probability can be solved with factorials. For example, the number of ways that five cards can be arranged is $5 \times 4 \times 3 \times 2 \times 1 = 5! = 120$. When you select the first card, you have five choices; when you select the second card, you have only four choices remaining, then three, two, and one. This approach is called combinatorial mathematics, or combinatorics. To calculate a factorial in MATLAB® use the factorial function. Thus

```
factorial(5)
ans =
     120
```

gives the same result as

```
5*4*3*2*1
ans =
     120
```

The value of a factorial quickly becomes very large. Ten factorial is 3,628,800. MATLAB® can handle up to 170! Anything larger gives `Inf` for an answer, because the maximum value for a real number is exceeded.

Table 3.3 Functions Used in Discrete Mathematics

factor(x)	Finds the prime factors of **x**.	**factor(12)** **ans =** **2 2 3**
gcd(x,y)	Finds the greatest common denominator of **x** and **y**.	**gcd(10,15)** **ans =** **5**

lcm(x,y)	Finds the least common multiple of **x** and **y**.	**lcm(2,5)**
		ans =
		   10
		lcm(2,10)
		ans =
		   10

rats(x)	Represents **x** as a fraction.	**rats(1.5)** **ans =** **3/2**
factorial(x)	Finds the value of **x** factorial (**x**!). A factorial is the product of all the integers less than **x**. For example, $6! = 6 \times 5 \times 4 \times 3 \times 2 \times 1 = 720$.	**factorial(6)** **ans =** **720**
nchoosek(n,k)	Finds the number of possible combinations of k items from a group of n items. For example, use this function to determine the number of possible subgroups of 3 chosen from a group of 10.	**nchoosek(10,3)** **ans =** **120**
primes(x)	Finds all the prime numbers less than **x**.	**primes(10)** **ans =** **2 3 5 7**

isprime(x)	Checks to see if **x** is a prime number. If it is, the function returns 1; if not, it returns 0.	**isprime(7)**
		ans =
		   1
		isprime(10)
		ans =
		   0

```
factorial(170)
     ans =
     7.2574e+306
factorial(171)
     ans =
           Inf
```

Factorials are used to calculate the number of permutations and combinations of possible outcomes. A permutation is the number of subgroups that can be formed when sampling from a larger group, *when the order matters*. Consider the following problem. How many different teams of two people can you form from a group of four? Assume that the order matters, since for this problem the first person chosen is the group leader. If we represent each person as a letter, the possibilities are as follows:

AB	BA	CA	DA
AC	BC	CB	DB
AD	BD	CD	DC

For the first member of the team, there are four choices, and for the second there are three choices, so the number of possible teams is $4 \times 3 = 12$. We could also express this as $4!/2!$. More generally, if you have a large group to choose from, call the group size n, and the size of the subgroup (team) m. Then the possible number of permutations is

$$\frac{n!}{(n-m)!}$$

If there are 100 people to choose from, the number of teams of two (where order matters) is

$$\frac{100!}{(100-2)!} = 9900$$

But, what if the order doesn't matter? In this case, team AB is the same as team BA, and we refer to all the possibilities as combinations instead of permutations. The possible number of combinations is

$$\frac{n!}{(n-m)! \times m!}$$

Although you could use MATLAB®'s factorial function to calculate the number of combinations, the nchoosek function will do it for you, and it offers some advantages when using larger numbers. If we want to know the number of possible teams of 2, chosen from a pool of 100 (100 choose 2),

```
nchoosek(100,2)
ans =
      4950
```

The nchoosek function allows us to calculate the number of combinations even if the pool size is greater than 170, which would not be possible using the factorial approach.

```
nchoosek(200,2)
ans =
    19900
        factorial(200)/(factorial(198)*factorial(2))
        ans =
              NaN
```

PRACTICE EXERCISES 3.3

1. Factor the number 322.
2. Find the greatest common denominator of 322 and 6.
3. Is 322 a prime number?
4. How many primes occur between 0 and 322?
5. Approximate π as a rational number.
6. Find 10! (10 factorial).
7. Find the number of possible groups containing 3 people from a group of 20, when order does not matter. (20 choose 3)

3.4 TRIGONOMETRIC FUNCTIONS

MATLAB® includes a complete set of the standard trigonometric functions and the hyperbolic trigonometric functions. Most of these functions assume that angles are expressed in radians. To convert radians to degrees or degrees to radians, we need to take advantage of the fact that π radians equals 180:

$$\text{degrees} = \text{radians}\left(\frac{180}{\pi}\right) \text{ and radians} = \text{degrees}\left(\frac{\pi}{180}\right)$$

KEY IDEA

Most trig functions require input in radians

The MATLAB® code to perform these conversions is

```
degrees = radians * 180/pi;
radians = degrees * pi/180;
```

To carry out these calculations, we need the value of π, so a constant, `pi`, is built into MATLAB®. However, since π cannot be expressed as a floating-point number, the constant `pi` in MATLAB® is only an approximation of the mathematical quantity π. Usually this is not important; however, you may notice some surprising results. For example, for

```
sin(pi)
ans =
      1.2246e-016
```

when you expect an answer of zero.

MATLAB® also includes a set of trigonometric functions that accept the angle in degrees so that you need not do the conversion to radians. These include `sind`, `cosd`, and `tand`.

You may access the help function from the menu bar for a complete list of trigonometric functions available in MATLAB®. Table 3.4 shows some of the more common ones.

Table 3.4 Some of the Available Trigonometric Functions

sin(x)	Finds the sine of **x** when **x** is expressed in radians.	**sin(0)** **ans = 0**
cos(x)	Finds the cosine of **x** when **x** is expressed in radians.	**cos(pi)** **ans = −1**
tan(x)	Finds the tangent of **x** when **x** is expressed in radians.	**tan(pi)** **ans =** **−1.2246** **e−016**
asin(x)	Finds the arcsine, or inverse sine, of **x**, where **x** must be between −1 and 1. The function returns an angle in radians between $\pi/2$ and $-\pi/2$.	**asin(−1)** **ans =** **−1.5708**
sinh(x)	Finds the hyperbolic sine of **x** when **x** is expressed in radians.	**sinh(pi)** **ans =** **11.5487**
asinh(x)	Finds the inverse hyperbolic sin of **x**.	**asinh(1)** **ans =** **0.8814**
sind(x)	Finds the sin of **x** when **x** is expressed in degrees.	**sind(90)** **ans =** **1**
asind(x)	Finds the inverse sin of **x** and reports the result in degrees.	**asind(1)** **ans =** **90**

HINT

Math texts often use the notation $\sin^{-1}(x)$ to indicate an inverse sine function, also called an arcsine. Students are often confused by this notation and try to create parallel MATLAB® code. Note, however, that

```
a = sin^-1(x)
```

is *not* a valid MATLAB® statement but instead should be

```
a = sin(x)
```

PRACTICE EXERCISES 3.4

Calculate the following (remember that mathematical notation is not necessarily the same as MATLAB® notation):

1. $\sin(2\theta)$ for $\theta = 3\pi$.
2. $\cos(\theta)$ for $0 \leq \theta \leq 2\pi$; let θ change in steps of 0.2π.
3. $\sin^{-1}(1)$.
4. $\cos^{-1}(x)$ for $-1 \leq x \leq 1$; let x change in steps of 0.2.
5. Find the cosine of 45°.
    a. Convert the angle from degrees to radians, and then use the **cos** function.
    b. Use the **cosd** function.
6. Find the angle whose sine is 0.5. Is your answer in degrees or radians?
7. Find the cosecant of 60. You may have to use the help function to find the appropriate syntax.

EXAMPLE 3.2

USING TRIGONOMETRIC FUNCTIONS

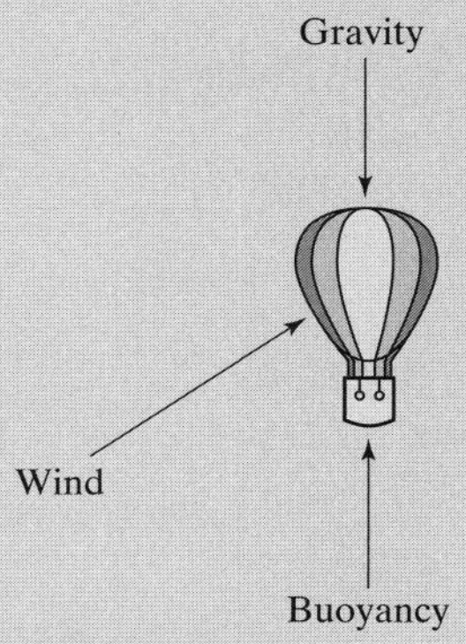

Gravity
Wind
Buoyancy

Figure 3.6
Force balance on a balloon.

A basic calculation in engineering is finding the resulting force on an object that is being pushed or pulled in multiple directions. Adding up forces is the primary calculation performed in both statics and dynamics classes. Consider a balloon that is acted upon by the forces shown in Figure 3.6.

To find the net force acting on the balloon, we need to add up the force due to gravity, the force due to buoyancy, and the force due to the wind. One approach is to find the force in the x direction and the force in the y direction for each individual force and then to recombine them into a final result.

The forces in the x and y directions can be found by trigonometry:

F = total force
F_x = force in the x direction
F_y = force in the y direction

We know from trigonometry that the sine is the opposite side over the hypotenuse, so

$$\sin(\theta) = F_y/F$$

and therefore,

$$F_y = F\sin(\theta)$$

Similarly, since the cosine is the adjacent side over the hypotenuse,

$$F_x = F\cos(\theta)$$

We can add up all the forces in the x direction and all the forces in the y direction and use these totals to find the resulting force:

$$F_{x\,total} = \Sigma F_{xi} \qquad F_{y\,total} = \Sigma F_{yi}$$

To find the magnitude and angle for F_{total}, we use trigonometry again. The tangent is the opposite side over the adjacent side. Therefore,

$$\tan(\theta) = \frac{F_{y\,total}}{F_{x\,total}}$$

We use an inverse tangent to write

$$\theta = \tan^{-1}\left(\frac{F_{y\,total}}{F_{x\,total}}\right)$$

(The inverse tangent is also called the *arctangent;* you'll see it on your scientific calculator as atan.)

Once we know θ, we can find F_{total}, using either the sine or the cosine. We have

$$F_{x\,total} = F_{total}\,\cos(\theta)$$

and rearranging terms gives

$$F_{total} = \frac{F_{x\,total}}{\cos(\theta)}$$

Now consider again the balloon shown in Figure 3.6. Assume that the force due to gravity on this particular balloon is 100 N, pointed downward. Assume further that the buoyant force is 200 N, pointed upward. Finally, assume that the wind is pushing on the balloon with a force of 50 N, at an angle of 30° from horizontal.
    Find the resulting force on the balloon.

1. State the Problem
   Find the resulting force on a balloon. Consider the forces due to gravity, buoyancy, and the wind.
2. Describe the Input and Output

   Input

Force	Magnitude	Direction
Gravity	100 N	−90
Buoyancy	200 N	+90
Wind	50 N	+30

   Output

   We'll need to find both the magnitude and the direction of the resulting force.

3. Develop a Hand Example

First find the x and y components of each force and sum the components:

Force	Horizontal Component	Vertical Component
Gravity	$F_x = F\cos(\theta)$	$F_y = F\sin(\theta)$
	$F_x = 100\cos(-90°) = 0$ N	$F_y = 100\sin(-90°) = -100$ N
Buoyancy	$F_x = F\cos(\theta)$	$F_y = F\sin(\theta)$
	$F_x = 200\cos(+90°) = 0$ N	$F_y = 200\sin(+90°) = +200$ N
Wind	$F_x = F\cos(\theta)$	$F_y = F\sin(\theta)$
	$F_x = 50\cos(+30°) = 43.301$ N	$F_y = 50\sin(+30°) = +25$ N
Sum	$F_{x\,total} = 0 + 0 + 43.301$	$F_{y\,total} = -100 + 200 + 25$
	$= 43.301$ N	$= 125$ N

Find the resulting angle:

$$\theta = \tan^{-1}\left(\frac{F_{y\,total}}{F_{x\,total}}\right)$$

$$\theta = \tan^{-1}\frac{125}{43.301} = 70.89°$$

Find the magnitude of the total force:

$$F_{total} = \frac{F_{x\,total}}{\cos(\theta)}$$

$$F_{total} = \frac{43.301}{\cos(70.89°)} = 132.29 \text{ N}$$

4. Develop a MATLAB® Solution

One solution is

```
%Example 3_2
clear, clc
%Define the input
Force =[100, 200, 50];
theta = [-90, +90, +30];
%convert angles to radians
theta = theta*pi/180;
%Find the x components
ForceX = Force.*cos(theta);
%Sum the x components
ForceX_total = sum(ForceX);
%Find and sum the y components in the same step
ForceY_total = sum(Force.*sin(theta));
%Find the resulting angle in radians
result_angle = atan(ForceY_total/ForceX_total);
%Find the resulting angle in degrees
result_degrees = result_angle*180/pi
%Find the magnitude of the resulting force
Force_total = ForceX_total/cos(result_angle)
```

which returns

```
result_degrees =
       70.8934

Force_total =
       132.2876
```

Notice that the values for the force and the angle were entered into an array. This makes the solution more general. Notice also that the angles were converted to radians. In the program listing, the output from all but the final calculations was suppressed. However, while developing the program, we left off the semicolons so that we could observe the intermediate results.

5. Test the Solution

Compare the MATLAB® solution with the hand solution. Now that you know it works, you can use the program to find the resultant of multiple forces. Just add the additional information to the definitions of the force vector `Force` and the angle vector `theta`. Note that we assumed a two-dimensional world in this example, but it would be easy to extend our solution to forces in all three dimensions.

3.5 DATA ANALYSIS FUNCTIONS

Analyzing data statistically in MATLAB® is particularly easy, partly because whole data sets can be represented by a single matrix and partly because of the large number of built-in data analysis functions.

3.5.1 Maximum and Minimum

Table 3.5 lists functions that find the minimum and maximum in a data set and the element at which those values occur.

Table 3.5 Maxima and Minima

max(x)	Finds the largest value in a **vector x**. For example, if $x = \begin{bmatrix} 1 & 5 & 3 \end{bmatrix}$, the maximum value is 5.	**x=[1, 5, 3];** **max(x)** **ans =** **5**
	Creates a row vector containing the maximum element from each column of a **matrix x**. For example, if $x = \begin{bmatrix} 1 & 5 & 3 \\ 2 & 4 & 6 \end{bmatrix}$, then the maximum value in column 1 is 2, the maximum value in column 2 is 5, and the maximum value in column 3 is 6.	**x=[1, 5, 3; 2, 4, 6];** **max(x)** **ans =** **2 5 6**

[a,b]=max(x)	Finds both the largest value in a **vector x** and its location in vector **x**. For $x = \begin{bmatrix} 1 & 5 & 3 \end{bmatrix}$ the maximum value is named **a** and is found to be 5. The location of the maximum value is element 2 and is named **b**.	**x=[1, 5, 3];**
		[a,b] = max(x)
		a =
		    5
		b =
		    2

	Creates a row vector containing the maximum element from each column of a matrix **x** and returns a row vector with the location of the maximum in each column of matrix **x**. For example, if $x = \begin{bmatrix} 1 & 5 & 3 \\ 2 & 4 & 6 \end{bmatrix}$, then the maximum value in column 1 is 2, the maximum value in column 2 is 5, and the maximum value in column 3 is 6. These maxima occur in row 2, row 1, and row 2, respectively.	**x=[1, 5, 3; 2, 4, 6];**
		[a,b] = max(x)
		a =
		    2 5 6
		b =
		    2 1 2

max(x,y)	Creates a matrix the same size as **x** and **y**. (Both **x** and **y** must have the same number of rows and columns.) Each element in the resulting matrix contains the maximum value from the corresponding positions in **x** and **y**. For example, if $x = \begin{bmatrix} 1 & 5 & 3 \\ 2 & 4 & 6 \end{bmatrix}$ and $y = \begin{bmatrix} 10 & 2 & 4 \\ 1 & 8 & 7 \end{bmatrix}$ then the resulting matrix will be $x = \begin{bmatrix} 10 & 5 & 4 \\ 2 & 8 & 7 \end{bmatrix}$	**x=[1, 5, 3; 2, 4, 6];**
		y=[10,2,4; 1, 8, 7];
		max(x,y)
		ans =
		    10 5 4
		    2 8 7

min(x)	Finds the smallest value in a **vector x**. For example, if $x = \begin{bmatrix} 1 & 5 & 3 \end{bmatrix}$ the minimum value is 1.	**x=[1, 5, 3];** **min(x)** **ans =** **1**
	Creates a row vector containing the minimum element from each column of a **matrix x**. For example, if $x = \begin{bmatrix} 1 & 5 & 3 \\ 2 & 4 & 6 \end{bmatrix}$, then the minimum value in column 1 is 1, the minimum value in column 2 is 4, and the minimum value in column 3 is 3.	**x=[1, 5, 3; 2, 4, 6];** **min(x)** **ans =** **1 4 3**

[a,b]=min(x)	Finds both the smallest value in a **vector x** and its location in vector **x**. For $x = \begin{bmatrix} 1 & 5 & 3 \end{bmatrix}$, the minimum value is named **a** and is found to be 1. The location of the minimum value is element 1 and is named **b**.	**x=[1, 5, 3];**
		[a,b]=min(x)
		a =
		    1
		b =
		    1

	Creates a row vector containing the minimum element from each column of a matrix **x** and returns a row vector with the location of the minimum in each column of matrix **x**. For example, if $x = \begin{bmatrix} 1 & 5 & 3 \\ 2 & 4 & 6 \end{bmatrix}$, then the minimum value in column 1 is 1, the minimum value in column 2 is 4, and the minimum value in column 3 is 3. These minima occur in row 1, row 2, and row 1, respectively.	**x=[1, 5, 3; 2, 4, 6];**
		[a,b]=min(x)
		a =
		    1 4 3
		b =
		    1 2 1

min(x,y)	Creates a matrix the same size as **x** and **y**. (Both **x** and **y** must have the same number of rows and columns.) Each element in the resulting matrix contains the minimum value from the corresponding positions in **x** and **y**. For example, if $x = \begin{bmatrix} 1 & 5 & 3 \\ 2 & 4 & 6 \end{bmatrix}$ and $y = \begin{bmatrix} 10 & 2 & 4 \\ 1 & 8 & 7 \end{bmatrix}$, then the resulting matrix will be $= \begin{bmatrix} 1 & 2 & 3 \\ 1 & 4 & 6 \end{bmatrix}$	**x=[1, 5, 3; 2, 4, 6];**
		y=[10,2,4; 1, 8, 7];
		min(x,y)
		ans =
		    1 2 3
		    1 4 6

All of the functions in this section work on the *columns* in two-dimensional matrices. MATLAB® is column dominant—in other words if there is a choice to make, MATLAB® will choose columns first over rows. If your data analysis requires you to evaluate data in rows, the data must be transposed. (In other words, the rows must become columns and the columns must become rows.) The transpose operator is a single quote ('). For example, if you want to find the maximum value in each *row* of the matrix

$$x = \begin{bmatrix} 1 & 5 & 3 \\ 2 & 4 & 6 \end{bmatrix}$$

use the command

```
max(x')
```

which returns

```
ans=
    5   6
```

HINT

A common mistake when finding the maximum or minimum value in a data set is to name the result max or min. This overwrites the function and it is no longer available for calculations. For example

```
max = max(x)
```

results in a variable named max for the answer. This is allowable MATLAB® code, but not wise. Trying to use the max function later in the program will result in an error. For example

```
another_max = max(y)
```

will return

```
??? Index exceeds matrix dimensions.
```

PRACTICE EXERCISES 3.5

Consider the following matrix:

$$x = \begin{bmatrix} 4 & 90 & 85 & 75 \\ 2 & 55 & 65 & 75 \\ 3 & 78 & 82 & 79 \\ 1 & 84 & 92 & 93 \end{bmatrix}$$

1. What is the maximum value in each column?
2. In which row does that maximum occur?
3. What is the maximum value in each row? (You'll have to transpose the matrix to answer this question.)
4. In which column does the maximum occur?
5. What is the maximum value in the entire table?

3.5.2 Mean and Median

MEAN
The average of all the values in the data set

MEDIAN
The middle value in a data set

There are several ways to find the "average" value in a data set. In statistics, the mean of a group of values is probably what most of us would call the average. The mean is the sum of all the values, divided by the total number of values. Another kind of average is the median, or the middle value. There are an equal number of values both larger and smaller than the median. The mode is the value that appears most often in a data set. MATLAB® provides functions for finding the mean, median, and the mode, as shown in Table 3.6. Recall that all of these functions are column dominant and will return an answer for each column in a two-dimensional matrix.

Table 3.6 Averages

mean(x)	Computes the mean value (or average value) of a **vector x**. For example if $x = \begin{bmatrix} 1 & 5 & 3 \end{bmatrix}$, the mean value is 3.	**x=[1, 5, 3];** **mean(x)** **ans =** **3.0000**
	Returns a row vector containing the mean value from each column of a **matrix x**. For example, if $x = \begin{bmatrix} 1 & 5 & 3 \\ 2 & 4 & 6 \end{bmatrix}$ then the mean value of column 1 is 1.5, the mean value of column 2 is 4.5, and the mean value of column 3 is 4.5.	**x=[1, 5, 3; 2, 4, 6];** **mean(x)** **ans =** **1.5 4.5 4.5**
median(x)	Finds the median of the elements of a **vector x**. For example, if $x = \begin{bmatrix} 1 & 5 & 3 \end{bmatrix}$, the median value is 3.	**x=[1, 5, 3];** **median(x)** **ans =** **3**
	Returns a row vector containing the median value from each column of a **matrix x**. For example, if $x = \begin{bmatrix} 1 & 5 & 3 \\ 2 & 4 & 6 \\ 3 & 8 & 4 \end{bmatrix}$, then the median value from column 1 is 2, the median value from column 2 is 5, and the median value from column 3 is 4.	**x=[1, 5, 3;** **2, 4, 6;** **3, 8, 4];** **median(x)** **ans =** **2 5 4**
mode(x)	Finds the value that occurs most often in an array. Thus, for the array $x = \begin{bmatrix} 1, & 2, & 3, & 3 \end{bmatrix}$ the mode is 3.	**x=[1,2,3,3]** **mode(x)** **ans =** **3**

3.5.3 Sums and Products

Often it is useful to add up (sum) all of the elements in a matrix or to multiply all of the elements together. MATLAB® provides a number of functions to calculate both sums and products, as shown in Table 3.7.

PRACTICE EXERCISES 3.6

Consider the following matrix:

$$x = \begin{bmatrix} 4 & 90 & 85 & 75 \\ 2 & 55 & 65 & 75 \\ 3 & 78 & 82 & 79 \\ 1 & 84 & 92 & 93 \end{bmatrix}$$

1. What is the mean value in each column?
2. What is the median for each column?
3. What is the mean value in each row?
4. What is the median for each row?
5. What is returned when you request the mode?
6. What is the mean for the entire matrix?

Table 3.7 Sums and Products

sum(x)	Sums the elements in **vector x**. For example, if $x = \begin{bmatrix} 1 & 5 & 3 \end{bmatrix}$, the sum is 9.	**x=[1, 5, 3];** **sum(x)** ans = 9
	Computes a row vector containing the sum of the elements in each column of a **matrix x**. For example, if $x = \begin{bmatrix} 1 & 5 & 3 \\ 2 & 4 & 6 \end{bmatrix}$ then the sum of column 1 is 3, the sum of column 2 is 9, and the sum of column 3 is 9.	**x=[1, 5, 3; 2, 4, 6];** **sum(x)** ans = 3 9 9
prod(x)	Computes the product of the elements of a **vector x**. For example, if $x = \begin{bmatrix} 1 & 5 & 3 \end{bmatrix}$ the product is 15.	**x=[1, 5, 3];** **prod(x)** ans = 15
	Computes a row vector containing the product of the elements in each column of a **matrix x**. For example, if $x = \begin{bmatrix} 1 & 5 & 3 \\ 2 & 4 & 6 \end{bmatrix}$, then the product of column 1 is 2, the product of column 2 is 20, and the product of column 3 is 18.	**x=[1, 5, 3; 2, 4, 6];** **prod(x)** ans = 2 20 18
cumsum(x)	Computes a vector of the same size as, and containing cumulative sums of the elements of, a **vector x**. For example, if $x = \begin{bmatrix} 1 & 5 & 3 \end{bmatrix}$, the resulting vector is $x = \begin{bmatrix} 1 & 6 & 9 \end{bmatrix}$.	**x=[1, 5, 3];** **cumsum(x)** ans = 1 6 9

	Computes a matrix containing the cumulative sum of the elements in each column of a **matrix x**. For example, if $x = \begin{bmatrix} 1 & 5 & 3 \\ 2 & 4 & 6 \end{bmatrix}$, the resulting matrix is $x = \begin{bmatrix} 1 & 5 & 3 \\ 3 & 9 & 9 \end{bmatrix}$.	**x=[1, 5, 3; 2, 4, 6];**
		cumsum(x)
		ans =
		  1 5 3
		  3 9 9

cumprod(x)	Computes a vector of the same size as, and containing cumulative products of the elements of, a **vector x**. For example, if $x = \begin{bmatrix} 1 & 5 & 3 \end{bmatrix}$, the resulting vector is $x = \begin{bmatrix} 1 & 5 & 15 \end{bmatrix}$.	**x=[1, 5, 3];** **cumprod(x)** ans = 1 5 15

	Computes a matrix containing the cumulative product of the elements in each column of a **matrix**. For example, if $x = \begin{bmatrix} 1 & 5 & 3 \\ 2 & 4 & 6 \end{bmatrix}$, the resulting matrix is $x = \begin{bmatrix} 1 & 5 & 3 \\ 2 & 20 & 18 \end{bmatrix}$.	**x=[1, 5, 3; 2, 4, 6];**
		cumprod(x)
		ans =
		  1 5 3
		  2 20 18

In addition to simply adding up all the elements, which returns a single value for each column in the array, the cumsum function (cumulative sum) adds all of the previous elements in an array and creates a new array of these intermediate totals. This is useful when dealing with the sequences of numbers in a series. Consider the harmonic series

$$\sum_{k-1}^{n} \frac{1}{k}$$

which is equivalent to

$$\frac{1}{1} + \frac{1}{2} + \frac{1}{3} + \frac{1}{4} + \cdots + \frac{1}{n}$$

We could use MATLAB® to create a sequence representing the first five values in the sequence as follows

```
k = 1:5;
sequence = 1./k
```

which gives us

```
sequence =
    1.0000    0.5000    0.3333    0.2500    0.2000
```

We could view the series as a sequence of fractions by changing the format to rational with the following code

```
format rat
sequence =
      1     1/2    1/3    1/4    1/5
```

Now we could use the `cumsum` function to find the value of the entire series for values of n from 1 to 5

```
format short
series = cumsum(sequence)
series =
    1.0000    1.5000    1.8333    2.0833    2.2833
```

Similarly the `cumprod` function finds the cumulative product of a sequence of numbers stored in an array.

3.5.4 Sorting Values

Table 3.8 lists several commands to sort data in a matrix into ascending or descending order. For example, if we define an array **x**

$$x = [1\ 6\ 3\ 9\ 4]$$

we can use the `sort` function to rearrange the values.

```
sort(x)
ans =
     1     3     4     6     9
```

The default is ascending order, but adding the string "descend" to the second field will force the function to list the values in descending order.

```
sort(x, 'descend')
ans =
     9     6     4     3     1
```

You can also use the sort command to rearrange entire matrices. This function is consistent with other MATLAB® functions, and sorts based on columns. Each column will be sorted independently. Thus

$$x = [1\ 3;\ 10\ 2;\ 3\ 1;\ 82\ 4;\ 5\ 5]$$

Table 3.8 Sorting Functions

sort(x)	Sorts the elements of a vector **x** into ascending order. For example, if $x = \begin{bmatrix} 1 & 5 & 3 \end{bmatrix}$, the resulting vector is $x = \begin{bmatrix} 1 & 3 & 5 \end{bmatrix}$.		**x=[1, 5, 3];** **sort(x)** **ans =** 1 3 5

	Sorts the elements in each column of a matrix **x** into ascending order. For example, if $x = \begin{bmatrix} 1 & 5 & 3 \\ 2 & 4 & 6 \end{bmatrix}$, the resulting matrix is $x = \begin{bmatrix} 1 & 4 & 3 \\ 2 & 5 & 6 \end{bmatrix}$.		**x=[1, 5, 3; 2, 4, 6];**
			sort(x)
			ans =
			   1 4 3
			   2 5 6

sort(x,'descend')	Sorts the elements in each column in descending order.		**x=[1, 5, 3; 2, 4, 6];**
			sort(x,'descend')
			ans =
			   2 5 6
			   1 4 3

sortrows(x)	Sorts the rows in a matrix in ascending order on the basis of the values in the first column, and keeps each row intact. For example, if $x = \begin{bmatrix} 3 & 1 & 2 \\ 1 & 9 & 3 \\ 4 & 3 & 6 \end{bmatrix}$, then using the **sortrows** command will move the middle row into the top position. The first column defaults to the basis for sorting.		**x=[3, 1, 3; 1, 9, 3; 4, 3, 6]**
			sortrows(x)
			ans =
			   1 9 3
			   3 1 2
			   4 3 6

sortrows(x,n)	Sorts the rows in a matrix on the basis of the values in column *n*. If *n* is negative, the values are sorted in descending order. If *n* is not specified, the default column used as the basis for sorting is column 1.		**sortrows(x,2)**
			ans =
			   3 1 2
			   4 3 6
			   1 9 3

gives

```
x =
    1    3
   10    2
    3    1
   82    4
    5    5
```

When we sort the array

```
sort(x)
```

each column is sorted in ascending order.

```
ans =
    1    1
    3    2
    5    3
   10    4
   82    5
```

The sortrows allows you to sort entire rows, based on the value in a specified column. Thus

sortrows(x,1)

sorts based on the first column, but maintains the relationship between values in columns one and two.

```
ans =
       1    3
       3    1
       5    5
      10    2
      82    4
```

Similarly you can sort based on values in the second column.

```
sortrows(x,2)
ans =
       3    1
      10    2
       1    3
       2    4
       5    5
```

These functions are particularly useful in analyzing data. Consider the results of the Men's 2006 Olympic 500-m speed skating event shown in Table 3.9.

The skaters were given a random number for this illustration, but once the race is over we'd like to sort the table in ascending order, based on the times in the second column.

```
skating_results = [1.0000   42.0930
                   2.0000   42.0890
                   3.0000   41.9350
                   4.0000   42.4970
                   5.0000   42.0020]

sortrows(skating_results,2)
ans =
       3.0000   41.9350
       5.0000   42.0020
       2.0000   42.0890
       1.0000   42.0930
       4.0000   42.4970
```

As you may remember, the winning time was posted by Apolo Anton Ohno, who in our example, is skater number 3.

Table 3.9 2006 Olympic Speed Skating Times

Skater Number	Time (min)
1	42.093
2	42.089
3	41.935
4	42.497
5	42.002

The `sortrows` function can also sort in descending order but uses a different syntax from the sort function. To sort in descending order, place a minus sign in front of the column number used for sorting. Thus

```
sortrows(skating_results, -2)
```

sorts the array in descending order, based on the second column. The result of this command is

```
ans =
        4.0000   42.4970
        1.0000   42.0930
        2.0000   42.0890
        5.0000   42.0020
        3.0000   41.9350
```

3.5.5 Determining Matrix Size

MATLAB® offers three functions (Table 3.10) that allow us to determine how big a matrix is: `size`, `length`, and `numel`. The `size` function returns the number of rows and columns in a matrix. The `length` function returns the larger of the matrix dimensions. The `numel` function returns the total number of elements in a matrix. For example, if

```
x = [1 2 3; 4 5 6];
size(x);
```

MATLAB® returns the following result

```
ans =
    2    3
```

This tells us that the **x** array has two rows and three columns. However, if we use the `length` function

```
length(x)
```

the result is

```
ans =
    3
```

Table 3.10 Size Functions

size(x)	Determines the number of rows and columns in matrix **x**. (If **x** is a multidimensional array, **size** determines how many dimensions exist and how big they are.)	x=[1, 5, 3; 2, 4, 6]; size(x) ans = 2 3

[a,b] = size(x)	Determines the number of rows and columns in matrix **x** and assigns the number of rows to **a** and the number of columns to **b**.	[a,b]=size(x)
		a =
		  2
		b =
		  3

length(x)	Determines the largest dimension of a matrix **x**.	x=[1, 5, 3; 2, 4, 6]; length(x) ans = 3
numel(x)	Determines the total number of elements in a matrix **x**.	x=[1, 5, 3; 2, 4, 6]; numel(x) ans = 6

because the largest of the array dimensions is 3.

Finally, if we use the `numel` function

```
numel(x)
```

the result is

```
ans =
    6
```

The `length` function is particularly useful when used with a loop structure, since it can easily determine how many times to execute the loop—based on the dimensions of an array.

EXAMPLE 3.3

WEATHER DATA

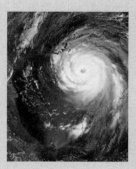

Figure 3.7
Satellite photo of a hurricane. (Courtesy of NASA/Jet Propulsion Laboratory.)

The National Weather Service collects massive amounts of weather data every day (Figure 3.7). Those data are available to all of us on the agency's online service at http://cdo.ncdc.noaa.gov/CDO/cdo. Analyzing large amounts of data can be confusing, so it's a good idea to start with a small data set, develop an approach that works, and then apply it to the larger data set that we are interested in.

We have extracted precipitation information from the National Weather Service for one location for all of 1999 and stored it in a file called Weather_Data.xls. (The .xls indicates that the data are in an Excel spreadsheet.) Each row represents a month, so there are 12 rows, and each column represents the day of the month (1 to 31), so there are 31 columns. Since not every month has the same number of days, data are missing for some locations in the last several columns. We place the number -99999 in those locations. The precipitation information is presented in hundredths of an inch. For example, on February 1 there was 0.61 inch of precipitation, and on April 1, 2.60 inches. A sample of the data is displayed in Table 3.11, with labels added for clarity; however, *the data in the file contain only numbers.*

Table 3.11 Precipitation Data from Asheville, North Carolina

1999	Day1	Day2	Day3	Day4	...	Day28	Day29	Day30	Day31
January	0	0	272	0		0	0	33	33
February	61	103	0	2		62	−99999	−99999	−99999
March	2	0	17	27		0	5	8	0
April	260	1	0	0		13	86	0	−99999
May	47	0	0	0		0	0	0	0
June	0	0	30	42		14	14	8	−99999
July	0	0	0	0		5	0	0	0
August	0	45	0	0		0	0	0	0
September	0	0	0	0		138	58	10	−99999
October	0	0	0	14		0	0	0	1
November	1	163	5	0		0	0	0	−99999
December	0	0	0	0		0	0	0	0

Use the data in the file to find the following:

a. The total precipitation in each month.
b. The total precipitation for the year.
c. The month and day on which the maximum precipitation during the year was recorded.

1. State the Problem
   Using the data in the file Weather_Data.xls, find the total monthly precipitation, the total precipitation for the year, and the day on which it rained the most.
2. Describe the Input and Output
   Input   The input for this example is included in a data file called Weather_Data.xls and consists of a two-dimensional matrix. Each row represents a month, and each column represents a day.

   Output   The output should be the total precipitation for each month, the total precipitation for the year, and the day on which the precipitation was a maximum. We have decided to present precipitation in inches, since no other units were specified in the statement of the problem.
3. Develop a Hand Example
   For the hand example, deal only with a small subset of the data. The information included in Table 3.11 is enough. The total for January, days 1 to 4, is

$$\text{total\_1} = (0 + 0 + 272 + 0)/100 = 2.72 \text{ inches}$$

The total for February, days 1 to 4, is

$$\text{total\_2} = (61 + 103 + 0 + 2)/100 = 1.66 \text{ inches}$$

Now add the months together to get the combined total. If our sample "year" is just January and February, then

$$\text{total} = \text{total\_1} + \text{total\_2} = 2.72 + 1.66 = 4.38 \text{ inches}$$

To find the day on which the maximum precipitation occurred, first find the maximum in the table, and then determine which row and which column it is in.

Working through a hand example allows you to formulate the steps required to solve the problem in MATLAB®.
4. Develop a MATLAB® Solution
   First we'll need to save the data file into MATLAB® as a matrix. Because the file is an Excel spreadsheet, the easiest approach is to use the Import Wizard. Double-click on the file in the current folder window to launch the Import Wizard.

   Once the Import Wizard has completed execution, the variable name `Sheet1` will appear in the workspace window. (See Figure 3.8; your version may name the variable `Weather_data` or `Sheet1`.)

   Because not every month has 31 days, there are a number of entries for nonexistent days. The value -99999 was inserted into those fields. You can double-click the variable name, `data`, in the workspace window, to edit this matrix and change the "phantom" values to 0 (see Figure 3.9).

   Now write the script M-file to solve the problem:

```
clc
%Example 3.3 - Weather Data
%In this example we will find the total precipitation
%for each month, and for the entire year, using a data file
```

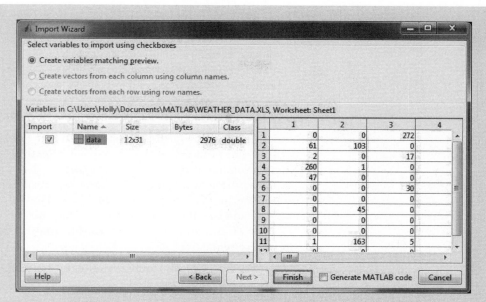

Figure 3.8
MATLAB® Import Wizard.

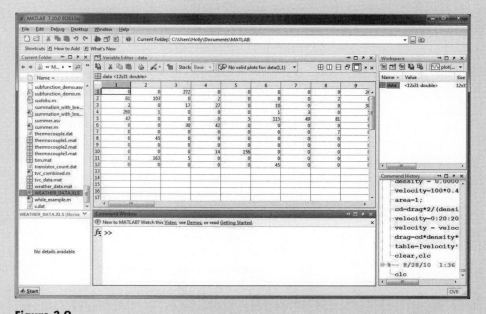

Figure 3.9
MATLAB® array editor. You can edit the array in this window and change all of the "phantom values" from − 99999 to 0.

```
%We will also find the month and day on which the
%precipitation was the maximum
weather_data=data;
%Use the transpose operator to change rows to columns
weather_data = weather_data';
%Find the sum of each column, which is the sum for each %month
```

```
monthly_total=sum(weather_data)/100
%Find the annual total
yearly_total = sum(monthly_total)
%Find the annual maximum and the day on which it occurs
[maximum_precip,month]=max(max(weather_data))
%Find the annual maximum and the month in which it occurs
[maximum_precip,day]=max(max(weather_data'))
```

Notice that the code did not start with our usual clear, clc commands, because that would clear the workspace, effectively deleting the data variable. Next we rename data to weather_data.

Next, the matrix weather_data is transposed, so that the data for each month are in a column instead of a row. That allows us to use the sum command to add up all the precipitation values for the month.

Now we can add up all the monthly totals to get the total for the year. An alternative syntax is

$$yearly\_total = sum(sum(weather\_data))$$

Finding the maximum daily precipitation is easy; what makes this example hard is determining the day and month on which the maximum occurred. The command

$$[maximum\_precip, month] = max(max(weather\_data))$$

is easier to understand if we break it up into two commands.

First,

$$[a,b] = max(weather\_data)$$

returns a matrix of maxima for each column, which in this case is the maximum for each month. This value is assigned to the variable name a. The variable b becomes a matrix of index numbers that represent the row in each column at which the maximum occurred. The result, then, is

```
a =
    Columns 1 through 9
     272    135     78    260    115    240    157    158    138
    Columns 10 through 12
     156    255     97
b =
    Columns 1 through 9
       3     18     27      1      6     25     12     24     28
    Columns 10 through 12
       5     26     14
```

Now when we execute the max command the second time, we determine the maximum precipitation for the entire data set, which is the maximum value in matrix a. Also, from matrix a, we find the index number for that maximum:

```
[c,d]=max(a)
c =
     272
d =
       1
```

These results tell us that the maximum precipitation occurred in column 1 of the a matrix, which means that it occurred in the first month.

Similarly, transposing the `weather_data` matrix (i.e., obtaining `weather_data'`) and finding the maximum twice allows us to find the day of the month on which the maximum occurred.

There are several things you should notice about the MATLAB® screen shown in Figure 3.10. In the **workspace window**, both `data` and `weather_data` are listed. The variable `data` is a 12×31 matrix, whereas `weather_data` is a 31×12 matrix. All of the variables created when the M-file was executed are now available to the command window. This makes it easy to perform additional calculations in the command window after the M-file has completed running. For example, notice that we forgot to change the `maximum_precip` value to inches from hundredths of an inch. Adding the command

$$\textbf{maximum\_precip} = \textbf{maximun\_precip}/100$$

would correct that oversight. Notice also that the Weather_Data.xls file is still in the current folder. Finally, notice that the **command history** window reflects only commands issued from the **command window**; it does not show commands executed from an M-file.

5. Test the Solution
   Open the Weather_Data.xls file, and confirm that the maximum precipitation occurred on January 3. Once you've confirmed that your M-file program works, you can use it to analyze other data. The National Weather Service maintains similar records for all of its recording stations.

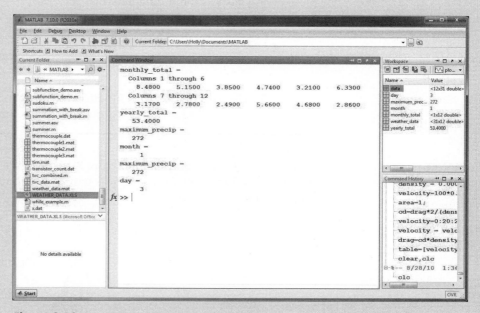

Figure 3.10
Results from the precipitation calculations.

STANDARD DEVIATION
A measure of the spread of values in a data set

3.5.6 Variance and Standard Deviation

The standard deviation and variance are measures of how much elements in a data set vary with respect to each other. Every student knows that the average score on a test is important, but you also need to know the high and low scores to get an idea of how well you did. Test scores, like many kinds of data that are important in engineering, are often distributed in a "bell"-shaped curve. In a normal (Gaussian) distribution of a large amount of data, approximately 68% of the data falls within one standard deviation (sigma) of the mean (± one sigma). If you extend the range to a two-sigma variation (± two sigma), approximately 95% of the data should fall inside these bounds, and if you go out to three sigma, over 99% of the data should fall in this range (Figure 3.11). Usually, measures such as the standard deviation and variance are meaningful only with large data sets.

PRACTICE EXERCISES 3.7

Consider the following matrix:

$$x = \begin{bmatrix} 4 & 90 & 85 & 75 \\ 2 & 55 & 65 & 75 \\ 3 & 78 & 82 & 79 \\ 1 & 84 & 92 & 93 \end{bmatrix}$$

1. Use the `size` function to determine the number of rows and columns in this matrix.
2. Use the `sort` function to sort each column in ascending order.
3. Use the `sort` function to sort each column in descending order.
4. Use the `sortrows` function to sort the matrix so that the first column is in ascending order, but each row still retains its original data. Your matrix should look like this:

$$x = \begin{bmatrix} 1 & 84 & 92 & 93 \\ 2 & 55 & 65 & 75 \\ 3 & 78 & 82 & 79 \\ 4 & 90 & 85 & 75 \end{bmatrix}$$

5. Use the `sortrows` function to sort the matrix from Exercise 4 in descending order, based on the third column.

VARIANCE
The standard deviation squared

Consider the data graphed in Figure 3.12. Both sets of data have the same average (mean) value of 50. However, it is easy to see that the first data set has more variation than the second.

Figure 3.11
Normal distribution.

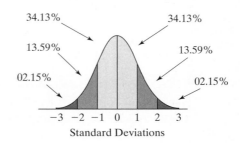

34.13%   34.13%
13.59%   13.59%
02.15%   02.15%

−3 −2 −1 0 1 2 3
Standard Deviations

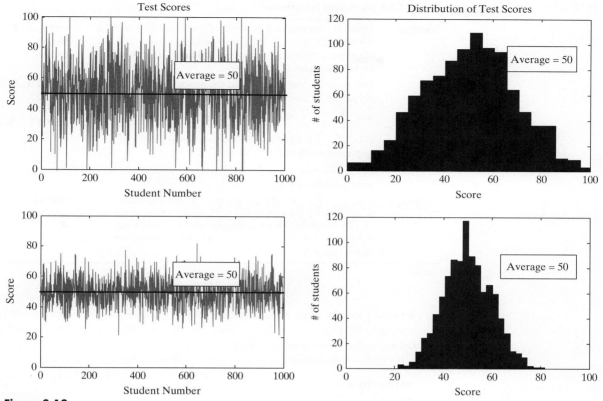

Figure 3.12
Test scores from two different tests.

The mathematical definition of variance is

$$\text{variance} = \sigma^2 = \frac{\displaystyle\sum_{k=1}^{N}(x_k - \mu)^2}{N - 1}$$

In this equation, the symbol μ represents the mean of the values x_k in the data set. Thus, the term $x_k - \mu$ is simply the difference between the actual value and the average value. The terms are squared and added together:

$$\sum_{k=1}^{N}(x_k - \mu)^2$$

Finally, we divide the summation term by the number of values in the data set (N), minus 1.

The standard deviation (σ), which is used more often than the variance, is the square root of the variance.

The MATLAB® function used to find the standard deviation is std. When we applied this function on the large data set shown in Figure 3.12, we obtained the following output:

```
std(scores1)
ans =
    20.3653
std(scores2)
ans =
    9.8753
```

Table 3.12 Statistical Functions

std(x)	Computes the standard deviation of the values in a vector **x**. For example, if $x = \begin{bmatrix} 1 & 5 & 3 \end{bmatrix}$, the standard deviation is 2. However, standard deviations are not usually calculated for small samples of data.	**x=[1, 5, 3];** **std(x)** **ans =** **2**
	Returns a row vector containing the standard deviation calculated for each column of a matrix **x**. For example, if $x = \begin{bmatrix} 1 & 5 & 3 \\ 2 & 4 & 6 \end{bmatrix}$ the standard deviation in column 1 is 0.7071, the standard deviation in column 2 is 0.7071, and standard deviation in column 3 is 2.1213. Again, standard deviations are not usually calculated for small samples of data.	**x=[1, 5, 3; 2, 4, 6];** **std(x)** **ans =** **0.7071 0.7071** **2.1213**
var(x)	Calculates the variance of the data in **x**. For example, if $x = \begin{bmatrix} 1 & 5 & 3 \end{bmatrix}$, the variance is 4. However, variance is not usually calculated for small samples of data. Notice that the standard deviation in this example is the square root of the variance.	**var(x)** **ans =** **4**

In other words, approximately 68% of the data in the first data set fall between the average, 50, and ± 20.3653. Similarly 68% of the data in the second data set fall between the same average, 50, and ± 9.8753.

The variance is found in a similar manner with the var function:

```
var(scores1)
ans =
   414.7454
var(scores2)
ans =
   97.5209
```

The syntax for calculating both standard deviation and variance is shown in Table 3.12.

PRACTICE EXERCISES 3.8

Consider the following matrix:

$$x = \begin{bmatrix} 4 & 90 & 85 & 75 \\ 2 & 55 & 65 & 75 \\ 3 & 78 & 82 & 79 \\ 1 & 84 & 92 & 93 \end{bmatrix}$$

1. Find the standard deviation for each column.
2. Find the variance for each column.
3. Calculate the square root of the variance you found for each column.
4. How do the results from Exercise 3 compare against the standard deviation you found in Exercise 1?

EXAMPLE 3.4

CLIMATOLOGIC DATA

Climatologists examine weather data over long periods of time, trying to find a pattern. Weather data have been kept reliably in the United States since the 1850s; however, most reporting stations have been in place only since the 1930s and 1940s (Figure 3.13). Climatologists perform statistical calculations on the data they collect. Although the data in Weather_Data.xls represent just one location for 1 year, we can use them to practice statistical calculations. Find the mean daily precipitation for each month and the mean daily precipitation for the year, and then find the standard deviation for each month and for the year.

1. State the Problem
   Find the mean daily precipitation for each month and for the year, on the basis of the data in Weather_Data.xls. Also, find the standard deviation of the data during each month and during the entire year.
2. Describe the Input and Output
   Input Use the Weather_Data.xls file as input to the problem.

   Output Find
   The mean daily precipitation for each month.
   The mean daily precipitation for the year.
   The standard deviation of the daily precipitation data for each month.
   The standard deviation of the daily precipitation data for the year.
3. Develop a Hand Example
   Use just the data for the first 4 days of the month:

   January average = $(0 + 0 + 272 + 0)/4 = 68$ hundredths of an inch of precipitation, or 0.68 inch.

   The standard deviation is found from the following equation:

   $$\sigma = \sqrt{\frac{\sum_{k=1}^{N}(x_k - \mu)^2}{N - 1}}$$

   Using just the first 4 days of January, first calculate the sum of the squares of the difference between the mean and the actual value:

   $$(0 - 68)^2 + (0 - 68)^2 + (272 - 68)^2 + (0 - 68)^2 = 55{,}488$$

   Divide by the number of data points minus 1:

   $$55{,}488/(4 - 1) = 18{,}496$$

   Finally, take the square root, to give 136 hundredths of an inch of precipitation, or 1.36 inches.

Figure 3.13
A hurricane over Florida. (Courtesy of NASA/Jet Propulsion Laboratory.)

(*continued*)

4. **Develop a MATLAB® Solution**
First we need to load the Weather_Data.xls file and edit out the −99999 entries. Although we could do that as described in Example 3.3, there is an easier way: The data from Example 3.3 could be saved to a file, so that they are available to use later. If we want to save the entire workspace, just type

```
save <filename>
```

where `filename` is a user-defined file name. If you just want to save one variable, type

```
save <filename> <variable_name>
```

which saves a single variable or a list of variables to a file. All we need to save is the variable `weather_data`, so the following command is sufficient:

save weather_data weather_data

This command saves the matrix `weather_data` into the **weather_data.mat** file. Check the current folder window to make sure that **weather_data.mat** has been stored (Figure 3.14).
Now the M-file we create to solve this example can load the data automatically:

```
clear, clc
% Example 3.4 Climatological Data
% In this example, we find the mean daily
% precipitation for each month
% and the mean daily precipitation for the year
% We also find the standard deviation of the data
%
% Changing the format to bank often makes the output
```

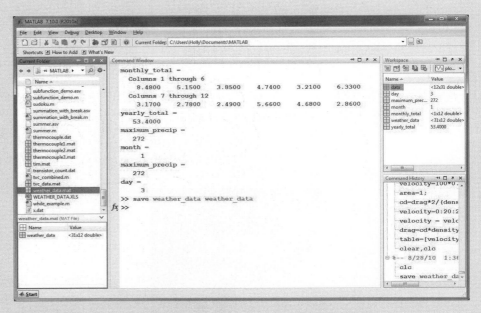

Figure 3.14
The current folder records the name of the saved file.

```
% easier to read
format bank
% By saving the variable weather_data from the last example, it is
% available to use in this problem
load weather_data
Average_daily_precip_monthly = mean(weather_data)
Average_daily_precip_yearly = mean(weather_data(:))
% Another way to find the average yearly precipitation
Average_daily_precip_yearly = mean(mean(weather_data))
% Now calculate the standard deviation
Monthly_Stdeviation = std(weather_data)
Yearly_Stdeviation = std(weather_data(:))
```

The results, shown in the command window, are

```
Average_daily_precip_monthly =
  Columns 1 through 3
    27.35 16.61 12.42
  Columns 4 through 6
    15.29 10.35 20.42
  Columns 7 through 9
    10.23 8.97 8.03
  Columns 10 through 12
    18.26 15.10 9.23
Average_daily_precip_yearly =
    14.35
Average_daily_precip_yearly =
    14.35
Monthly_Stdeviation =
  Columns 1 through 3
    63.78 35.06 20.40
  Columns 4 through 6
    48.98 26.65 50.46
  Columns 7 through 9
    30.63 30.77 27.03
  Columns 10 through 12
    42.08 53.34 21.01
Yearly_Stdeviation =
    39.62
```

The mean daily precipitation for the year was calculated in two equivalent ways. The mean of each month was found, and then the mean (average) of the monthly values was found. This works out to be the same as taking the mean of all the data at once. Some new syntax was introduced in this example. The command

```
weather_data(:)
```

converts the two-dimensional matrix `weather_data` into a one-dimensional matrix, thus making it possible to find the mean in one step.

The situation is different for the standard deviation of daily precipitation for the year. Here, we need to perform just one calculation:

```
std(weather_data(:))
```

(continued)

Otherwise you would find the standard deviation of the standard deviation—not what you want at all.

5. Test the Solution

First, check the results to make sure they make sense. For example, the first time we executed the M-file, the `weather_data` matrix still contained -99999 values. That resulted in mean values less than 1. Since it isn't possible to have negative rainfall, checking the data for reasonability alerted us to the problem. Finally, although calculating the mean daily rainfall for one month by hand would serve as an excellent check, it would be tedious. You can use MATLAB® to help you by calculating the mean without using a predefined function. The command window is a convenient place to perform these calculations:

```
load weather_data
sum(weather_data(:,1))        %Find the sum of all the rows in
                              %column one of matrix weather_data
ans =
    848.00
ans/31
ans =
    27.35
```

Compare these results with those for January (month 1).

HINT

Use the colon operator to change a two-dimensional matrix into a single column:

```
A = X(:)
```

3.6 RANDOM NUMBERS

Random numbers are often used in engineering calculations to simulate measured data. Measured data rarely behave exactly as predicted by mathematical models, so we can add small values of random numbers to our predictions to make a model behave more like a real system. Random numbers are also used to model games of chance. Two different types of random numbers can be generated in MATLAB®: uniform random numbers and Gaussian random numbers (often called a normal distribution).

3.6.1 Uniform Random Numbers

Uniform random numbers are generated with the `rand` function. These numbers are evenly distributed between 0 and 1. (Consult the help function for more details.) Table 3.13 lists several MATLAB® commands for generating random numbers.

We can create a set of random numbers over other ranges by modifying the numbers created by the `rand` function. For example, to create a set of 100 evenly distributed numbers between 0 and 5, first create a set over the default range with the command

```
r = rand(100,1);
```

This results in a 100×1 matrix of values. Now we just need to multiply by 5 to expand the range to 0 to 5:

```
r = r * 5;
```

Table 3.13 Random-Number Generators

rand(n)	Returns an $n \times n$ matrix. Each value in the matrix is a random number between 0 and 1.	**rand(2)** **ans =** 0.9501 0.6068 0.2311 0.4860
rand(m,n)	Returns an $m \times n$ matrix. Each value in the matrix is a random number between 0 and 1.	**rand(3,2)** **ans =** 0.8913 0.0185 0.7621 0.8214 0.4565 0.4447
randn(n)	Returns an $n \times n$ matrix. Each value in the matrix is a Gaussian (or normal) random number with a mean of 0 and a variance of 1.	**randn(2)** **ans =** −0.4326 0.1253 −1.6656 0.2877
randn(m,n)	Returns an $m \times n$ matrix. Each value in the matrix is a Gaussian (or normal) random number with a mean of 0 and a variance of 1.	**randn(3,2)** **ans =** −1.1465 −0.0376 1.1909 0.3273 1.1892 0.1746

If we want to change the range to 5 to 10, we can add 5 to every value in the array:

```
r = r + 5;
```

The result will be random numbers varying from 5 to 10. We can generalize these results with the equation

$$x = (\text{max} - \text{min}) \cdot \text{random\_number\_set} + \text{min}$$

3.6.2 Gaussian Random Numbers

Gaussian random numbers have the normal distribution shown in Figure 3.11. There is no absolute upper or lower bound to a data set of this type; we are just less and less likely to find data, the farther away from the mean we get. Gaussian random-number sets are described by specifying their average and the standard deviation of the data set.

MATLAB® generates Gaussian values with a mean of 0 and a variance of 1.0, using the randn function. For example,

```
randn(3)
```

returns a 3×3 matrix

```
ans =
 -0.4326    0.2877    1.1892
 -1.6656   -1.1465   -0.0376
  0.1253    1.1909    0.3273
```

If we need a data set with a different average or a different standard deviation, we start with the default set of random numbers and then modify it. Since the default standard deviation is 1, we must *multiply* by the required standard deviation for the new data set. Since the default mean is 0, we'll need to *add* the new mean:

$$x = \text{standard\_deviation} \cdot \text{random\_data\_set} + \text{mean}$$

For example, to create a sequence of 500 Gaussian random variables with a standard deviation of 2.5 and a mean of 3, type

```
x = randn(1,500)*2.5 + 3;
```

Notice that both `rand` and `randn` can accept either one or two input values. If only one is specified the result is a square matrix. If two values are specified they represent the number of rows and the number of columns in the resulting matrix.

PRACTICE EXERCISES 3.9

1. Create a 3×3 matrix of evenly distributed random numbers.
2. Create a 3×3 matrix of normally distributed random numbers.
3. Create a 100×5 matrix of evenly distributed random numbers. Be sure to suppress the output.
4. Find the maximum, the standard deviation, the variance, and the mean for each column in the matrix that you created in Exercise 3.
5. Create a 100×5 matrix of normally distributed random numbers. Be sure to suppress the output.
6. Find the maximum, the standard deviation, the variance, and the mean for each column in the matrix you created in Exercise 5.
7. Explain why your results for Exercises 4 and 6 are different.

EXAMPLE 3.5

NOISE

Random numbers can be used to simulate the noise we hear as static on the radio. By adding this noise to data files that store music, we can study the effect of static on recordings.

MATLAB® has the ability to play music files by means of the `sound` function. To demonstrate this function, it also has a built-in music file with a short segment of Handel's Messiah. In this example, we will use the `randn` function to create noise, and then we'll add the noise to the music clip.

Music is stored in MATLAB® as an array with values from −1 to 1. To convert this array into music, the `sound` function requires a sample frequency. The handel.mat file contains both an array representing the music and the value of the

Figure 3.15
Utah Symphony
Orchestra.

sample frequency. To hear the Messiah, you must first load the file, using the command

```
load handel
```

Notice that two new variables—y and Fs—were added to the workspace window when the handel file was loaded. To play the clip, type

```
sound(y, Fs)
```

Experiment with different values of Fs to hear the effect of different sample frequencies on the music. (Clearly, the sound must be engaged on your computer, or you won't be able to hear the playback.)

1. State the Problem
   Add a noise component to the recording of Handel's Messiah included with MATLAB®.

2. Describe the Input and Output
   Input   MATLAB® data file of Handel's Messiah, stored as the built-in file handel

   Output An array representing the Messiah, with static added
                 A graph of the first 200 elements of the data file

3. Develop a Hand Example
   Since the data in the music file vary between -1 and $+1$, we should add noise values of a smaller order of magnitude. First we'll try values centered on 0 and with a standard deviation of 0.1.

4. Develop a MATLAB® Solution

```
%Example 3.5
%Noise
load handel          %Load the music data file
sound(y,Fs)          %Play the music data file
pause                %Pause to listen to the music
% Be sure to hit enter to continue after playing the music
% Add random noise
noise=randn(length(y),1)*0.10;
sound(y+noise,Fs)
```

This program allows you to play the recording of the Messiah, both with and without the added noise. You can adjust the multiplier on the noise line to observe the effect of changing the magnitude of the added static. For example:

```
noise=randn(length(y),1)*0.20
```

5. Test the Solution
   In addition to playing back the music both with and without the added noise, we could plot the results. Because the file is quite large (73,113 elements), we'll just plot the first 200 points:

```
% Plot the first 200 data points in each file
t=1:length(y);
noisy = y + noise;
plot(t(1,1:200),y(1:200,1),t(1,1:200),noisy(1:200,1),':')
title('Handel"s Messiah')
xlabel('Element Number in Music Array')
ylabel('Frequency')
```

(continued)

Figure 3.16
Handel's *Messiah*. The solid line represents the original data, and the dotted line is the data to which we've added noise.

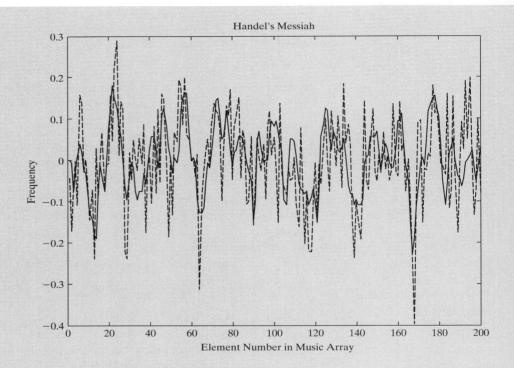

These commands tell MATLAB® to plot the index number of the data on the x-axis and the value stored in the music arrays on the y-axis. Plotting is introduced in more detail in a later chapter.

In Figure 3.16, the solid line represents the original data, and the dotted line the data to which we've added noise. As expected, the noisy data has a bigger range and doesn't always follow the same pattern as the original.

3.7 COMPLEX NUMBERS

MATLAB® includes several functions used primarily with complex numbers. Complex numbers consist of two parts: a real and an imaginary component. For example,

$$5 + 3i$$

COMPLEX NUMBER
A number with both real and imaginary components

is a complex number. The real component is 5, and the imaginary component is 3. Complex numbers can be entered into MATLAB® in two ways: as an addition problem, such as

```
A = 5 + 3i    or    A = 5+3*i
```

or with the complex function, as in

```
A = complex(5,3)
```

which returns

```
A =
   5.0000 + 3.0000i
```

As is standard in MATLAB®, the input to the `complex` function can be either two scalars or two arrays of values. Thus, if x and y are defined as

```
x = 1:3;
y = [-1,5,12];
```

then the `complex` function can be used to define an array of complex numbers as follows:

```
complex(x,y)
ans =
  1.0000 - 1.0000i 2.0000 + 5.0000i 3.0000 +12.0000i
```

The `real` and `imag` functions can be used to separate the real and imaginary components of complex numbers. For example, for A = 5 + 3*i, we have

```
real(A)
ans =
   5
imag(A)
ans =
   3
```

The `isreal` function can be used to determine whether a variable is storing a complex number. It returns a 1 if the variable is real and a 0 if it is complex. Since A is a complex number, we get

```
isreal(A)
ans =
   0
```

Thus, the `isreal` function is false and returns a value of 0.

The complex conjugate of a complex number consists of the same real component, but an imaginary component of the opposite sign. The `conj` function returns the complex conjugate:

```
conj(A)
ans =
   5.0000 - 3.0000i
```

The transpose operator also returns the complex conjugate of an array, in addition to converting rows to columns and columns to rows. Thus, we have

```
A'
ans =
   5.0000 - 3.0000i
```

Of course, in this example A is a scalar. We can create a complex array B by using A and performing both addition and multiplication operations:

```
B = [A, A+1, A*3]
B =
   5.0000 + 3.0000i 6.0000 + 3.0000i 15.0000 + 9.0000i
```

The transpose of B is

```
B'
ans =
    5.0000 - 3.0000i
    6.0000 - 3.0000i
   15.0000 - 9.0000i
```

Complex numbers are often thought of as describing a position on an x–y plane. The real part of the number corresponds to the x-value, and the imaginary component corresponds to the y-value, as shown in Figure 3.17a. Another way to think about this point is to describe it with polar coordinates—that is, with a radius and an angle (Figure 3.17b).

MATLAB® includes functions to convert complex numbers from Cartesian to polar form.

When the absolute-value function is used with a complex number, it calculates the radius, using the Pythagorean theorem:

```
abs(A)
ans =
    5.8310
```

$$\text{radius} = \sqrt{(\text{real component})^2 + (\text{imaginary component})^2}$$

Since, in this example, the real component is 5, and the imaginary component is 3,

$$\text{radius} = \sqrt{5^2 + 3^2} = 5.8310$$

We could also calculate the radius in MATLAB®, using the real and imag functions described earlier:

```
sqrt(real(A).^2 + imag(A).^2)
ans =
    5.8310
```

Similarly, the angle is found with the angle function:

```
angle(A)
ans =
    0.5404
```

Figure 3.17
(a) Complex number represented in a Cartesian coordinate system. (b) A complex number can also be described with polar coordinates.

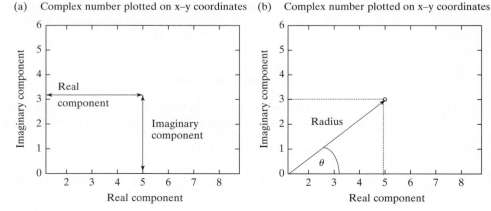

(a) Complex number plotted on x–y coordinates (b) Complex number plotted on x–y coordinates

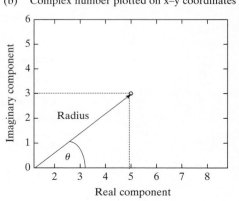

The result is expressed in radians. Both functions, `abs` and `angle`, will accept scalars or arrays as input. Recall that B is a 1×3 array of complex numbers:

```
B =
   5.0000 + 3.0000i 6.0000 + 3.0000i 15.0000 + 9.0000i
```

The `abs` function returns the radius if the number is represented in polar coordinates:

```
abs(B)
ans =
    5.8310 6.7082 17.4929
```

The angle from the horizontal can be found with the `angle` function:

```
angle(B)
ans =
    0.5404 0.4636 0.5404
```

The MATLAB® functions commonly used with complex numbers are summarized in Table 3.14.

Table 3.14 Functions Used with Complex Numbers

abs(x)	Computes the absolute value of a complex number, using the Pythagorean theorem. This is equivalent to the radius if the complex number is represented in polar coordinates. For example, if $x = 3 + 4i$, the absolute value is $\sqrt{3^2 + 4^2} = 5$	**x=3+4i;** **abs(x)** **ans =** **5**
angle(x)	Computes the angle from the horizontal in radians when a complex number is represented in polar coordinates.	**x=3+4i;** **angle(x)** **ans =** **0.9273**
complex(x,y)	Generates a complex number with a real component x and an imaginary component y.	**x=3;** **y=4;** **complex(x,y)** **ans =** **3.0000 + 4.0000i**
real(x)	Extracts the real component from a complex number.	**x=3+4i;** **real(x)** **ans =** **3**
imag(x)	Extracts the imaginary component from a complex number.	**x=3+4i;** **imag(x)** **ans =** **4**
isreal(x)	Determines whether the values in an array are real. If they are real, the function returns a 1; if they are complex, it returns a 0.	**x=3+4i;** **isreal(x)** **ans =** **0**
conj(x)	Generates the complex conjugate of a complex number.	**x=3+4i;** **conj(x)** **ans =** **3.0000 - 4.0000i**

PRACTICE EXERCISES 3.10

1. Create the following complex numbers:
   a. $A = 1 + i$
   b. $B = 2 - 3i$
   c. $C = 8 + 2i$
2. Create a vector D of complex numbers whose real components are 2, 4, and 6 and whose imaginary components are −3, 8, and −16.
3. Find the magnitude (absolute value) of each of the vectors you created in Exercises 1 and 2.
4. Find the angle from the horizontal of each of the complex numbers you created in Exercises 1 and 2.
5. Find the complex conjugate of vector D.
6. Use the transpose operator to find the complex conjugate of vector D.
7. Multiply A by its complex conjugate, and then take the square root of your answer. How does this value compare against the magnitude (absolute value) of A?

3.8 COMPUTATIONAL LIMITATIONS

KEY IDEA

There is a limit to how small or how large a number can be handled by computer programs

The variables stored in a computer can assume a wide range of values. On the majority of computers, the range extends from about 10^{-308} to 10^{308}, which should be enough to accommodate most computations. MATLAB® includes functions to identify the largest real numbers and the largest integers the program can process (Table 3.15).

The value of `realmax` corresponds roughly to 2^{1024}, since computers actually perform their calculations in binary (base-2) arithmetic. Of course, it is possible to formulate a problem in which the result of an expression is larger or smaller than the permitted maximum. For example, suppose that we execute the following commands:

```
x = 2.5e200;
y = 1.0e200;
z = x*y
```

Table 3.15 Computational Limits

realmax	Returns the largest possible floating-point number used in MATLAB®.	**realmax** ans = **1.7977e+308**
realmin	Returns the smallest possible floating-point number used in MATLAB®.	**realmin** ans = **2.2251e-308**
intmax	Returns the largest possible integer number used in MATLAB®.	**intmax** ans = **2147483647**
intmin	Returns the smallest possible integer number used in MATLAB®.	**intmin** ans = **−2147483648**

OVERFLOW
A calculational result that is too large for the computer program to handle

UNDERFLOW
A calculational result that is too small for the computer program to distinguish from zero

MATLAB® responds with

```
z =
    Inf
```

because the answer (2.5**e**400) is outside the allowable range. This error is called **exponent overflow**, because the exponent of the result of an arithmetic operation is too large to store in the computer's memory.

Exponent underflow is a similar error, caused by the exponent of the result of an arithmetic operation being too *small* to store in the computer's memory. Using the same allowable range, we obtain an exponent underflow with the following commands:

```
x = 2.5e-200;
y = 1.0e200
z = x/y
```

Together, these commands return

```
z = 0
```

The result of an exponent underflow is zero.

We also know that division by zero is an invalid operation. If an expression results in a division by zero, the result of the division is infinity:

```
z = y/0
z =
        Inf
```

KEY IDEA
Careful planning can help you avoid calculational overflow or underflow

MATLAB® may print a warning telling you that division by zero is not possible.

In performing calculations with very large or very small numbers, it may be possible to reorder the calculations to avoid an underflow or an overflow. Suppose, for example, that you would like to perform the following string of multiplications:

$$(2.5 \times 10^{200}) \times (2 \times 10^{200}) \times (1 \times 10^{-100})$$

The answer is 5×10^{300}, within the bounds allowed by MATLAB®. However, consider what happens when we enter the problem into MATLAB®:

```
2.5e200*2e200*1e-100
ans =
        Inf
```

Because MATLAB® executes the problem from left to right, the first multiplication yields a value outside the allowable range (5×10^{400}), resulting in an answer of infinity. However, by rearranging the problem to

```
2.5e200*1e-100*2e200
ans =
5.0000e+300
```

we avoid the overflow and find the correct answer.

3.9 SPECIAL VALUES AND MISCELLANEOUS FUNCTIONS

Most, but not all, functions require an input argument. Although used as if they were scalar constants, the functions listed in Table 3.16 do *not* require any input.

Table 3.16 Special Functions

pi	Mathematical constant π.	pi **ans =** **3.1416**
I	Imaginary number.	i **ans =** **0 + 1.0000i**
J	Imaginary number.	i **ans =** **0 + 1.0000i**
Inf	Infinity, which often occurs during a calculational overflow or when a number is divided by zero.	5/0 **Warning: Divide by zero.** **ans =** **Inf**

NaN	Not a number.
	Occurs when a calculation is undefined.
		0/0
		Warning: Divide by zero.
		ans =
		  NaN
		inf/inf
		ans =
		  NaN

clock	Current time.
	Returns a six-member array [year month day hour minute second].
	When the clock function was called on July 19, 2008, at 5:19 p.m. and 30.0 seconds, MATLAB® returned the output shown at the right.
	
	The fix and clock functions together result in a format that is easier to read. The fix function rounds toward zero. A similar result could be obtained by setting format bank.
		clock
		ans =
		  1.0e+003 *
		2.0080 0.0070 0.0190
		0.0170 0.0190 0.0300
		fix(clock)
		ans =
		  2008 7 19
		  17 19 30

date	Current date. Similar to the clock function. However, it returns the date in a "string format."	date **ans =** **19-Jul-2008**
eps	The distance between 1 and the next-larger double-precision floating-point number.	eps **ans =** **2.2204e-016**

MATLAB® allows you to redefine these special values as variable names; however, doing so can have unexpected consequences. For example, the following MATLAB® code is allowed, even though it is not wise:

```
pi = 12.8;
```

From this point on, whenever the variable pi is called, the new value will be used. Similarly, you can redefine any function as a variable name, such as

```
sin = 10;
```

To restore sin to its job as a trigonometric function (or to restore the default value of pi), you must clear the workspace with

```
clear
```

or you may clear each variable independently with

```
clear sin
clear pi
```

Now check to see the result by issuing the command for π.

    pi

This command returns

    pi =
        3.1416

HINT

The function i is the most common of these functions to be unintentionally renamed by MATLAB® users.

The NaN function stands for "not a number." It is returned when a user attempts a calculation where the result is undefined—for example 0/0. It can also be useful as a placeholder in an array.

PRACTICE EXERCISES 3.11

1. Use the `clock` function to add the time and date to your work sheet.
2. Use the `date` function to add the date to your work sheet.
3. Convert the following calculations to MATLAB® code and explain your results:
   a. 322! (Remember that, to a mathematician, the symbol ! means factorial.)
   b. $5 * 10^{500}$
   c. $1/5 * 10^{500}$
   d. 0/0

SUMMARY

In this chapter, we explored a number of predefined MATLAB® functions, including the following:

- General mathematical functions, such as
  - exponential functions
  - logarithmic functions
  - roots
- Rounding functions
- Functions used in discrete mathematics, such as
  - factoring functions
  - prime-number functions
- Trigonometric functions, including
  - standard trigonometric functions
  - inverse trigonometric functions
  - hyperbolic trigonometric functions
  - trigonometric functions that use degrees instead of radians
- Data analysis functions, such as
  - maxima and minima
  - averages (mean and median)

○ sums and products
○ sorting
○ standard deviation and variance
- Random-number generation for both
○ uniform distributions
○ Gaussian (normal) distributions
- Functions used with complex numbers

We explored the computational limits inherent in MATLAB® and introduced special values, such as pi, that are built into the program.

MATLAB® SUMMARY

The following MATLAB® summary lists and briefly describes all of the special characters, commands, and functions that were defined in this chapter:

Special Characters and Functions	
eps	smallest difference recognized
i	imaginary number
clock	returns the time
date	returns the date
Inf	infinity
intmax	returns the largest possible integer number used in MATLAB®
intmin	returns the smallest possible integer number used in MATLAB®
j	imaginary number
NaN	not a number
pi	mathematical constant π
realmax	returns the largest possible floating-point number used in MATLAB®
realmin	returns the smallest possible floating-point number used in MATLAB®

Commands and Functions	
abs	computes the absolute value of a real number or the magnitude of a complex number
angle	computes the angle when complex numbers are represented in polar coordinates
asin	computes the inverse sine (arcsine)
asind	computes the inverse sine and reports the result in degrees
ceil	rounds to the nearest integer toward positive infinity
complex	creates a complex number
conj	creates the complex conjugate of a complex number
cos	computes the cosine
cumprod	computes a cumulative product of the values in an array
cumsum	computes a cumulative sum of the values in an array
erf	calculates the error function
exp	computes the value of e^x
factor	finds the prime factors
factorial	calculates the factorial
fix	rounds to the nearest integer toward zero
floor	rounds to the nearest integer toward minus infinity
gcd	finds the greatest common denominator
help	opens the help function
helpwin	opens the windowed help function

Commands and Functions

imag	extracts the imaginary component of a complex number
isprime	determines whether a value is prime
isreal	determines whether a value is real or complex
lcm	finds the least common multiple
length	determines the largest dimension of an array
log	computes the natural logarithm or the logarithm to the base e ($\log_e$)
log10	computes the common logarithm or the logarithm to the base 10 ($\log_{10}$)
log2	computes the logarithm to the base 2 ($\log_2$)
max	finds the maximum value in an array and determines which element stores the maximum value
mean	computes the average of the elements in an array
median	finds the median of the elements in an array
min	finds the minimum value in an array and determines which element stores the minimum value
mode	finds the most common number in an array
nchoosek	finds the number of possible combinations when a subgroup of k values is chosen from a group of n values.
nthroot	find the real nth root of the input matrix
numel	determines the total number of elements in an array
primes	finds the prime numbers less than the input value
prod	multiplies the values in an array
rand	calculates evenly distributed random numbers
randn	calculates normally distributed (Gaussian) random numbers
rats	converts the input to a rational representation (i.e., a fraction)
real	extracts the real component of a complex number
rem	calculates the remainder in a division problem
round	rounds to the nearest integer
sign	determines the sign (positive or negative)
sin	computes the sine, using radians as input
sind	computes the sine, using angles in degrees as input
sinh	computes the hyperbolic sine
size	determines the number of rows and columns in an array
sort	sorts the elements of a vector
sortrows	sorts the rows of a vector on the basis of the values in the first column
sound	plays back music files
sqrt	calculates the square root of a number
std	determines the standard deviation
sum	sums the values in an array
tan	computes the tangent, using radians as input
var	computes the variance

KEY TERMS

argument	variation	real numbers
average	mean	seed
complex numbers	median	standard deviation
discrete mathematics	nesting	underflow
function	normal random variation	uniform random
function input	overflow	number
Gaussian random	rational numbers	variance

PROBLEMS

Elementary Math Functions

3.1 Find the cube root of -5, both by using the `nthroot` function and by raising -5 to the $1/3$ power. Explain the difference in your answers. Prove that both results are indeed correct answers by cubing them and showing that they equal -5.

3.2 MATLAB® contains functions to calculate the natural logarithm (`log`), the logarithm to the base 10 (`log10`), and the logarithm to the base 2 (`log2`). However, if you want to find a logarithm to another base—for example, base b—you'll have to do the math yourself with the formula

$$\log_b(x) = \frac{\log_e(x)}{\log_e(b)}$$

What is the $\log_b$ of 10 when b is defined from 1 to 10 in increments of 1?

3.3 Populations tend to expand exponentially, that is,

$$P = P_0 e^{rt}$$

where

P = current population
P_0 = original population
r = continuous growth rate, expressed as a fraction
t = time.

If you originally have 100 rabbits that breed at a continuous growth rate of 90% $(r = 0.9)$ per year, find how many rabbits you will have at the end of 10 years.

3.4 Chemical reaction rates are proportional to a rate constant k that changes with temperature according to the Arrhenius equation

$$k = k_0 e^{-Q/RT}$$

For a certain reaction,

$$Q = 8000 \text{ cal/mol}$$

$$R = 1.987 \text{ cal/mol K}$$

$$k_0 = 1200 \text{ min}^{-1}$$

Find the values of k for temperatures from 100 K to 500 K, in 50° increments. Create a table of your results.

3.5 Consider the air-conditioning requirements of the large home shown in Figure P3.5.

The interior of the house is warmed by waste heat from lighting and electrical appliances, by heat leaking in from the outdoors, and by heat generated by the people in the home. An air-conditioner must be able to remove all this thermal energy in order to keep the inside temperature from rising. Suppose there are 20 light bulbs emitting 100 J/s of energy each and four appliances emitting 500 J/s each. Suppose also that heat leaks in from the outside at a rate of 3000 J/s.

(a) How much heat must the air-conditioner be able to remove from the home per second?

Figure P3.5

Air conditioning must remove heat from a number of sources.

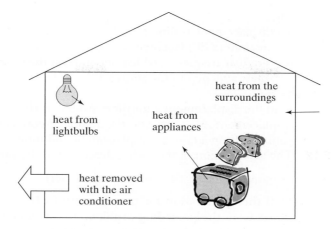

(b) One particular air-conditioning unit can handle 2000 J/s. How many of these units are needed to keep the home at a constant temperature?

3.6. **(a)** If you have four people, how many different ways can you arrange them in a line?

(b) If you have 10 different tiles, how many different ways can you arrange them?

3.7. **(a)** If you have 12 people, how many different committees of two people each can you create? Remember that a committee of Bob and Alice is the same as a committee of Alice and Bob.

(b) How many different soccer teams of 11 players can you form from a class of 30 students? (Combinations—order does not matter).

(c) Since each player on a soccer team is assigned a particular role, order does matter. Recalculate the possible number of different soccer teams that can be formed when order is taken into account.

3.8 There are 52 *different* cards in a deck. How many different hands of 5 cards each are possible? Remember, every hand can be arranged 120 (5!) different ways.

3.9 Very large prime numbers are used in cryptography. How many prime numbers are there between 10,000 and 20,000? (These aren't big enough primes to be useful in ciphers.) (*Hint*: Use the `primes` function and the `length` command.)

Trigonometric Functions

3.10 Sometimes it is convenient to have a table of sine, cosine, and tangent values instead of using a calculator. Create a table of all three of these trigonometric functions for angles from 0 to 2π, with a spacing of 0.1 radian. Your table should contain a column for the angle and then for the sine, cosine, and tangent.

3.11 The displacement of the oscillating spring shown in Figure P3.11 can be described by

$$x = A \cos(\omega t)$$

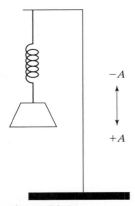

Figure P3.11
An oscillating spring.

where
x = displacement at time t
A = maximum displacement
ω = angular frequency, which depends on the spring constant and the mass attached to the spring
t = time.

Find the displacement x for times from 0 to 10 seconds when the maximum displacement A is 4 cm, and the angular frequency is 0.6 radian/s. Present your results in a table of displacement and time values.

3.12 The acceleration of the spring described in the preceding exercise is

$$a = -A\omega^2 \cos(\omega t)$$

Find the acceleration for times from 0 to 10 seconds, using the constant values from the preceding problem. Create a table that includes the time, the displacement from corresponding values in the previous exercise, and the acceleration.

3.13 You can use trigonometry to find the height of a building as shown in Figure P3.13. Suppose you measure the angle between the line of sight and the horizontal line connecting the measuring point and the building. You can calculate the height of the building with the following formulas:

$$\tan(\theta) = h/d$$

$$h = d \tan(\theta)$$

Assume that the distance to the building along the ground is 120 m and the angle measured along the line of sight is 30° ±3°. Find the maximum and minimum heights the building can be.

3.14 Consider the building from the previous exercise.

(a) If it is 200 feet tall and you are 20 feet away, at what angle from the ground will you have to tilt your head to see the top of the building? (Assume that your head is even with the ground.)

(b) How far is it from your head to the top of the building?

Figure P3.13
You can determine the height of a building with trigonometry.

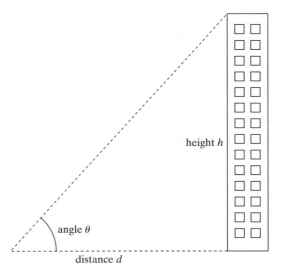

Data Analysis Functions

3.15 Consider the following table of data representing temperature readings in a reactor:

Thermocouple 1	Thermocouple 2	Thermocouple 3
84.3	90.0	86.7
86.4	89.5	87.6
85.2	88.6	88.3
87.1	88.9	85.3
83.5	88.9	80.3
84.8	90.4	82.4
85.0	89.3	83.4
85.3	89.5	85.4
85.3	88.9	86.3
85.2	89.1	85.3
82.3	89.5	89.0
84.7	89.4	87.3
83.6	89.8	87.2

Your instructor may provide you with a file named thermocouple.dat, or you may need to enter the data yourself.

Use MATLAB® to find

(a) The maximum temperature measured by each thermocouple.
(b) The minimum temperature measured by each thermocouple.

3.16 The range of an object shot at an angle θ with respect to the x-axis and an initial velocity v_0 (Figure P3.16) is given by

$$\text{Range} = \frac{v_0^2}{g}\sin(2\theta)$$

for $0 \leq \theta \leq \pi/2$ and neglecting air resistance. Use $g = 9.81 \text{ m/s}^2$ and an initial velocity v_0 of 100 m/s. Show that the maximum range is obtained at approximately $\theta = \pi/4$ by computing the range in increments of $\pi/100$ between $0 \leq \theta \leq \pi/2$. You won't be able to find the exact angle that results in the maximum range, because your calculations are at evenly spaced angles of $\pi/100$ radian.

3.17 The vector

G=[68, 83, 61, 70, 75, 82, 57, 5, 76, 85, 62, 71, 96, 78, 76, 68, 72, 75, 83, 93]

represents the distribution of final grades in a dynamics course. Compute the mean, median, mode, and standard deviation of G. Which better represents the "most typical grade," the mean, median, or mode? Why? Use MATLAB® to determine the number of grades in the array (don't just count them) and to sort them into ascending order.

3.18 Generate 10,000 Gaussian random numbers with a mean of 80 and standard deviation of 23.5. (You'll want to suppress the output so that you don't overwhelm the command window with data.) Use the `mean` function to confirm that your array actually has a mean of 80. Use the `std` function to confirm that your standard deviation is actually 23.5.

3.19 Use the `date` function to add the current date to your homework.

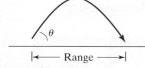

Figure P3.16
The range depends on the launch angle and the launch velocity.

Random Numbers

3.20 Many games require the player to roll two dice. The number on each die can vary from 1 to 6.

(a) Use the rand function in combination with a rounding function to create a simulation of one roll of one die.

(b) Use your results from part (a) to create a simulation of the value rolled with a second die.

(c) Add your two results to create a value representing the total rolled during each turn.

(d) Use your program to determine the values rolled in a favorite board game, or use the game shown in Figure P3.20.

3.21 Suppose you are designing a container to ship sensitive medical materials between hospitals. The container needs to keep the contents within a specified temperature range. You have created a model predicting how the container responds to the exterior temperature, and you now need to run a simulation.

(a) Create a normal distribution (Gaussian distribution) of temperatures with a mean of 70°F and a standard deviation of 2°, corresponding to a 2-hour duration. You'll need a temperature for each time value from 0 to 120 minutes. (That's 121 values.)

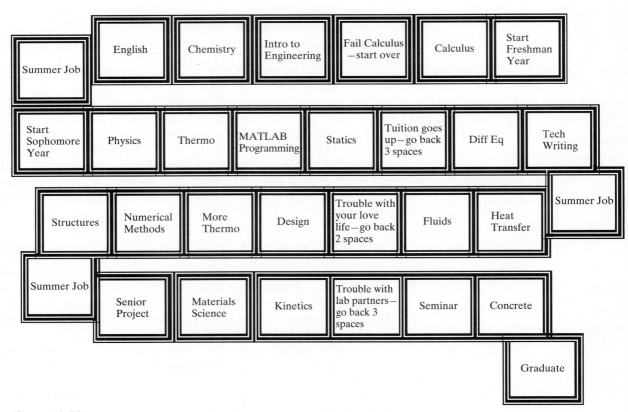

Figure P3.20
The college game.

(b) Plot the data on an x–y plot. Don't worry about labels. Recall that the MATLAB® function for plotting is plot(x,y).

(c) Find the maximum temperature, the minimum temperature, and the times at which they occur.

Complex Numbers

3.22 Consider the circuit shown in Figure P3.22, which includes the following:
- A sinusoidally varying voltage source, V.
- An inductor, with an inductance, L.
- A capacitor, with a capacitance, C.
- A resistor, with a resistance, R.

We can find the current, I, in the circuit by using Ohm's law (generalized for alternating currents),

$$V = IZ_T$$

where Z_T is the total impedance in the circuit. (Impedance is the AC corollary to resistance.)

Assume that the impedance for each component is as follows:

$$Z_L = 0 + 5j \text{ ohms}$$

$$Z_C = 0 - 15j \text{ ohms}$$

$$R = Z_R = 5 + 0j \text{ ohms}$$

$$Z_T = Z_C + Z_L + R$$

and that the applied voltage is

$$V = 10 + 0j \text{ volts}$$

(Electrical engineers usually use j instead of i for imaginary numbers.)

Find the current, I, in the circuit. You should expect a complex number as a result. Enter the complex values of impedance into your calculations using the **complex** function.

3.23 Impedance is related to the inductance, L, and the capacitance, C, by the following equations

$$Z_C = \frac{1}{\omega C j}$$

$$Z_L = \frac{1}{\omega L j}$$

For a circuit similar to the one shown in Figure P3.22 assume the following:

Figure P3.22
A simple circuit illustrating a sinusoidally varying voltage source, V.

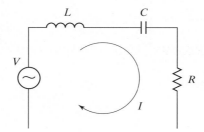

$$C = 1\,\mu F \text{ (microfarads)}$$

$$L = 200 \text{ mH (millihenries)}$$

$$R = 5 \text{ ohms}$$

$$f = 15 \text{ kHz (kilohertz)}$$

$$\omega = 2\pi f$$

$$V = 10 \text{ volts}$$

(a) Find the impedance for the capacitor (Z_C) and for the inductor (Z_L).

(b) Find the total impedance

$$Z_T = Z_C + Z_L + R$$

(c) Find the current by solving Ohm's law for I.

$$V = IZ_T$$

(d) Electrical engineers often describe complex parameters using polar coordinates, that is, the parameter has both an angle and a magnitude. (Imagine plotting a point on the complex plane, where the x-axis represents the real part of the number, and the y-axis represents the imaginary part of the number.) Use the abs function to find the magnitude of the current found in part c, and use the angle function to find the corresponding angle.

4 Manipulating MATLAB® Matrices

Objectives

After reading this chapter, you should be able to:

- Manipulate matrices
- Extract data from matrices

- Solve problems with two matrix variables of different sizes
- Create and use special matrices

4.1 MANIPULATING MATRICES

As you solve more and more complicated problems with MATLAB®, you'll find that you will need to combine small matrices into larger matrices, extract information from large matrices, create very large matrices, and use matrices with special properties.

4.1.1 Defining Matrices

In MATLAB®, you can define a matrix by typing in a list of numbers enclosed in square brackets. You can separate the numbers by spaces or by commas, at your discretion. (You can even combine the two techniques in the same matrix definition.) To indicate a new row, you can use a semicolon. For example,

```
A = [3.5];
B = [1.5, 3.1]; or B = [1.5 3.1];
C = [-1, 0, 0; 1, 1, 0; 0, 0, 2];
```

You can also define a matrix by listing each row on a separate line, as in the following set of MATLAB® commands:

```
C = row[-1,  0, 0;
          1,  1, 0;
          1, -1, 0;
          0,  0, 2]
```

You don't even need to enter the semicolon to indicate a new row. MATLAB® interprets

```
C =    [-1,  0, 0
        1,  1, 0
        1, -1, 0
        0,  0, 2]
```

as a 4×3 matrix. You could also enter a column matrix in this manner:

```
A = [
       1
       2
       3 ]
```

ELLIPSIS
A set of three periods used to indicate that a row is continued on the next line

If there are too many numbers in a row to fit on one line, you can continue the statement on the next line, but a comma and an ellipsis (...) are required at the end of the line, indicating that the row is to be continued. You can also use the ellipsis to continue other long assignment statements in MATLAB®.

If we want to define **F** with 10 values, we can use either of the following statements:

```
F = [1, 52, 64, 197, 42, -42, 55, 82, 22, 109]; or
F = [1, 52, 64, 197, 42, -42, ...
        55, 82, 22, 109];
```

MATLAB® also allows you to define a matrix in terms of another matrix that has already been defined. For example, the statements

```
B = [1.5, 3.1];
S = [3.0, B]
```

return

```
S =
       3.0    1.5    3.1
```

Similarly,

```
T = [ 1, 2, 3; S]
```

returns

```
T =
       1    2    3
       3   1.5   3.1
```

INDEX
A number used to identify elements in an array

We can change values in a matrix, or include additional values, by using an index number to specify a particular element. This process is called **indexing into an array**. Thus, the command

```
S(2) = -1.0;
```

changes the second value in the matrix **S** from 1.5 to -1. If we type the matrix name

```
S
```

into the command window, then MATLAB® returns

```
S =
       3.0   -1.0   3.1
```

We can also extend a matrix by defining new elements. If we execute the command

```
S(4) = 5.5;
```

we extend the matrix **S** to four elements instead of three. If we define element

```
S(8) = 9.5;
```

matrix **S** will have eight values, and the values of $S(5)$, $S(6)$, and $S(7)$ will be set to 0. Thus,

```
S
```

returns

```
S =
      3.0    -1.0    3.1    5.5    0    0    0    9.5
```

4.1.2 Using the Colon Operator

The colon operator is very powerful in defining new matrices and modifying existing ones. First, we can use it to define an evenly spaced matrix. For example,

```
H = 1:8
```

returns

```
H =
      1    2    3    4    5    6    7    8
```

The default spacing is 1. However, when colons are used to separate three numbers, the middle value becomes the spacing. Thus,

```
time = 0.0 : 0.5 : 2.0
```

returns

```
time =
      0    0.5000    1.0000    1.5000    2.0000
```

The colon operator can also be used to extract data from matrices, a feature that is very useful in data analysis. When a colon is used in a matrix reference in place of a specific index number, the colon represents the entire row or column.

Suppose we define M as

```
M = [1 2 3 4 5;
     2 3 4 5 6;
     3 4 5 6 7];
```

We can extract column 1 from matrix M with the command

```
x = M(:, 1)
```

which returns

```
x =
      1
      2
      3
```

We read this syntax as "all the rows in column 1." We can extract any of the columns in a similar manner. For instance,

```
y = M(:, 4)
```

returns

```
y =
        4
        5
        6
```

and can be interpreted as "all the rows in column 4." Similarly, to extract a row,

```
z = M(1,:)
```

returns

```
z =
        1    2    3    4    5
```

and is read as "row 1, all the columns."

We don't have to extract an entire row or an entire column. The colon operator can also be used to mean "from row to row" or "from column to column." To extract the two bottom rows of the matrix M, type

```
w = M(2:3,:)
```

which returns

```
w =
        2    3    4    5    6
        3    4    5    6    7
```

and reads "rows 2 to 3, all the columns." Similarly, to extract just the four numbers in the lower right-hand corner of matrix M,

```
w = M(2:3, 4:5)
```

returns

```
w =
        5    6
        6    7
```

and reads "rows 2 to 3 in columns 4 to 5."

In MATLAB®, it is valid to have a matrix that is empty. For example, each of the following statements will generate an empty matrix:

```
a = [ ];
b = 4:-1:5;
```

Finally, using the matrix name with a single colon, such as

```
M(:)
```

transforms the matrix into one long column.

```
M =
   1
   2
   3
   2
   3
   4
   3
   4
   5
   4
   5
   6
   5
   6
   7
```

The matrix was formed by first listing column 1, then adding column 2 onto the end, tacking on column 3, and so on. Actually, the computer does not store two-dimensional arrays in a two-dimensional pattern. Rather, it "thinks" of a matrix as one long list, just like the matrix **M** at the left. There are two ways you can extract a single value from an array: by using a single index number or by using the row, column notation. To find the value in row 2, column 3, use the following commands:

```
M
M =
        1    2    3    4    5
        2    3    4    5    6
        3    4    5    6    7
M(2, 3)
ans =
             4
```

Alternatively, you can use a single index number. The value in row 2, column 3 of matrix M is element number 8. (Count down column 1, then down column 2, and finally down column 3 to the correct element.) The associated MATLAB® command is

```
M(8)
ans = 4
```

HINT

You can use the word "end" to identify the final row or column in a matrix, even if you don't know how big it is. For example,

```
M(1,end)
```

returns

```
M(1,end)
ans =
        5
```

and

```
M(end, end)
```

returns

```
ans =
        7
```

as does

```
M(end)
ans =
        7
```

PRACTICE EXERCISES 4.1

Create MATLAB® variables to represent the following matrices, and use them in the exercises that follow:

$$a = \begin{bmatrix} 12 & 17 & 3 & 6 \end{bmatrix} \qquad b = \begin{bmatrix} 5 & 8 & 3 \\ 1 & 2 & 3 \\ 2 & 4 & 6 \end{bmatrix} \qquad c = \begin{bmatrix} 22 \\ 17 \\ 4 \end{bmatrix}$$

1. Assign to the variable x1 the value in the second column of matrix a. This is sometimes represented in mathematics textbooks as element $a_{1,2}$ and could be expressed as x1 = $a_{1,2}$.
2. Assign to the variable x2 the third column of matrix b.
3. Assign to the variable x3 the third row of matrix b.
4. Assign to the variable x4 the values in matrix b along the diagonal (i.e., elements $b_{1,1}$, $b_{2,2}$, and $b_{3,3}$).
5. Assign to the variable x5 the first three values in matrix a as the first row and all the values in matrix b as the second through the fourth row.
6. Assign to the variable x6 the values in matrix c as the first column, the values in matrix b as columns 2, 3, and 4, and the values in matrix a as the last row.
7. Assign to the variable x7 the value of element 8 in matrix b, using the single-index-number identification scheme.
8. Convert matrix b to a column vector named x8.

EXAMPLE 4.1

USING TEMPERATURE DATA

The data collected by the National Weather Service are extensive but are not always organized in exactly the way we would like (Figure 4.1). Take, for example, the summary of the 1999 Asheville, North Carolina, Climatological Data. We'll use these data to practice manipulating matrices—both extracting elements and recombining elements to form new matrices.

Figure 4.1
Temperature data collected from a weather satellite were used to create this composite false-color image. (Courtesy of NASA/Jet Propulsion Laboratory.)

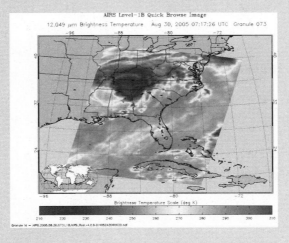

The numeric information has been extracted from the table and is in an Excel file called **Asheville_1999.xls** (Appendix D, available online). Use MATLAB® to confirm that the reported values on the annual row are correct for the mean maximum temperature and the mean minimum temperature, as well as for the annual high temperature and the annual low temperature. Combine these four columns of data into a new matrix called **temp_data**.

1. State the Problem

   Calculate the annual mean maximum temperature, the annual mean minimum temperature, the highest temperature reached during the year, and the lowest temperature reached during the year for 1999 in Asheville, North Carolina.

2. Describe the Input and Output

   Input   Import a matrix from the Excel file **Asheville_1999.xls**.

   Output   Find the following four values: annual mean maximum temperature
                                                  annual mean minimum temperature
                                                  highest temperature
                                                  lowest temperature

   Create a matrix composed of the mean maximum temperature values, the mean minimum temperature values, the highest monthly temperatures, and the lowest monthly temperatures. Do not include the annual data.

3. Develop a Hand Example

   Using a calculator, add the values in column 2 of the table and divide by 12.

4. Develop a MATLAB® Solution

   First import the data from Excel, then save them in the current directory as **Asheville_1999**. Save the variable **Asheville_1999** as the file **Asheville_1999.mat**. This makes it available to be loaded into the workspace from our M-file program:

```
% Example 4.1
% In this example, we extract data from a large matrix and
% use the data analysis functions to find the mean high
% and mean low temperatures for the year and to find the
% high temperature and the low temperature for the year
%
clear, clc
% load the data matrix from a file
load asheville_1999
% extract the mean high temperatures from the large matrix
mean_max = asheville_1999(1:12,2);
% extract the mean low temperatures from the large matrix
mean_min = asheville_1999(1:12,3);
% Calculate the annual means
annual_mean_max = mean(mean_max)
annual_mean_min = mean(mean_min)
% extract the high and low temperatures from the large
% matrix
high_temp = asheville_1999(1:12,8);
low_temp = asheville_1999(1:12,10);
% Find the max and min temperature for the year
```

(continued)

```
max_high = max(high_temp)
min_low = min(low_temp)
%  Create a new matrix with just the temperature
%  information
new_table =[mean_max, mean_min, high_temp, low_temp]
```

The results are displayed in the command window:

```
annual_mean_max =
    68.0500
annual_mean_min =
    46.3250
max_high =
    96
min_low =
    9
new_table =
     51.4000    31.5000    78.0000     9.0000
     52.6000    32.1000    66.0000    16.0000
     52.7000    32.5000    76.0000    22.0000
     70.1000    48.2000    83.0000    34.0000
     75.0000    51.5000    83.0000    40.0000
     80.2000    60.9000    90.0000    50.0000
     85.7000    64.9000    96.0000    56.0000
     86.4000    63.0000    94.0000    54.0000
     79.1000    54.6000    91.0000    39.0000
     67.6000    45.5000    78.0000    28.0000
     62.2000    40.7000    76.0000    26.0000
     53.6000    30.5000    69.0000    15.0000
```

5. Test the Solution
   Compare the results against the bottom line of the table from the Asheville, North Carolina, Climatological Survey. It is important to confirm that the results are accurate before you start to use any computer program to process data.

4.2 PROBLEMS WITH TWO VARIABLES

All of the calculations we have done thus far have used only one variable. Of course, most physical phenomena can vary with many different factors. In this section, we consider how to perform the same calculations when the variables are represented by vectors.

Consider the following MATLAB® statements:

```
x = 3;
y = 5;
A = x * y
```

Since **x** and **y** are scalars, it's an easy calculation: $x \cdot y = 15$, or

```
A =
    15
```

Now, let's see what happens if **x** is a matrix and **y** is still a scalar:

```
x = 1:5;
```

returns five values of **x**. Because **y** is still a scalar with only one value (5),

```
A = x * y
```

returns

```
A =
         5    10    15    20    25
```

This is still a review. But what happens if **y** is now a vector? Then

```
y = 1:3;
A = x * y
```

returns an error statement:

```
??? Error using = => *
Inner matrix dimensions must agree.
```

This error statement reminds us that the asterisk is the operator for matrix multiplication, which is not what we want. We want the dot-asterisk operator (.*), which will perform an element-by-element multiplication. However, the two vectors, **x** and **y**, will need to be the same length for this to work. Thus,

```
y = linspace(1,3,5)
```

creates a new vector **y** with five evenly spaced elements:

```
y =
         1.0000    1.5000    2.0000    2.5000    3.0000
A = x .* y
A =
         1    3    6    10    15
```

However, although this solution works, the result is probably not what you really want. You can think of the results as the diagonal on a matrix (Table 4.1).

What if we want to know the result for element 3 of vector **x** and element 5 of vector **y**? This approach obviously doesn't give us all the possible answers. We want a two-dimensional matrix of answers that corresponds to all the combinations of **x** and **y**. In order for the answer **A**, to be a two-dimensional matrix, the input vectors must be two-dimensional matrices. MATLAB® has a built-in function called meshgrid that will help us accomplish this—and **x** and **y** don't even have to be the same size.

First, let's change **y** back to a three-element vector:

Table 4.1 Results of an Element-by-Element Calculation

			x			
		1	**2**	**3**	**4**	**5**
	1.0	1				
	1.5		3			
Y	**2.0**			6		
	2.5				10	
	3.0			?		15

```
y = 1:3;
```

Then, we'll use `meshgrid` to create a new two-dimensional version of both **x** and **y** that we'll call `new_x` and `new_y`:

```
[new_x, new_y]=meshgrid(x,y)
```

KEY IDEA

Use the meshgrid function to map two one-dimensional variables into two-dimensional variables of equal size

The `meshgrid` command takes the two input vectors and creates two two-dimensional matrices. Each of the resulting matrices has the same number of rows and columns. The number of columns is determined by the number of elements in the **x** vector, and the number of rows is determined by the number of elements in the **y** vector. This operation is called *mapping the vectors into a two-dimensional array*:

```
new_x =
        1    2    3    4    5
        1    2    3    4    5
        1    2    3    4    5
new_y =
        1    1    1    1    1
        2    2    2    2    2
        3    3    3    3    3
```

Notice that all the rows in `new_x` are the same and all the columns in `new_y` are the same. Now, it's possible to multiply `new_x` by `new_y` and get the two-dimensional grid of results we really want:

```
A = new_x.*new_y
A =
        1    2    3     4     5
        2    4    6     8    10
        3    6    9    12    15
```

PRACTICE EXERCISES 4.2

Using Meshgrid

1. The area of a rectangle (Figure 4.2) is length times width (area = length × width). Find the areas of rectangles with lengths of 1, 3, and 5 cm and with widths of 2, 4, 6, and 8 cm. (You should have 12 answers.)
2. The volume of a circular cylinder is, volume = $\pi r^2 h$. Find the volume of cylindrical containers with radii from 0 to 12 m and heights from 10 to 20 m. Increment the radius dimension by 3 m and the height by 2 m as you span the two ranges.

Figure 4.2

Dimensions of a rectangle and a circular cylinder.

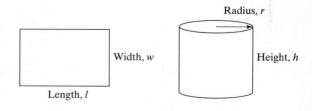

EXAMPLE 4.2

DISTANCE TO THE HORIZON

You've probably experienced standing on the top of a hill or a mountain and feeling like you can see forever. How far can you really see? It depends on the height of the mountain and the radius of the earth, as shown in Figure 4.3. The distance to the horizon is quite different on the moon than on the earth, because the radius is different for each.

Using the Pythagorean theorem, we see that

$$R^2 + d^2 = (R + h)^2$$

and solving for d yields, $d = \sqrt{h^2 + 2Rh}$.

From this last expression, find the distance to the horizon on the earth and on the moon, for mountains from 0 to 8000 m. (Mount Everest is 8850 m tall.) The radius of the earth is 6378 km and the radius of the moon is 1737 km.

1. **State the Problem**
   Find the distance to the horizon from the top of a mountain on the moon and on the earth.
2. **Describe the Input and Output**

   Input

Radius of the moon	1737 km
Radius of the earth	6378 km
Height of the mountains	0 to 8000 m

   Output

   Distance to the horizon, in kilometers.

3. **Develop a Hand Example**

   $$d = \sqrt{h^2 + 2Rh}$$

   Using the radius of the earth and an 8000-m mountain yields

   $$d = \sqrt{(8 \text{ km})^2 + 2 \times 6378 \text{ km} \times 8 \text{ km}} = 319 \text{ km}$$

4. **Develop a MATLAB® Solution**

   ```
   %Example 4.2
   %Find the distance to the horizon
   %Define the height of the mountains
   ```

Figure 4.3
Distance to the horizon.

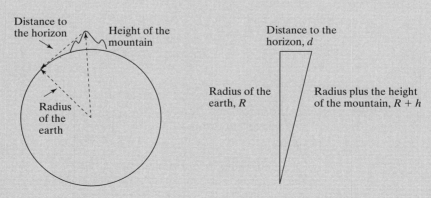

(*continued*)

```
%in meters
clear, clc
format bank
%Define the height vector
height=0:1000:8000;
%Convert meters to km
height=height/1000;
%Define the radii of the moon and earth
radius = [1737 6378];
%Map the radii and heights onto a 2D grid
  [Radius,Height]=meshgrid(radius,height);
%Calculate the distance to the horizon
distance=sqrt(Height.^2 + 2*Height.*Radius)
```

Executing the preceding M-file returns a table of the distances to the horizon on both the moon and the earth:

```
distance =
            0              0
        58.95         112.95
        83.38         159.74
       102.13         195.65
       117.95         225.92
       131.89         252.60
       144.50         276.72
       156.10         298.90
       166.90         319.55
```

5. Test the Solution
   Compare the MATLAB® solution with the hand solution. The distance to the horizon from near the top of Mount Everest (8000 m) is over 300 km and matches the value calculated in MATLAB®.

EXAMPLE 4.3

FREE FALL

The general equation for the distance that a freely falling body has traveled (neglecting air friction) is

$$d = \frac{1}{2}gt^2$$

where

d = distance
g = acceleration due to gravity
t = time.

When a satellite orbits a planet, it is in free fall. Many people believe that when the space shuttle enters orbit, it leaves gravity behind; gravity, though, is what keeps the shuttle in orbit. The shuttle (or any satellite) is actually falling toward the earth

Figure 4.4
The space shuttle is constantly falling toward the earth. (Courtesy of NASA/Jet Propulsion Laboratory.)

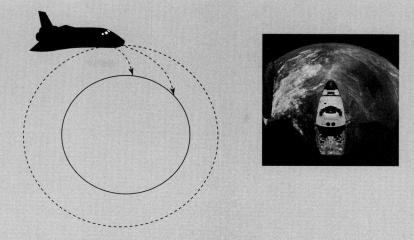

(Figure 4.4). If it is going fast enough horizontally, it stays in orbit; if it's going too slowly, it hits the ground.

The value of the constant g, the acceleration due to gravity, depends on the mass of the planet. On different planets, g has different values (Table 4.2).

Find how far an object would fall at times from 0 to 100 seconds on each planet in our solar system and on our moon.

1. **State the Problem**
   Find the distance traveled by a freely falling object on planets with different gravities.
2. **Describe the Input and Output**
   Input   Value of g, the acceleration due to gravity, on each of the planets and the moon

$$\text{Time} = 0 \text{ to } 100 \text{ s}$$

   Output Distances calculated for each planet and the moon.

3. **Develop a Hand Example**
$$d = 1/2 \ gt^2, \text{ so on Mercury at 100 seconds:}$$
$$d = 1/2 \times 3.7 \text{ m/s}^2 \times 100^2 \text{ s}^2$$
$$d = 18{,}500 \text{ m}$$

Table 4.2 Acceleration Due to Gravity in Our Solar System

Mercury	$g = 3.7$ m/s^2
Venus	$g = 8.87$ m/s^2
Earth	$g = 9.8$ m/s^2
Moon	$g = 1.6$ m/s^2
Mars	$g = 3.7$ m/s^2
Jupiter	$g = 23.12$ m/s^2
Saturn	$g = 8.96$ m/s^2
Uranus	$g = 8.69$ m/s^2
Neptune	$g = 11.0$ m/s^2
Pluto	$g = .58$ m/s^2

(*continued*)

4. Develop a MATLAB® Solution

```
%Example 4.3
%Free fall
clear, clc
%Try the problem first with only two planets, and a coarse
% grid
format bank
%Define constants for acceleration due to gravity on
%Mercury and Venus
acceleration_due_to_gravity = [3.7, 8.87];
time=0:10:100;  %Define time vector
%Map acceleration_due_to_gravity and time into 2D matrices
 [g,t]=meshgrid(acceleration_due_to_gravity, time);
%Calculate the distances
distance=1/2*g.*t.^2
```

Executing the preceding M-file returns the following values of distance traveled on Mercury and on Venus.

```
distance =
               0            0
          185.00       443.50
          740.00      1774.00
         1665.00      3991.50
         2960.00      7096.00
         4625.00     11087.50
         6660.00     15966.00
         9065.00     21731.50
        11840.00     28384.00
        14985.00     35923.50
        18500.00     44350.00
```

5. Test the Solution

Compare the MATLAB® solution with the hand solution. We can see that the distance traveled on Mercury at 100 seconds is 18,500 m, which corresponds to the hand calculation.

    The M-file included the calculations for just the first two planets and was performed first to work out any programming difficulties. Once we've confirmed that the program works, it is easy to redo with the data for all the planets:

```
%Redo the problem with all the data
clear, clc
format bank
%Define constants
acceleration_due_to_gravity = [3.7, 8.87, 9.8, 1.6, 3.7,
23.12 8.96, 8.69, 11.0, 0.58];
time=0:10:100;
%Map acceleration_due_to_gravity and time into 2D matrices
 [g,t]=meshgrid(acceleration_due_to_gravity,time);
%Calculate the distances
d=1/2*g.*t.^2
```

Figure 4.5
Results of the distance calculations for an object falling on each of the planets.

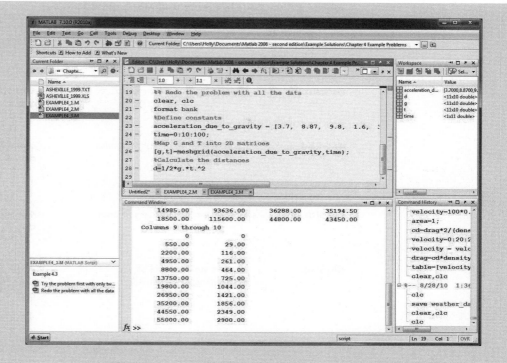

There are several important things to notice about the results shown in Figure 4.5. First, look at the workspace window—`acceleration_due_to_gravity` is a 1×10 matrix (one value for each of the planets and the moon), and `time` is a 1×11 matrix (11 values of time). However, both g and t are 11×10 matrices— the result of the `meshgrid` operation. The results shown in the command window were formatted with the `format bank` command to make the output easier to read; otherwise there would have been a common scale factor.

HINT

As you create a MATLAB® program in the editing window, you may want to comment out those parts of the code which you know work and then uncomment them later. Although you can do this by adding one % at a time to each line, it's easier to select **text** from the menu bar. Just highlight the part of the code you want to comment out, and then choose **comment** from the **text** drop-down menu. To delete the comments, highlight and select **uncomment** from the **text** drop-down menu (text → uncomment). You can also access this menu by right-clicking in the edit window.

4.3 SPECIAL MATRICES

MATLAB® contains a group of functions that generate special matrices; we present some of these functions in Table 4.3.

Table 4.3 Functions to Create and Manipulate Matrices

zeros(m)	Creates an *m* × *m* matrix of zeros.	**zeros(3)**
		ans =
		    0 0 0
		    0 0 0
		    0 0 0

zeros(m,n)	Creates an *m* × *n* matrix of zeros.	**zeros(2,3)**
		ans =
		    0 0 0
		    0 0 0

ones(m)	Creates an *m* × *m* matrix of ones.	**ones(3)**
		ans =
		    1 1 1
		    1 1 1
		    1 1 1

ones(m,n)	Creates an *m* × *n* matrix of ones.	**ones(2,3)**
		ans =
		    1 1 1
		    1 1 1

diag(A)	Extracts the diagonal of a two-dimensional matrix **A**.	**A=[1 2 3; 3 4 5; 1 2 3];**
		diag(A)
		ans =
		    1
		    4
		    3

	For any vector **A**, creates a square matrix with **A** as the diagonal. Check the **help** function for other ways the **diag** function can be used.	**A=[1 2 3];**
		diag(A)
		ans =
		    1 0 0
		    0 2 0
		    0 0 3

fliplr	Flips a matrix into its mirror image, from right to left.	**A=[1 0 0; 0 2 0; 0 0 3];**
		fliplr(A)
		ans =
		    0 0 1
		    0 2 0
		    3 0 0

flipud	Flips a matrix vertically.	**flipud(A)**
		ans =
		    0 0 3
		    0 2 0
		    1 0 0

magic(m)	Creates an *m* × *m* "magic" matrix.	**magic(3)**
		ans =
		    8 1 6
		    3 5 7
		    4 9 2

4.3.1 Matrix of Zeros

It is sometimes useful to create a matrix of all zeros. When the **zeros** function is used with a single scalar input argument, a square matrix is generated:

```
A = zeros(3)
A =
        0    0    0
        0    0    0
        0    0    0
```

If we use two scalar arguments, the first value specifies the number of rows and the second the number of columns:

```
B = zeros(3,2)
B =
        0    0
        0    0
        0    0
```

4.3.2 Matrix of Ones

The ones function is similar to the zeros function, but creates a matrix of ones:

```
A = ones(3)
A =
        1    1    1
        1    1    1
        1    1    1
```

As with the zeros function, if we use two inputs, we can control the number of rows and columns:

```
B = ones(3,2)
B =
        1    1
        1    1
        1    1
```

The zeros and ones functions are useful for creating matrices with "placeholder" values that will be filled in later. For example, if you wanted a vector of five numbers, all of which were equal to π, you might first create a vector of ones:

```
a = ones(1,5)
```

This gives

```
a =
        1    1    1    1    1
```

Then, multiply by π.

```
b = a*pi
```

The result is

```
b =
        3.1416    3.1416    3.1416    3.1416    3.1416
```

The same result could be obtained by adding π to a matrix of zeros. For example,

```
a = zeros(1,5);
b = a+pi
```

gives

```
b =
        3.1416    3.1416    3.1416    3.1416    3.1416
```

A placeholder matrix is especially useful in MATLAB® programs with a loop structure, because it can reduce the time required to execute the loop.

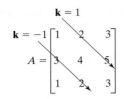

Figure 4.6
Each diagonal in a matrix can be described by means of the parameter **k**.

4.3.3 Diagonal Matrices

We can use the `diag` function to extract the diagonal from a matrix. For example, if we define a square matrix

    A = [1 2 3; 3 4 5; 1 2 3];

then using the function

    diag(A)

extracts the main diagonal and gives the following results:

    ans =
        1.00
        4.00
        3.00

Other diagonals can be extracted by defining a second input, k, to `diag`. Positive values of k specify diagonals in the upper right-hand corner of the matrix, and negative values specify diagonals in the lower left-hand corner (see Figure 4.6).
    Thus, the command

    diag(A,1)

returns

    ans =
        2
        5

If, instead of using a two-dimensional matrix as input to the `diag` function, we use a vector such as

    B = [1 2 3];

then, MATLAB® uses the vector for the values along the diagonal of a new matrix and fills in the remaining elements with zeros:

    diag(B)
    ans =
        1   0   0
        0   2   0
        0   0   3

By specifying a second parameter, we can move the diagonal to any place in the matrix:

    diag(B,1)
    ans =
        0   1   0   0
        0   0   2   0
        0   0   0   3
        0   0   0   0

4.3.4 Magic Matrices

MATLAB® includes a matrix function called `magic` that generates a matrix with unusual properties. At the present time, there does not seem to be any practical use for magic matrices—except that they are interesting. In a magic matrix, the sums of the columns are the same, as are the sums of the rows. An example is

```
A = magic(4)
A =
    16     2     3    13
     5    11    10     8
     9     7     6    12
     4    14    15     1
sum(A)
ans =
    34    34    34    34
```

To find the sums of the rows, we need to transpose the matrix:

```
sum(A')
ans =
   34    34    34    34
```

Not only are the sums of all the columns and rows the same, but the sums of the diagonals are the same. The diagonal from left to right is

```
diag(A)
ans =
   16
   11
    6
    1
```

The sum of the diagonal is the same number as the sums of the rows and columns:

```
sum(diag(A))
ans =
   34
```

Finally, to find the diagonal from lower left to upper right, we first have to "flip" the matrix and then find the sum of the diagonal:

```
fliplr(A)
ans =
   13     3     2    16
    8    10    11     5
   12     6     7     9
    1    15    14     4
diag(ans)
ans =
   13
   10
    7
    4
sum(ans)
ans =
   34
```

Figure 4.7
"Melancholia" by Albrecht Dürer, 1514. (Courtesy of the Library of Congress.)

Figure 4.8
Albrecht Dürer included the date of the woodcut (1514) in the magic square. (Courtesy of the Library of Congress.)

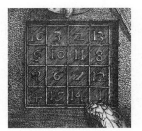

Figure 4.7 shows one of the earliest documented examples of a magic square—Albrecht Dürer's woodcut "Melancholia," created in 1514. Scholars believe the square was a reference to alchemical concepts popular at the time. The date 1514 is included in the two middle squares of the bottom row (see Figure 4.8).

Magic squares have fascinated both professional and amateur mathematicians for centuries. For example, Benjamin Franklin experimented with magic squares. You can create magic squares of any size greater than 2×2 in MATLAB®. MATLAB®'s solution is not the only one; other magic squares are possible.

PRACTICE EXERCISES 4.3

1. Create a 3 × 3 matrix of zeros.
2. Create a 3 × 4 matrix of zeros.
3. Create a 3 × 3 matrix of ones.
4. Create a 5 × 3 matrix of ones.
5. Create a 4 × 6 matrix in which all the elements have a value of pi.
6. Use the `diag` function to create a matrix whose diagonal has values of 1, 2, 3.
7. Create a 10 × 10 magic matrix.
   a. Extract the diagonal from this matrix.
   b. Extract the diagonal that runs from lower left to upper right from this matrix.
   c. Confirm that the sums of the rows, columns, and diagonals are all the same.

SUMMARY

This chapter concentrated on manipulating matrices, a capability that allows the user to create complicated matrices by combining smaller ones. It also lets you extract portions of an existing matrix. The colon operator is especially useful for these operations. The colon operator should be interpreted as "all of the rows" or "all of the columns" when used in place of a row or column designation. It should be interpreted as "from _ to _" when it is used between row or column numbers. For example,

```
A(:,2:3)
```

should be interpreted as "all the rows in matrix **A**, and all the columns from 2 to 3." When used alone as the sole index, as in **A(:)**, it creates a matrix that is a single column from a two-dimensional representation. The computer actually stores all array information as a list, making both single-index notation and row-column notation useful alternatives for specifying the location of a value in a matrix.

The `meshgrid` function is extremely useful, since it can be used to map vectors into two-dimensional matrices, making it possible to perform array calculations with vectors of unequal size.

MATLAB® contains a number of functions that make it easy to create special matrices:

- **zeros**, which is used to create a matrix composed entirely of zeros
- **ones**, which is used to create a matrix composed entirely of ones
- **diag**, which can be used to extract the diagonal from a matrix or, if the input is a vector, to create a square matrix
- **magic**, which can be used to create a matrix with the unusual property that all the rows and columns add up to the same value, as do the diagonals.

In addition, a number of functions were included that allow the user to "flip" the matrix either from left to right or from top to bottom.

MATLAB® SUMMARY

The following MATLAB® summary lists and briefly describes all of the special characters, commands, and functions that were defined in this chapter.

Special Characters	
:	colon operator
...	ellipsis, indicating continuation on the next line
[]	empty matrix

Commands and Functions	
meshgrid	maps vectors into a two-dimensional array
zeros	creates a matrix of zeros
ones	creates a matrix of ones
diag	extracts the diagonal from a matrix
fliplr	flips a matrix into its mirror image, from left to right
flipud	flips a matrix vertically
magic	creates a "magic" matrix

KEY TERMS

elements   magic matrices   subscripts
index numbers   mapping

PROBLEMS

Manipulating Matrices

4.1 Create the following matrices, and use them in the exercises that follow:

$$a = \begin{bmatrix} 15 & 3 & 22 \\ 3 & 8 & 5 \\ 14 & 3 & 82 \end{bmatrix} \quad b = \begin{bmatrix} 1 \\ 5 \\ 6 \end{bmatrix} \quad c = \begin{bmatrix} 12 & 18 & 5 & 2 \end{bmatrix}$$

(a) Create a matrix called **d** from the third column of matrix **a**.

(b) Combine matrix **b** and matrix **d** to create matrix **e**, a two-dimensional matrix with three rows and two columns.

(c) Combine matrix **b** and matrix **d** to create matrix **f**, a one-dimensional matrix with six rows and one column.

(d) Create a matrix **g** from matrix **a** and the first three elements of matrix **c**, with four rows and three columns.

(e) Create a matrix **h** with the first element equal to $a_{1,3}$, the second element equal to $c_{1,2}$, and the third element equal to $b_{2,1}$.

4.2 Load the file **thermo_scores.dat** provided by your instructor, or enter the matrix at the top of page 137 and name it **thermo_scores**. (Enter only the numbers.)

(a) Extract the scores and student number for student 5 into a row vector named **student_5**.

(b) Extract the scores for Test 1 into a column vector named **test_1**.

(c) Find the standard deviation and variance for each test.

(d) Assuming that each test was worth 100 points, find each student's final total score and final percentage. (Be careful not to add in the student number.)

(e) Create a table that includes the final percentages and the scores from the original table.

Student No.	Test 1	Test 2	Test 3
1	68	45	92
2	83	54	93
3	61	67	91
4	70	66	92
5	75	68	96
6	82	67	90
7	57	65	89
8	5	69	89
9	76	62	97
10	85	52	94
11	62	34	87
12	71	45	85
13	96	56	45
14	78	65	87
15	76	43	97
16	68	76	95
17	72	65	89
18	75	67	88
19	83	68	91
20	93	90	92

(f) Sort the matrix on the basis of the final percentage, from high to low (in descending order), keeping the data in each row together. (You may need to consult the **help** function to determine the proper syntax.)

4.3 Consider the following table:

Time (h)	Thermocouple 1 °F	Thermocouple 2 °F	Thermocouple 3 °F
0	84.3	90.0	86.7
2	86.4	89.5	87.6
4	85.2	88.6 min	88.3
6	87.1 max	88.9	85.3
8	83.5	88.9	80.3 min
10	84.8	90.4 max	82.4
12	85.0	89.3	83.4
14	85.3	89.5	85.4
16	85.3	88.9	86.3
18	85.2	89.1	85.3
20	82.3 min	89.5	89.0
22	84.7	89.4	87.3
24	83.6	89.8	87.2

(a) Create a column vector named **times** going from 0 to 24 in 2-hour increments.

(b) Your instructor may provide you with the thermocouple temperatures in a file called **thermocouple.dat**, or you may need to create a matrix named **thermocouple** yourself by typing in the data.

(c) Combine the **times** vector you created in part (a) with the data from **thermocouple** to create a matrix corresponding to the table in this problem.

(d) Recall that both the **max** and **min** functions can return not only the maximum values in a column, but also the element number where those values occur. Use this capability to determine the values of **times** at which the maxima and minima occur in each column.

4.4 Suppose that a file named **sensor.dat** contains information collected from a set of sensors. Your instructor may provide you with this file, or you may need to enter it by hand from the following data:

Time (s)	Sensor 1	Sensor 2	Sensor 3	Sensor 4	Sensor 5
0.0000	70.6432	68.3470	72.3469	67.6751	73.1764
1.0000	73.2823	65.7819	65.4822	71.8548	66.9929
2.0000	64.1609	72.4888	70.1794	73.6414	72.7559
3.0000	67.6970	77.4425	66.8623	80.5608	64.5008
4.0000	68.6878	67.2676	72.6770	63.2135	70.4300
5.0000	63.9342	65.7662	2.7644	64.8869	59.9772
6.0000	63.4028	68.7683	68.9815	75.1892	67.5346
7.0000	74.6561	73.3151	59.7284	68.0510	72.3102
8.0000	70.0562	65.7290	70.6628	63.0937	68.3950
9.0000	66.7743	63.9934	77.9647	71.5777	76.1828
10.0000	74.0286	69.4007	75.0921	77.7662	66.8436
11.0000	71.1581	69.6735	62.0980	73.5395	58.3739
12.0000	65.0512	72.4265	69.6067	79.7869	63.8418
13.0000	76.6979	67.0225	66.5917	72.5227	75.2782
14.0000	71.4475	69.2517	64.8772	79.3226	69.4339
15.0000	77.3946	67.8262	63.8282	68.3009	71.8961
16.0000	75.6901	69.6033	71.4440	64.3011	74.7210
17.0000	66.5793	77.6758	67.8535	68.9444	59.3979
18.0000	63.5403	66.9676	70.2790	75.9512	66.7766
19.0000	69.6354	63.2632	68.1606	64.4190	66.4785

Each row contains a set of sensor readings, with the first row containing values collected at 0 seconds, the second row containing values collected at 1.0 seconds, and so on.

(a) Read the data file and print the number of sensors and the number of seconds of data contained in the file. (*Hint*: Use the **size** function—don't just count the two numbers.)

(b) Find both the maximum value and the minimum value recorded on each sensor. Use MATLAB® to determine at what times they occurred.

(c) Find the mean and standard deviation for each sensor and for all the data values collected. Remember, column 1 does not contain sensor data; it contains time data.

4.5 The American National Oceanic and Atmospheric Administration (NOAA) measures the intensity of a hurricane season with the accumulated cyclone energy (ACE) index. The ACE for a season is the sum of the ACE for each tropical storm with winds exceeding 35 knots (65 km/h). The maximum sustained winds (measured in knots) in the storm are measured or approximated every six hours. The values are squared and summed over the duration of the storm. The total is divided by 10,000, to make the parameter easier to use.

$$\text{ACE} = \frac{\Sigma\, v_{max}^2}{10^4}$$

This parameter is related to the energy of the storm, since kinetic energy is proportional to velocity squared. However, it does not take into account the size of the storm, which would be necessary for a true total energy estimate. Reliable

Atlantic Basin Hurricane Seasons, 1950–2010

Year	ACE Index	# Tropical Storms	# Hurricanes Cat. 1–5	# Major Hurricanes Cat. 3–5
1950	243	13	11	8
1951	137	10	8	5
1952	87	7	6	3
1953	104	14	6	4
1954	113	11	8	2
1955	199	12	9	6
1956	54	8	4	2
1957	84	8	3	2
1958	121	10	7	5
1959	77	11	7	2
1960	88	7	4	2
1961	205	11	8	7
1962	36	5	3	1
1963	118	9	7	2
1964	170	12	6	6
1965	84	6	4	1
1966	145	11	7	3
1967	122	8	6	1
1968	35	7	4	0
1969	158	17	12	5
1970	34	10	5	2
1971	97	13	6	1
1972	28	4	3	0
1973	43	7	4	1
1974	61	7	4	2
1975	73	8	6	3
1976	81	8	6	2
1977	25	6	5	1
1978	62	11	5	2
1979	91	8	5	2
1980	147	11	9	2
1981	93	11	7	3
1982	29	5	2	1
1983	17	4	3	1
1984	71	12	5	1

(continued)

Year	ACE Index	# Tropical Storms	# Hurricanes Cat. 1–5	# Major Hurricanes Cat. 3–5
1985	88	11	7	3
1986	36	6	4	0
1987	34	7	3	1
1988	103	12	5	3
1989	135	11	7	2
1990	91	14	8	1
1991	34	8	4	2
1992	75	6	4	1
1993	39	8	4	1
1994	32	7	3	0
1995	228	19	11	5
1996	166	13	9	6
1997	40	7	3	1
1998	182	14	10	3
1999	177	12	8	5
2000	116	14	8	3
2001	106	15	9	4
2002	65	12	4	2
2003	175	16	7	3
2004	225	14	9	6
2005	248	28	15	7
2006	79	10	5	2
2007	72	15	6	2
2008	145	16	8	5
2009	51	9	3	2
2010	165	19	12	5

storm data have been collected in the Atlantic Ocean since 1950, and are included here. This data may also be available to you from your instructor as an EXCEL worksheet, ace.xlsx, and was extracted from the *Accumulated Cyclone Energy* article in Wikipedia. (http://en.wikipedia.org/wiki/Accumulated_cyclone_energy). It was collected by the National Oceanic and Atmospheric Administration (http://www.aoml.noaa.gov/hrd/tcfaq/E11.html).

(a) Import the data into MATLAB®, and name the array **ace_data**.

(b) Extract the data from each column, into individual arrays. You should have arrays named
   - **years**
   - **ace**
   - **tropical_storms**
   - **hurricanes**
   - **major_hurricanes**

(c) Use the max function to determine which year had the highest
   - ACE value
   - Number of tropical storms
   - Number of hurricanes
   - Number of major hurricanes

(d) Determine the **mean** and the **median** values for each column in the array, except for the year.

(e) Use the sortrows function to rearrange the **ace_data** array based on the ACE value, sorted from high to low.

The data presented in this problem is updated regularly. Similar data is available for the eastern Pacific and central Pacific oceans.

Problems with Two Variables

4.6 The area of a triangle is, area = ½ base × height (see Figure P4.6). Find the area of a group of triangles whose base varies from 0 to 10 m and whose height varies from 2 to 6 m. Choose an appropriate spacing for your calculational variables. Your answer should be a two-dimensional matrix.

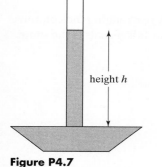

Figure P4.6
The area of a triangle.

4.7 A barometer (see Figure P4.7) is used to measure atmospheric pressure and is filled with a high-density fluid. In the past, mercury was used, but because of its toxic properties it has been replaced with a variety of other fluids. The pressure, P, measured by a barometer is the height of the fluid column, h, times the density of the liquid, ρ, times the acceleration due to gravity, g, or

$$P = h\rho g$$

This equation could be solved for the height:

$$h = \frac{P}{\rho g}$$

Find the height to which the liquid column will rise for pressures from 0 to 100 kPa for two different barometers. Assume that the first uses mercury, with a density of 13.56 g/cm^3 (13,560 kg/m^3) and the second uses water, with a density of 1.0 g/cm^3 (1000 kg/m^3). The acceleration due to gravity is 9.81 m/s^2. Before you start calculating, be sure to check the units in this calculation. The metric measurement of pressure is a pascal (Pa), equal to 1 kg/m s^2. A kPa is 1000 times as big as a Pa. Your answer should be a two-dimensional matrix.

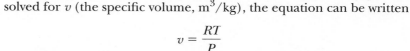

Figure P4.7
Barometer.

4.8 The ideal gas law, $Pv = RT$, describes the behavior of many gases. When solved for v (the specific volume, m^3/kg), the equation can be written

$$v = \frac{RT}{P}$$

Find the specific volume for air, for temperatures from 100 to 1000 K and for pressures from 100 kPa to 1000 kPa. The value of R for air is 0.2870 kJ/(kg K). In this formulation of the ideal gas law, R is different for every gas. There are other formulations in which R is a constant, and the molecular weight of the gas must be included in the calculation. You'll learn more about this equation in chemistry classes and thermodynamics classes. Your answer should be a two-dimensional matrix.

Special Matrices

4.9 Create a matrix of zeros the same size as each of the matrices **a**, **b**, and **c** from Problem 4.1. (Use the **size** function to help you accomplish this task.)

4.10 Create a 6 × 6 magic matrix.

(a) What is the sum of each of the rows?
(b) What is the sum of each of the columns?
(c) What is the sum of each of the diagonals?

4.11 Extract a 3 × 3 matrix from the upper left-hand corner of the magic matrix you created in Problem 4.9. Is this also a magic matrix?

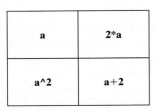

Figure P4.12
Create a matrix out of other matrices.

4.12 Create a 5 × 5 magic matrix named **a**.

(a) Is **a** times a constant such as 2 also a magic matrix?

(b) If you square each element of **a**, is the new matrix a magic matrix?

(c) If you add a constant to each element, is the new matrix **a** magic matrix?

(d) Create a 10 × 10 matrix out of the following components (see Figure P4.12):
- The matrix **a**
- 2 times the matrix **a**
- A matrix formed by squaring each element of **a**
- 2 plus the matrix **a**

Is your result a magic matrix? Does the order in which you arrange the components affect your answer?

4.13 Albrecht Durer's magic square (Figure 4.8) is not exactly the same as the 4 × 4 magic square created with the command

magic(4)

(a) Recreate Durer's magic square in MATLAB® by rearranging the columns.

(b) Prove that the sum of all the rows, columns, and diagonals is the same.

5 Plotting

Objectives

After reading this chapter, you should be able to:

- Create and label two-dimensional plots
- Adjust the appearance of your plots
- Divide the plotting window into subplots
- Create three-dimensional plots
- Use the interactive MATLAB® plotting tools

INTRODUCTION

Large tables of data are difficult to interpret. Engineers use graphing techniques to make the information easier to understand. With a graph, it is easy to identify trends, pick out highs and lows, and isolate data points that may be measurement or calculation errors. Graphs can also be used as a quick check to determine whether a computer solution is yielding expected results.

5.1 TWO-DIMENSIONAL PLOTS

The most useful plot for engineers is the *x–y* plot. A set of ordered pairs is used to identify points on a two-dimensional graph; the points are then connected by straight lines. The values of *x* and *y* may be measured or calculated. Generally, the independent variable is given the name *x* and is plotted on the *x*-axis, and the dependent variable is given the name *y* and is plotted on the *y*-axis.

5.1.1 Basic Plotting

Simple x–y Plots
Once vectors of *x*-values and *y*-values have been defined, MATLAB® makes it easy to create plots. Suppose a set of time versus distance data were obtained through measurement.

We can store the time values in a vector called **x** (the user can define any convenient name) and the distance values in a vector called **y**:

```
x = [0:2:18];
y = [0, 0.33, 4.13, 6.29, 6.85, 11.19, 13.19, 13.96, 16.33,
    18.17];
```

To plot these points, use the `plot` command, with **x** and **y** as arguments:

```
plot(x,y)
```

Time, s	Distance, ft
0	0
2	0.33
4	4.13
6	6.29
8	6.85
10	11.19
12	13.19
14	13.96
16	16.33
18	18.17

A graphics window automatically opens, which MATLAB® calls Figure 1. The resulting plot is shown in Figure 5.1. (Slight variations in scaling of the plot may occur, depending on the size of the graphics window.)

Titles, Labels, and Grids

Good engineering practice requires that we include axis labels and a title in our plot. The following commands add a title, *x*- and *y*-axis labels, and a background grid:

```
plot(x,y)
xlabel('Time, sec')
ylabel('Distance, ft')
grid on
```

Figure 5.1
Simple plot of time versus distance created in MATLAB®.

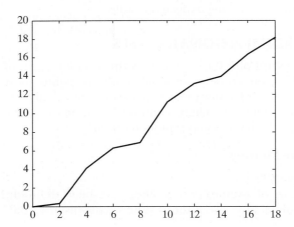

Figure 5.2
Adding a grid, a title, and labels makes a plot easier to interpret.

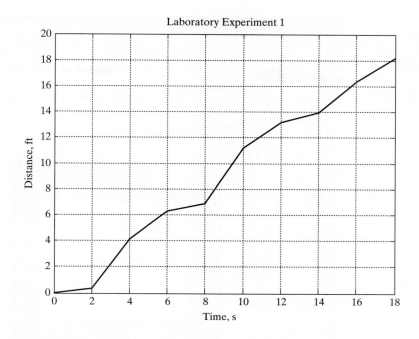

These commands generate the plot in Figure 5.2. As with any MATLAB® commands, they could also be combined onto one or two lines, separated by commas:

```
plot(x,y) , title('Laboratory Experiment 1')
xlabel('Time, sec' ), ylabel('Distance, ft'), grid
```

STRING
A list of characters enclosed by single quotes

As you type the preceding commands into MATLAB®, notice that the text color changes to red when you enter a single quote ('). This alerts you that you are starting a string. The color changes to purple when you type the final single quote ('), indicating that you have completed the string. Paying attention to these visual aids will help you avoid coding mistakes. MATLAB® 6 used different color cues, but the idea is the same.

If you are working in the command window, the graphics window will open on top of the other windows (see Figure 5.3). To continue working, either click in the command window or minimize the graphics window. You can also resize the graphics window to whatever size is convenient for you or add it to the MATLAB® desktop by selecting the docking arrow underneath the exit icon in the upper right-hand corner of the figure window.

HINT

Once you click in the command window, the figure window is hidden behind the current window. To see the changes to your figure, you will need to select the figure from the Windows task bar at the bottom of the screen, or open the Window menu from the main MATLAB® desktop and select the window of interest.

Figure 5.3
The graphics window opens on top of the command window. You can resize it to a convenient shape, or dock it with the MATLAB® desktop.

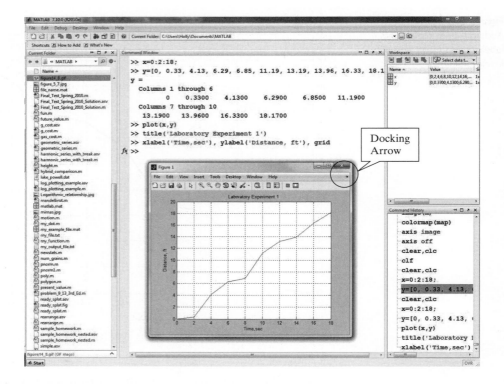

HINT

You must create a graph *before* you add the title and labels. If you specify the title and labels first, they are erased when the plot command executes.

HINT

Because a single quote is used to end the string used in xlabel, ylabel, and title commands, MATLAB® interprets an apostrophe (as in the word *it's*) as the end of the string. Entering the single quote twice, as in xlabel('Holly"s Data'), will allow you to use apostrophes in your text. (Don't use a double quote, which is a different character.)

Creating Multiple Plots

If you are working in an M-file when you request a plot, and then you continue with more computations, MATLAB® will generate and display the graphics window and then return immediately to execute the rest of the commands in the program. If you request a second plot, the graph you created will be overwritten. There are two possible solutions to this problem: Use the pause command to temporarily halt the execution of your M-file program so that you can examine the figure, or create a second figure, using the figure function.

The pause command stops the program execution until any key is pressed. If you want to pause for a specified number of seconds, use the pause (n) command, which will cause execution to pause for n seconds before continuing.

Table 5.1 Basic Plotting Functions

plot	Creates an x–y plot	**plot(x,y)**
title	Adds a title to a plot	**title('My Graph')**
xlabel	Adds a label to the x-axis	**xlabel('Independent Variable')**
ylabel	Adds a label to the y-axis	**ylabel('Dependent Variable')**
grid	Adds a grid to the graph	**grid** **grid on** **grid off**
pause	Pauses the execution of the program, allowing the user to view the graph	**pause**
figure	Determines which figure will be used for the current plot	**figure** **figure(2)**
hold	Freezes the current plot, so that an additional plot can be overlaid	**hold on** **hold off**

The figure command allows you to open a new figure window. The next time you request a plot, it will be displayed in this new window. For example,

```
figure(2)
```

opens a window named "Figure 2," which then becomes the window used for subsequent plotting. Executing figure without an input parameter causes a new window to open, numbered consecutively one up from the current window. For example, if the current figure window is named "Figure 2," executing figure will cause "Figure 3" to open. The commands used to create a simple plot are summarized in Table 5.1.

Plots with More than One Line

A plot with more than one line can be created in several ways. By default, the execution of a second plot statement will erase the first plot. However, you can layer plots on top of one another by using the hold on command. Execute the following statements to create a plot with both functions plotted on the same graph, as shown in Figure 5.4:

```
x = 0:pi/100:2*pi;
y1 = cos(x*4);
plot(x,y1)
```

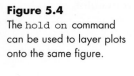

Figure 5.4
The hold on command can be used to layer plots onto the same figure.

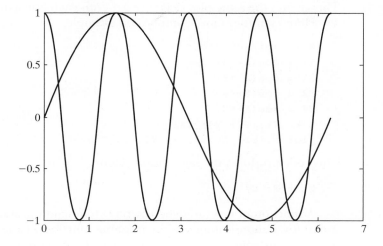

```
y2 = sin(x);
hold on;
plot(x, y2)
```

Semicolons are optional on both the `plot` statement and the `hold on` statement. MATLAB® will continue to layer the plots until the `hold off` command is executed:

```
hold off
```

Another way to create a graph with multiple lines is to request both lines in a single `plot` command. MATLAB® interprets the input to `plot` as alternating *x* and *y* vectors, as in

```
plot(X1, Y1, X2, Y2)
```

where the variables X1, Y1 form an ordered set of values to be plotted and X2, Y2 form a second ordered set of values. Using the data from the previous example,

```
plot(x, y1, x, y2)
```

produces the same graph as Figure 5.4, with one exception: The two lines are different colors. MATLAB® uses a default plotting color (blue) for the first line drawn in a `plot` command. In the `hold on` approach, each line is drawn in a separate plot command and thus is the same color. By requesting two lines in a single command, such as `plot(x,y1,x,y2)`, the second line defaults to green, allowing the user to distinguish between the two plots.

If the `plot` function is called with a single matrix argument, MATLAB® draws a separate line for each column of the matrix. The *x*-axis is labeled with the row index vector, 1:*k*, where *k* is the number of rows in the matrix. This produces an evenly spaced plot, sometimes called a line plot. If `plot` is called with two arguments, one a vector and the other a matrix, MATLAB® successively plots a line for each row in the matrix. For example, we can combine y1 and y2 into a single matrix and plot against x:

```
Y = [y1; y2];
plot(x,Y)
```

This creates the same plot as Figure 5.4, with each line a different color.

Here's another more complicated example:

```
X = 0:pi/100:2*pi;
Y1 = cos(X)*2;
Y2 = cos(X)*3;
Y3 = cos(X)*4;
Y4 = cos(X)*5;
Z = [Y1; Y2; Y3; Y4];
plot(X, Y1, X, Y2, X, Y3, X, Y4)
```

This code produces the same result (Figure 5.5) as

```
plot(X, Z)
```

A function of two variables, the `peaks` function produces sample data that are useful for demonstrating certain graphing functions. (The data are created by scaling and translating Gaussian distributions.) Calling `peaks` with a single argument n

Figure 5.5
Multiple plots on the same graph.

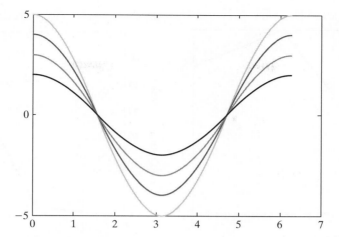

will create an $n \times n$ matrix. We can use peaks to demonstrate the power of using a matrix argument in the plot function. The command

```
plot(peaks(100))
```

results in the impressive graph in Figure 5.6. The input to the plot function created by peaks is a 100×100 matrix. Notice that the x-axis goes from 1 to 100, the index numbers of the data. You undoubtedly can't tell, but there are 100 lines drawn to create this graph—one for each column.

Plots of Complex Arrays

If the input to the plot command is a single array of complex numbers, MATLAB®
plots the real component on the x-axis and the imaginary component on the y-axis. For example, if

```
A = [0+0i,1+2i, 2+5i, 3+4i]
```

then

```
plot(A)
title('Plot of a Single Complex Array')
xlabel('Real Component')
ylabel('Imaginary Component')
```

returns the graph shown in Figure 5.7a.

Figure 5.6
The peaks function, plotted with a single argument in the plot command.

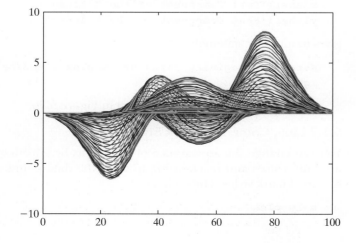

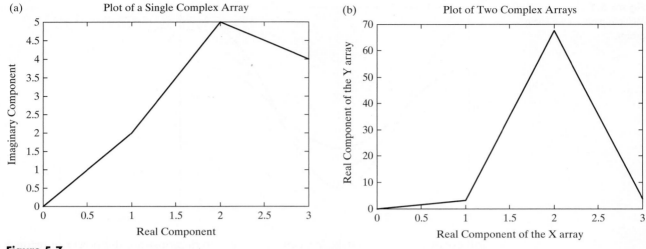

Figure 5.7
(a) Complex numbers are plotted with the real component on the *x*-axis and the imaginary component on the *y*-axis when a single array is used as input. (b) When two complex arrays are used in the `plot` function, the imaginary components are ignored.

If we attempt to use two arrays of complex numbers in the `plot` function, the imaginary components are ignored. The real portion of the first array is used for the *x*-values, and the real portion of the second array is used for the *y*-values. To illustrate, first create another array called B by taking the sine of the complex array A:

```
B = sin(A)
```

returns

```
B =

    0 3.1658 + 1.9596i 67.4789 -30.8794i 3.8537 -27.0168i
```

and

```
plot(A,B)
title('Plot of Two Complex Arrays')
xlabel('Real Component of the X array')
ylabel('Real Component of the Y array')
```

gives us an error statement.

```
Warning: Imaginary parts of complex X and/or Y arguments
ignored.
```

The data are still plotted, as shown in Figure 5.7b.

5.1.2 Line, Color, and Mark Style

You can change the appearance of your plots by selecting user-defined line styles and line colors and by choosing to show the data points on the graph with user-specified mark styles. The command

```
help plot
```

Table 5.2 Line, Mark, and Color Options

Line Type	Indicator	Point Type	Indicator	Color	Indicator
solid	-	point	.	blue	b
dotted	:	circle	o	green	g
dash-dot	-.	x-mark	x	red	r
dashed	- -	plus	+	cyan	c
		star	*	magenta	m
		square	s	yellow	y
		diamond	d	black	k
		triangle down	v	white	w
		triangle up	^		
		triangle left	<		
		triangle right	>		
		pentagram	p		
		hexagram	h		

returns a list of the available options. You can select solid (the default), dashed, dotted, and dash-dot line styles, and you can choose to show the points. The choices among marks include plus signs, stars, circles, and x-marks, among others. There are seven different color choices. (See Table 5.2 for a complete list.)

The following commands illustrate the use of line, color, and mark styles:

```
x = [1:10];
y = [58.5, 63.8, 64.2, 67.3, 71.5, 88.3, 90.1, 90.6,
     89.5,90.4];
plot(x,y,':ok')
```

The resulting plot (Figure 5.8a) consists of a dashed line, together with data points marked with circles. The line, the points, and the circles are drawn in black.

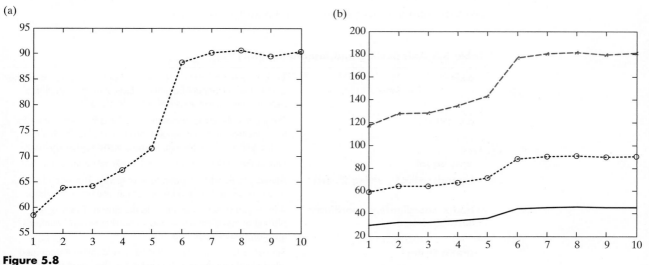

Figure 5.8
(a) Adjusting the line, mark, and color style. (b) Multiple plots with varying line styles and point styles.

The indicators were listed inside a string, denoted with single quotes. The order in which they are entered is arbitrary and does not affect the output.

To specify line, mark, and color styles for multiple lines, add a string containing the choices after each pair of data points. If the string is not included, the defaults are used. For example,

```
plot(x,y,':ok',x,y*2,'--xr',x,y/2,'-b')
```

results in the graph shown in Figure 5.8b.

The `plot` command offers additional options to control the way the plot appears. For example, the line width can be controlled. Plots intended for overhead presentations may look better with thicker lines. Use the `help` function to learn more about controlling the appearance of the plot, or use the interactive controls described in Section 5.5.

5.1.3 Axis Scaling and Annotating Plots

MATLAB® automatically selects appropriate x-axis and y-axis scaling. Sometimes, it is useful for the user to be able to control the scaling. Control is accomplished with the `axis` function, shown in Table 5.3. Executing the `axis` function without any input

```
axis
```

freezes the scaling of the plot. If you use the `hold on` command to add a second line to your graph, the scaling cannot change. To return control of the scaling to MATLAB®, simply re-execute the `axis` function.

The `axis` function also accepts input defining the x-axis and y-axis scaling. The argument is a single matrix, with four values representing:

- The minimum x value shown on the x-axis
- The maximum x value shown on the x-axis
- The minimum y value shown on the y-axis
- The maximum y value shown on the y-axis

Thus, the command

```
axis([-2, 3, 0, 10])
```

fixes the plot axes to x from -2 to $+3$ and y from 0 to 10.

Table 5.3 Axis Scaling and Annotating Plots

axis	When the **axis** function is used without inputs, it freezes the axis at the current configuration. Executing the function a second time returns axis control to MATLAB®.
axis(v)	The input to the **axis** command must be a four-element vector that specifies the minimum and maximum values for both the x- and y-axes—for example, **[xmin, xmax,ymin,ymax]**.
axis equal	Forces the scaling on the x- and y-axis to be the same.
legend('string1', 'string 2', etc)	Allows you to add a legend to your graph. The legend shows a sample of the line and lists the string you have specified.
text(x_coordinate,y_coordinate, 'string')	Allows you to add a text box to the graph. The box is placed at the specified x- and y-coordinates and contains the string value specified.
gtext('string')	Similar to text. The box is placed at a location determined interactively by the user by clicking in the figure window.

Figure 5.9

Final version of the sample graph, annotated with a legend, a text box, a title, *x* and *y* labels, and a modified axis.

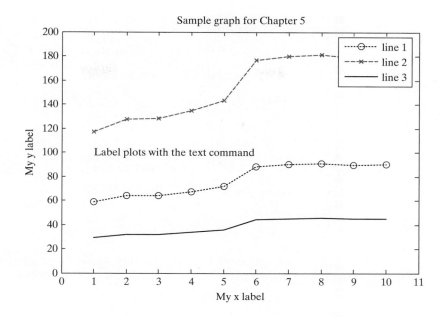

It is often useful to create plots where the scaling is the same on the *x*- and *y*-axis. This is accomplished with the command

```
axis equal
```

MATLAB® offers several additional functions, also listed in Table 5.3, that allow you to annotate your plots. The legend function requires the user to specify a legend in the form of a string for each line plotted, and displays it in the upper right-hand corner of the plot. The text function allows you to add a text box to your plot, which is useful for describing features on the graph. It requires the user to specify the location of the lower left-hand corner of the box in the plot window as the first two input fields, with a string specifying the contents of the text box in the third input field. The use of both legend and text is demonstrated in the following code, which modifies the graph from Figure 5.8b.

```
legend('line 1', 'line 2', 'line3')
text(1,100,'Label plots with the text command')
```

We added a title, *x* and *y* labels, and adjusted the axis with the following commands:

```
xlabel('My x label'), ylabel('My y label')
title('Example graph for Chapter 5'
axis([0,11,0,200])
```

The results are shown in Figure 5.9.

HINT

You can use Greek letters in your titles and labels by putting a backslash (\) before the name of the letter. For example,

```
title('\alpha \beta \gamma')
```

creates the plot title

$$\alpha\beta\gamma$$

To create a superscript, use a caret. Thus,

```
title('x ^2')
```

gives

$$x^2$$

To create a subscript, use an underscore.

```
title('x_5')
```

gives

$$x_5$$

If your expression requires a group of characters as either a subscript or a superscript, enclose them in curly braces. For example,

```
title('k^{-1}')
```

which returns

$$k^{-1}$$

Finally, to create a title with more than one line of text, you'll need to use a cell array. You can learn more about cell arrays in a later chapter, but the syntax is:

```
title({'First line of text'; 'Second line of text'})
```

MATLAB® has the ability to create other more complicated mathematical expressions for use as titles, axis labels, and other text strings, using the TeX markup language. To learn more, consult the help feature. (Search on "text properties.")

PRACTICE EXERCISES 5.1

1. Plot x versus y for $y = \sin(x)$. Let x vary from 0 to 2π in increments of 0.1π.
2. Add a title and labels to your plot.
3. Plot x versus y_1 and y_2 for $y_1 = \sin(x)$ and $y_2 = \cos(x)$. Let x vary from 0 to 2π in increments of 0.1π. Add a title and labels to your plot.
4. Re-create the plot from Exercise 3, but make the $\sin(x)$ line dashed and red. Make the $\cos(x)$ line green and dotted.
5. Add a legend to the graph in Exercise 4.
6. Adjust the axes so that the x-axis goes from -1 to $2\pi + 1$ and the y-axis from -1.5 to $+1.5$.
7. Create a new vector, $a = \cos(x)$. Let x vary from 0 to 2π in increments of 0.1π. Plot just a without specifying the x values (plot(a)) and observe the result. Compare this result with the graph produced by plotting x versus a.

EXAMPLE 5.1

USING THE CLAUSIUS–CLAPEYRON EQUATION

The Clausius–Clapeyron equation can be used to find the saturation vapor pressure of water in the atmosphere, for different temperatures. The saturation water vapor pressure is useful to meteorologists because it can be used to calculate relative humidity, an important component of weather prediction, when the actual partial pressure of water in the air is known.

The following table presents the results of calculating the saturation vapor pressure of water in the atmosphere for various air temperatures with the use of the Clausius–Clapeyron equation:

Air Temperature, °F	Saturation Vapor Pressure, mbar
−60.0000	0.0698
−50.0000	0.1252
−40.0000	0.2184
−30.0000	0.3714
−20.0000	0.6163
−10.0000	1.0000
0	1.5888
10.0000	2.4749
20.0000	3.7847
30.0000	5.6880
40.0000	8.4102
50.0000	12.2458
60.0000	17.5747
70.0000	24.8807
80.0000	34.7729
90.0000	48.0098
100.0000	65.5257
110.0000	88.4608
120.0000	118.1931

Let us present these results graphically as well.

The Clausius–Clapeyron equation is

$$\ln\,(P^0/6.11) = \left(\frac{\Delta H_v}{R_{\text{air}}}\right) * \left(\frac{1}{273} - \frac{1}{T}\right)$$

where

P^0 = saturation vapor pressure for water, in mbar, at temperature T
ΔH_v = latent heat of vaporization for water, 2.453×10^6 J/kg
R_{air} = gas constant for moist air, 461 J/kg
T = temperature in kelvins.

1. State the Problem
   Find the saturation vapor pressure at temperatures from −60°F to 120°F, using the Clausius–Clapeyron equation.

(continued)

2. Describe the Input and Output

Input

$$\Delta H_v = 2.453 \times 10^6 \text{ J/kg}$$
$$R_{air} = 461 \text{ J/kg}$$
$$T = -60°F \text{ to } 120°F$$

Since the number of temperature values was not specified, we'll choose to recalculate every 10°F.

Output

Table of temperature versus saturation vapor pressures
Graph of temperature versus saturation vapor pressures

3. Develop a Hand Example

Change the temperatures from degree Fahrenheit to kelvin:

$$T_k = \frac{(T_f + 459.6)}{1.8}$$

Solve the Clausius–Clapeyron equation for the saturation vapor pressure (P^0):

$$\ln\left(\frac{P^0}{6.11}\right) = \left(\frac{\Delta H_v}{R_{air}}\right) \times \left(\frac{1}{273} - \frac{1}{T}\right)$$

$$P^0 = 6.11 * \exp\left(\left(\frac{\Delta H_v}{R_{air}}\right) \times \left(\frac{1}{273} - \frac{1}{T}\right)\right)$$

Notice that the expression for the saturation vapor pressure, P^0, is an exponential equation. We would thus expect the graph to have the shape shown in Figure 5.10.

4. Develop a MATLAB® Solution

```
%Example 5.1
%Using the Clausius-Clapeyron equation, find the
%saturation vapor pressure for water at different
%temperatures
%
  TF=[-60:10:120];        %Define temp matrix in F
  TK=(TF + 459.6)/1.8;    %Convert temp to K
  Delta_H=2.45e6;         %Define latent heat of
  R_air = 461;            %vaporization
                          %Define ideal gas constant
                          %for air
%
%Calculate the vapor pressures
  Vapor_Pressure=6.11*exp((Delta_H/R_air)*(1/273 - 1./TK));
  %Display the results in a table
    my_results = [TF',Vapor_Pressure']
%
%Create an x-y plot
  plot(TF,Vapor_Pressure)
  title('Clausius-Clapeyron Behavior')
```

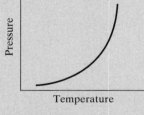

Figure 5.10
A sketch of the predicted equation behavior.

```
xlabel('Temperature, F')
ylabel('Saturation Vapor Pressure, mbar')
```

The resulting table is

```
my_results =

        -60.0000      0.0698
        -50.0000      0.1252
        -40.0000      0.2184
        -30.0000      0.3714
        -20.0000      0.6163
        -10.0000      1.0000
               0      1.5888
         10.0000      2.4749
         20.0000      3.7847
         30.0000      5.6880
         40.0000      8.4102
         50.0000     12.2458
         60.0000     17.5747
         70.0000     24.8807
         80.0000     34.7729
         90.0000     48.0098
        100.0000     65.5257
        110.0000     88.4608
        120.0000    118.1931
```

A figure window opens to display the graphical results, shown in Figure 5.11.

Figure 5.11
A plot of the Clausius–Clapeyron equation.

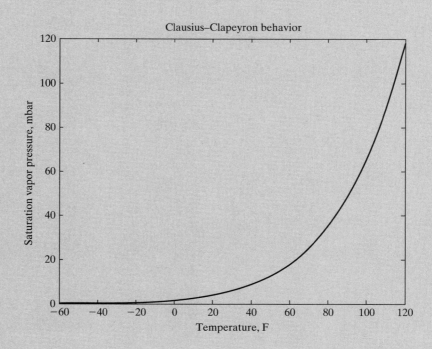

5. Test the Solution

The plot follows the expected trend. It is almost always easier to determine whether computational results make sense if a graph is produced. Tabular data are extremely difficult to interpret.

EXAMPLE 5.2

BALLISTICS

The range of an object (see Figure 5.12) shot at an angle θ with respect to the x-axis and an initial velocity v_0 is given by

$$R(\theta) = \frac{v^2}{g} \sin(2\theta) \quad \text{for } 0 \leq \theta \leq \frac{\pi}{2} (\text{neglecting air resistance})$$

Use $g = 9.9$ m/s^2 and an initial velocity of 100 m/s. Show that the maximum range is obtained at $\theta = \pi/4$ by computing and plotting the range for values of θ from

$$0 \leq \theta \leq \frac{\pi}{2}$$

in increments of 0.05.

Repeat your calculations with an initial velocity of 50 m/s, and plot both sets of results on a single graph.

1. State the Problem

Calculate the range as a function of the launch angle.

2. Describe the Input and Output

Input

$g = 9.9$ m/s^2
$\theta = 0$ to $\pi/2$, incremented by 0.05
$v_0 = 50$ m/s and 100 m/s

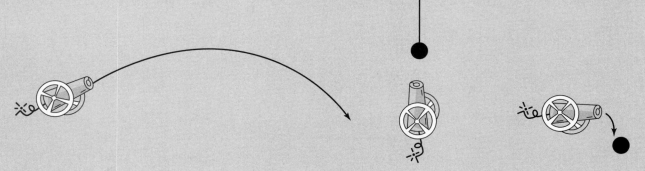

Figure 5.12
The range is zero, if the cannon is perfectly vertical or perfectly horizontal.

Output

Range R

Present the results as a plot.

3. Develop a Hand Example

If the cannon is pointed straight up, we know that the range is zero, and if the cannon is horizontal, the range is also zero (see Figure 5.12).

    This means that the range must increase with the cannon angle up to some maximum and then decrease. A sample calculation at 45° ($\pi/4$ radians) shows that

$$R(\theta) = \frac{v^2}{g} \sin(2\theta)$$

$$R\left(\frac{\pi}{4}\right) = \frac{100^2}{9.9}\sin\left(\frac{2\pi}{4}\right) = 1010 \text{ m when the initial velocity is 100 m/s}$$

4. Develop a MATLAB® Solution

```
%Example 5.2
%The program calculates the range of a ballistic projectile
%
%Define the constants
   g = 9.9;
   v1 = 50;
   v2 = 100;
%Define the angle vector
   angle = 0:0.05:pi/2;
%Calculate the range
   R1 = v1^2/g*sin(2*angle);
   R2 = v2^2/g*sin(2*angle);
%Plot the results
   plot(angle,R1,angle,R2,':')
   title('Cannon Range')
   xlabel('Cannon Angle')
   ylabel('Range, meters')
   legend('Initial Velocity=50 m/s', 'Initial Velocity=100 m/s')
```

Notice that in the `plot` command, we requested MATLAB® to print the second set of data as a dashed line. A title, labels, and a legend were also added. The results are plotted in Figure 5.13.

5. Test the Solution

Compare the MATLAB® results with those from the hand example. Both graphs start and end at zero. The maximum range for an initial velocity of 100 m/s is approximately 1000 m, which corresponds well to the calculated value of 1010 m. Notice that both solutions peak at the same angle, approximately 0.8 radian. The numerical value for $\pi/4$ is 0.785 radian, confirming the hypothesis presented in the problem statement that the maximum range is achieved by pointing the cannon at an angle of $\pi/4$ radians (45°).

(continued)

Figure 5.13
The predicted range of a projectile.

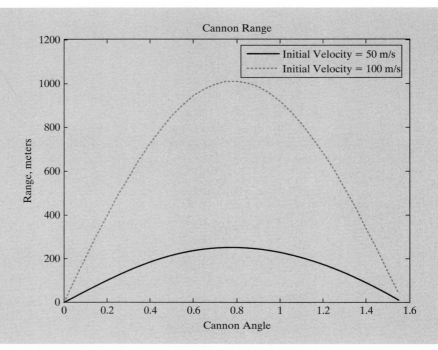

HINT

To clear a figure, use the `clf` command. To close the active figure window, use the `close` command, and to close all open figure windows use `close all`.

A function similar to `text` is `gtext`, which allows the user to interactively place a text box in an existing plot. The `gtext` function requires a single input, the string to be displayed.

```
gtext('This string will display on the graph')
```

Once executed, a crosshair appears on the graph. The user positions the crosshair to the appropriate position. The text is added to the graph when any key on the keyboard is depressed, or a mouse button is selected.

5.2 SUBPLOTS

The `subplot` command allows you to subdivide the graphing window into a grid of m rows and n columns. The function

```
subplot(m,n,p)
```

splits the figure into an $m \times n$ matrix. The variable p identifies the portion of the window where the next plot will be drawn. For example, if the command

```
subplot(2,2,1)
```

is used, the window is divided into two rows and two columns, and the plot is drawn in the upper left-hand window (Figure 5.14).

$p = 1$	$p = 2$
$p = 3$	$p = 4$

Figure 5.14
Subplots are used to subdivide the figure window into an $m \times n$ matrix.

Figure 5.15
The subplot command allows the user to create multiple graphs in the same figure window.

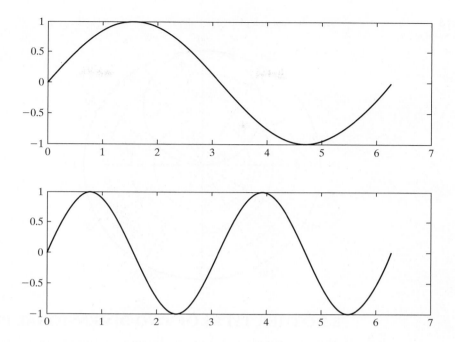

The windows are numbered from left to right, top to bottom. Similarly, the following commands split the graph window into a top plot and a bottom plot:

```
x = 0:pi/20:2*pi;
subplot(2,1,1)
plot(x,sin(x))
subplot(2,1,2)
plot(x,sin(2*x)
```

The first graph is drawn in the top window, since p = 1. Then the subplot command is used again to draw the next graph in the bottom window. Figure 5.15 shows both graphs.

Titles are added above each subwindow as the graphs are drawn, as are x- and y-axis labels and any annotation desired. The use of the subplot command is illustrated in several of the sections that follow.

PRACTICE EXERCISES 5.2

1. Subdivide a figure window into two rows and one column.
2. In the top window, plot $y = \tan(x)$ for $-1.5 \le x \le 1.5$. Use an increment of 0.1.
3. Add a title and axis labels to your graph.
4. In the bottom window, plot $y = \sinh(x)$ for the same range.
5. Add a title and labels to your graph.
6. Try the preceding exercises again, but divide the figure window vertically instead of horizontally.

Figure 5.16
A polar plot of the sine function.

The sine function plotted in polar coordinates is a circle.

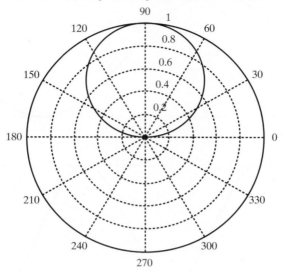

5.3 OTHER TYPES OF TWO-DIMENSIONAL PLOTS

Although simple *x–y* plots are the most common type of engineering plot, there are many other ways to represent data. Depending on the situation, these techniques may be more appropriate than an *x–y* plot.

5.3.1 Polar Plots

MATLAB® provides plotting capability with polar coordinates:

```
polar(theta, r)
```

generates a polar plot of angle theta (in radians) and radial distance *r*.
    For example, the code

```
x = 0:pi/100:pi;
y = sin(x);
polar(x,y)
```

generates the plot in Figure 5.16. A title was added in the usual way:

```
title('The sine function plotted in polar coordinates is a
circle.')
```

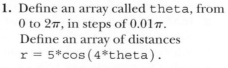
PRACTICE EXERCISES 5.3

1. Define an array called theta, from 0 to 2π, in steps of 0.01π.
   Define an array of distances
   r = 5*cos(4*theta).

   Make a polar plot of theta versus r.

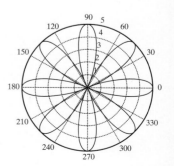

2. Use the `hold` on command to freeze the graph.
Assign `r = 4*cos(6*theta)` and plot. Add a title.

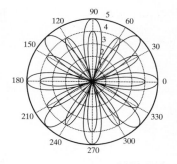

3. Create a new figure.
Use the `theta` array from the preceding exercises.
Assign `r = 5 − 5*sin(theta)` and create a new polar plot.

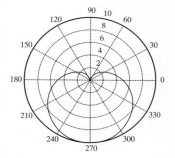

4. Create a new figure.
Use the `theta` array from the preceding exercises.
Assign `r = sqrt(5^2*cos(2*theta))` and create a new polar plot.

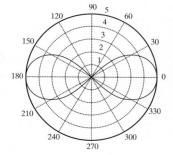

5. Create a new figure.
Define a theta array such that `theta = pi/2:4/5*pi:4.5pi;`
Create a six-member array of ones called `r`.
Create a new polar plot of theta versus r.

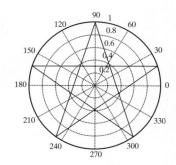

Table 5.4 Rectangular and Logarithmic Plots

plot(x,y)	Generates a linear plot of the vectors x and y
semilogx(x,y)	Generates a plot of the values of x and y, using a logarithmic scale for x and a linear scale for y
semilogy(x,y)	Generates a plot of the values of x and y, using a linear scale for x and a logarithmic scale for y
loglog(x,y)	Generates a plot of the vectors x and y, using a logarithmic scale for both x and y

KEY IDEA

Logarithmic plots are especially useful if the data vary exponentially

5.3.2 Logarithmic Plots

For most plots that we generate, the *x*- and *y*-axes are divided into equally spaced intervals; these plots are called *linear* or *rectangular* plots. Occasionally, however, we may want to use a logarithmic scale on one or both of the axes. A logarithmic scale (to the base 10) is convenient when a variable ranges over many orders of magnitude, because the wide range of values can be graphed without compressing the smaller values. Logarithmic plots are also useful for representing data that vary exponentially. Appendix B discusses in more detail when to use the various types of logarithmic scaling.

The MATLAB® commands for generating linear and logarithmic plots of the vectors x and y are listed in Table 5.4.

Remember that the logarithm of a negative number or of zero does not exist. If your data include these values, MATLAB® will issue a warning message and will not plot the points in question. However, it will generate a plot based on the remaining points.

Each command for logarithmic plotting can be executed with one argument, as we saw in plot(y) for a linear plot. In these cases, the plots are generated with the values of the indices of the vector y used as x values.

As an example, plots of $y = 5x^2$ were created using all four scaling approaches, as shown in Figure 5.17. The linear (rectangular) plot, semilog plot along the *x*-axis, semilog plot along the *y*-axis, and log–log plot are all shown on one figure, plotted with the subplot function in the following code:

```
x = 0:0.5:50;
y = 5*x.^2;
subplot(2,2,1)
plot(x,y)
    title('Polynomial - linear/linear')
    ylabel('y'), grid
subplot(2,2,2)
semilogx(x,y)
    title('Polynomial - log/linear')
    ylabel('y'), grid
subplot(2,2,3)
semilogy(x,y)
    title('Polynomial - linear/log')
    xlabel('x'), ylabel('y'), grid
subplot(2,2,4)
loglog(x,y)
    title('Polynomial - log/log')
    xlabel('x'), ylabel('y'), grid
```

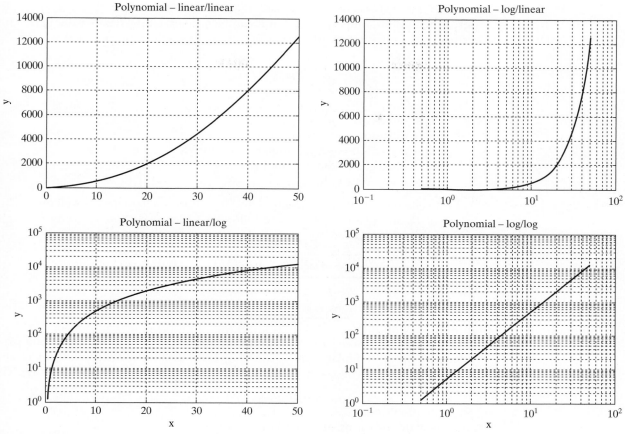

Figure 5.17
Linear and logarithmic plots, displayed using the subplot function.

The indenting is intended to make the code easier to read—MATLAB® ignores white space. As a matter of style, notice that only the bottom two subplots have *x*-axis labels.

EXAMPLE 5.3

RATES OF DIFFUSION

Metals are often treated to make them stronger and therefore wear longer. One problem with making a strong piece of metal is that it becomes difficult to form it into a desired shape. A strategy that gets around this problem is to form a soft metal into the shape you desire and then harden the surface. This makes the metal wear well without making it brittle.

A common hardening process is called *carburizing*. The metal part is exposed to carbon, which diffuses into the part, making it harder. This is a very slow process if

(*continued*)

performed at low temperatures, but it can be accelerated by heating the part. The diffusivity is a measure of how fast diffusion occurs and can be modeled as

$$D = D_0 \exp\left(\frac{-Q}{RT}\right)$$

where

D = diffusivity, cm^2/s
D_0 = diffusion coefficient, cm^2/s
Q = activation energy, J/mol, 8.314 J/mol K
R = ideal gas constant, J/mol K
T = temperature, K.

As iron is heated, it changes structure and its diffusion characteristics change. The values of D_0 and Q are shown in the following table for carbon diffusing through each of iron's structures:

Type of Metal	D_0 (cm^2/s)	Q (J/mol K)
alpha Fe (BCC)	0.0062	80,000
gamma Fe (FCC)	0.23	148,000

Create a plot of diffusivity versus inverse temperature $(1/T)$, using the data provided. Try the rectangular, semilog, and log–log plots to see which you think might represent the results best. Let the temperature vary from room temperature (25°C) to 1200°C.

1. State the Problem
   Calculate the diffusivity of carbon in iron.
2. Describe the Input and Output

   Input

   For C in alpha iron, $D_0 = 0.0062$ cm^2/s and $Q = 80,000$ J/mol K
   For C in gamma iron, $D_0 = 0.23$ cm^2/s and $Q = 148,000$ J/mol K
   $R = 8.314$ J/mol K
   T varies from 25°C to 1200°C

   Output

   Calculate the diffusivity and plot it.
3. Develop a Hand Example
   The diffusivity is given by

$$D = D_0 \exp\left(\frac{-Q}{RT}\right)$$

At room temperature, the diffusivity for carbon in alpha iron is

$$D = 0.0062 \exp\left(\frac{-80,000}{8.314 \times (25 + 273)}\right)$$

$$D = 5.9 \times 10^{-17}$$

(Notice that the temperature had to be changed from Celsius to Kelvin.)

4. Develop a MATLAB® Solution

```
% Example 5.3
% Calculate the diffusivity of carbon in iron
  clear, clc
% Define the constants
  D0alpha = 0.0062;
  D0gamma = 0.23;
  Qalpha = 80000;
  Qgamma = 148000;
  R = 8.314;
  T = 25:5:1200;
% Change T from C to K
  T = T+273;
% Calculate the diffusivity
  Dalpha = D0alpha*exp(-Qalpha./(R*T));
  Dgamma = D0gamma*exp(-Qgamma./(R*T));
% Plot the results
  subplot(2,2,1)
  plot(1./T,Dalpha, 1./T,Dgamma)
  title('Diffusivity of C in Fe')
  xlabel('Inverse Temperature, K^{-1}'),
  ylabel('Diffusivity, cm^2/s')
  grid on

  subplot(2,2,2)
  semilogx(1./T,Dalpha, 1./T,Dgamma)
  title('Diffusivity of C in Fe')
  xlabel('Inverse Temperature, K^{-1}'),
  ylabel('Diffusivity, cm^2/s')
  grid on

  subplot(2,2,3)
  semilogy(1./T,Dalpha, 1./T,Dgamma)
  title('Diffusivity of C in Fe')
  xlabel('Inverse Temperature, K^{-1}'),
  ylabel('Diffusivity, cm^2/s')
  grid on

  subplot(2,2,4)
  loglog(1./T,Dalpha, 1./T,Dgamma)
  title('Diffusivity of C in Fe')
  xlabel('Inverse Temperature, K^{-1}'),
  ylabel('Diffusivity, cm^2/s')
  grid on
```

Subplots were used in Figure 5.18, so that all four variations of the plot are in the same figure. Notice that x-labels were added only to the bottom two graphs, to reduce clutter, and that a legend was added only to the first plot. The `semilogy` plot resulted in straight lines and allows a user to read values off the graph easily over a wide range of both temperatures and diffusivities. This is the plotting scheme usually used in textbooks and handbooks to present diffusivity values.

(*continued*)

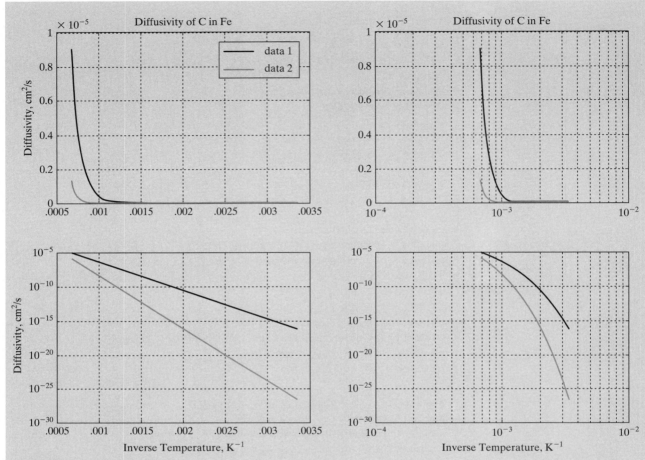

Figure 5.18
Diffusivity data plotted on different scales. The data follows a straight line when the $\log_{10}$ of the diffusivity is plotted on the y-axis versus the inverse temperature on the x-axis.

5. Test the Solution

Compare the MATLAB® results with those from the hand example. We calculated the diffusivity to be

$$5.9 \times 10^{-17} \text{ cm}^2/\text{s at } 25°C$$

for carbon in alpha iron. To check our answer, we'll need to change 25°C to kelvins and take the inverse:

$$\frac{1}{(25 + 273)} = 3.36 \times 10^{-3}$$

From the semilogy graph (lower left-hand corner), we can see that the diffusivity for alpha iron is approximately 10^{-17}.

PRACTICE EXERCISE 5.4

Create appropriate x and y arrays to use in plotting each of the expressions that follow. Use the `subplot` command to divide your figures into four sections, and create each of these four graphs for each expression:

- Rectangular
- Semilogx
- Semilogy
- Loglog
  1. $y = 5x + 3$
  2. $y = 3x^2$
  3. $y = 12e^{(x+2)}$
  4. $y = 1/x$

Physical data usually are plotted so that they fall on a straight line. Which of the preceding types of plot results in a straight line for each problem?

5.3.3 Bar Graphs and Pie Charts

Bar graphs, histograms, and pie charts are popular forms for reporting data. Some of the commonly used MATLAB® functions for creating bar graphs and pie charts are listed in Table 5.5.

Examples of some of these graphs are shown in Figure 5.19. The graphs make use of the `subplot` function to allow four plots in the same figure window:

```
clear, clc
x = [1,2,5,4,8];
y = [x;1:5];
subplot(2,2,1)
  bar(x),title('A bar graph of vector x')
subplot(2,2,2)
  bar(y),title('A bar graph of matrix y')
subplot(2,2,3)
  bar3(y),title('A three-dimensional bar graph')
subplot(2,2,4)
  pie(x),title('A pie chart of x')
```

Table 5.5 Bar Graphs and Pie Charts

bar(x)	When **x** is a vector, bar generates a vertical bar graph. When x is a two-dimensional matrix, bar groups the data by row.
barh(x)	When **x** is a vector, barh generates a horizontal bar graph. When **x** is a two-dimensional matrix, barh groups the data by row.
bar3(x)	Generates a three-dimensional bar chart
bar3h(x)	Generates a three-dimensional horizontal bar chart
pie(x)	Generates a pie chart. Each element in the matrix is represented as a slice of the pie.
pie3(x)	Generates a three-dimensional pie chart. Each element in the matrix is represented as a slice of the pie.
hist(x)	Generates a histogram

Figure 5.19
Sample bar graphs and pie charts. The `subplot` function was used to divide the window into quadrants.

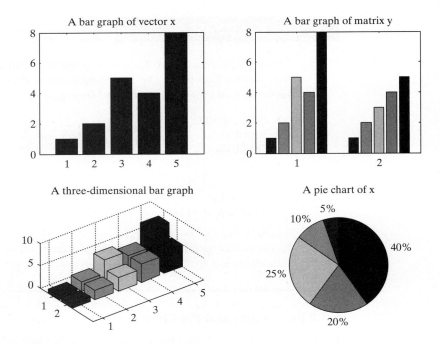

5.3.4 Histograms

A histogram is a special type of graph that is particularly useful for the statistical analysis of data. It is a plot showing the distribution of a set of values. In MATLAB®, the histogram computes the number of values falling into 10 bins (categories) that are equally spaced between the minimum and maximum values. For example, if we define a matrix **x** as the set of grades from the Introduction to Engineering final, the scores could be represented in a histogram, shown in Figure 5.20 and generated with the following code:

```
x = [100,95,74,87,22,78,34,35,93,88,86,42,55,48];
hist(x)
```

Figure 5.20
A histogram of grade data.

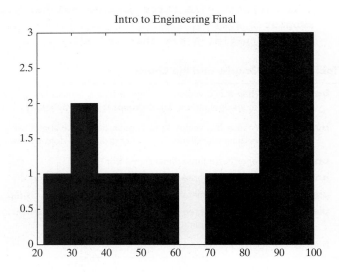

The default number of bins is 10, but if we have a large data set, we may want to divide the data up into more bins. For example, to create a histogram with 25 bins, the command would be

```
hist(x, 25)
```

If you set the hist function equal to a variable, as in

```
A = hist(x)
```

the data used in the plot are stored in **A**:

```
A =
    1  2  1  1  1  0  1  1  3  3
```

EXAMPLE 5.4

WEIGHT DISTRIBUTIONS

The average 18-year-old American male weighs 152 pounds. A group of 100 young men were weighed and the data stored in a file called **weight.dat**. Create a graph to represent the data.

1. State the Problem
   Use the data file to create a line graph and a histogram. Which is a better representation of the data?
2. Describe the Input and Output

   Input   **weight.dat**, an ASCII data file that contains weight data

   Output   A line plot of the data
              A histogram of the data

3. Develop a Hand Example
   Since this is a sample of actual weights, we would expect the data to approximate a normal random distribution (a Gaussian distribution). The histogram should be bell shaped.
4. Develop a MATLAB® Solution
   The following code generates the plots shown in Figure 5.21:

```
% Example 5.4
% Using Weight Data
%
load weight.dat
% Create the line plot of weight data
subplot(1,2,1)
plot(weight)
title('Weight of Freshman Class Men')
xlabel('Student Number')
ylabel('Weight, lb')
grid on
% Create the histogram of the data
subplot(1,2,2)
hist(weight)
```

(continued)

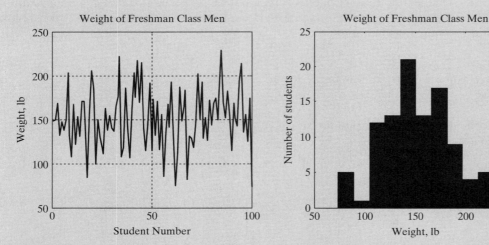

Figure 5.21
Histograms and line plots are two different ways to visualize numeric information.

```
xlabel('Weight, lb')
ylabel('Number of students')
title('Weight of Freshman Class Men')
```

5. Test the Solution
The graphs match our expectations. The weight appears to average about 150 lb and varies in what looks like a normal distribution. We can use MATLAB® to find the average and the standard deviation of the data, as well as the maximum and minimum weights in the data set. The MATLAB® code

```
average_weight = mean(weight)
standard_deviation = std(weight)
maximum_weight = max(weight)
minimum_weight = min(weight)
```

returns

```
average_weight =
   151.1500
standard_deviation =
   32.9411
maximum_weight =
   228
minimum_weight =
   74
```

5.3.5 *X–Y* Graphs with Two *Y*-Axes

Sometimes, it is useful to overlay two x–y plots onto the same figure. However, if the orders of magnitude of the y-values are quite different, it may be difficult to see how the data behave. Consider, for example, a graph of $\sin(x)$ and e^x drawn

Figure 5.22
MATLAB® allows the y-axis to be scaled differently on the left-hand and right-hand sides of the figure. In the top graph, both lines were drawn using the same scaling. In the bottom graph, the sine curve was drawn using the scaling on the left axis, while the exponential curve was drawn using the scaling on the right axis.

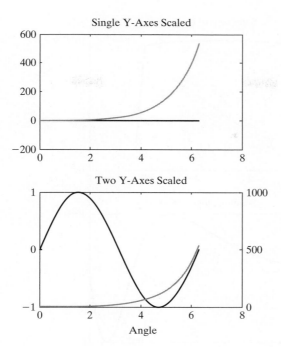

on the same figure. The results, obtained with the following code, are shown in Figure 5.22:

```
x = 0:pi/20:2*pi;
y1 = sin(x);
y2 = exp(x);
subplot(2,1,1)
plot(x,y1,x,y2)
```

The plot of $\sin(x)$ looks like it runs straight along the line $x = 0$, because of the scale. The plotyy function allows us to create a graph with two y-axes, the one on the left for the first set of ordered pairs and the one on the right for the second set of ordered pairs:

```
subplot(2,1,2)
plotyy(x,y1,x,y2)
```

Titles and labels were added in the usual way. The y-axis was not labeled, because the results are dimensionless.

The plotyy function can create a number of different types of plots by adding a string with the name of the plot type after the second set of ordered pairs. In Figure 5.23, the plots were created with the following code and have a logarithmically scaled axis:

```
subplot(2,1,1)
plotyy(x,y1,x,y2, 'semilogy')
subplot(2,1,2)
plotyy(x,y1,x,y2,'semilogx')
```

For other problems you may need to add y-axis labels. The left-hand y-axis is easy—just add the label in the usual way

```
ylabel('Left y-axis label')
```

Figure 5.23

The `plotyy` function can generate several types of graphs, including semilogx, semilogy, and loglog.

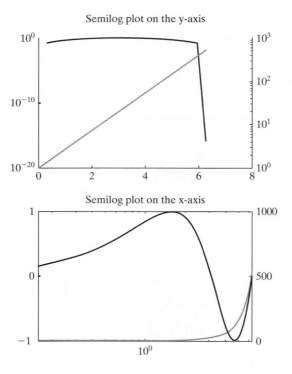

The right-hand *y*-axis label is trickier. You can add it using MATLAB®'s interactive controls, described in a later section, or you can use handle graphics. This involves giving the plot a name, and then using the name to switch to the second axis set (which corresponds to the *y*-axis on the right-hand side of the figure). Here is the code

```
a = plotyy(x,y1,x,y2)
ylabel(a(2),'Right y-axis label')
```

EXAMPLE 5.5

PERIODIC PROPERTIES OF THE ELEMENTS

The properties of elements in the same row or column in the periodic table usually display a recognizable trend as we move across a row or down a column. For example, the melting point usually goes down as we move down a column, because the atoms are farther apart and the bonds between the atoms are therefore weaker. Similarly, the radius of the atoms goes up as we move down a column, because there are more electrons in each atom and correspondingly bigger orbitals. It is instructive to plot these trends against atomic weight on the same graph.

1. State the Problem
   Plot the melting point and the atomic radius of the Group I elements against the atomic weight, and comment on the trends you observe.

Table 5.6 Group I Elements and Selected Physical Properties

Element	Atomic Number	Melting Point, °C	Atomic Radius, pm
Lithium	3	181	0.1520
Sodium	11	98	0.1860
Potassium	19	63	0.2270
Rubidium	37	34	0.2480
Cesium	55	28.4	0.2650

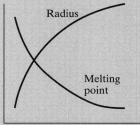

Figure 5.24
Sketch of the predicted data behavior.

2. Describe the Input and Output

    Input   The atomic weights, melting points, and atomic radii of the Group I elements are listed in Table 5.6.

    Output   Plot with both melting point and atomic radius on the same graph.

3. Develop a Hand Example
    We would expect the graph to look something like the sketch shown in Figure 5.24.

4. Develop a MATLAB® Solution
    The following code produces the plot shown in Figure 5.25:

```
% Example 5.5
clear, clc
% Define the variables
atomic_number = [ 3, 11, 19, 37, 55];
melting_point = [181, 98, 63, 34, 28.4];
atomic_radius = [0.152, 0.186, 0.227, 0.2480, 0.2650];
% Create the plot with two lines on the same scale
subplot(1,2,1)
plot(atomic_number,melting_point,atomic_number,atomic_radius)
title('Periodic Properties')
```

Figure 5.25
In the left-hand figure, the two sets of values were plotted using the same scale. Using two y-axes allows us to plot data with different units on the same graph, as shown in the right-hand figure.

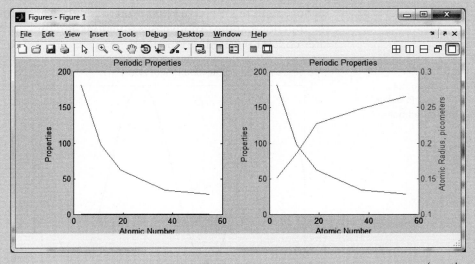

(*continued*)

```
xlabel('Atomic Number')
ylabel('Properties')
% Create the second plot with two different y scales
subplot(1,2,2)
h=plotyy(atomic_number,melting_point,atomic_number,atomic_
radius)
title('Periodic Properties')
xlabel('Atomic Number')
ylabel('Melting Point, C')
ylabel(h(2),'Atomic Radius, picometers')
```

On the second graph, which has two different y scales, we used the plotyy function instead of the plot function. This forced the addition of a second scale, on the right-hand side of the plot. We needed it because atomic radius and melting point have different units and the values for each have different magnitudes. Notice that in the first plot it is almost impossible to see the atomic-radius line; it is on top of the x-axis because the numbers are so small.

5. Test the Solution

Compare the MATLAB® results with those from the hand example. The trend matches our prediction. Clearly, the graph with two y-axes is the superior representation, because we can see the property trends.

5.3.6 Function Plots

The fplot function allows you to plot a function without defining arrays of corresponding x- and y-values. For example,

```
fplot('sin(x)',[-2*pi,2*pi])
```

creates a plot (Figure 5.26) of x versus $\sin(x)$ for x-values from -2π to 2π. MATLAB® automatically calculates the spacing of x-values to create a smooth curve. Notice that the first argument in the fplot function is a string containing the function and the second argument is an array. For more complicated functions that may be inconvenient to enter as a string, you may define an anonymous function and enter the function handle. Anonymous functions and function handles are described in a later chapter devoted to functions.

Figure 5.26

Function plots do not require the user to define arrays of ordered pairs.

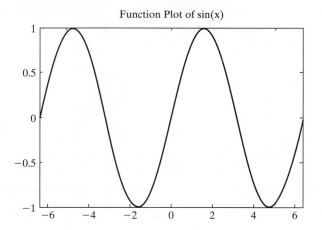

Function Plot of sin(x)

PRACTICE EXERCISE 5.5 ▌▌

Create a plot of the functions that follow, using `fplot`. You'll need to select an appropriate range for each plot. Don't forget to title and label your graphs.

1. $f(t) = 5t^2$
2. $f(t) = 5 \sin^2(t) + t \cos^2(t)$
3. $f(t) = te^t$
4. $f(t) = \ln(t) + \sin(t)$

HINT

■ The correct MATLAB® syntax for the mathematical expression $\sin^2(t)$ is
■ `sin(t).^2`.

5.4 THREE-DIMENSIONAL PLOTTING

MATLAB® offers a variety of three-dimensional plotting commands, several of which are listed in Table 5.7.

5.4.1 Three-Dimensional Line Plot

The `plot3` function is similar to the `plot` function, except that it accepts data in three dimensions. Instead of just providing **x** and **y** vectors, the user must also provide a **z** vector. These ordered triples are then plotted in three-space and connected with straight lines. For example,

```
clear, clc
x = linspace(0,10*pi,1000);
y = cos(x);
z = sin(x);
plot3(x,y,z)
grid
xlabel('angle'), ylabel('cos(x)') zlabel('sin(x)') title('A
Spring')
```

Table 5.7 Three-Dimensional Plots

plot3(x,y,z)	Creates a three-dimensional line plot
comet3(x,y,z)	Generates an animated version of `plot3`
mesh(z) or mesh(x,y,z)	Creates a meshed surface plot
surf(z) or surf(x,y,z)	Creates a surface plot; similar to the mesh function
shading interp	Interpolates between the colors used to illustrate surface plots
shading flat	Colors each grid section with a solid color
colormap(map_name)	Allows the user to select the color pattern used on surface plots
contour(z) or contour(x,y,z)	Generates a contour plot
surfc(z) or surfc(x,y,z)	Creates a combined surface plot and contour plot
pcolor(z) or pcolor(x,y,z)	Creates a pseudo color plot

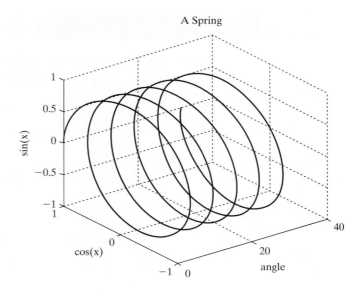

The title, labels, and grid are added to the graph in Figure 5.27 in the usual way, with the addition of `zlabel` for the z-axis.

The coordinate system used with `plot3` is oriented using the right-handed coordinate system familiar to engineers.

KEY IDEA

The axes used for three-dimensional plotting correspond to the right-hand rule

HINT

Just for fun, re-create the plot shown in Figure 5.27, but this time with the `comet3` function:

```
comet3(x,y,z)
```

This plotting function "draws" the graph in an animation sequence. If your animation runs too quickly, add more data points. For two-dimensional line graphs, use the `comet` function.

5.4.2 Surface Plots

Surface plots allow us to represent data as a surface. We will be experimenting with two types of surface plots: `mesh` plots and `surf` plots.

Mesh Plots

There are several ways to use `mesh` plots. They can be used to good effect with a single two-dimensional $m \times n$ matrix. In this application, the value in the matrix represents the **z**-value in the plot. The **x**- and **y**-values are based on the matrix dimensions. Take, for example, the following very simple matrix:

```
z = [1, 2, 3, 4,  5,  6,  7,  8,  9, 10;
     2, 4, 6, 8, 10, 12, 14, 16, 18, 20;
     3, 4, 5, 6,  7,  8,  9, 10, 11, 12];
```

Figure 5.28

Simple mesh created with a single two-dimensional matrix.

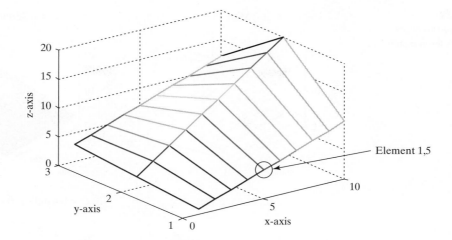

The code

```
mesh(z)
xlabel('x-axis')
ylabel('y-axis')
zlabel('z-axis')
```

generates the graph in Figure 5.28.

The graph is a "mesh" created by connecting the points defined in z into a rectilinear grid. Notice that the *x*-axis goes from 0 to 10 and *y* goes from 1 to 3. The matrix index numbers were used for the axis values. For example, note that $z_{1,5}$ —the value of *z* in row 1, column 5—is equal to 5. This element is circled in Figure 5.28.

The mesh function can also be used with three arguments: mesh(x,y,z). In this case, x is a list of *x*-coordinates, y is a list of *y*-coordinates, and z is a list of *z*-coordinates.

```
x = linspace(1,50,10)
y = linspace(500,1000,3)
z = [1, 2, 3, 4,  5,  6,  7,  8,  9, 10;
     2, 4, 6, 8, 10, 12, 14, 16, 18, 20;
     3, 4, 5, 6,  7,  8,  9, 10, 11, 12]
```

The x vector must have the same number of elements as the number of columns in the z vector and the y vector must have the same number of elements as the number of rows in the z vector. The command

```
mesh(x,y,z)
```

creates the plot in Figure 5.29a. Notice that the *x*-axis varies from 0 to 50, with data plotted from 1 to 50. Compare this scaling with that in Figure 5.28, which used the **z** matrix index numbers for the *x*- and *y*-axes.

Surf Plots

Surf plots are similar to mesh plots, but surf creates a three-dimensional colored surface instead of a mesh. The colors vary with the value of **z**.

The surf command takes the same input as mesh: either a single input— for example, surf(z),in which case it uses the row and column indices as *x*- and

Figure 5.29
Mesh and surf plots are
created with three input
arguments.

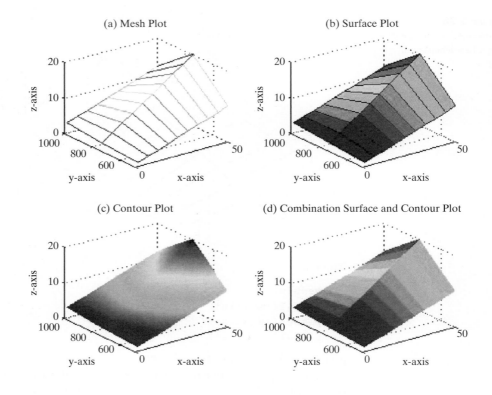

(a) Mesh Plot

(b) Surface Plot

(c) Contour Plot

(d) Combination Surface and Contour Plot

y-coordinates—or three matrices. Figure 5.29b was generated with the same commands as those used to generate Figure 5.29a, except that `surf` replaced `mesh`.

The shading scheme for surface plots is controlled with the shading command. The default, shown in Figure 5.29b, is "faceted." Interpolated shading can create interesting effects. The plot shown in Figure 5.29c was created by adding

    `shading interp`

to the previous list of commands. Flat shading without the grid is generated when

    `shading flat`

is used, as shown in Figure 5.29d.

KEY IDEA
The colormap function
controls the colors used on
surface plots

The color scheme used in surface plots can be controlled with the `colormap` function. For example,

    `colormap(gray)`

forces a grayscale representation for surface plots. This may be appropriate if you'll be making black-and-white copies of your plots. Other available `colormaps` are

autumn	bone	hot
spring	colorcube	hsv
summer	cool	pink
winter	copper	prism
jet (default)	flag	white

Figure 5.30
Surface and contour plots are different ways of visualizing the same data.

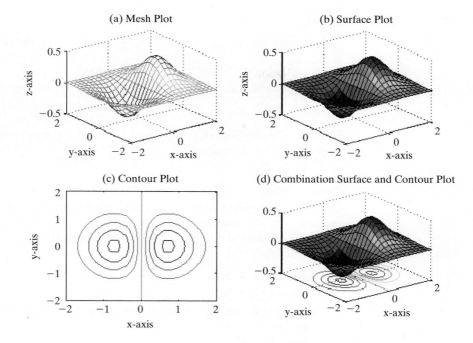

(a) Mesh Plot

(b) Surface Plot

(c) Contour Plot

(d) Combination Surface and Contour Plot

Use the `help` command to see a description of the various options:

```
help colormap
```

Another Example
A more complicated surface can be created by calculating the values of Z:

```
x= [-2:0.2:2];
y= [-2:0.2:2];
[X,Y] = meshgrid(x,y);
Z = X.*exp(-X.^2 - Y.^2);
```

In the preceding code, the `meshgrid` function is used to create the two-dimensional matrices X and Y from the one-dimensional vectors x and y. The values in Z are then calculated. The following code plots the calculated values:

```
subplot(2,2,1)
mesh(X,Y,Z)
title('Mesh Plot'), xlabel('x-axis'), ylabel('y-axis'),
zlabel('z-axis')

subplot(2,2,2)

surf(X,Y,Z)
title('Surface Plot'), xlabel('x-axis'), ylabel('y-axis'),
zlabel('z-axis')
```

Either the x, y vectors or the X, Y matrices can be used to define the *x*- and *y*-axes. Figure 5.30a is a `mesh` plot of the given function, and Figure 5.30b is a `surf` plot of the same function.

HINT

If a single vector is used in the `meshgrid` function, the program interprets it as

```
[X,Y] = meshgrid(x,x)
```

You could also use the vector definition as input to `meshgrid`:

```
[X,Y] = meshgrid(-2:0.2:2)
```

Both of these lines of code would produce the same result as the commands listed in the example.

Contour Plots

Contour plots are two-dimensional representations of three-dimensional surfaces, much like the familiar contour maps used by many hikers. The `contour` command was used to create Figure 5.30c, and the `surfc` command was used to create Figure 5.30d:

```
subplot(2,2,3)
contour(X,Y,Z)
xlabel('x-axis'), ylabel('y-axis'), title('Contour Plot')
subplot(2,2,4)
surfc(X,Y,Z)
xlabel('x-axis'), ylabel('y-axis')
title('Combination Surface and Contour Plot')
```

Pseudo Color Plots

Pseudo color plots are similar to contour plots, except that instead of lines outlining a specific contour, a two-dimensional shaded map is generated over a grid. MATLAB® includes a sample function called `peaks` that generates the **x**, **y**, and **z** matrices of an interesting surface that looks like a mountain range:

```
[x,y,z] = peaks;
```

With the following code, we can use this surface to demonstrate the use of pseudo color plots, shown in Figure 5.31:

```
subplot(2,2,1)
pcolor(x,y,z)
```

The grid is deleted when interpolated shading is used:

```
subplot(2,2,2)
pcolor(x,y,z)
shading interp
```

You can add contours to the image by overlaying a contour plot:

```
subplot(2,2,3)
pcolor(x,y,z)
shading interp
hold on
contour(x,y,z,20,'k')
```

The number 20 specifies that 20 contour lines are drawn, and the `'k'` indicates that the lines should be black. If we hadn't specified black lines, they would

Figure 5.31
A variety of contour plots is available in MATLAB®.

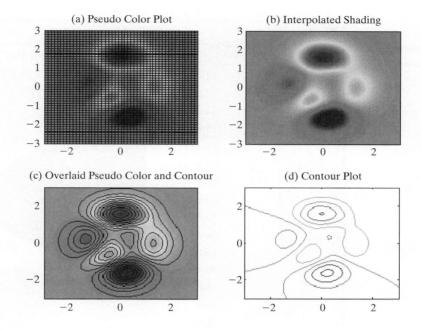

(a) Pseudo Color Plot
(b) Interpolated Shading
(c) Overlaid Pseudo Color and Contour
(d) Contour Plot

have been the same color as the pseudo color plot and would have disappeared into the image. Finally, a simple contour plot was added to the figure for comparison:

```
subplot(2,2,4)
contour(x,y,z)
```

Additional options for using all the three-dimensional plotting functions are included in the help window.

5.5 EDITING PLOTS FROM THE MENU BAR

KEY IDEA
When you interactively edit a plot, your changes will be lost if you rerun the program

In addition to controlling the way your plots look by using MATLAB® commands, you can edit a plot once you've created it. The plot in Figure 5.32 was created with the sphere command, which is one of several sample functions, like peaks, used to demonstrate plotting.

```
sphere
```

In the figure, the **Insert menu** has been selected. Notice that you can insert labels, titles, legends, text boxes, and so on, all by using this menu. The **Tools menu** allows you to change the way the plot looks, by zooming in or out, changing the aspect ratio, etc. The figure toolbar, underneath the menu toolbar, offers icons that allow you to do the same thing.

The plot in Figure 5.32 doesn't really look like a sphere; it's also missing labels and a title, and the meaning of the colors may not be clear. We edited this plot by first adjusting the shape:

- Select **Edit** → **Axes Properties** from the menu toolbar.
- From the **Property Editor—Axes window**, select More Properties → **Data Aspect Ratio Mode**.
- Set the mode to manual (see Figure 5.33).

Figure 5.32
MATLAB® offers interactive tools, such as the insert tool, that allow the user to adjust the appearance of graphs.

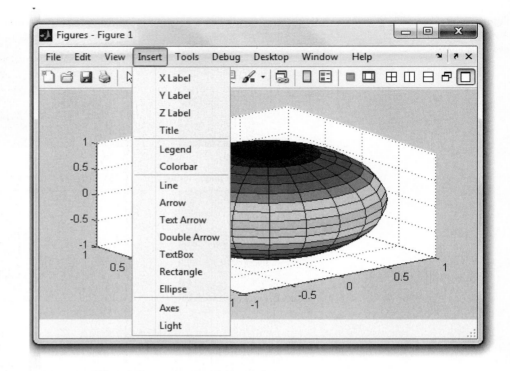

Figure 5.33
MATLAB® allows you to edit plots by using commands from the toolbar.

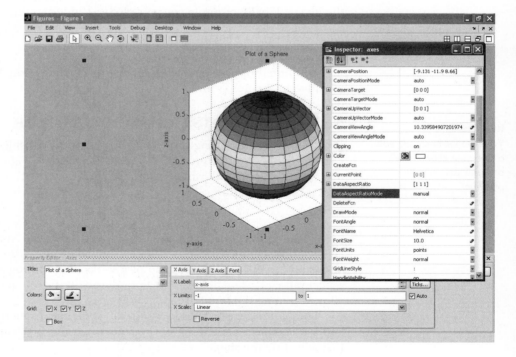

Figure 5.34
Edited plot of a sphere.

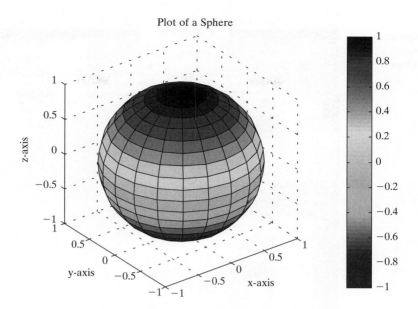

Similarly, labels, a title, and a color bar were added (Figure 5.34) using the Property Editor. They could also have been added by using **the Insert menu** option on the menu bar. Editing your plot in this manner is more interactive and allows you to fine-tune the plot's appearance. The only problem with editing a figure interactively is that if you run your MATLAB® program again, you will lose all of your improvements.

HINT

You can force a plot to space the data equally on all the axes by using the `axis equal` command. This approach has the advantage that you can program **axis equal** into an M-file and retain your improvements.

5.6 CREATING PLOTS FROM THE WORKSPACE WINDOW

A great feature of MATLAB® 7 is its ability to create plots interactively from the workspace window. In this window, select a variable, then select the drop-down menu on the **plotting icon** (shown in Figure 5.35). MATLAB® will list the plotting options it "thinks" are reasonable for the data stored in your variable. Simply select the appropriate option, and your plot is created in the current **figure window**. If you don't like any of the suggested types of plot, choose **More plots** from the drop-down menu, and a new window will open with the complete list of available plotting options for you to choose from. This is especially useful, because it may suggest options that had not occurred to you. For example, Figure 5.35 shows a scatter plot of the **x and y** matrices highlighted in the figure. The matrices were created by loading the seamount data set, which is built into MATLAB®.

Figure 5.35
Plotting from the workspace window, using the interactive plotting feature.

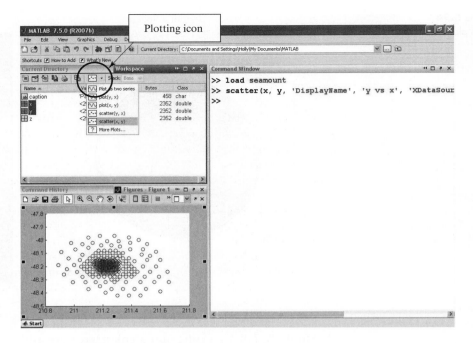

If you want to plot more than one variable, highlight the first, then hold down the *Ctrl* key and select the additional variables. To annotate your plots, use the interactive editing process described in Section 5.5. The interactive environment is a rich resource. You'll get the most out of it by exploring and experimenting.

5.7 SAVING YOUR PLOTS

There are several ways to save plots created in MATLAB®:

- If you created the plot with programming code stored in an M-file, simply rerunning the code will re-create the figure.
- You can also save the figure from the file menu, using the Save As . . . option. You'll be presented with several choices:
  1. You may save the figure as a **.fig** file, which is a MATLAB®-specific file format. To retrieve the figure, just double-click on the file name in the current folder. You can do the same thing programatically with the code

     ```
     open <figurename.fig>
     ```

  2. You may save the figure in a number of different standard graphics formats, such as jpeg (**.jpg**) and enhanced metafile (**.emf**). These versions of the figure can be inserted into other documents, such as a Word document.
  3. You can select Edit from the menu bar, then select **copy figure**, and paste the figure into another document.
  4. You can use the file menu to create an M-file that will re-create the figure.

PRACTICE EXERCISE 5.6

Create a plot of $y = \cos(x)$. Practice saving the file and inserting it into a Word document.

SUMMARY

The most commonly used graph in engineering is the x–y plot. This two-dimensional plot can be used to graph data or to visualize mathematical functions. No matter what a graph represents, it should always include a title and x- and y-axis labels. Axis labels should be descriptive and should include units, such as ft/s or kJ/kg.

MATLAB® includes extensive options for controlling the appearance of your plots. The user can specify the color, line style, and marker style for each line on a graph. A grid can be added to the graph, and the axis range can be adjusted. Text boxes and a legend can be employed to describe the graph. The subplot function is used to divide the plot window into an $m \times n$ grid. Inside each of these subwindows, any of the MATLAB® plots can be created and modified.

In addition to x–y plots, MATLAB® offers a variety of plotting options, including polar plots, pie charts, bar graphs, histograms, and x–y graphs with two y-axes. The scaling on x–y plots can be modified to produce logarithmic plots on either or both x- and y-axes. Engineers often use logarithmic scaling to represent data as a straight line.

The function `fplot` allows the user to plot a function without defining a vector of x- and y-values. MATLAB® automatically chooses the appropriate number of points and spacing to produce a smooth graph. Additional function-plotting capability is available in the symbolic toolbox.

The three-dimensional plotting options in MATLAB® include a line plot, a number of surface plots, and contour plots. Most of the options available in two-dimensional plotting also apply to these three-dimensional plots. The `meshgrid` function is especially useful in creating three-dimensional surface plots.

Interactive tools allow the user to modify existing plots. These tools are available from the figure menu bar. Plots can also be created with the interactive plotting option from the workspace window. The interactive environment is a rich resource. You'll get the most out of it by exploring and experimenting.

Figures created in MATLAB® can be saved in a variety of ways, either to be edited later or to be inserted into other documents. MATLAB® offers both proprietary file formats that minimize the storage space required to store figures and standard file formats suitable to import into other applications.

MATLAB® SUMMARY

The following MATLAB® summary lists all the special characters, commands, and functions that were defined in this chapter:

Special Characters					
Line Type	**Indicator**	**Point Type**	**Indicator**	**Color**	**Indicator**
solid	-	point	.	blue	**b**
dotted	:	circle	**o**	green	**g**
dash-dot	-.	x-mark	**x**	red	**r**
dashed	- -	plus	**+**	cyan	**c**
		star	*****	magenta	**m**
		square	**s**	yellow	**y**
		diamond	**d**	black	**k**

(continued)

Special Characters (continued)

Line Type	Indicator	Point Type	Indicator	Color	Indicator
		triangle down	**v**	white	**w**
		triangle up	**^**		
		triangle left	**<**		
		triangle right	**>**		
		pentagram	**p**		
		hexagram	**h**		

Commands and Functions

autumn	optional colormap used in surface plots
axis	freezes the current axis scaling for subsequent plots or specifies the axis dimensions
axis equal	forces the same scale spacing for each axis
bar	generates a bar graph
bar3	generates a three-dimensional bar graph
barh	generates a horizontal bar graph
bar3h	generates a horizontal three-dimensional bar graph
bone	optional colormap used in surface plots
clf	clear figure
close	close the current figure window
close all	close all open figure windows
colorcube	optional colormap used in surface plots
colormap	color scheme used in surface plots
comet	draws an x–y plot in a pseudo animation sequence
comet3	draws a three-dimensional line plot in a pseudo animation sequence
contour	generates a contour map of a three-dimensional surface
cool	optional colormap used in surface plots
copper	optional colormap used in surface plots
figure	opens a new figure window
flag	optional colormap used in surface plots
fplot	creates an x–y plot based on a function
gtext	similar to text; the box is placed at a location determined interactively by the user by clicking in the figure window
grid	adds a grid to the current plot only
grid off	turns the grid off
grid on	adds a grid to the current and all subsequent graphs in the current figure
hist	generates a histogram
hold off	instructs matlab® **to** erase figure contents before adding new information
hold on	instructs matlab® **not to** erase figure contents before adding new information
hot	optional colormap used in surface plots
hsv	optional colormap used in surface plots
jet	default colormap used in surface plots
legend	adds a legend to a graph
linspace	creates a linearly spaced vector

(continued)

Commands and Functions	
loglog	generates an x–y plot with both axes scaled logarithmically
mesh	generates a mesh plot of a surface
meshgrid	places each of two vectors into separate two-dimensional matrices, the size of which is determined by the source vectors
pause	pauses the execution of a program until any key is hit
pcolor	creates a pseudo color plot similar to a contour map
peaks	creates a sample matrix used to demonstrate graphing functions
pie	generates a pie chart
pie3	generates a three-dimensional pie chart
pink	optional colormap used in surface plots
plot	creates an x–y plot
plot3	generates a three-dimensional line plot
plotyy	creates a plot with two y-axes
polar	creates a polar plot
prism	optional colormap used in surface plots
semilogx	generates an x–y plot with the x-axis scaled logarithmically
semilogy	generates an x–y plot with the y-axis scaled logarithmically
shading flat	shades a surface plot with one color per grid section
shading inte:	shades a surface plot by interpolation
sphere	sample function used to demonstrate graphing
spring	optional colormap used in surface plots
subplot	divides the graphics window into sections available for plotting
summer	optional colormap used in surface plots
surf	generates a surface plot
surfc	generates a combination surface and contour plot
text	adds a text box to a graph
title	adds a title to a plot
white	optional colormap used in surface plots
winter	optional colormap used in surface plots
xlabel	adds a label to the x-axis
ylabel	adds a label to the y-axis
zlabel	adds a label to the z-axis

PROBLEMS

Two-Dimensional (x–y) Plots

5.1 Create plots of the following functions from $x = 0$ to 10.

(a) $y = e^x$
(b) $y = \sin(x)$
(c) $y = ax^2 + bx + c$, where $a = 5$, $b = 2$, and $c = 4$
(d) $y = \sqrt{x}$

Each of your plots should include a title, an x-axis label, a y-axis label, and a grid.

5.2 Plot the following set of data:
$$y = \begin{bmatrix} 12, & 14, & 12, & 22, & 8, & 9 \end{bmatrix}$$
Allow MATLAB® to use the matrix index number as the parameter for the x-axis.

5.3 Plot the following functions on the same graph for x values from $-\pi$ to π, selecting spacing to create a smooth plot:
$$y_1 = \sin(x)$$
$$y_2 = \sin(2x)$$
$$y_3 = \sin(3x)$$
(*Hint:* Recall that the appropriate MATLAB® syntax for $2x$ is **2*x**.)

5.4 Adjust the plot created in Problem 5.3 so that:
- Line 1 is red and dashed.
- Line 2 is blue and solid.
- Line 3 is green and dotted.

Do not include markers on any of the graphs. In general, markers are included only on plots of measured data, not for calculated values.

5.5 Adjust the plot created in Problem 5.4 so that the x-axis goes from -6 to $+6$.
- Add a legend.
- Add a text box describing the plots.

x–y Plotting with Projectiles

Use the following information in Problems 5.6 through 5.10:
The distance a projectile travels when fired at an angle θ is a function of time and can be divided into horizontal and vertical distances according to the formulas
$$\text{horizontal}(t) = tV_0 \cos(\theta)$$
and
$$\text{vertical}(t) = tV_0 \sin(\theta) - \tfrac{1}{2}gt^2$$
where

horizontal	=	distance traveled in the x direction
vertical	=	distance traveled in the y direction
V_0	=	initial velocity
g	=	acceleration due to gravity, 9.8 m/s^2
t	=	time, s.

5.6 Suppose the projectile just described is fired at an initial velocity of 100 m/s and a launch angle of $\pi/4$ ($45°$). Find the distance traveled both horizontally and vertically (in the x and y directions) for times from 0 to 20 s with a spacing of .01 seconds.
(a) Graph horizontal distance versus time.
(b) In a new figure window, plot vertical distance versus time (with time on the x-axis).
Don't forget a title and labels.

5.7 In a new figure window, plot horizontal distance on the x-axis and vertical distance on the y-axis.

5.8 Replot horizontal distance on the x-axis and vertical distance on the y-axis using the comet function. If the plot draws too quickly or too slowly on your computer, adjust the number of time values used in your calculations.

5.9   Calculate three new vectors for each of the vertical (v_1, v_2, v_3) and horizontal (h_1, h_2, h_3) distances traveled, assuming launch angles of $\pi/2$, $\pi/4$, and $\pi/6$.

   • In a new figure window, graph horizontal distance on the *x*-axis and vertical distance on the *y*-axis, for all three cases. (You'll have three lines.)
   • Make one line solid, one dashed, and one dotted. Add a legend to identify which line is which.

5.10   Re-create the plot from Problem 5.9. This time, create a matrix `theta` of the three angles, $\pi/2$, $\pi/4$, and $\pi/6$. Use the `meshgrid` function to create a mesh of `theta` and the time vector (`t`). Then use the two new meshed variables you create to recalculate vertical distance (`v`) and horizontal distance (`h`) traveled. Each of your results should be a 2001 × 3 matrix. Use the `plot` command to plot `h` on the *x*-axis and `v` on the *y*-axis.

5.11   A tensile testing machine such as the one shown in Figure P5.11 is used to determine the behavior of materials as they are deformed. In the typical test, a specimen is stretched at a steady rate. The force (load) required to deform the material is measured, as is the resulting deformation. An example set of data measured in one such test is shown in Table P5.11. These data

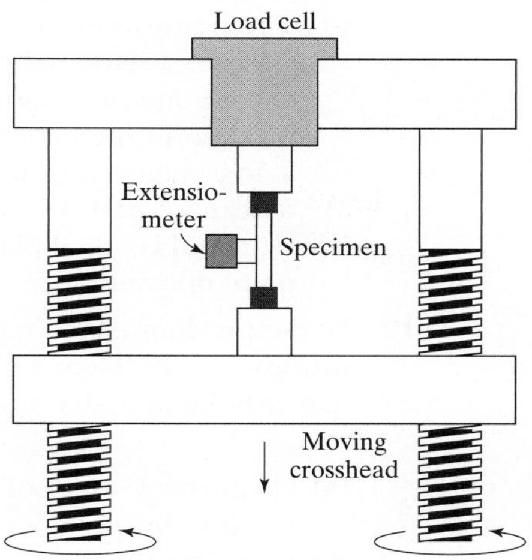

Figure P5.11
A tensile testing machine is used to measure stress and strain and to characterize the behavior of materials as they are deformed.

Table P5.11 Tensile Testing Data

load, lbf	length, inches
0	2
1650	2.002
3400	2.004
5200	2.006
6850	2.008
7750	2.010
8650	2.020
9300	2.040
10100	2.080
10400	2.120

(From William Callister, *Materials Science and Engineering, An Introduction*, 5th ed., p. 149.)

can be used to calculate the applied stress and the resulting strain with the following equations.

$$\sigma = \frac{F}{A} \quad \text{and} \quad \varepsilon = \frac{l - l_0}{l_0}$$

where

σ = stress in $lb_f/in.^2$ (psi)
F = applied force in lb_f
A = sample cross-sectional area in $in.^2$
ε = strain in in./in.
l = sample length
l_0 = original sample length

(a) Use the provided data to calculate the stress and the corresponding strain for each data pair. The tested sample was a rod of diameter 0.505 in., so you'll need to find the cross-sectional area to use in your calculations.

(b) Create an x–y plot with strain on the x-axis and stress on the y-axis. Connect the data points with a solid black line, and use circles to mark each data point.

(c) Add a title and appropriate axis labels.

(d) The point where the graph changes from a straight line with a steep slope to a flattened curve is called the yield stress or yield point. This corresponds to a significant change in the material behavior. Before the yield point the material is elastic, returning to its original shape if the load is removed—much like a rubber band. Once the material has been deformed past the yield point, the change in shape becomes permanent and is called plastic deformation. Use a text box to mark the yield point on your graph.

5.12 In the previous chapter, the accumulated cyclone energy index (ACE) was introduced (Problem 4.5). Use that data to solve the following problems. It may also be available to you as an EXCEL spreadsheet, named ace_data.xlsx.

(a) Create an x–y plot of the year (on the x-axis) versus the ACE index values (on the y-axis.)

(b) Calculate the mean ACE value, and use it to draw the mean value on your graph. (*Hint:* You just need two points, one at the first year and another at the final year).

(c) Use the `filter` function to find a running weighted average of the ACE data, over a 10-year period, using the following syntax, assuming you have named the data extracted from the ACE column, `ace`.

```
running_avg_ace = filter(ones(1,10)/10,1,ace);
```

Create a plot of the year (on the x-axis) versus the ACE value and the weighted average on the y-axis. (You will have two lines.) From your graph, do you think hurricane intensity is increasing? You can find out more about the `filter` function by searching the `help` documentation.

Using Subplots

5.13 In Problem 5.1, you created four plots. Combine these into one figure with four subwindows, using the `subplot` function of MATLAB®.

5.14 In Problems 5.6, 5.7, and 5.9, you created a total of four plots. Combine these into one figure with four subwindows, using the `subplot` function of MATLAB®.

Polar Plots

5.15 Create a vector of angles from 0 to 2π. Use the `polar` plotting function to create graphs of the functions that follow. Remember, polar plots expect the angle and the radius as the two inputs to the `polar` function. Use the `subplot` function to put all four of your graphs in the same figure.

(a) $r = \sin^2(\theta) + \cos^2(\theta)$
(b) $r = \sin(\theta)$
(c) $r = e^{\theta/5}$
(d) $r = \sinh(\theta)$

5.16 In Practice Exercises 5.3, you created a number of interesting shapes in polar coordinates. Use those exercises as a help in creating the following figures:

(a) Create a "flower" with three petals.
(b) Overlay your figure with eight additional petals, half the size of the three original ones.
(c) Create a heart shape.
(d) Create a six-pointed star.
(e) Create a hexagon.

Logarithmic Plots

5.17 When interest is compounded continuously, the following equation represents the growth of your savings:

$$P = P_0 e^{rt}$$

In this equation,
P = current balance
P_0 = initial balance
r = growth constant, expressed as a decimal fraction
t = time invested.
Determine the amount in your account at the end of each year if you invest $1000 at 8% (0.08) for 30 years. (Make a table.)

Create a figure with four subplots. Plot time on the x-axis and current balance P on the y-axis.

(a) In the first quadrant, plot t versus P in a rectangular coordinate system.
(b) In the second quadrant, plot t versus P, scaling the x-axis logarithmically.
(c) In the third quadrant, plot t versus P, scaling the y-axis logarithmically.
(d) In the fourth quadrant, plot t versus P, scaling both axes logarithmically.

Which of the four plotting techniques do you think displays the data best?

5.18 According to Moore's law (an observation made in 1965 by Gordon Moore, a cofounder of Intel Corporation; see Figure P5.18), the number of transistors that would fit per square inch on a semiconductor integrated circuit doubles approximately every 2 years. Although Moore's law is often reported as predicting doubling every 18 months, this is incorrect. A colleague of Moore took into account the fact that transistor performance is also improving, and when combined with the increased number of transistors results in doubling of *performance* every 18 months. The year 2005 was the 40th

Figure P5.18
Gordon Moore, a pioneer of the semiconductor industry. (Copyright © 2005 Intel Corporation.)

anniversary of the law. Over the last 40 years, Moore's projection has been consistently met. In 1965, the then state-of-the-art technology allowed for 30 transistors per square inch. Moore's law says that transistor density can be predicted by $d(t) = 30 \ (2^{t/2})$, where t is measured in years.

(a) Letting $t = 0$ represent the year 1965 and $t = 46$ represent 2011, use this model to calculate the predicted number of transistors per square inch for the 46 years from 1965 to 2011. Let t increase in increments of 2 years. Display the results in a table with two columns—one for the year and one for the number of transistors.

(b) Using the subplot feature, plot the data in a linear x–y plot, a semilog x plot, a semilog y plot, and a log–log plot. Be sure to title the plots and label the axes.

5.19 The total transistor count on integrated circuits produced over the last 35 years is shown in Table P5.19. Create a semilog plot (with the y-axis scaled

Table P5.19 Exponential Increase in Transistor Count on Integrated Circuits*

Processor	Transistor Count	Date of Introduction	Manufacturer
Intel 4004	2300	1971	Intel
Intel 8008	2500	1972	Intel
Intel 8080	4500	1974	Intel
Intel 8088	29000	1979	Intel
Intel 80286	134000	1982	Intel
Intel 80386	275000	1985	Intel
Intel 80486	1200000	1989	Intel
Pentium	3100000	1993	Intel
AMD K5	4300000	1996	AMD
Pentium II	7500000	1997	Intel
AMD K6	8800000	1997	AMD
Pentium III	9500000	1999	Intel
AMD K6-III	21300000	1999	AMD
AMD K7	22000000	1999	AMD
Pentium 4	42000000	2000	Intel
Barton	54300000	2003	AMD
AMD K8	105900000	2003	AMD
Itanium 2	220000000	2003	Intel
Itanium 2 with 9MB cache	592000000	2004	Intel
Cell	241000000	2006	Sony/IBM/Toshiba
Core 2 Duo	291000000	2006	Intel
Core 2 Quad	582000000	2006	Intel
G80	681000000	2006	NVIDIA
POWER6	789000000	2007	IBM
Dual-Core Itanium 2	1700000000	2006	Intel
Quad-Core Itanium Tukwila (processor)[1]	2000000000	2008	Intel
8-Core Xeon Nehalem-EX	2300000000	2010	Intel
10-Core Xeon Westmere-EX	2600000000	2011	Intel

*Data from *Wikipedia*, http://en.wikipedia.org/wiki/Transistor_count.

logarithmically) of the actual data, using circles only to indicate the data points (no lines). Include a second line representing the predicted values using Moore's law, based on the 1971 count as the starting point. Add a legend to your plot.

5.20 Many physical phenomena can be described by the Arrhenius equation. For example, reaction-rate constants for chemical reactions are modeled as

$$k = k_0 e^{(-Q/RT)}$$

where

k_0 = constant with units that depend upon the reaction
Q = activation energy, kJ/kmol
R = ideal gas constant, kJ/kmol K
T = temperature in K.

For a certain chemical reaction, the values of the constants are

$$Q = 1000 \, \text{J/mol}$$
$$k_0 = 10 \, \text{s}^{-1}$$
$$R = 8.314 \, \text{J/mol K}$$

for T from 300 K to 1000 K. Find the values of k. Create the following two graphs of your data in a single figure window:

(a) Plot T on the x-axis and k on the y-axis.
(b) Plot your results as the $\log_{10}$ of k on the y-axis and $1/T$ on the x-axis.

Bar Graphs, Pie Charts, and Histograms

5.21 Let the vector

$$G = [68, 83, 61, 70, 75, 82, 57, 5, 76, 85, 62, 71, 96, 78, 76, 68, 72, 75, 83, 93]$$

represent the distribution of final grades in an engineering course.

(a) Use MATLAB® to sort the data and create a bar graph of the scores.
(b) Create a histogram of the scores.

5.22 In the engineering class mentioned in Problem 5.21, there are
2 A's
4 B's
8 C's
4 D's
2 E's

(a) Create a vector of the grade distribution

$$\text{grades} = [2, 4, 8, 4, 2]$$

Create a pie chart of the grades vector. Add a legend listing the grade names (A, B, C, etc.)
(b) Use the **menu** text option instead of a legend to add a text box to each slice of pie, and save your modified graph as a **.fig** file.
(c) Create a three-dimensional pie chart of the same data. Earlier versions of MATLAB® had trouble with legends for many three-dimensional figures, so don't be surprised if your legend doesn't match the pie chart.

5.23 The inventory of a certain type of screw in a warehouse at the end of each month is listed in the following table:

	2009	2010
January	2345	2343
February	4363	5766
March	3212	4534
April	4565	4719
May	8776	3422
June	7679	2200
July	6532	3454
August	2376	7865
September	2238	6543
October	4509	4508
November	5643	2312
December	1137	4566

Plot the data in a bar graph.

5.24 Use the `randn` function to create 1000 values in a normal (Gaussian) distribution of numbers with a mean of 70 and a standard deviation of 3.5. Create a histogram of the data set you calculated.

Graphs with Two y-Axes

5.25 In the introduction to Problems 5.6 through 5.9, we learned that the equations for the distance traveled by a projectile as a function of time are

$$\text{Horizontal}(t) = tV_0 \cos(\theta)$$
$$\text{Vertical}(t) = tV_0 \sin(\theta) - \tfrac{1}{2}gt^2$$

For time from 0 to 20 s, plot both the horizontal distance versus time and the vertical distance versus time on the same graph, using separate y-axes for each line. Assume a launch angle of 45° ($\pi/4$ radians) and an initial velocity of 100 m/s. Assume also that the acceleration due to gravity, g, is 9.8 m/s. Be sure to label both y-axes.

5.26 If the equation modeling the vertical distance traveled by a projectile as a function of time is

$$\text{Vertical}(t) = tV_0 \sin(\theta) - 1/2 \, gt^2$$

then, from calculus, the velocity in the vertical direction is

$$\text{Velocity}(t) = V_0 \sin(\theta) - gt$$

Create a vector t from 0 to 20 s, and calculate both the vertical position and the velocity in the vertical direction, assuming a launch angle θ of $\pi/4$ radians and an initial velocity of 100 m/s. Plot both quantities on the same graph with separate y-axes. Be sure to label both y-axes.

The velocity should be zero at the point where the projectile is the highest in the vertical direction. Does your graph support this prediction?

5.27 For many metals, deformation changes their physical properties. In a process called *cold work*, metal is intentionally deformed to make it stronger.

The following data tabulate both the strength and ductility of a metal that has been cold worked to different degrees:

Percent Cold Work	Yield Strength, MPa	Ductility, %
10	275	43
15	310	30
20	340	23
25	360	17
30	375	12
40	390	7
50	400	4
60	407	3
68	410	2

Plot these data on a single *x–y* plot with two *y*-axes. Be sure to label both *y*-axes.

Three-Dimensional Line Plots

5.28 Create a vector **x** of values from 0 to 20 π, with a spacing of $\pi/100$. Define vectors **y** and **z** as

$$y = x \sin(x)$$

and

$$z = x \cos(x)$$

(a) Create an *x–y* plot of **x** and **y**.
(b) Create a polar plot of **x** and **y**.
(c) Create a three-dimensional line plot of **x**, **y**, and **z**. Don't forget a title and labels.

5.29 Figure out how to adjust your input to `plot3` in Problem 5.28 so as to create a graph that looks like a tornado (see Figure P5.29). Use `comet3` instead of `plot3` to create the graph.

Figure P5.29
Tornado plot.

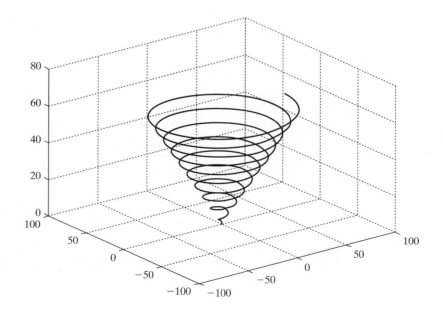

Three-Dimensional Surface and Contour Plots

5.30 Create x and y vectors from -5 to $+5$ with a spacing of 0.5. Use the mesh-grid function to map x and y onto two new two-dimensional matrices called X and Y. Use your new matrices to calculate vector Z, with magnitude

$$Z = \sin\left(\sqrt{X^2 + Y^2}\right)$$

(a) Use the mesh plotting function to create a three-dimensional plot of Z.

(b) Use the surf plotting function to create a three-dimensional plot of Z. Compare the results you obtain with a single input (Z) with those obtained with inputs for all three dimensions (X, Y, Z).

(c) Modify your surface plot with interpolated shading. Try using different colormaps.

(d) Generate a contour plot of Z.

(e) Generate a combination surface and contour plot of Z.

6 User-Defined Functions

Objectives

After reading this chapter, you should be able to:

- Create and use your own MATLAB® functions with both single and multiple inputs and outputs
- Store and access your own functions in toolboxes

- Create and use anonymous functions
- Create and use function handles
- Create and use subfunctions and nested subfunctions

INTRODUCTION

The MATLAB® programming language is built around functions. A *function* is a piece of computer code that accepts an input argument from the user and provides output to the program. Functions allow us to program efficiently, enabling us to avoid rewriting the computer code for calculations that are performed frequently. For example, most computer programs contain a function that calculates the sine of a number. In MATLAB®, sin is the function name used to call up a series of commands that perform the necessary calculations. The user needs to provide an angle, and MATLAB® returns a result. It isn't necessary for the programmer to know how MATLAB® calculates the value of sin(x).

6.1 CREATING FUNCTION M-FILES

We have already explored many of MATLAB®'s built-in functions, but you may wish to define your own functions—those that are used commonly in your programming. User-defined functions are stored as M-files and can be accessed by MATLAB® if they are in the current folder or on MATLAB®'s search path.

6.1.1 Syntax

Both built-in MATLAB® functions and user-defined MATLAB® functions have the same structure. Each consists of a name, user-provided input, and calculated output. For example, the function

    cos(x)

- is named `cos`,
- takes the user input inside the parentheses (in this case, **x**), and
- calculates a result.

The user does not see the calculations performed, but just accepts the answer. User-defined functions work the same way. Imagine that you have created a function called `my_function`. Using

    my_function(x)

in a program or from the command window will return a result, as long as **x** is defined and the logic in the function definition works.

User-defined functions are created in M-files. Each must start with a function-definition line that contains:

- The word `function`
- A variable that defines the function output
- A function name
- A variable used for the input argument

For example,

    function output = my_function(x)

is the first line of the user-defined function called `my_function`. It requires one input argument, which the program will call x, and will calculate one output argument, which the program will call `output`. The function name and the names of the input and output variables are arbitrary and are selected by the programmer. Here's an example of an appropriate first line for a function called `calculation`:

    function result = calculation(a)

In this case, the function name is `calculation`, the input argument will be called a in any calculations performed in the function program, and the output will be called `result`. Although any valid MATLAB® names can be used, it is good programming practice to use meaningful names for all variables and for function names.

HINT

Students are often confused about the use of the word *input* as it refers to a function. We use it here to describe the input argument—the value that goes inside the parentheses when we call a function. In MATLAB®, input arguments are different from the `input` command.

Here's an example of a very simple MATLAB® function that calculates the value of a particular polynomial:

    function output = poly(x)
    %This function calculates the value of a third-order

```
%polynomial
output = 3*x.^3 + 5*x.^2 - 2*x +1;
```

The function name is `poly`, the input argument is x, and the output variable is named `output`.

Before this function can be used, it must be saved into the current folder. The file name *must be the same* as the function name in order for MATLAB® to find it. All of the MATLAB® naming conventions we learned for naming variables apply to naming user-defined functions. In particular,

- The function name must start with a letter.
- It can consist of letters, numbers, and the underscore.
- Reserved names cannot be used.
- Any length is allowed, although long names are not good programming practice.

Once the M-file has been saved, the function is available for use from the command window, from a script M-file, or from another function. You cannot execute a function M-file directly from the M-file itself. This makes sense, since the input parameters have not been defined until you call the function from the command window or a script M-file. Consider the `poly` function just created. If, in the command window, we type

poly(4)

then MATLAB® responds with

```
ans =
        265
```

If we set **a** equal to 4 and use **a** as the input argument, we get the same result:

```
a = 4;
poly(a)

ans =
        265
```

If we define a vector, we get a vector of answers. Thus,

```
y = 1:5;
poly(y)
```

gives

```
ans =
        7     41    121    265    491
```

If, however, you try to execute the function by selecting the save-and-run icon from the function menu bar, the following error message is displayed:

```
???Input argument "x" is undefined.
Error in ==> poly at 3
output = 3*x.^3 + 5*x.^2 - 2*x +1;
```

The value of **x** must be passed to the function when it is used—either in the command window or from within a script M-file program.

HINT

While you are creating a function, it may be useful to allow intermediate calculations to print to the command window. However, once you complete your "debugging," make sure that all your output is suppressed. If you don't, you'll see extraneous information in the command window.

PRACTICE EXERCISES 6.1

Create MATLAB® functions to evaluate the following mathematical functions (make sure you select meaningful function names) and test them. To test your functions you'll need to call them from the command window, or use them in a script M-file program. Remember, each function requires its own M-file.

1. $y(x) = x^2$
2. $y(x) = e^{1/x}$
3. $y(x) = \sin(x^2)$

Create MATLAB® functions for the following unit conversions (you may need to consult a textbook or the Internet for the appropriate conversion factors). Be sure to test your functions, either from the command window, or by using them in a script M-file program.

4. Inches to feet
5. Calories to joules
6. Watts to BTU/hr
7. Meters to miles
8. Miles per hour (mph) to ft/s

EXAMPLE 6.1

Figure 6.1
Trigonometric functions require angles to be expressed in radians. Trigonometry is regularly used in engineering drawings.

CONVERTING BETWEEN DEGREES AND RADIANS

Engineers usually measure angles in degrees, yet most computer programs and many calculators require that the input to trigonometric functions be in radians. Write and test a function **DR** that changes degrees to radians and another function **RD** that changes radians to degrees. Your functions should be able to accept both scalar and matrix input.

1. State the Problem
   Create and test two functions, **DR** and **RD**, to change degrees to radians and radians to degrees (see Figure 6.1).
2. Describe the Input and Output

Input	A vector of degree values
	A vector of radian values
Output	A table converting degrees to radians
	A table converting radians to degrees

3. Develop a Hand Example

$$degrees = radians \times 180/\pi$$
$$radians = degrees \times \pi/180$$

Degrees to Radians	
Degrees	**Radians**
0	0
30	$30(\pi/180) = \pi/6 = 0.524$
60	$60(\pi/180) = \pi/3 = 1.047$
90	$90(\pi/180) = \pi/2 = 1.571$

4. Develop a MATLAB® Solution

```
%Example 6.1
%
clear, clc
%Define a vector of degree values
degrees = 0:15:180;
% Call the DR function, and use it to find radians
radians = DR(degrees);
%Create a table to use in the output
degrees_radians = [degrees;radians]'
%Define a vector of radian values
radians = 0:pi/12:pi;
%Call the RD function, and use it to find degrees
degrees = RD(radians);
radians_degrees = [radians;degrees]'
```

The functions called by the program are

```
function output = DR(x)
%This function changes degrees to radians
output = x*pi/180;
```

and

```
function output = RD(x)
%This function changes radians to degrees
output = x*180/pi;
```

Remember that in order for the script M-file to find the functions, they must be in the current folder and must be named **DR.m** and **RD.m**. The program generates the following results in the command window:

```
degrees_radians =
       0      0.000
      15      0.262
      30      0.524
      45      0.785
      60      1.047
```

(continued)

75	1.309
90	1.571
105	1.833
120	2.094
135	2.356
150	2.618
165	2.880
180	3.142

radians_degrees =

0.000	0.000
0.262	15.000
0.524	30.000
0.785	45.000
1.047	60.000
1.309	75.000
1.571	90.000
1.833	105.000
2.094	120.000
2.356	135.000
2.618	150.000
2.880	165.000
3.142	180.000

5. Test the Solution

   Compare the MATLAB® solution with the hand solution. Since the output is a table, it is easy to see that the conversions generated by MATLAB® correspond to those calculated by hand.

EXAMPLE 6.2

ASTM GRAIN SIZE

Figure 6.2
Typical microstructures of iron (400×). (From *Metals Handbook*, 9th ed., Vol. 1, American Society of Metals, Metals Park, Ohio, 1978.)

You may not be used to thinking of metals as crystals, but they are. If you look at a polished piece of metal under a microscope, the structure becomes clear, as seen in Figure 6.2. As you can see, every crystal (called a grain in metallurgy) is a different size and shape. The size of the grains affects the metal's strength; the finer the grains, the stronger the metal.

Because it is difficult to determine an "average" grain size, a standard technique has been developed by ASTM (formerly known as the American Society for Testing and Materials, but now known just by its initials). A sample of metal is examined under a microscope at a magnification of 100, and the number of grains in 1 square inch is counted. The parameters are related by

$$N = 2^{n-1}$$

where n is the ASTM grain size and N is the number of grains per square inch at 100×. The equation can be solved for n to give

$$n = \frac{(\log(N) + \log(2))}{\log(2)}$$

This equation is not hard to use, but it's awkward. Instead, let's create a MATLAB® function called `grain_size`.

1. **State the Problem**
   Create and test a function called `grain_size` to determine the ASTM grain size of a piece of metal.

2. **Describe the Input and Output**
   To test the function, we'll need to choose an arbitrary number of grains. For example:

   Input 16 grains per square inch at 100×

   Output ASTM grain size

3. **Develop a Hand Example**

$$n = \frac{(\log(N) + \log(2))}{\log(2)}$$

$$n = \frac{(\log(16) + \log(2))}{\log(2)} = 5$$

4. **Develop a MATLAB® Solution**
   The function, created in a separate M-file, is

```
function output = grain_size(N)
%Calculates the ASTM grain size n
output = (log10(N) + log10(2))./log10(2);
```

   which was saved as `grain_size.m` in the current folder. To use this function, we can call it from the command window:

```
grain_size(16)
ans =
        5
```

5. **Test the Solution**
   The MATLAB® solution is the same as the hand solution. It might be interesting to see how the ASTM grain size varies with the number of grains per square inch. We could use the function with an array of values and plot the results in Figure 6.3.

```
%Example 6.2
%ASTM Grain Size
N = 1:100;
n = grain_size(N);
plot(N,n)
title('ASTM Grain Size')
xlabel('Number of grains per square inch at 100x')
ylabel('ASTM Grain Size')
grid
```

As expected, the grain size increases as the number of grains per square inch increases.

(continued)

Figure 6.3
A plot of a function's behavior is a good way to help determine whether you have programmed it correctly.

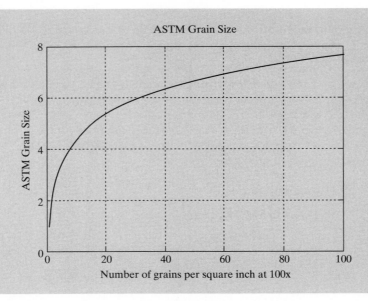

6.1.2 Comments

As with any computer program, you should comment your code liberally so that it is easy to follow. However, in a MATLAB® function, the comments on the line immediately following the very first line serve a special role. These lines are returned when the `help` function is queried from the command window. Consider, for example, the following function:

```
function results = f(x)
%This function converts seconds to minutes
results = x./60;
```

Querying the `help` function from the command window

```
help f
```

returns

```
This function converts seconds to minutes
```

6.1.3 Functions with Multiple Inputs and Outputs

Just as the predefined MATLAB® functions may require multiple inputs and may return multiple outputs, more complicated user-defined functions can be written. Recall, for example, the remainder function. This predefined function calculates the remainder in a division problem and requires the user to input the dividend and the divisor. For the problem $\frac{5}{3}$, the correct syntax is

```
rem(5,3)
```

which gives

```
ans =
        2
```

Similarly, a user-defined function could be written to multiply two vectors together:

```
function output = g(x,y)
% This function multiplies x and y together
% x and y must be the same size matrices
a = x .*y;
output = a;
```

When **x** and **y** are defined in the command window and the function **g** is called, a vector of output values is returned:

```
x = 1:5;
y = 5:9;
g(x,y)
ans =
      5    12    21    32    45
```

You can use the comment lines to let users know what kind of input is required and to describe the function. In this example, an intermediate calculation (**a**) was performed, but the only output from this function is the variable we've named `output`. This output can be a matrix containing a variety of numbers, but it's still only one variable.

You can also create functions that return more than one output variable. Many of the predefined MATLAB® functions return more than one result. For example, `max` returns both the maximum value in a matrix and the element number at which the maximum occurs. To achieve the same result in a user-defined function, make the output a matrix of answers instead of a single variable, as in

```
function [dist, vel, accel] = motion(t)
% This function calculates the distance, velocity, and
% acceleration of a particular car for a given value of t
% assuming all 3 parameters are initially 0.
accel = 0.5 .*t;
vel = t.^2/4;
dist = t.^3/12;
```

Once saved as `motion` in the current folder, you can use the function to find values of `distance`, `velocity`, and `acceleration` at specified times:

```
[distance, velocity, acceleration] = motion(10)

distance =
        83.33
velocity =
        25
acceleration =
        5
```

If you call the `motion` function without specifying all three outputs, only the first output will be returned:

```
motion(10)
ans =
        83.333
```

Remember, all variables in MATLAB® are matrices, so it's important in the preceding example to use the .* operator, which specifies element-by-element multiplication. For example, using a vector of time values from 0 to 30 in the motion function

```
time = 0:10:30;
[distance, velocity, acceleration] = motion(time)
```

returns three vectors of answers:

```
distance =
        0     83.33    666.67    2250.00
velocity =
        0     25.00    100.00     225.00
acceleration =
        0      5.00     10.00      15.00
```

It's easier to see the results if you group the vectors together, as in

```
results = [time',distance',velocity',acceleration']
```

which returns

```
results =

        0          0          0          0
    10.00      83.33      25.00       5.00
    20.00     666.67     100.00      10.00
    30.00    2250.00     225.00      15.00
```

Because time, distance, velocity, and acceleration were row vectors, the transpose operator was used to convert them into columns.

PRACTICE EXERCISES 6.2

Assuming that the matrix dimensions agree, create and test MATLAB® functions to evaluate the following simple mathematical functions with multiple input vectors and a single output vector:

1. $z(x, y) = x + y$
2. $z(a, b, c) = ab^c$
3. $z(w, x, y) = we^{(x/y)}$
4. $z(p, t) = p/\sin(t)$

Assuming that the matrix dimensions agree, create and test MATLAB® functions to evaluate the following simple mathematical functions with a single input vector and multiple output vectors:

5. $f(x) = \cos(x)$
   $f(x) = \sin(x)$
6. $f(x) = 5x^2 + 2$
   $f(x) = \sqrt{5x^2 + 2}$
7. $f(x) = \exp(x)$
   $f(x) = \ln(x)$

Assuming that the matrix dimensions agree, create, and test MATLAB® functions to evaluate the following simple mathematical functions with multiple input vectors and multiple output vectors:

8. $f(x, y) = x + y$
   $f(x, y) = x - y$
9. $f(x, y) = ye^x$
   $f(x, y) = xe^y$

EXAMPLE 6.3

HOW GRAIN SIZE AFFECTS METAL STRENGTH: A FUNCTION WITH THREE INPUTS

Metals composed of small crystals are stronger than metals composed of fewer large crystals. The metal yield strength (the amount of stress at which the metal starts to permanently deform) is related to the average grain diameter by the *Hall–Petch equation*:

$$\sigma = \sigma_0 + Kd^{-1/2}$$

where the symbols σ_0 and K represent constants that are different for every metal.

Create a function called `HallPetch` that requires three inputs—σ_0, K, and d—and calculates the value of yield strength. Call this function from a MATLAB® program that supplies values of σ_0 and K, then plots the value of yield strength for values of d from 0.1 to 10 mm.

1. **State the Problem**
   Create a function called `HallPetch` that determines the yield strength of a piece of metal, using the Hall–Petch equation. Use the function to create a plot of yield strength versus grain diameter.

2. **Describe the Input and Output**

   Input $K = 9600 \text{ psi}/\sqrt{\text{mm}}$
   $\sigma_0 = 12{,}000 \text{ psi}$
   $d = 0.1 \text{ to } 10 \text{ mm}$

   Output Plot of yield strength versus diameter

3. **Develop a Hand Example**
   The Hall–Petch equation is

   $$\sigma = \sigma_0 + Kd^{-1/2}$$

   Substituting values of 12,000 psi and 9600 psi/$\sqrt{\text{mm}}$ for σ_0 and K, respectively, then

   $$\sigma = 12{,}000 + 9600d^{-1/2}$$

   For $d = 1$ mm,

   $$\sigma = 12{,}000 + 9600 = 21{,}600$$

(continued)

4. Develop a MATLAB® Solution
   The desired function, created in a separate M-file, is

```
function output = HallPetch(sigma0,k,d)
%Hall-Petch equation to determine the yield
%strength of metals
output = sigma0 + K*d.^(-0.5);
```

and was saved as **HallPetch.m** in the current folder:

```
%Example 6.3
clear,clc
format compact
s0 = 12000
K = 9600
%Define the values of grain diameter
diameter = 0.1:0.1:10;
yield = HallPetch(s0,K,d);
%Plot the results
figure(1)
plot(diameter,yield)
title('Yield strengths found with the Hall-Petch equation')
xlabel('diameter, mm')
ylabel('yield strength, psi')
```

The graph shown in Figure 6.4 was generated by the program.

5. Test the Solution
   We can use the graph to compare the results to the hand solution.

Figure 6.4
Yield strengths predicted with the Hall–Petch equation. Small grain diameters correspond to large values of the yield strength.

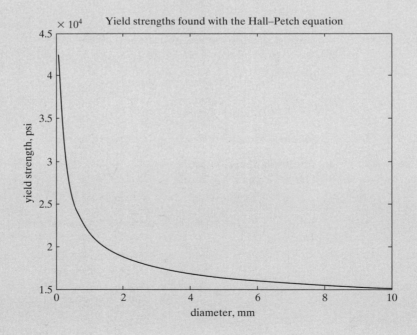

Yield strengths found with the Hall–Petch equation

EXAMPLE 6.4

KINETIC ENERGY: A FUNCTION WITH TWO INPUTS

The kinetic energy of a moving object (Figure 6.5) is

$$KE = {}^1\!/_2\, mv^2.$$

Create and test a function called KE to find the kinetic energy of a moving car if you know the mass m and the velocity v of the vehicle.

1. **State the Problem**
   Create a function called KE to find the kinetic energy of a car.
2. **Describe the Input and Output**
   Input   Mass of the car, in kilograms
                  Velocity of the car, in m/s

   Output   Kinetic energy, in joules
3. **Develop a Hand Example**
   If the mass is 1000 kg, and the velocity is 25 m/s, then

   $$KE = {}^1\!/_2 \times 1000\ \text{kg} \times (25\ \text{m/s})^2 = 312{,}500\ \text{J} = 312.5\ \text{kJ}$$

4. **Develop a MATLAB® Solution**

   ```
   function output = ke(mass,velocity)
   output = 1/2*mass*velocity.^2;
   ```

5. **Test the Solution**

   ```
   v = 25;
   m = 1000;
   ke(m,v)
   ans =
           312500
   ```

   This result matches the hand example, confirming that the function works correctly and can now be used in a larger MATLAB® program.

Figure 6.5
Race cars store a significant amount of kinetic energy. (Rick Graves/Getty Images.)

6.1.4 Functions with No Input or No Output

Although most functions need at least one input and return at least one output value, in some situations no inputs or outputs are required. For example, consider this function, which draws a star in polar coordinates:

```
function [] = star( )
theta = pi/2:0.8*pi:4.8*pi;
r = ones(1,6);
polar(theta,r)
```

The square brackets on the first line indicate that the output of the function is an empty matrix (i.e., no value is returned). The empty parentheses tell us that no input is expected. If, from the command window, you type

```
star
```

then no values are returned, but a figure window opens showing a star drawn in polar coordinates (see Figure 6.6).

HINT

You may ask yourself if the **star** function is really an example of a function that does not return an output; after all, it does draw a star. But the output of a function is defined as a *value* that is returned when you call the function. If we ask MATLAB® to perform the calculation

```
A = star
```

an error statement is generated, because the **star** function does not return anything! Thus, there is nothing to set **A** equal to.

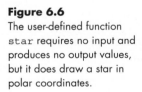

Figure 6.6
The user-defined function
star requires no input and
produces no output values,
but it does draw a star in
polar coordinates.

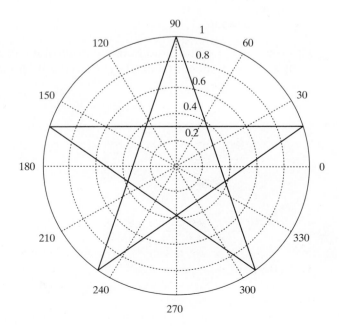

There are numerous built-in MATLAB® functions that do not require any input. For example,

```
A = clock
```

returns the current time:

```
A =
  1.0e+003 *
  Columns 1 through 4
    2.0050    0.0030    0.0200    0.0150
  Columns 5 through 6
    0.0250    0.0277
```

Also,

```
A = pi
```

returns the value of the mathematical constant π:

```
A =

  3.1416
```

However, if we try to set the MATLAB® function `tic` equal to a variable name, an error statement is generated, because `tic` does not return an output value:

```
A = tic
???Error using ==> tic
Too many output arguments.
```

(The `tic` function starts a timer going for later use in the **toc** function.)

6.1.5 Determining the Number of Input and Output Arguments

There may be times when you want to know the number of input arguments or output values associated with a function. MATLAB® provides two built-in functions for this purpose.

The `nargin` function determines the number of input arguments in either a user-defined function or a built-in function. The name of the function must be specified as a string, as, for example, in

```
nargin('sin')
ans =

    1
```

The remainder function, `rem`, requires two inputs; thus,

```
nargin('rem')
ans =
    2
```

When `nargin` is used inside a user-defined function, it determines how many input arguments were actually entered. This allows a function to have a variable number of inputs. Recall graphing functions such as `surf`. When `surf` has a single matrix input, a graph is created, using the matrix index numbers as the *x*- and *y*-coordinates. When there are three inputs, *x*, *y*, and *z*, the graph is based on the specified *x*- and *y*-values. The `nargin` function allows the programmer to determine how to create the plot, based on the number of inputs.

The surf function is an example of a function with a variable number of inputs. If we use nargin from the command window to determine the number of declared inputs, there isn't one correct answer. The nargin function returns a negative number to let us know that a variable number of inputs are possible:

```
nargin('surf')
ans =
   -1
```

The nargout function is similar to nargin, but it determines the number of outputs from a function:

```
nargout('sin')
ans =
   1
```

The number of outputs is determined by how many matrices are returned, not how many values are in the matrix. We know that **size** returns the number of rows and columns in a matrix, so we might expect nargout to return 2 when applied to size. However,

```
nargout('size')
ans =
   1
```

returns only one matrix, which has just two elements, as for example, in

```
x = 1:10;
size(x)
ans =
  1 10
```

An example of a function with multiple outputs is max:

```
nargout('max')
ans =
    2
```

When used inside a user-defined function, nargout determines how many outputs have been requested by the user. Consider this example, in which we have rewritten the function from Section 6.1.4 to create a star:

```
function A = star1( )
theta = pi/2:0.8*pi:4.8*pi;
r = ones(1,6);
polar(theta,r)
if nargout==1
    A = 'Twinkle twinkle little star';
end
```

If we use nargout from the command window, as in

```
nargout('star1')
ans =
    1
```

MATLAB® tells us that one output is specified. If we call the function simply as

```
star1
```

nothing is returned to the command window, although the plot is drawn. If we call the function by setting it equal to a variable, as in

```
x = star1
x =
Twinkle twinkle little star
```

a value for **x** is returned, based on the `if` statement embedded in the function, which used `nargout` to determine the number of output values.

`If` statements are introduced in Chapter 8.

6.1.6 Local Variables

The variables used in function M-files are known as *local variables*. The only way a function can communicate with the workspace is through input arguments and the output it returns. Any variables defined within the function exist only for the function to use. For example, consider the g function previously described:

```
function output = g(x,y)
% This function multiplies x and y together
% x and y must be the same size matrices
a = x .*y;
output = a;
```

LOCAL VARIABLE
A variable that only has meaning inside a program or function

The variables a, x, y, and `output` are local variables. They can be used for additional calculations inside the g function, but they are not stored in the workspace. To confirm this, clear the workspace and the command window and then call the g function:

```
clear, clc
g(10,20)
```

The function returns

```
g(10,20)
ans =
        200
```

Notice that the only variable stored in the workspace window is `ans`, which is characterized as follows:

Name	Value	Size	Bytes	Class
ans	200	1 × 1	8	double array

Just as calculations performed in the command window or from a script M-file cannot access variables defined in functions, functions cannot access the variables defined in the workspace. This means that functions must be completely self-contained: The only way they can get information from your program is through the input arguments, and the only way they can deliver information is through the function output.

Consider a function written to find the distance an object falls due to gravity:

```
function result = distance(t)
%This function calculates the distance a falling object
%travels due to gravity
g = 9.8 %meters per second squared
result = 1/2*g*t.^2;
```

The value of g must be included *inside* the function. It doesn't matter whether g has or has not been used in the main program. How g is defined is hidden to the distance function unless g is specified inside the function.

Of course, you could also pass the value of g to the function as an input argument:

```
function result = distance(g,t)
%This function calculates the distance a falling object
%travels due to gravity
result = 1/2*g*t.^2;
```

HINT

The same matrix names can be used in both a function and the program that references it. However, they do not *have* to be the same. Since variable names are local to either the function or the program that calls the function, the variables are completely separate. As a beginning programmer, you would be wise to use different variable names in your functions and your programs—just so you don't confuse *yourself*.

6.1.7 Global Variables

KEY IDEA
It is usually a bad idea to define global variables

Unlike local variables, global variables are available to all parts of a computer program. In general, *it is a bad idea* to define global variables. However, MATLAB® protects users from unintentionally using a global variable by requiring that it be identified both in the command-window environment (or in a script M-file) and in the function that will use it.

GLOBAL VARIABLE
A variable that is available from multiple programs

Consider the distance function once again:

```
function result = distance(t)
%This function calculates the distance a falling object
%travels due to gravity
global G
result = 1/2*G*t.^2;
```

The `global` command alerts the function to look in the workspace for the value of **G**. **G** must also have been defined in the command window (or script M-file) as a global variable:

```
global G
G = 9.8;
```

This approach allows you to change the value of **G** without needing to redefine the distance function or providing the value of **G** as an input argument to the distance function.

HINT

As a matter of style, always make the names of global variables uppercase. MATLAB® doesn't care, but it is easier to identify global variables if you use a consistent naming convention.

HINT

It may seem like a good idea to use global variables because they can simplify your programs. However, consider this example of using global variables in your everyday life: It would be easier to order a book from an online book-seller if you had posted your credit card information on a site where any retailer could just look it up. Then the bookseller wouldn't have to ask you to type in the number. However, this might produce some unintended consequences (like other people using your credit card without your permission or knowledge!). When you create a global variable, it becomes available to other functions and can be changed by those functions, sometimes leading to unintended consequences.

6.1.8 Accessing M-File Code

The functions provided with MATLAB® are of two types. One type is built in, and the code is not accessible for us to review. The other type consists of M-files, stored in toolboxes provided with the program. We can see these M-files (or the M-files we've written) with the `type` command. For example, the `sphere` function creates a three-dimensional representation of a sphere; thus,

```
type sphere
```

or

```
type('sphere')
```

returns the contents of the **sphere.m** file:

```
function [xx,yy,zz] = sphere(varargin)
%SPHERE Generate sphere.
%    [X,Y,Z] = SPHERE(N) generates three (N+1)-by-(N+1)
%    matrices so that SURF(X,Y,Z) produces a unit sphere.
%
%    [X,Y,Z] = SPHERE uses N = 20.
%
%    SPHERE(N) and just SPHERE graph the sphere as a SURFACE
%    and do not return anything.
%
%    SPHERE(AX,(. . .) plots into AX instead of GCA.
%
%    See also ELLIPSOID, CYLINDER.
%    Clay M. Thompson 4-24-91, CBM 8-21-92.
%    Copyright 1984-2002 The MathWorks, Inc.
%    $Revision: 5.8.4.1 $ $Date: 2002/09/26 01:55:25 $

%    Parse possible Axes input
error(nargchk(0,2,nargin));
[cax,args,nargs] = axescheck(varargin{:});

n = 20;
if nargs > 0, n = args{1}; end
% -pi <= theta <= pi is a row vector.
% -pi/2 <= phi <= pi/2 is a column vector.
```

```
theta = (-n:2:n)/n*pi;
phi = (-n:2:n)'/n*pi/2;
cosphi = cos(phi); cosphi(1) = 0; cosphi(n+1) = 0;
sintheta = sin(theta); sintheta(1) = 0; sintheta(n+1) = 0;

x = cosphi*cos(theta);
y = cosphi*sintheta;
z = sin(phi)*ones(1,n+1);

if nargout == 0
    cax = newplot(cax);
    surf(x,y,z,'parent',cax)
else
    xx = x; yy = y; zz = z;
end
```

HINT

Notice that the sphere function uses varargin to indicate that it will accept a variable number of input arguments. The function also makes use of the nargin and nargout functions. Studying this function may give you ideas on how to program your own function M-files. The sphere function also uses an if/else structure, which is introduced in a subsequent chapter of this text.

6.2 CREATING YOUR OWN TOOLBOX OF FUNCTIONS

When you call a function in MATLAB®, the program first looks in the current folder to see if the function is defined. If it can't find the function listed there, it starts down a predefined search path, looking for a file with the function name. To view the path the program takes as it looks for files, select

KEY IDEA
Group your functions together into toolboxes

File → Set Path

from the menu bar or type

```
pathtool
```

in the command window (Figure 6.7).

As you create more and more functions to use in your programming, you may wish to modify the path to look in a directory where you've stored your own personal tools. For example, suppose you have stored the degrees-to-radians and radians-to-degrees functions created in Example 6.1 in a directory called **My_functions**.

You can add this directory (folder) to the path by selecting **Add Folder** from the list of option buttons in the Set Path dialog window, as shown in Figure 6.7. You'll be prompted to either supply the folder location or browse to find it, as shown in Figure 6.8.

MATLAB® now first looks into the current folder for function definitions and then works down the modified search path, as shown in Figure 6.9.

Once you've added a folder to the path, the change applies only to the current MATLAB® session, unless you save your changes permanently. Clearly, you should never make permanent changes to a public computer. However, if someone else has made changes you wish to reverse, you can select the default button as shown in Figure 6.9 to return the search path to its original settings.

Figure 6.7
The path tool allows you to change where MATLAB® looks for function definitions.

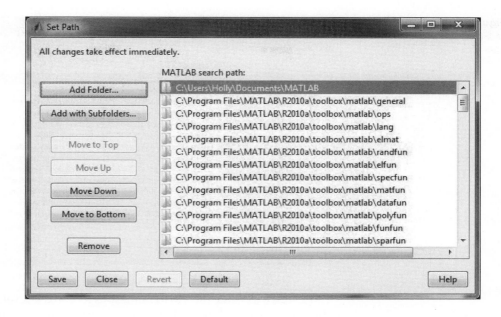

Figure 6.8
The Browse for Folder window.

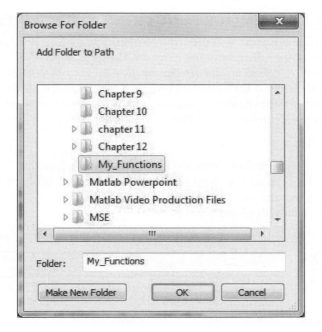

The path tool allows you to change the MATLAB® search path interactively; however, the addpath function allows you to insert the logic to add a search path to any MATLAB® program. Consult

help addpath

if you wish to modify the path in this way.

Figure 6.9
Modified MATLAB® search path.

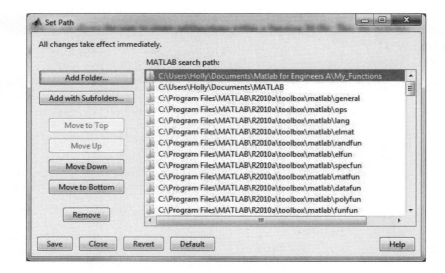

MATLAB® provides access to numerous toolboxes developed at The MathWorks or by the user community. For more information, see the firm's website, www. mathworks.com.

6.3 ANONYMOUS FUNCTIONS AND FUNCTION HANDLES

KEY IDEA

Anonymous functions may be included in M-file programs with other commands or may be defined from the command window

Normally, if you go to the trouble of creating a function, you will want to store it for use in other programming projects. However, MATLAB® includes a simpler kind of function, called an *anonymous function*. New to MATLAB® 7, anonymous functions are defined in the command window or in a script M-file and are available—much as are variable names—only until the workspace is cleared. To create an anonymous function, consider the following example:

```
ln = @(x) log(x)
```

- The @ symbol alerts MATLAB® that **ln** is a function.
- Immediately following the @ symbol, the input to the function is listed in parentheses.
- Finally, the function is defined.

The function name appears in the variable window, listed as a function_handle:

Name	Value	Size	Bytes	Class
ln	@(x) log(x)	1×1	16	function_handle

HINT

Think of a function handle as a nickname for the function.

Anonymous functions can be used like any other function—for example,

```
ln(10)
ans =
    2.3026
```

Once the workspace is cleared, the anonymous function no longer exists. Anonymous functions can be saved as .mat files, just like any variable, and can be restored with the `load` command. For example to save the anonymous function `ln`, type:

```
save my_ln_function ln
```

A file named **my_ln_function.mat** is created, which contains the anonymous `ln` function. Once the workspace is cleared, the `ln` function no longer exists, but it can be reloaded from the .mat file

```
load my_ln_function
```

It is possible to assign a function handle to any M-file function. Earlier in this chapter we created an M-file function called distance.m.

```
function result = distance(t)
result = 1/2*9.8*t.^2;
```

The command

```
distance_handle = @(t) distance(t)
```

assigns the handle **distance_handle** to the distance function.

Anonymous functions and the related function handles are useful in functions that require other functions as input (function functions).

6.4 FUNCTION FUNCTIONS

MATLAB®'s function functions have an odd, but descriptive name. They are functions that require other functions as input. One example of a MATLAB® built-in function function is the function plot, `fplot`. This function requires two inputs: a function or a function handle, and a range over which to plot. We can demonstrate the use of `fplot` with the function handle `ln`, defined as

```
ln = @(x) log(x)
```

The function handle can now be used as input to the `fplot` function:

```
fplot(ln,[0.1, 10])
```

The result is shown in Figure 6.10. We could also use the `fplot` function without the function handle. We just need to insert the function syntax directly, as a string:

```
fplot('log(x)',[0.1, 10])
```

The advantage to using function handles isn't obvious from this example, but consider instead this anonymous function describing a particular fifth-order polynomial:

```
poly5 = @(x) -5*x.^5 + 400*x.^4 + 3*x.^3 + 20*x.^2 - x + 5;
```

Entering the equation directly into the `fplot` function would be awkward. Using the function handle is considerably simpler.

```
fplot(poly5,[-30,90])
```

The results are shown in Figure 6.11.

A wide variety of MATLAB® functions accept function handles as input. For example, the `fzero` function finds the value of x where $f(x)$ is equal to 0. It accepts

Figure 6.10
Function handles can be used as input to a function function, such as `fplot`.

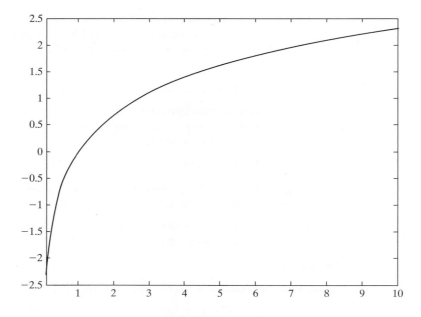

Figure 6.11
This fifth-order polynomial was plotted using the `fplot` function function, with a function handle as input.

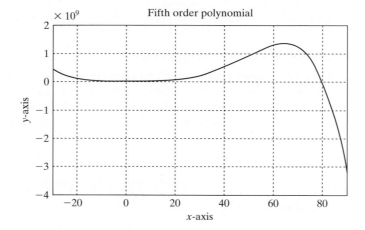

a function handle and a rough guess for *x*. From Figure 6.11, we see that our fifth-order polynomial probably has a zero between 75 and 85, so a rough guess for the zero point might be *x* = 75.

```
fzero(poly5,75)
ans =
       80.0081
```

6.5 SUBFUNCTIONS

More complicated functions can be created by grouping functions together in a single file as subfunctions. These subfunctions can be called only from the primary function, so they have limited utility. Subfunctions can be used to modularize your code and to make the primary function easier to read.

HINT

■
■ You should not attempt to create code using subfunctions until you have
■ mastered function M-files containing a single function.
■

Each MATLAB® function M-file has *one* primary function. The name of the M-file must be the same as the *primary* function name. Thus, the primary function stored in the M-file my_function.m must be named my_function. Subfunctions are added after the primary function, and can have any legitimate MATLAB® variable name. Figure 6.12 shows a very simple example of a function that both adds and subtracts two vectors. The primary function is named subfunction_demo. The file includes two subfunctions: add and subtract.

Notice in the editing window that the contents of each function are identified with a gray bracket. Each code section can be either collapsed or expanded, to make the contents easier to read, by clicking on the + or − sign included with the bracket. MATLAB® uses the term "folding" for this functionality. You can also access folding from the "Text" menu on the menu bar.

When could you use subfunctions effectively? Imagine that your instructor has assigned three homework problems, each requiring you to create and test a function.

- Problem 1 Create and test a function called square to square values of *x*. Assume *x* varies between −3 and +3.
- Problem 2 Create and test a function called cold_work to find the percent cold work experienced by a metallic rod, as it is drawn into a wire. Cold work is described by the following equation

$$\% \, \text{Cold Work} = \frac{r_i^2 - r_f^2}{r_i^2} \times 100$$

where r_i is the initial radius of the rod, and r_f is the final radius of the rod. To test your function let $r_i = 0.5$ cm and let $r_f = 0.25$ cm.
- Problem 3 Create and test a function called potential_energy to determine the potential energy change of a given mass. The change in potential energy is given by

$$\Delta \text{PE} = m \times g \times \Delta z$$

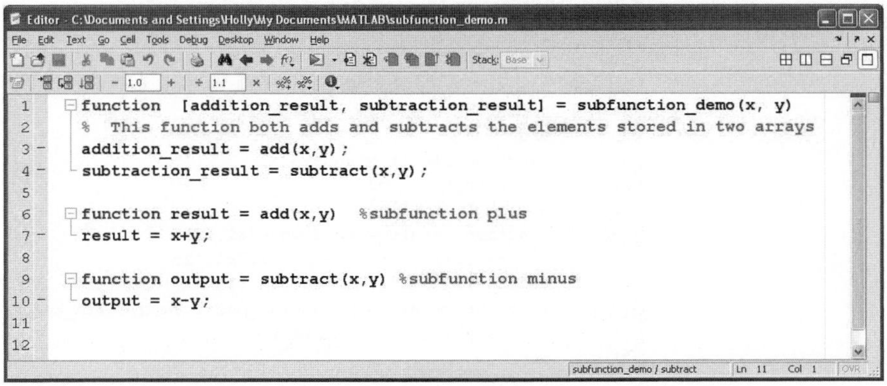

Figure 6.12
MATLAB® allows the user to create subfunctions within a function M-file. This file includes the primary function, subfunction_demo, and two subfunctions add and subtract.

Your function should have three inputs: m, g, and Δz. Use the following data to test your function.

$$m = \begin{bmatrix} 1 & 2 & 3 \end{bmatrix} \text{ kg (The array represents three different masses.)}$$
$$g = 9.8 \text{ m/s}^2$$
$$\Delta z = 5 \text{ m}$$

To complete the assignment you would need to create four M-files: one for each function and one to call and test the functions. We can use subfunctions to reduce the number of M-files to one, as shown in Figure 6.13.

Note the primary function has no input and no output. To execute the primary function, type the function name at the command prompt:

```
sample_homework
```

or select the save and run icon.

When the primary function executes, it calls the subfunctions, and the results are displayed in the command window, as follows:

```
Problem 1
The squares of the input values are listed below
       9    4    1    0    1    4    9
Problem 2
The percent cold work is
```

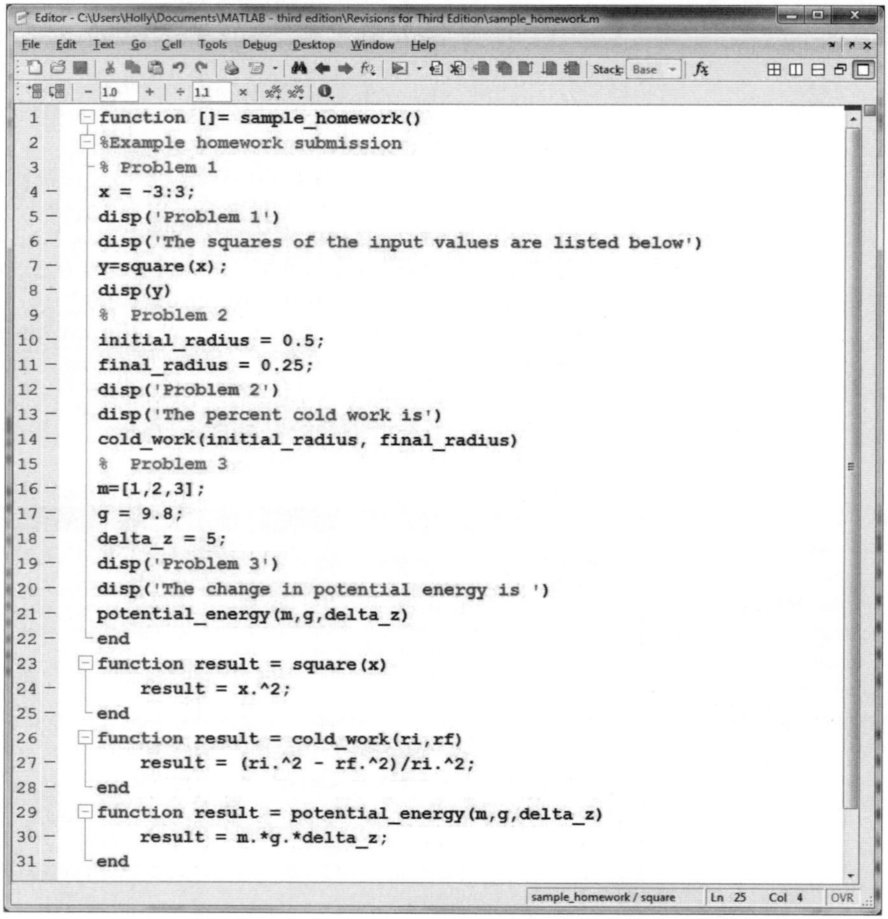

Figure 6.13
This M-file is an example of a function with sequential subfunctions.

```
 1   function []= sample_homework()
 2   %Example homework submission
 3   % Problem 1
 4   x = -3:3;
 5   disp('Problem 1')
 6   disp('The squares of the input values are listed below')
 7   y=square(x);
 8   disp(y)
 9   % Problem 2
10   initial_radius = 0.5;
11   final_radius = 0.25;
12   disp('Problem 2')
13   disp('The percent cold work is')
14   cold_work(initial_radius, final_radius)
15   % Problem 3
16   m=[1,2,3];
17   g = 9.8;
18   delta_z = 5;
19   disp('Problem 3')
20   disp('The change in potential energy is ')
21   potential_energy(m,g,delta_z)
22   end
23   function result = square(x)
24       result = x.^2;
25   end
26   function result = cold_work(ri,rf)
27       result = (ri.^2 - rf.^2)/ri.^2;
28   end
29   function result = potential_energy(m,g,delta_z)
30       result = m.*g.*delta_z;
31   end
```

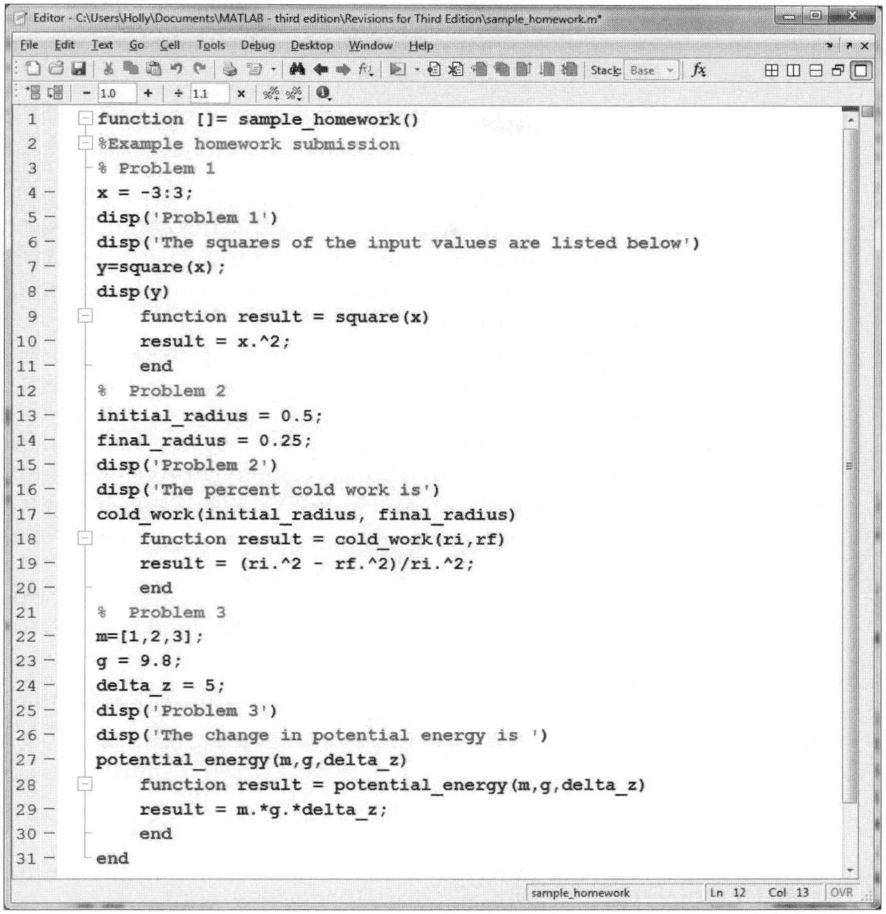

Figure 6.14
This function M-file includes nested subfunctions.

```
Editor - C:\Users\Holly\Documents\MATLAB - third edition\Revisions for Third Edition\sample_homework.m*

File   Edit   Text   Go   Cell   Tools   Debug   Desktop   Window   Help

1    ┌ function []= sample_homework()
2    │ %Example homework submission
3    ├ % Problem 1
4 -  │ x = -3:3;
5 -  │ disp('Problem 1')
6 -  │ disp('The squares of the input values are listed below')
7 -  │ y=square(x);
8 -  │ disp(y)
9    │ ┌   function result = square(x)
10 - │ │      result = x.^2;
11 - │ └      end
12   │ %  Problem 2
13 - │ initial_radius = 0.5;
14 - │ final_radius = 0.25;
15 - │ disp('Problem 2')
16 - │ disp('The percent cold work is')
17 - │ cold_work(initial_radius, final_radius)
18 - │ ┌   function result = cold_work(ri,rf)
19 - │ │      result = (ri.^2 - rf.^2)/ri.^2;
20 - │ └      end
21   │ %  Problem 3
22 - │ m=[1,2,3];
23 - │ g = 9.8;
24 - │ delta_z = 5;
25 - │ disp('Problem 3')
26 - │ disp('The change in potential energy is ')
27 - │ potential_energy(m,g,delta_z)
28 - │ ┌   function result = potential_energy(m,g,delta_z)
29 - │ │      result = m.*g.*delta_z;
30 - │ └      end
31 - └ end

                                           sample_homework      Ln 12    Col 13    OVR
```

```
ans =
    0.7500
Problem 3
The change in potential energy is
ans =
  49 98 147
```

In this example, the four functions (primary and three subfunctions) are listed sequentially. An alternate approach is to list the subfunction *within* the primary function, usually placed near the portion of the code from which it is called. This is called *nesting*. When functions are nested, we need to indicate the end of each individual function with the end command (see Figure 6.14).

SUMMARY

MATLAB® contains a wide variety of built-in functions. However, you will often find it useful to create your own MATLAB® functions. The most common type of user-defined MATLAB® function is the function M-file, which must start with a function-definition line that contains

- the word `function`,
- a variable that defines the function output,
- a function name, and
- a variable used for the input argument.

For example,

$$\text{function output} = \text{my\_function}(x)$$

The function name must also be the name of the M-file in which the function is stored. Function names follow the standard MATLAB® naming rules.

Like the built-in functions, user-defined functions can accept multiple inputs and can return multiple results.

Comments immediately following the function-definition line can be accessed from the command window with the `help` command.

Variables defined within a function are local to that function. They are not stored in the workspace and cannot be accessed from the command window. Global variables can be defined with the `global` command used in both the command window (or script M-file) and a MATLAB® function. Good programming style suggests that you define global variables with capital letters. In general, however, it is not wise to use global variables.

Groups of user-defined functions, called "toolboxes," may be stored in a common directory and accessed by modifying the MATLAB® search path. This is accomplished interactively with the path tool, either from the menu bar, as in

$$\text{File} \rightarrow \text{Set Path}$$

or from the command line, with

$$\text{pathtool}$$

MATLAB® provides access to numerous toolboxes developed at The MathWorks or by the user community.

Another type of function is the anonymous function, which is defined in a MATLAB® session or in a script M-file and exists only during that session. Anonymous functions are especially useful for very simple mathematical expressions or as input to the more complicated function functions.

MATLAB® SUMMARY

The following MATLAB® summary lists and briefly describes all of the special characters, commands, and functions that were defined in this chapter:

Special Characters	
@	identifies a function handle, such as that used with anonymous functions
%	comment

Commands and Functions	
addpath	adds a directory to the MATLAB® search path
fminbnd	a function function that accepts a function handle or function definition as input and finds the function minimum between two bounds
Fplot	a function function that accepts a function handle or function definition as input and creates the corresponding plot between two bounds

Commands and Functions

Fzero	a function function that accepts a function handle or function definition as input and finds the function zero point nearest a specified value
function	identifies an M-file as a function
global	defines a variable that can be used in multiple sections of code
meshgrid	maps two input vectors onto two two-dimensional matrices
nargin	determines the number of input arguments in a function
nargout	determines the number of output arguments from a function
pathtool	opens the interactive path tool
varargin	indicates that a variable number of arguments may be input to a function

KEY TERMS

anonymous	folding	in-line
argument	function	input argument
comments	function function	local variable
directory	function handle	M-file
file name	function name	nesting
folder	global variable	toolbox

PROBLEMS

Function M-Files

As you create functions in this section, be sure to comment them appropriately. Remember that, although many of these problems could be solved without a function, the objective of this chapter is to learn to write and use functions. Each of these functions (except for the anonymous functions) must be created in its own M-file and then called from the command window or a script M-file program.

6.1 As described in Example 6.2, metals are actually crystalline materials. Metal crystals are called grains. When the average grain size is small, the metal is strong; when it is large, the metal is weaker. Since every crystal in a particular sample of metal is a different size, it isn't obvious how we should describe the average crystal size. The American Society for Testing and Materials (ASTM) has developed the following correlation for standardizing grain-size measurements:

$$N = 2^{n-1}$$

The ASTM grain size (n) is determined by looking at a sample of a metal under a microscope at a magnification of $100 \times$ (100 power). The number of grains in a 1-square-inch area (actual dimensions of 0.01 in $\times$ 0.01 in) is estimated (N) and used in the preceding equation to find the ASTM grain size.

(a) Write a MATLAB® function called num_grains to find the number of grains in a 1-square-inch area (N) at $100 \times$ magnification when the ASTM grain size is known.

(b) Use your function to find the number of grains for ASTM grain sizes $n = 10$ to 100.

(c) Create a plot of your results.

6.2 Perhaps the most famous equation in physics is

$$E = mc^2$$

which relates energy E to mass m. The speed of light in a vacuum, c, is the property that links the two together. The speed of light in a vacuum is 2.9979×10^8 m/s.

(a) Create a function called `energy` to find the energy corresponding to a given mass in kilograms. Your result will be in joules, since $1 \text{ kg m}^2/\text{s}^2 = 1 \text{ J}$.

(b) Use your function to find the energy corresponding to masses from 1 kg to 10^6 kg. Use the `logspace` function (consult `help logspace`) to create an appropriate mass vector.

(c) Create a plot of your results. Try using different logarithmic plotting approaches (e.g., `semilogy`, `semilogx`, and `loglog`) to determine the best way to graph your results.

6.3 The future-value-of-money formula relates how much a current investment will be worth in the future, assuming a constant interest rate.

$$FV = PV \times (1 + I)^n$$

where

FV is the future value
PV is the present value or investment
I is the interest rate expressed as a fractional amount per compounding period—i.e., 5% is expressed as .05
n is the number of compounding periods.

(a) Create a MATLAB® function called `future_value` with three inputs: the investment (present value), the interest rate expressed as a fraction, and the number of compounding periods.

(b) Use your function to determine the value of a $1000 investment in 10 years, assuming the interest rate is 0.5% per month, and the interest is compounded monthly.

6.4 In freshman chemistry, the relationship between moles and mass is introduced:

$$n = \frac{m}{MW}$$

where

n = number of moles of a substance
m = mass of the substance
MW = molecular weight (molar mass) of the substance.

(a) Create a function M-file called `nmoles` that requires two vector inputs—the mass and molecular weight—and returns the corresponding number of moles. Because you are providing vector input, it will be necessary to use the `meshgrid` function in your calculations.

(b) Test your function for the compounds shown in the following table, for masses from 1 to 10 g:

Compound	Molecular Weight (Molar Mass)
Benzene	78.115 g/mol
Ethyl alcohol	46.07 g/mol
Refrigerant R134a (tetrafluoroethane)	102.3 g/mol

Your result should be a 10 × 3 matrix.

6.5 By rearranging the preceding relationship between moles and mass, you can find the mass if you know the number of moles of a compound:

$$m = n \times \text{MW}$$

(a) Create a function M-file called `mass` that requires two vector inputs—the number of moles and the molecular weight—and returns the corresponding mass. Because you are providing vector input, it will be necessary to use the `meshgrid` function in your calculations.

(b) Test your function with the compounds listed in the previous problem, for values of n from 1 to 10.

6.6 The distance to the horizon increases as you climb a mountain (or a hill). The expression

$$d = \sqrt{2rh + h^2}$$

where
   d = distance to the horizon
   r = radius of the earth
   h = height of the hill

can be used to calculate that distance. The distance depends on how high the hill is and on the radius of the earth (or another planetary body).

(a) Create a function M-file called `distance` to find the distance to the horizon. Your function should accept two vector inputs—radius and height—and should return the distance to the horizon. Don't forget that you'll need to use `meshgrid` because your inputs are vectors.

(b) Create a MATLAB® program that uses your distance function to find the distance in miles to the horizon, both on the earth and on Mars, for hills from 0 to 10,000 feet. Remember to use consistent units in your calculations. Note that

   • Earth's diameter = 7926 miles
   • Mars' diameter = 4217 miles

Report your results in a table. Each column should represent a different planet, and each row a different hill height.

6.7 A rocket is launched vertically. At time $t = 0$, the rocket's engine shuts down. At that time, the rocket has reached an altitude of 500 m and is rising at a velocity of 125 m/s. Gravity then takes over. The height of the rocket as a function of time is

$$h(t) = -\frac{9.8}{2} t^2 + 125t + 500 \text{ for } t > 0$$

(a) Create a function called `height` that accepts time as an input and returns the height of the rocket. Use your function in your solutions to parts b and c.

(b) Plot `height` versus time for times from 0 to 30 seconds. Use an increment of 0.5 second in your time vector.

(c) Find the time when the rocket starts to fall back to the ground. (The `max` function will be helpful in this exercise.)

6.8 The distance a freely falling object travels is

$$x = \frac{1}{2} gt^2$$

where

g = acceleration due to gravity, 9.8 m/s^2
t = time in seconds
x = distance traveled in meters.

If you have taken calculus, you know that we can find the velocity of the object by taking the derivative of the preceding equation. That is,

$$\frac{dx}{dt} = v = gt$$

We can find the acceleration by taking the derivative again:

$$\frac{dv}{dt} = a = g$$

(a) Create a function called `free_fall` with a single input vector `t` that returns values for distance `x`, velocity `v`, and acceleration `g`.

(b) Test your function with a time vector that ranges from 0 to 20 seconds.

6.9 Create a function called `polygon` that draws a polygon with any number of sides. Your function should require a single input: the number of sides desired. It should not return any value to the command window but should draw the requested polygon in polar coordinates.

Creating Your Own Toolbox

6.10 This problem requires you to generate temperature-conversion tables. Use the following equations, which describe the relationships between temperatures in degrees Fahrenheit (T_F), degrees Celsius (T_C), kelvins (T_K), and degrees Rankine (T_R), respectively:

$$T_F = T_R - 459.67°R$$

$$T_F = \frac{9}{5}T_C + 32°F$$

$$T_R = \frac{9}{5}T_K$$

You will need to rearrange these expressions to solve some of the problems.

(a) Create a function called `F_to_K` that converts temperatures in Fahrenheit to Kelvin. Use your function to generate a conversion table for values from 0°F to 200°F.

(b) Create a function called `C_to_R` that converts temperatures in Celsius to Rankine. Use your function to generate a conversion table from 0°C to 100°C. Print 25 lines in the table. (Use the `linspace` function to create your input vector.)

(c) Create a function called `C_to_F` that converts temperatures in Celsius to Fahrenheit. Use your function to generate a conversion table from 0°C to 100°C. Choose an appropriate spacing.

(d) Group your functions into a folder (directory) called `my_temp_conversions`. Adjust the MATLAB® search path so that it finds your folder. (Don't save any changes on a public computer!)

Anonymous Functions and Function Handles

6.11 Barometers have been used for almost 400 years to measure pressure changes in the atmosphere. The first known barometer was invented by Evangelista Torricelli (1608–1647), a student of Galileo during his final years in Florence, Italy. The height of a liquid in a barometer is directly proportional to the atmospheric pressure, or

$$P = \rho g h$$

where P is the pressure, ρ is the density of the barometer fluid, and h is the height of the liquid column. For mercury barometers, the density of the fluid is 13,560 kg/m^3. On the surface of the earth, the acceleration due to gravity, g, is approximately 9.8 m/s^2. Thus, the only variable in the equation is the height of the fluid column, h, which should have the unit of meters.

(a) Create an anonymous function P that finds the pressure if the value of h is provided. The units of your answer will be

$$\frac{\text{kg m}}{\text{m}^3 \text{s}^2}\text{m} = \frac{\text{kg}}{\text{m}}\frac{1}{\text{s}^2} = \text{Pa}$$

(b) Create another anonymous function to convert pressure in Pa (Pascals) to pressure in atmospheres (atm). Call the function `Pa_to_atm`. Note that

$$1 \text{ atm} = 101{,}325 \text{ Pa}$$

(c) Use your anonymous functions to find the pressure for fluid heights from 0.5 m to 1.0 m of mercury.

(d) Save your anonymous functions as `.mat` files

6.12 The energy required to heat water at constant pressure is approximately equal to

$$E = m C_p \Delta T$$

where

m = mass of the water, in grams
C_p = heat capacity of water, 1 cal/g K
ΔT = change in temperature, K.

(a) Create an anonymous function called `heat` to find the energy required to heat 1 gram of water if the change in temperature is provided as the input.

(b) Your result will be in calories:

$$g\frac{\text{cal}}{\text{g K}}\frac{1}{}K = \text{cal}$$

Joules are the unit of energy used most often in engineering. Create another anonymous function `cal_to_J` to convert your answer from part (a) into joules. (There are 4.2 J/cal.)

(c) Save your anonymous functions as `.mat` files.

6.13. (a) Create an anonymous function called `my_function`, equal to

$$-x^2 - 5x - 3 + e^x$$

(b) Use the `fplot` function to create a plot from $x = -5$ to $x = +5$. Recall that the `fplot` function can accept a function handle as input.

(c) Use the `fminbnd` function to find the minimum function value in this range. The `fminbnd` function is an example of a function function, since it requires a function or function handle as input. The syntax is

fminbnd(function_handle, xmin, xmax)

Three inputs are required: the function handle, the minimum value of x, and the maximum value of x. The function searches between the minimum value of x and the maximum value of x for the point where the function value is a minimum.

6.14 In Problem 6.7, you created an M-file function called `height` to evaluate the height of a rocket as a function of time. The relationship between time, t, and height, $h(t)$, is:

$$h(t) = -\frac{9.8}{2}t^2 + 125t + 500 \text{ for } t > 0$$

(a) Create a function handle to the `height` function called `height_handle`.

(b) Use `height_handle` as input to the `fplot` function, and create a graph from 0 to 60 seconds.

(c) Use the `fzero` function to find the time when the rocket hits the ground (i.e., when the function value is zero). The `fzero` function is an example of a function function, since it requires a function or function handle as input. The syntax is

fzero(function_handle, x_guess)

The `fzero` function requires two inputs—a function handle and your guess as to the time value where the function is close to zero. You can select a reasonable `x_guess` value by inspecting the graph created in part (b).

Subfunctions

6.15 In Problem 6.10 you were asked to create and use three different temperature-conversion functions, based on the following conversion equations:

$$T_F = T_R - 459.67°R$$

$$T_F = \frac{9}{5}T_C + 32°F$$

$$T_R = \frac{9}{5}T_K$$

Recreate Problem 6.10 using nested subfunctions. The primary function should be called `temperature_conversions` and should include the subfunctions

```
F_to_K
C_to_R
C_to_F
```

Within the primary function use the subfunctions to:

(a) Generate a conversion table for values from 0°F to 200°F. Include a column for temperature in Fahrenheit and Kelvin.

(b) Generate a conversion table from 0°C to 100°C. Print 25 lines in the table. (Use the `linspace` function to create your input vector.) Your table should include a column for temperature in Celsius and Rankine.

(c) Generate a conversion table from 0°C to 100°C. Choose an appropriate spacing. Include a column for temperature in Celsius and Fahrenheit.

Recall that you will need to call your primary function from the command window or from a script M-file.

7

User-Controlled Input and Output

Objectives

After reading this chapter, you should be able to:

- Prompt the user for input to an M-file program
- Create output with the `disp` function
- Create formatted output by using `fprintf`

- Create formatted output for use in other functions with the `sprintf` function
- Use graphical techniques to provide program input
- Use the cell mode to modify and run M-file programs

INTRODUCTION

So far, we have explored the use of MATLAB® in two modes: in the command window as a scratch pad and in the editing window to write simple programs (script M-files). The programmer has been the user. Now we move on to more complicated programs, written in the editing window, where the programmer and the user may be different people. That will make it necessary to use input and output commands to communicate with the user, instead of rewriting the actual code to solve similar problems. MATLAB® offers built-in functions to allow a user to communicate with a program as it executes. The `input` command pauses the program and prompts the user for input; the `disp` and `fprintf` commands provide output to the command window.

7.1 USER-DEFINED INPUT

Although we have written programs in script M-files, we have assumed that the programmer (you) and the user are the same person. To run the program with different input values, we actually changed some of the code. We can create more general programs by allowing the user to input values of a matrix from the keyboard while the program is running. The `input` function allows us to do this. It displays a text string in the command window and then waits for the user to provide the requested input. For example,

```
z = input('Enter a value')
```

displays

```
Enter a value
```

in the command window. If the user enters a value such as

```
5
```

the program assigns the value 5 to the variable **z**. If the `input` command does not end with a semicolon, the value entered is displayed on the screen:

```
z =
      5
```

The same approach can be used to enter a one- or two-dimensional matrix. The user must provide the appropriate brackets and delimiters (commas and semicolons). For example,

```
z = input('Enter values for z in brackets')
```

KEY IDEA

The `input` function can be used to communicate with the program user

requests the user to input a matrix such as

```
[1, 2, 3; 4, 5, 6]
```

and responds with

```
z =
      1 2 3
      4 5 6
```

This input value of z can then be used in subsequent calculations by the script M-file.

Data entered with `input` does not need to be numeric information. Suppose we prompt the user with the command

```
x = input('Enter your name in single quotes')
```

and enter

```
'Holly'
```

when prompted. Because we haven't used a semicolon at the end of the `input` command, MATLAB® will respond

```
x =
  Holly
```

Notice in the workspace window that **x** is listed as a 1×5 character array:

Name	Value	Size	Bytes	Class
abc x	'Holly'	1 × 5	6	char

If you are entering a string (in MATLAB®, strings are character arrays), you must enclose the characters in single quotes. However, an alternative form of the input command alerts the function to expect character input without the single quotes by specifying string input in the second field:

```
x = input('Enter your name', 's')
```

Now you need only enter the characters, such as

```
Ralph
```

and the program responds with

```
x =
   Ralph
```

PRACTICE EXERCISES 7.1

1. Create an M-file to calculate the area A of a triangle:
$$A = \frac{1}{2} \text{ base height}$$
Prompt the user to enter the values for the base and for the height.
2. Create an M-file to find the volume V of a right circular cylinder:
$$V = \pi r^2 h$$
Prompt the user to enter the values of r and h.
3. Create a vector from 0 to n, allowing the user to enter the value of n.
4. Create a vector that starts at a, ends at b, and has a spacing of c. Allow the user to input all of these parameters.

EXAMPLE 7.1

FREELY FALLING OBJECTS

Consider the behavior of a freely falling object under the influence of gravity (see Figure 7.1).

Figure 7.1
The Leaning Tower of Pisa. (Courtesy of Tim Galligan.)

The position of the object is described by

$$d = \frac{1}{2}gt^2$$

where d = distance the object travels
  g = acceleration due to gravity
  t = elapsed time.
We shall allow the user to specify the value of g—the acceleration due to gravity—and a vector of time values.

1. **State the Problem**
   Find the distance traveled by a freely falling object and plot the results.
2. **Describe the Input and Output**
   Input   Value of g, the acceleration due to gravity, provided by the user
           Time, provided by the user
   Output DistancesPlot of distance versus time
3. **Develop a Hand Example**

$$d = \frac{1}{2}gt^2, \text{ so, on the moon at 100 seconds,}$$

$$d = \frac{1}{2} \times 1.6 \text{ m/s}^2 \times 100^2 \text{ s}^2$$

$$d = 8000 \text{ m}$$

4. **Develop a MATLAB® Solution**

```
%Example 7.1
%Free fall
clear, clc
%Request input from the user
g = input('What is the value of acceleration due to
  gravity?')
start = input('What starting time would you like?')
finish = input('What ending time would you like?')
incr = input('What time increments would you like
  calculated?')
time = start:incr:finish;
%Calculate the distance
distance = 1/2*g*time.^2;
%Plot the results
loglog(time,distance)
title('Distance Traveled in Free Fall')
xlabel('time, s'),ylabel('distance, m')
%Find the maximum distance traveled
final_distance = max(distance)
```

The interaction in the command window is:

```
What is the value of acceleration due to gravity? 1.6
g =
   1.6000
```

(continued)

Figure 7.2
Distance traveled when the acceleration is 1.6 m/s. Notice that the figure is a loglog plot.

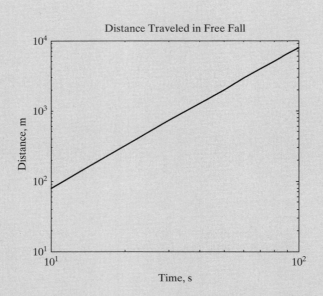

```
What starting time would you like? 0
start =
    0
What ending time would you like? 100
finish =
    100
What time increments would you like calculated? 10
incr =
    10
final_distance =
    8000
```

The results are plotted in Figure 7.2.

5. Test the Solution

Compare the MATLAB® solution with the hand solution. Since the user can control the input, we entered the data used in the hand solution. MATLAB® tells us that the final distance traveled is 8000 m, which, since we entered 100 seconds as the final time, corresponds to the distance traveled after 100 seconds.

7.2 OUTPUT OPTIONS

There are several ways to display the contents of a matrix. The simplest is to enter the name of the matrix, without a semicolon. The name will be repeated, and the values of the matrix will be displayed, starting on the next line. For example, first define a matrix **x**:

```
x = 1:5;
```

Because there is a semicolon at the end of the assignment statement, the values in **x** are not repeated in the command window. However, if you want to display **x** later in your program, simply type in the variable name

```
x
```

which returns

```
X =
     1        2        3        4        5
```

MATLAB® offers two other approaches to displaying results: the disp function and the fprintf function.

7.2.1 Display Function

The display (disp) function can be used to display the contents of a matrix without printing the matrix name. It accepts a single array as input. Thus,

```
disp(x)
```

returns

```
1    2    3    4    5
```

The display command can also be used to display a string (text enclosed in single quotation marks). For example,

```
disp('The values in the x matrix are:');
```

returns

```
The values in the x matrix are:
```

When you enter a string as input into the disp function, you are really entering an array of character information. Try entering the following on the command line:

```
'The values in the x matrix are:'
```

MATLAB® responds

```
ans =
'The values in the x matrix are:'
```

The workspace window lists ans as a 1×32 character array.

Name	Size	Bytes	Class
abc ans	1×32	90	char array

Character arrays store character information in arrays similar to numerical arrays. Characters can be letters, numbers, punctuation, and even some nondisplayed characters. Each character, including spaces, is an element in the character array.

When we execute the two display functions

```
disp('The values in the x matrix are:');
disp(x)
```

MATLAB® responds

```
The values in the x matrix are:
1 2 3 4 5
```

Notice that the two disp outputs are displayed on separate lines. You can get around this feature by creating a combined matrix of your two outputs, using the num2str (number to string) function. The process is called concatenation and creates a single character array. Thus,

```
disp(['The values in the x array are:' num2str(x)])
```

returns

```
The values in the x array are: 1 2 3 4 5
```

The num2str function changes an array of numbers into an array of characters. In the preceding example, we used num2str to transform the **x** matrix to a character array, which was then combined with the first string (by means of square brackets, [ ]) to make a bigger character array. You can see the resulting matrix by typing

```
A = ['The values in the x array are: ' num2str(x)]
```

which returns

```
A =
    The values in 1 2 3 4 5 the x array are:
```

Checking in the workspace window, we see that **A** is a 1×45 matrix. The workspace window also tells us that the matrix contains character data instead of numeric information. This is evidenced both by the icon in front of **A** and in the class column.

Name	Size	Bytes	Class
ab A	1 × 45	90	char array

HINT

If you want to include an apostrophe in a string, you need to enter the apostrophe twice. If you don't do this, MATLAB® will interpret the apostrophe as terminating the string. An example of the use of two apostrophes is

```
disp('The moon''s gravity is 1/6th that of the earth')
```

You can use a combination of the **input** and **disp** functions to mimic a conversation. Try creating and running the following M-file:

```
disp('Hi There');
disp('I'm your MATLAB program');
name = input('Who are you?','s');
disp(['Hi',name]);
answer = input('Don''t you just love computers?','s');
disp([answer,'?']);
disp('Computers are very useful');
disp('You''ll use them a lot in college!!');
disp('Good luck with your studies')
pause(2);
disp('Bye bye')
```

This interaction made use of the pause function. If you execute pause without any input, the program waits until the user hits the Enter key. If a value is used as input to the pause function, the program waits for the specified number of seconds, and then continues.

7.2.2 Formatted Output—The fprintf Function

The fprintf function (formatted print function) gives you even more control over the output than you have with the disp function. In addition to displaying both text and matrix values, you can specify the format to be used in displaying the values, and you can specify when to skip to a new line. If you are a C programmer, you will be familiar with the syntax of this function. With few exceptions, the MATLAB® fprintf function uses the same formatting specifications as the C fprintf function. This is hardly surprising, since MATLAB® was written in C. (It was originally written in Fortran and then later rewritten in C.)

The general form of the fprintf command contains two arguments, one a string and the other a list of matrices:

```
fprintf(format-string, var,. . .)
```

Consider the following example:

```
cows = 5;
fprintf('There are %f cows in the pasture', cows)
```

The string, which is the first argument inside the fprintf function, contains a placeholder (%) where the value of the variable (in this case, cows) will be inserted. The placeholder also contains formatting information. In this example, the %f tells MATLAB® to display the value of cows in a default fixed-point format. The default format displays six places after the decimal point:

```
There are 5.000000 cows in the pasture
```

Besides defaulting to a fixed-point format, MATLAB® allows you to specify an exponential format, %e, or lets you allow MATLAB® to choose whichever is shorter, fixed point or exponential (%g). It also lets you display character information (%c) or a string of characters (%s). The decimal format (%d) is especially useful if the number you wish to display is an integer.

```
fprintf('There are %d cows in the pasture', cows)
There are 5 cows in the pasture
```

Table 7.1 illustrates the various formats supported by fprintf, and the related sprintf functions.

KEY IDEA
The fprintf function allows you to control how numbers are displayed

MATLAB® does not automatically start a new line after an fprintf function is executed. If you tried out the preceding fprintf command example, you probably noticed that the command prompt is on the same line as the output:

```
There are 5.000000 cows in the pasture>>
```

Table 7.1 Type Field Format

Type Field	Result
%f	fixed-point notation
%e	exponential notation
%d	decimal notation—does not include trailing zeros if the value displayed is an integer. If the number includes a fractional component, it is displayed using exponential notation.
%g	whichever is shorter, **%f** or **%e**
%c	character information (displays one character at a time)
%s	string of characters (displays the entire string)

Additional type fields are described in the help feature.

If we execute another command, the results will appear on the same line instead of moving down. Thus, if we issue the new commands

```
cows = 6;
fprintf('There are %f cows in the pasture', cows);
```

from an M-file, MATLAB® continues the command window display on the same line:

```
There are 5.000000 cows in the pasture There are 6.000000 cows
in the pasture
```

To cause MATLAB® to start a new line, you'll need to use \n, called a linefeed, at the end of the string. For example, the code

```
cows = 5;
fprintf('There are %f cows in the pasture \n', cows)
cows = 6;
fprintf('There are %f cows in the pasture \n', cows)
```

returns the following output:

```
There are 5.000000 cows in the pasture
There are 6.000000 cows in the pasture
```

HINT

The backslash (\) and forward slash (/) are different characters. It's a common mistake to confuse them—and then the linefeed command doesn't work! Instead, the output to the command window will be

```
There are 5.000000 cows in the pasture /n
```

Other special format commands are listed in Table 7.2. The tab (\t) is especially useful for creating tables in which everything lines up neatly.

You can further control how the variables are displayed by using the optional width field and precision field with the format command. The width field controls the minimum number of characters to be printed. It must be a positive decimal integer. The precision field is preceded by a period (.) and specifies the number of decimal places after the decimal point for exponential and fixed-point types. For example, %8.2f specifies that the minimum total width available to display your result is eight digits, two of which are after the decimal point. Thus, the code

```
voltage = 3.5;
fprintf('The voltage is %8.2f millivolts \n',voltage);
```

Table 7.2 Special Format Commands

Format Command	Resulting Action
\n	Linefeed
\r	carriage return (similar to linefeed)
\t	tab
\b	backspace

returns

> **The voltage is   3.50 millivolts**

Notice the empty space before the number 3.50. This occurs because we reserved six spaces (eight total, two after the decimal) for the portion of the number to the left of the decimal point.

Often when you use the `fprintf` function, your variable will be a matrix—for example,

> **x = 1:5;**

MATLAB® will repeat the string in the `fprintf` command until it uses all the values in the matrix. Thus,

> **fprintf('%8.2f \n',x);**

returns

> **1.00**
> **2.00**
> **3.00**
> **4.00**
> **5.00**

If the variable is a two-dimensional matrix, MATLAB® uses the values one column at a time, going down the first column, then the second, and so on. Here's a more complicated example:

> **feet = 1:3;**
> **inches = feet.*12;**

Combine these two matrices:

> **table = [feet;inches]**

MATLAB® then returns

> **table =**
> **   1   2   3**
> **   12 24 36**

Now we can use the `fprintf` function to create a table that is easier to interpret. For instance,

> **fprintf('%4.0f %7.2f \n',table)**

sends the following output to the command window:

> **1   12.00**
> **2   24.00**
> **3   36.00**

Why don't the two outputs look the same? The `fprintf` statement we created uses two values at a time. It goes through the `table` array one *column* at a time to find the numbers it needs. Thus, the first two numbers used in the `fprintf` output are from the first column of the `table` array.

The `fprintf` function can accept a variable number of matrices after the string. It uses all of the values in each of these matrices, in order, before moving on to the next matrix. As an example, suppose we wanted to use the feet and inches matrices without combining them into the table matrix. Then we could type

```
fprintf('%4.0f %7.2f \n', feet, inches)
   1     2.00
   3    12.00
  24    36.00
```

The function works through the values of `feet` first and then uses the values in `inches`. It is unlikely that this is what you really want the function to do (in this example it wasn't), so the output values are almost always grouped into a single matrix to use in `fprintf`.

The `fprintf` command gives you considerably more control over the form of your output than MATLAB®'s simple format commands. It does, however, require some care and forethought to use.

In addition to creating formatted output for display in the command window, the `fprintf` function can be used to send formatted output to a file. First, you'll need to create and open an output file and assign it a file identifier (nickname). You do this with the `fopen` function

```
file_id = fopen('my_output_file.txt', 'wt');
```

The first field is the name of the file, and the second field makes it possible for us to write data to the file (hence the string 'wt'). Once the file has been identified and opened for writing, we use the `fprintf` function, adding the file identifier as the first field in the function input.

```
fprintf(file_id, 'Some example output is %4.2f \n', pi*1000)
```

This form of the function sends the result of the formatted string

```
Some example output is 3141.59
```

to `my_output_file.txt`. To the command window the function sends a count of the number of bytes transferred to the file.

```
ans =
    32
```

HINT

A common mistake new programmers make when using `fprintf` is to forget to include the field type identifier, such as `f`, in the placeholder sequence. The `fprintf` function won't work, but no error message is returned either.

HINT

If you want to include a percentage sign in an `fprintf` statement, you need to enter the % twice. If you don't, MATLAB® will interpret the % as a placeholder for data. For example,

```
fprintf('The interest rate is %5.2f %% \n', 5)
```

results in

```
The interest rate is 5.00 %
```

EXAMPLE 7.2

FREE FALL: FORMATTED OUTPUT

Let's redo Example 7.1, but this time let's create a table of results instead of a plot, and let's use the `disp` and `fprintf` commands to control the appearance of the output.

1. **State the Problem**
   Find the distance traveled by a freely falling object.
2. **Describe the Input and Output**
   Input   Value of g, the acceleration due to gravity, provided by the user
   Time t, provided by the user
   Output   Distances calculated for each planet and the moon
3. **Develop a Hand Example**

$$d = \frac{1}{2}gt^2, \text{ so, on the moon at 100 seconds,}$$

$$d = \frac{1}{2} \times 1.6 \text{ m/s}^2 \times 100^2 \text{ s}^2$$

$$d = 8000 \text{ m}$$

4. **Develop a MATLAB® Solution**

```
%Example 7.2
%Free Fall
clear, clc
%Request input from the user
g = input('What is the value of acceleration due to
  gravity?')
start = input('What starting time would you like?')
finish = input('What ending time would you like?')
incr = input('What time increments would you like
  calculated?')
time = start:incr:finish;
%Calculate the distance
distance = 1/2*g*time.^2;
%Create a matrix of the output data
table = [time;distance];
%Send the output to the command window
fprintf('For an acceleration due to gravity of %5.1f seconds
  \n the following data were calculated \n', g)
disp('Distance Traveled in Free Fall')
disp('time, s distance, m')
fprintf('%8.0f %10.2f\n',table)
```

This M-file produces the following interaction in the command window:

```
What is the value of acceleration due to gravity? 1.6
g =
  1.6000
What starting time would you like? 0
start =
  0
```

(*continued*)

```
What ending time would you like? 100
finish =
   100
What time increments would you like calculated? 10
incr =
   10
For an acceleration due to gravity of 1.6 seconds the following
  data were calculated
Distance Traveled in Free Fall
time, s  distance, m
     0           0.00
    10          80.00
    20         320.00
    30         720.00
    40        1280.00
    50        2000.00
    60        2880.00
    70        3920.00
    80        5120.00
    90        6480.00
   100        8000.00
```

5. Test the Solution

Compare the MATLAB® solution with the hand solution. Since the output is a table, it is easy to see that the distance traveled at 100 seconds is 8000 m. Try using other data as input, and compare your results with the graph produced in Example 7.1.

PRACTICE EXERCISES 7.2

In an M-file,
1. Use the disp command to create a title for a table that converts inches to feet.
2. Use the disp command to create column headings for your table.
3. Create an inches vector from 0 to 120 with an increment of 10.
4. Calculate the corresponding values of feet.
5. Group the inch vector and the feet vector together into a table matrix.
6. Use the fprintf command to send your table to the command window.

7.2.3 Formatted Output—The sprintf Function

KEY IDEA

The sprintf function is similar to fprintf and is useful for annotating plots

The sprintf function is similar to fprintf, but instead of just sending the result of the formatted string to the command window, sprintf assigns it a name and sends it to the command window.

```
a = sprintf('Some example output is %4.2f \n', pi*1000) =
a =
     Some example output is 3141.59
```

When would this be useful? In Example 7.3, the **sprintf** function is used to specify the contents of a text box, which is shown as an annotation on a graph.

EXAMPLE 7.3

PROJECTILE MOTION: ANNOTATING A GRAPH

Recall from earlier examples that the equation describing the range of a projectile fired from a cannon is

$$R(\theta) = \frac{v^2}{g}\sin(2\theta)$$

where

$R(\theta)$ is the range in meters

v is the initial projectile velocity in m/s

θ is the launch angle

g is the acceleration due to gravity, 9.9 m/s^2

Plot the angle on the x-axis versus the range on the y-axis and add a text box indicating the value of the maximum range.

1. **State the Problem**

   Find and plot the distance traveled by projectile, as a function of launch angle. Annotate a plot, indicating the maximum range.

2. **Describe the Input and Output**

   Input   Acceleration due to gravity, $g = 9.9$ m/s^2

   Launch angle

   Initial projectile velocity, 100 m/s

   Output   An annotated graph indicating the maximum range.

3. **Develop a Hand Example**

   We know from physics and from previous examples that the maximum range occurs at a launch angle of 45°. Substituting into the provided equation,

   $$R = (45°) = \frac{100^2 \text{m}^2/\text{s}^2}{9.9 \text{ m/s}^2}\sin(2 * 45°)$$

   Since the angle is specified in degrees, you must either set your calculator to accept degrees into the sine function or else convert 45° to the corresponding number of radians $(\pi/4)$. After you have done so, the result is

   $$R(45°) = 1010 \text{ m}$$

4. **Develop a MATLAB® Solution**

```
% Example 7.3
% Find the maximum projectile range
% Create an annotated graph of the results
% Define the input parameters
  g=9.9;    %Acceleration due to gravity
  velocity = 100; %Initial velocity, m/s^2
  theta = [0:5:90]   %Launch angle in degrees
% Calculate the range
  range = velocity^2/g*sind(2*theta);
% Calculate the maximum range
  m = max(range);
% Create the input for the textbox
  tinput=sprintf('Max range was %4.0f me \n',m);
```

(continued)

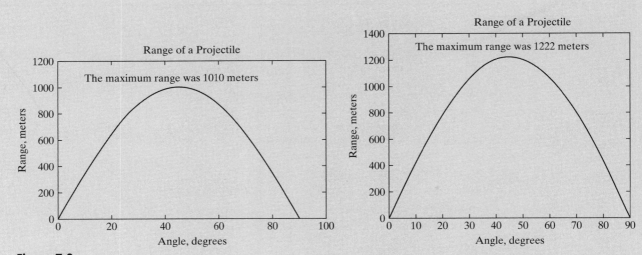

Figure 7.3
The contents of the text box change, depending on the input to the program, and are controlled by the sprintf Function.

```
% Plot the results
plot(theta,range)
title('Range of a Projectile')
xlabel('Angle, degrees'), ylabel('Range, meters')
text(10,m,tinput)
```

There are several things to notice about this program. First, we took advantage of the sind function to calculate the value of sine, using degrees as input. Second, the location of the text box will always start on the graph at 10° (measured on the x-axis), but the y location depends on the maximum range. This M-file produces the graph shown in Figure 7.3a.

5. Test the Solution
Compare the MATLAB® solution with the hand solution. The text box used to annotate the graph lists the maximum range as 1010 m, the same value calculated by hand. We could also test the program with a different initial velocity, for example, 110 m/s. The result is shown in Figure 7.3.

7.3 GRAPHICAL INPUT

MATLAB® offers a technique for entering ordered pairs of x- and y-values graphically. The ginput command allows the user to select points from a figure window and converts the points into the appropriate x- and y-coordinates. In the statement

```
[x,y] = ginput(n)
```

MATLAB® requests the user to select n points from the figure window. If the value of n is not included, as in

```
[x,y] = ginput
```

MATLAB® accepts points until the return key is pressed.

This technique is useful for picking points off a graph. Consider the graph in Figure 7.4.

Figure 7.4
The `ginput` function allows the user to pick points off a graph.

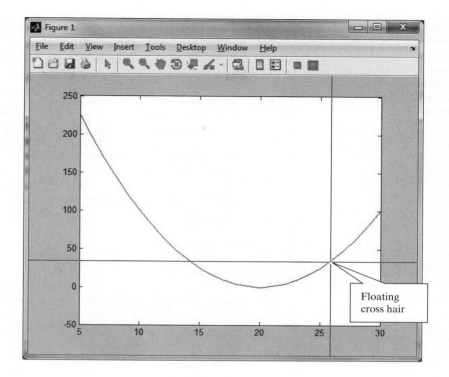

The figure was created by defining *x* from 5 to 30 and calculating *y*:

```
x = 5:30;
y = x.^2 - 40.*x + 400;
plot(x,y)
axis([5,30,-50,250])
```

The axis values were defined so that the graph would be easier to trace.
Once the `ginput` function has been executed, as in

```
[a,b] = ginput
```

MATLAB® adds a floating cross hair to the graph, as shown in Figure 7.4. After this cross hair is positioned to the user's satisfaction, right-clicking and then selecting Return (Enter) sends the values of the *x*- and *y*-coordinates to the program:

```
a =
  24.4412
b =
  19.7368
```

7.4 MORE CELL MODE FEATURES

KEY IDEA
Cell mode allows you to create reports in HTML, Word, and PowerPoint

A useful feature to use in conjunction with cell mode is Publish. It allows the user to publish an M-file program to an HTML file. MATLAB® runs the program and creates a report showing the code in each cell, as well as the calculational results that were sent to the command window. Any figures created are also included in the report. Figure 7.5 shows part of an M-file created to solve the homework problems from a previous chapter. It was created using cell mode, as can be seen from the cell dividers. A portion of the report created using the publish feature is shown in Figure 7.6.

Figure 7.5
M-Files such as this script, which was used to solve homework problems from a previous chapter, can be published using MATLAB®'s publish feature.

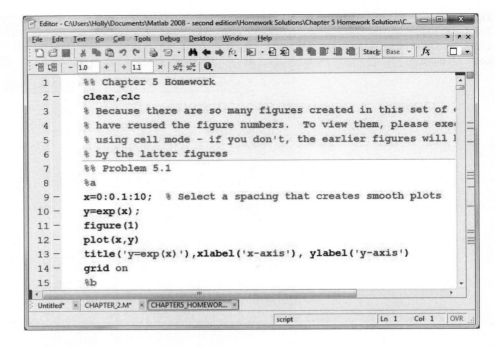

If you prefer a report in a different format, such as Word, PowerPoint or pdf, you can use the menu bar option

File → Publish Configuration for ...

to publish the results in your choice of several different formats. You'll need to select "edit publish configurations" and then the "output file format" setting, and

Figure 7.6
HTML report created from a MATLAB® M-file using the **Publish** feature.

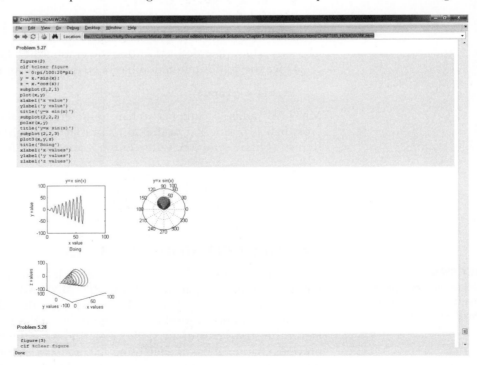

change it from html to your desired format, as shown in Figure 7.7. The publish feature does not work well if you have programmed user interactions such as prompts for data input into the file. During the publishing process, the M-file program is executed, but no values are available for the user input. This results in an error message, which is included in the published version of the file. The publish feature can be used to publish M-file programs that do not contain cells. The result is equivalent to a program that consists of only one cell.

The cell toolbar also includes a set of value-manipulation tools, as shown in Figure 7.8. Whatever number is closest to the cursor (in Figure 7.8, it's the number 2)

Figure 7.7
Change the output file format in the edit configuration window to create reports in a number of popular formats, including Word documents and pdf files.

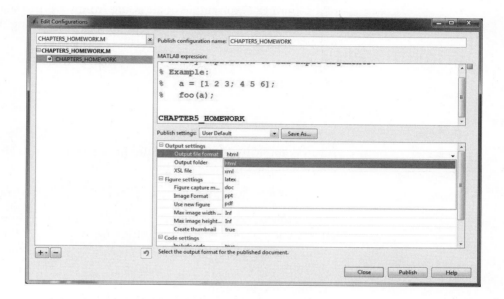

Figure 7.8
Value manipulation tools allow the user to experiment with changing values in calculations.

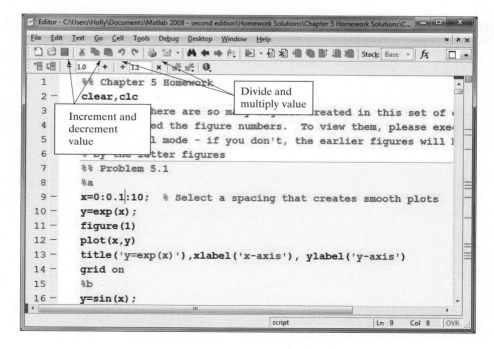

can be adjusted by the factor shown on the toolbar by selecting the appropriate icon ($-$,$+$, $\div$, or $\times$). When this feature is used in combination with the `evaluate cell` tool, you can repeat a set of calculations multiple times while easily adjusting a variable of interest.

EXAMPLE 7.4

INTERACTIVELY ADJUSTING PARAMETERS

On the basis of an energy balance calculation, you know that the change in enthalpy of a 1-kmol (29-kg) sample of air going from state 1 to state 2 is 8900 kJ. You'd like to know the final temperature, but the equation relating the change in enthalpy to temperature, namely,

$$\Delta h = \int_1^2 C_{\mathrm{p}} \mathrm{d}T$$

where

$$C_{\mathrm{p}} = a + bT + cT^2 + dT^3$$

is too complicated to solve for the final temperature. However, using techniques learned in calculus, we find that

$$\Delta h = a(T_2 - T_1) + \frac{b}{2}(T_2^2 - T_1^2) + \frac{c}{3}(T_2^3 - T_1^3) + \frac{d}{4}(T_2^4 - T_1^4)$$

If we know the starting temperature (T_1) and the values of a, b, c, and d, we can guess values of the final temperature (T_2) until we get the correct value of Δh. The interactive ability to modify variable values in the cell mode makes solving this problem easy.

1. State the Problem
   Find the final temperature of air when you know the starting temperature and the change in internal energy.
2. Describe the Input and Output

   Input   Used in the equation for C_{p}, these values of a, b, c, and d will give a heat capacity value in kJ/kmol K:

   $$a = 28.90$$
   $$b = 0.1967 \times 10^{-2}$$
   $$c = 0.4802 \times 10^{-5}$$
   $$d = -1.966 \times 10^{-9}$$
   $$\Delta h = 8900 \text{ kJ}$$
   $$T_1 = 300 \text{ K}$$

   Output   For every guessed value of the final temperature, an estimate of Δh should print to the screen.

3. Develop a Hand Example
   If we guess a final temperature of 400 K, then

   $$\Delta h = a(T_2 - T_1) + \frac{b}{2}(T_2^2 - T_1^2) + \frac{c}{3}(T_2^3 - T_1^3) + \frac{d}{4}(T_2^4 - T_1^4)$$

   $$\Delta h = 28.9(400 - 300) + \frac{0.1967 \times 10^{-2}}{2}(400^2 - 300^2) + \frac{0.4802 \times 10^{-5}}{3}$$
   $$\times (400^3 - 300^3) + \cdots \frac{-1.966 \times 10^{-9}}{4}(400^4 - 300^4)$$

which gives

$$\Delta h = 3009.47$$

4. Develop a MATLAB® Solution

```
%% Example 7.4
% Interactively Adjusting Parameters
clear,clc
a = 28.90;
b = 0.1967e-2;
c = 0.4802e-5;
d = -1.966e-9;
T1 = 300
%% guess T2 and adjust
T2 = 400
format bank
delta_h = a*(T2-T1) + b*(T2.^2 - T1.^2)/2 + c*(T2.^3-T1.^3)/
3 + d*(T2.^4-T1.^4)/4
```

Run the program once, and MATLAB® returns

```
T1 = 300.00
T2 = 400.00
delta_h = 3009.47
```

Now position the cursor near the T2=400 statement, as shown in Figure 7.9. (In this example, the edit window was docked with the MATLAB® desktop.)

Figure 7.9
The original guess gives us an idea of how far away we are from the final answer.

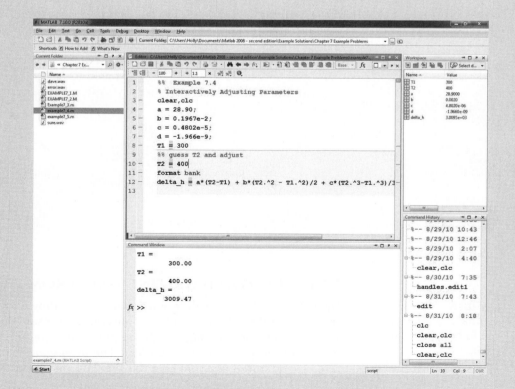

Figure 7.10
Adjust the value closest to the cursor by selecting one of the Increment/Decrement icons and adjusting the step size shown on the cell-mode toolbar.

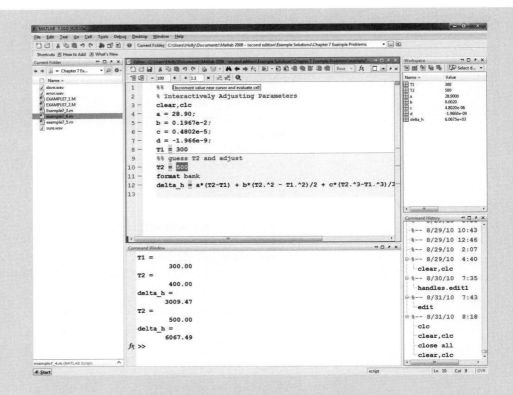

By selecting the Increment Value icon, with the value set at 100, we can quickly try several different temperatures (see Figure 7.10). Once we're close, we can change the increment and zero in the answer.

A T_2 value of 592 K gave a calculated Δh value of 8927, which is fairly close to our goal. We could get closer if we believed that the added accuracy was justified.

5. Test the Solution

   Substitute the calculated value of T_2 into the original equation, and check the results with a calculator:

$$\Delta h = 28.9(592 - 300) + \frac{0.1967 \times 10^{-2}}{2}(592^2 - 300^2)$$

$$+ \frac{0.4802 \times 10^{-5}}{3}(592^3 - 300^3) + \frac{-1.966 \times 10^{-9}}{4}(592^4 - 300^4)$$

$$\Delta h = 8927.46$$

7.5 READING AND WRITING DATA FROM FILES

KEY IDEA

MATLAB® can import data from files using a variety of formats

Data are stored in many different formats, depending on the devices and programs that created the data and on the application. For example, sound might be stored in a .wav file, and an image might be stored in a .jpg file. Many applications store data in Excel spreadsheets (.xls files). The most generic of these files is the ASCII file, usually stored as a .dat or a .txt file. You may want to import these data into

MATLAB® to analyze in a MATLAB® program, or you might want to save your data in one of these formats to make the file easier to export to another application.

7.5.1 Importing Data

Import Wizard

If you select a data file from the current folder and double-click on the file name, the Import Wizard launches. The Import Wizard determines what kind of data is in the file and suggests ways to represent the data in MATLAB®. Table 7.3 is a list of some of the data types MATLAB® recognizes. Not every possible data format is supported by MATLAB®. You can find a complete list by typing

```
doc fileformats
```

in the command window.

The Import Wizard can be used for simple ASCII files and for Excel spreadsheet files. Many of the other formats can also be imported with the Import Wizard, which can be launched from the command line, using the `uiimport` function:

```
uiimport(' filename.extension ')
```

For example, it is easy to record sound files using a variety of software tools, or to find existing files on the Internet. To import a sound file, such as one called `decision.wav`, type

```
uiimport(' decision.wav ')
```

The Import Wizard then opens, as shown in Figure 7.11.

Either technique for launching the Import Wizard (double-clicking on the file name in the current folder window, or using the uiimport function in the command window) requires an interaction with the user (through the Wizard). If you want to load a data file from a MATLAB® program, you'll need a different approach.

Table 7.3 Some of the Data File Types Supported by MATLAB®

File Type	Extension	Remark
Text	.mat	MATLAB® workspace
	.dat	ASCII data
	.txt	ASCII data
	.csv	Comma-separated values ASCII data
Other common scientific	.cdf	common data format
data formats	.fits	flexible image transport system data
	.hdf	hierarchical data format
Spreadsheet data	.xls, xlxx	Excel spreadsheet
	.wk1	Lotus 123
Image data	.tiff	tagged image file format
	.bmp	bit map
	.jpeg or jpg	joint photographics expert group
	.gif	graphics interchange format
Audio data	.au, snd	audio
	.wav	Microsoft wave file
Movie	.avi	audio/video interleaved file
	mpg	motion picture experts group

Figure 7.11
The Import Wizard launches when the uiimport command is executed.

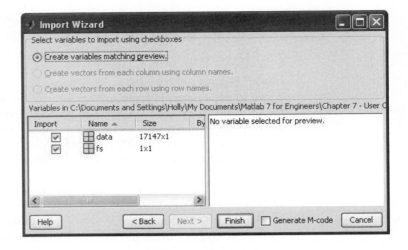

Import Commands

You can bypass the Wizard interactions by using one of the functions that are especially designed to read each of the supported file formats. For example, to read in a .wav file, use the wavread function:

```
[data,fs] = wavread('decision.wav')
```

Clearly, you need to understand what kind of data to expect, so that you can name the created variables appropriately. Recall that you can find a list of import functions by typing

```
doc fileformats
```

To use the files you have imported, you'll need to use a function appropriate to the data. In the case of a .wav file, the sound function is appropriate, so the code to play the decision.wav file is

```
sound(data,fs)
```

You should be aware that data storage formats are constantly changing, which can affect MATLAB®'s ability to interpret them. For example, some but not all .wav files use a data compression algorithm not supported by MATLAB®.

7.5.2 Exporting Data

The easiest way to find the appropriate function for writing a file is to use the help tutorial to find the correct function to read it and then to follow the links to the write function. For example, to read an Excel spreadsheet file (.xls), we'd use xlsread:

```
xlsread('filename.xls')
```

At the end of the tutorial page, we are referred to the correct function for writing an Excel file, namely,

```
xlswrite('filename.xls', M)
```

where M is the array you want to store in the Excel spreadsheet.

7.6 DEBUGGING YOUR CODE

A software bug is a problem that exists in the code you have written. It can be a mistake that results in the code not working at all (a coding error), or it can be a logic error that results in a wrong answer. The term "bug" has its genesis in electronics, where actual insects sometimes caused equipment failure. Perhaps the most famous example is the moth (Figure 7.12) found in the innards of one of the earliest computers, the Harvard Mark II Aiken Relay Calculator, in 1947.

MATLAB® includes a number of tools to help you debug your code, including the error bar and more comprehensive tools that allow you to step through the code.

7.6.1 Error Bar

Whenever you use an M-file, notice that along the right-hand side of the figure window a vertical bar appears, that marks locations where there are actual errors or where MATLAB® has issued warnings. The portion of the code that concerns MATLAB® is highlighted. For example, in Figure 7.13 there are several places marked with a light orange highlight, which indicates a warning. If you run your cursor over the highlight (either in the code or along the bar), a message appears with a suggested fix for the problem. Not every warning corresponds to a real problem. For example, the warnings issued for the program in Figure 7.13 resulted from lines of code without semicolons at the end of the line. In this particular case we wanted the code to report answers to the command window; in other cases you might want to suppress the output. You can edit which error messages are shown by selecting

```
File → Preferences → Code Analyser
```

If the errors shown on the error bar are marked in red, they will cause the M-file to stop executing. In Figure 7.14, the code was adjusted to introduce such an error. On line 22 the right-hand parentheses are missing, as indicated by the error message. You can walk through the warnings and actual error messages by clicking on the square at the top of the error bar.

Figure 7.12
The moth found trapped between in a relay in Harvard's Mark II Aiken Relay Calculator. This is often erroneously reported to be the first use of the term "bug" as a synonym for a computer problem. This page from the computer log book is currently on exhibit in the Smithsonian Institute's National Museum of American History. (Image courtesy of the Naval Surface Warfare Center, Dalgren, VA, 1988.)

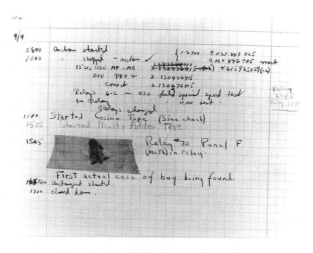

Figure 7.13
The error bar on the right-hand side of the screen identifies lines of code with potential errors. Locations in the code with potential errors are indicated with a light orange highlight.

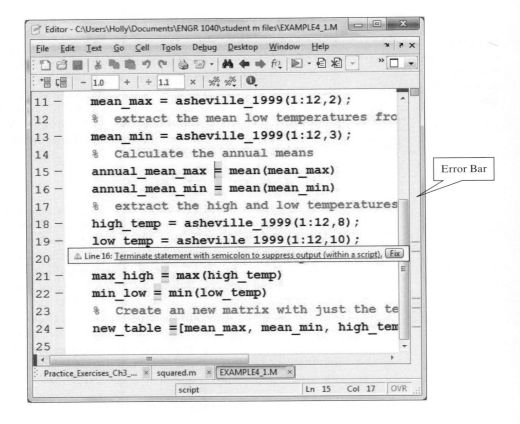

Figure 7.14
M-file with an error on line 22.

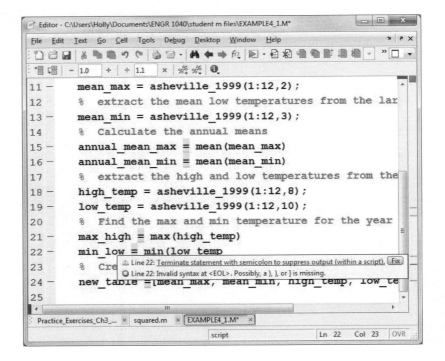

7.6.2 Debugging Toolbar

When trying to find logic errors in a piece of code, it is often useful to run sections of the program, then to stop, evaluate what has happened, and continue. Using cell mode is one way to accomplish this, but a more comprehensive approach is offered by the debugging toolbar. It allows you to set breakpoints (places in the code where the execution stops while you evaluate results) and to step through the code one line at a time. Breakpoints can't be enabled until all of the syntax errors have been resolved.

To set a breakpoint, click next to the line number on the left-hand side of the editing window, or select the set/clear breakpoint icon on the toolbar. A red circle should appear, as shown in Figure 7.15. If the circle is gray, syntax errors still exist in the program, or you have not saved the most recent version of the code. When you run the program, the execution will pause at the breakpoint, and mark the location with a green arrow. To continue, select the continue icon from the breakpoint toolbar.

You can also choose to step through the code one line at a time, using the step icon. If your code includes calls to user-defined functions, you can step into the function and then step through the function code one line at a time, using the step in icon. To leave the user-defined function, select the step out icon. For example, Figure 7.16 shows an M-file program that calls the user-defined function, RD. Both M-files are displayed in the editing window by selecting the arrange documents icon. Notice that the line where we "stepped out" of the main program and into the function is marked with a white arrow.

Figure 7.15

Breakpoints enable the user to move through the code in small pieces.

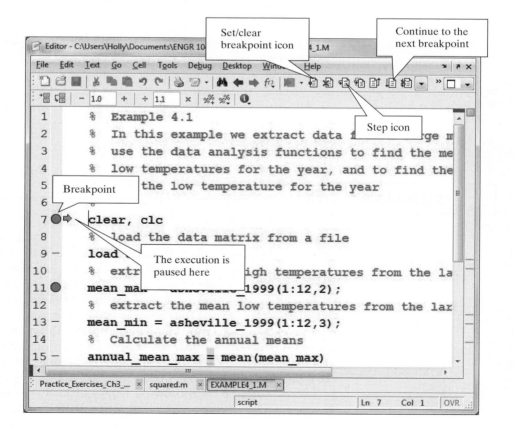

Figure 7.16
The step in icon makes it possible to step through user-defined functions one line at a time, as they are called by the main program.

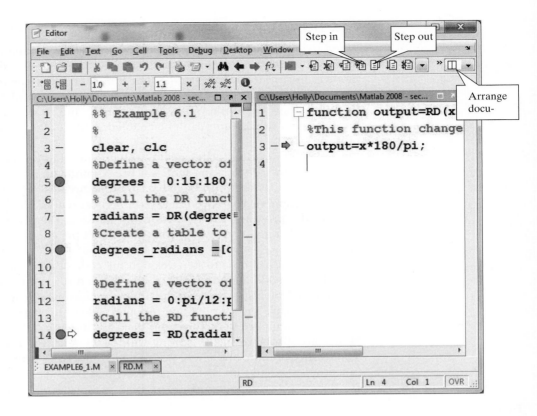

While you are executing the M-file using breakpoints to pause the code, the command window prompt is

K>>

The prompt returns to the standard symbol

>>

when you have completed the process.

SUMMARY

MATLAB® provides functions that allow the user to interact with an M-file program and allow the programmer to control the output to the command window.

The input function pauses the program and sends a prompt determined by the programmer to the command window. Once the user has entered a value or values and hits the return key, program execution continues.

The display (disp) function allows the programmer to display the contents of a string or a matrix in the command window. Although the disp function is adequate for many display tasks, the fprintf function gives the programmer considerably more control over the way results are displayed. The programmer can combine text and calculated results on the same line and specify the number format used. The sprintf function behaves exactly the same way as the fprintf function. However, the result of sprintf is assigned a variable name and can be used with other functions that require strings as input. For example, the functions used to annotate graphs such as `title`, `text`, and `xlabel` all accept strings as input and therefore will accept the result of the `sprintf` function as input.

For applications in which graphical input is required, the `ginput` command allows the user to provide input to a program by selecting points from a graphics window.

Cell mode includes a number of useful features, past just dividing up M-files into convenient sections. The `publish` tool creates a report containing both the M-file code and results as well as any figures generated when the program executes. The Increment and Decrement icons on the cell toolbar allow the user to automatically change the value of a parameter each time the code is executed, making it easy to test the result of changing a variable.

MATLAB® includes functions that allow the user to import and export data in a number of popular file formats. A complete list of these formats is available in the `help` tutorial on the File Formats page (doc fileformats). The `fprintf` function can also be used to export formatted output to a text file.

The error bar, located on the right-hand side of the M-file window, identifies lines of code with potential errors. Warnings are indicated in orange and errors that will cause the execution of the code to terminate are shown in red. More extensive debugging tools are available from the debugging toolbar.

MATLAB® SUMMARY

The following MATLAB® summary lists all the special characters, commands, and functions that were defined in this chapter:

Special Characters	
'	begins and ends a string
%	placeholder used in the `fprintf` command
%f	fixed-point, or decimal, notation
%d	signed integer notation
%e	exponential notation
%g	either fixed point or exponential notation
%s	string notation
%%	cell divider
\n	linefeed
\r	carriage return (similar to linefeed)
\t	tab
\b	backspace

Comma j261d Functions	
disp	displays a string or a matrix in the command window
fprintf	creates formatted output which can be sent to the command window or to a file
ginput	allows the user to pick values from a graph
input	allows the user to enter values
num2str	changes a number to a string
pause	pauses the program
sound	plays MATLAB® data through the speakers
sprintf	similar to **fprintf** creates formatted output which is assigned to a variable name and stored as a character array
uiimport	launches the Import Wizard
wavread	reads wave files
xlsimport	imports Excel data files
xlswrite	exports data as an Excel file

KEY TERMS

cell	formatted output	string
cell mode	precision field	width field
character array		

PROBLEMS

Input Function

7.1 Create an M-file that prompts the user to enter a value of x and then calculates the value of $\sin(x)$.

7.2 Create an M-file that prompts the user to enter a matrix and then use the max function to determine the largest value entered. Use the following matrix to test your program:

```
[1, 5, 3, 8, 9, 22]
```

7.3 The volume of a cone is

$$V = \tfrac{1}{3} \times \text{area of the base} \times \text{height}$$

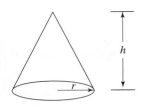

Figure P7.3
Volume of a cone.

Prompt the user to enter the area of the base and the height of the cone (Figure P7.3). Calculate the volume of the cone.

Disp Function

7.4 One of the first computer programs many students write is called "Hello, World." The only thing the program does is print this message to the computer screen. Write a "Hello, World" program in an M-file, using the disp function.

7.5 Use two separate input statements to prompt a user to enter his or her first and last names. Use the disp function to display those names on one line. (You'll need to combine the names and some spaces into an array.)

7.6 Prompt the user to enter his or her age. Then use the disp function to report the age back to the command window. If, for example, the user enters 5 when prompted for her age, your display should read

```
Your age is 5
```

This output requires combining both character data (a string) and numeric data in the disp function—which can be accomplished by using the num2str function.

7.7 Prompt the user to enter an array of numbers. Use the length function to determine how many values were entered, and use the disp function to report your results to the command window.

fprintf

7.8 Repeat Problem 7.7, and use fprintf to report your results.

7.9   Use `fprintf` to create the multiplication tables from 1 to 13 for the number 6. Your table should look like this.

> 1 times 6 is 6
> 2 times 6 is 12
> 3 times 6 is 18
>
> $\vdots$

7.10   Before calculators were readily available (about 1974), students used tables to determine the values of mathematical functions like sine, cosine, and log. Create such a table for sine, using the following steps:

- Create a vector of angle values from 0 to 2π in increments of $\pi/10$.
- Calculate the sine of each of the angles, and group your results into a table that includes the angle and the sine.
- Use `disp` to create a title for the table and a second `disp` command to create column headings.
- Use the `fprintf` function to display the numbers. Display only two values past the decimal point.

7.11   Very small dimensions—those on the atomic scale—are often measured in angstroms. An angstrom is represented by the symbol Å and corresponds to a length of 10^{-10} m. Create an inches-to-angstroms conversion table as follows for values of inches from 1 to 10:

- Use `disp` to create a title and column headings.
- Use `fprintf` to display the numerical information.
- Because the length represented in angstroms is so big, represent your result in scientific notation, showing two values after the decimal point. This corresponds to three significant figures (one before and two after the decimal point).

7.12   Use your favorite Internet search engine and World Wide Web browser to identify recent currency conversions for British pounds sterling, Japanese yen, and the European euro to US dollars. Use the conversion tables to create the following tables (use the `disp` and `fprintf` commands in your solution, which should include a title, column labels, and formatted output):

(a) Generate a table of conversions from yen to dollars. Start the yen column at 5 and increment by 5 yen. Print 25 lines in the table.

(b) Generate a table of conversions from the euros to dollars. Start the euro column at 1 euro and increment by 2 euros. Print 30 lines in the table.

(c) Generate a table with four columns. The first should contain dollars, the second the equivalent number of euros, the third the equivalent number of pounds, and the fourth the equivalent number of yen. Let the dollar column vary from 1 to 10.

Problems Combining the input, disp, and fprintf Commands

7.13   This problem requires you to generate temperature conversion tables. Use the following equations, which describe the relationships between temperatures in degrees Fahrenheit (T_F), degrees Celsius (T_C), kelvins (T_K), and degrees Rankine (T_R), respectively:

$$T_F = T_R - 459.67°R$$
$$T_F = \frac{9}{5}T_C + 32°F$$
$$T_R = \frac{9}{5}T_K$$

You will need to rearrange these expressions to solve some of the problems.

(a) Generate a table of conversions from Fahrenheit to Kelvin for values from 0°F to 200°F. Allow the user to enter the increments in degrees F between lines. Use `disp` and `fprintf` to create a table with a title, column headings, and appropriate spacing.

(b) Generate a table of conversions from Celsius to Rankine. Allow the user to enter the starting temperature and the increment between lines. Print 25 lines in the table. Use `disp` and `fprintf` to create a table with a title, column headings, and appropriate spacing.

(c) Generate a table of conversions from Celsius to Fahrenheit. Allow the user to enter the starting temperature, the increment between lines, and the number of lines for the table. Use `disp` and `fprintf` to create a table with a title, column headings, and appropriate spacing.

7.14 Engineers use both English and SI (Système International d'Unités) units on a regular basis. Some fields use primarily one or the other, but many combine the two systems. For example, the rate of energy input to a steam power plant from burning fossil fuels is usually measured in Btu/hour. However, the electricity produced by the same plant is usually measured in joules/s (watts). Automobile engines, by contrast, are often rated in horsepower or in ft lb$_f$/s. Here are some conversion factors relating these different power measurements:

$$1 \text{ kW} = 3412.14 \text{ Btu/h} = 737.56 \text{ ft lb}_f/s$$

$$1 \text{ hp} = 550 \text{ ft lb}_f/s = 2544.5 \text{ Btu/h}$$

(a) Generate a table of conversions from kW to hp. The table should start at 0 kW and end at 15 kW. Use the `input` function to let the user define the increment between table entries. Use `disp` and `fprintf` to create a table with a title, column headings, and appropriate spacing.

(b) Generate a table of conversions from ft lb$_f$/s to Btu/h. The table should start at 0 ft lb$_f$/s but let the user define the increment between table entries and the final table value. Use `disp` and `fprintf` to create a table with a title, column headings, and appropriate spacing.

(c) Generate a table that includes conversions from kW to Btu/h, hp, and ft lb$_f$/s. Let the user define the initial value of kW, the final value of kW, and the number of entries in the table. Use `disp` and `fprintf` to create a table with a title, column headings, and appropriate spacing.

ginput

7.15 At time $t = 0$, when a rocket's engine shuts down, the rocket has reached an altitude of 500 m and is rising at a velocity of 125 m/s. At this point, gravity takes over. The height of the rocket as a function of time is

$$h(t) = -\frac{9.8}{2}t^2 + 125t + 500 \text{ for } t > 0$$

Plot the height of the rocket from 0 to 30 seconds, and

• Use the `ginput` function to estimate the maximum height the rocket reaches and the time when the rocket hits the ground.

• Use the `disp` command to report your results to the command window.

7.16 The `ginput` function is useful for picking distances off a graph. Demonstrate this feature by doing the following:

- Create a graph of a circle by defining an array of angles from 0 to 2π, with a spacing of $\pi/100$.
- Use the `ginput` function to pick two points on the circumference of the circle.
- Use `hold on` to keep the figure from refreshing, and plot a line between the two points you picked.
- Use the data from the points to calculate the length of the line between them. (*Hint:* Use the Pythagorean theorem in your calculation.)

7.17 In recent years, the price of gasoline has increased dramatically. Automobile companies have responded with more fuel-efficient cars, in particular, hybrid models. But will you save money by purchasing a hybrid such as the Toyota Camry rather than a Camry with a standard engine? The hybrid vehicles are considerably more expensive, but get better gas mileage. Consider the vehicle prices and gas efficiencies shown in Table P7.17.

Table P7.17 A Comparison of Standard and Hybrid Vehicles

Year	Model	Base MSRP	Gas Efficiency, in-town/highway
2008	Toyota Camry	$18,720	21/31 mpg
2008	Toyota Camry Hybrid	$25,350	33/34 mpg
2008	Toyota Highlander 4WD	$28,750	17/23 mpg
2008	Toyota Highlander 4WD Hybrid	$33,700	27/25 mpg (hybrids may actually get better mileage in town than on the road)
2008	Ford Escape 2WD	$19,140	24/28 mpg
2008	Ford Escape 2WD Hybrid	$26,495	34/30 mpg

One way to compare two vehicles is to find the "cost to own."

Cost to own = Purchase cost + Upkeep + Gasoline cost

Assume for this exercise that the upkeep costs are the same, so in our comparison we'll set them equal to zero.

(a) What do you think the cost of gasoline will be over the next several years? Prompt the user to enter an estimate of gasoline cost in dollars/gallon.

(b) Find the "cost to own" as a function of the number of miles driven for a pair of vehicles from the table, based on the fuel price estimate from part a. Plot your results on an *x–y* graph. The point where the two lines cross is the break-even point.

(c) Use the `ginput` function to pick the break-even point off the graph.

(d) Use `sprintf` to create a string identifying the break-even point, and use the result to create a text-box annotation on your graph. Position the text box using the `gtext` function.

Cell Mode

7.18 Publish your program and results from Problem 7.17 to HTML, using the **publish to HTML** feature from the cell toolbar. Unfortunately, because this chapter's assignment requires interaction with the user, the published results will include errors.

7.19 Revisit Problem 7.17, which compares the cost to own for hybrids versus standard-engine vehicles.

(a) Instead of allowing the user to enter an estimate of fuel cost, assume that gasoline will cost $2.00 per gallon for the next several years.

(b) Use the incremental value adjustment tool on the cell-mode toolbar to change the value of the gasoline cost, until the break-even point occurs at less than 100,000 miles.

8

Logical Functions and Selection Structures

Objectives

After reading this chapter, you should be able to:

- Understand how MATLAB® interprets relational and logical operators
- Use the find function
- Understand the appropriate uses of the if/else family of commands
- Understand the switch/ case structure

INTRODUCTION

One way to think of a computer program (not just MATLAB®) is to consider how the statements that compose it are organized. Usually, sections of computer code can be categorized as **sequences**, **selection structures**, and **repetition structures** (see Figure 8.1). So far, we have written code that contains sequences but none of the other structures:

- A sequence is a list of commands that are executed one after another.
- A selection structure allows the programmer to execute one command (or set of commands) if some criterion is true and a second command (or set of commands) if the criterion is false. A selection statement provides the means of choosing between these paths, based on a **logical condition**. The conditions that are evaluated often contain both **relational** and **logical** operators or functions.
- A repetition structure, or loop, causes a group of statements to be executed multiple times. The number of times a loop is executed depends on either a counter or the evaluation of a logical condition.

Figure 8.1
Programming structures used in MATLAB®.

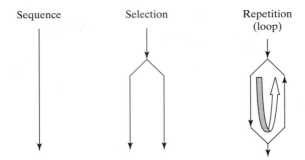

8.1 RELATIONAL AND LOGICAL OPERATORS

The selection and repetition structures used in MATLAB® depend on relational and logical operators. MATLAB® has six relational operators for comparing two matrices of equal size, as shown in Table 8.1.

Comparisons are either true or false, and most computer programs (including MATLAB®) use the number 1 for true and 0 for false. (MATLAB® actually takes any number that is not 0 to be true.) If we define two scalars

```
x = 5;
y = 1;
```

and use a relational operator such as <, the result of the comparison

```
x<y
```

is either true or false. In this case, **x** is not less than **y**, so MATLAB® responds

```
ans =
     0
```

indicating that the comparison is false. MATLAB® uses this answer in selection statements and in repetition structures to make decisions.

Of course, variables in MATLAB® usually represent entire matrices. If we redefine x and y, we can see how MATLAB® handles comparisons between matrices. For example,

```
x = 1:5;
y = x -4;
x<y
```

KEY IDEA
Relational operators compare values

Table 8.1 Relational Operators

Relational Operator	Interpretation
<	less than
<=	less than or equal to
>	greater than
>=	greater than or equal to
==	equal to
~=	not equal to

Table 8.2 Logical Operators

Logical Operator	Interpretation
&	and
~	not
|	or
xor	exclusive or

returns

```
ans =
    0   0   0   0   0
```

MATLAB® compares corresponding elements and creates an answer matrix of zeros and ones. In the preceding example, **x** was greater than **y** for every comparison of elements, so every comparison was false and the answer was a string of zeros. If, instead, we have

```
x = [ 1, 2, 3, 4, 5];
y = [-2, 0, 2, 4, 6];
x<y
```

then

```
ans =
    0   0   0   0   1
```

The results tell us that the comparison was false for the first four elements, but true for the last. For a comparison to be true for an entire matrix, it must be true for *every* element in the matrix. In other words, all the results must be ones.

KEY IDEA

Logical operators are used to combine comparison statements

MATLAB® also allows us to combine comparisons with the logical operators *and*, *not*, and *or* (see Table 8.2).

The code

```
x = [ 1, 2, 3, 4, 5];
y = [-2, 0, 2, 4, 6];
z = [ 8, 8, 8, 8, 8];
z>x & z>y
```

returns

```
ans =
    1   1   1   1   1
```

because **z** is greater than both **x** and **y** for every element. The statement

```
x>y | x>z
```

is read as "**x** is greater than **y** or **x** is greater than **z**" and returns

```
ans =
    1   1   1   0   0
```

This means that the condition is true for the first three elements and false for the last two.

These relational and logical operators are used in both selection structures and loops to determine what commands should be executed.

8.2 FLOWCHARTS AND PSEUDOCODE

With the addition of selection and repetition structures to your group of programming tools, it becomes even more important to plan your program before you start coding. Two common approaches are to use flowcharts and pseudocode. A flowchart is a graphical approach to creating your coding plan, and pseudocode is a verbal description of your plan. You may want to use either or both for your programming projects.

For simple programs, pseudocode may be the best (or at least the simplest) planning approach:

- Outline a set of statements describing the steps you will take to solve a problem.
- Convert these steps into comments in an M-file.
- Insert the appropriate MATLAB® code into the file between the comment lines.

Here's a really simple example: Suppose you've been asked to create a program to convert mph to ft/s. The output should be a table, complete with a title and column headings. Here's an outline of the steps you might follow:

- Define a vector of mph values.
- Convert mph to ft/s.
- Combine the mph and ft/s vectors into a matrix.
- Create a table title.
- Create column headings.
- Display the table.

Once you've identified the steps, put them into a MATLAB® M-file as comments:

```
%Define a vector of mph values
%Convert mph to ft/s
%Combine the mph and ft/s vectors into a matrix
%Create a table title
%Create column headings
%Display the table
```

Now you can insert the appropriate MATLAB® code into the M-file

```
%Define a vector of mph values
  mph = 0:10:100;
%Convert mph to ft/s
  fps = mph*5280/3600;
%Combine the mph and ft/s vectors into a matrix
  table = [mph;fps]
%Create a table title
  disp('Velocity Conversion Table')
%Create column headings
  disp('   mph    f/s')
%Display the table
  fprintf('%8.0f   %8.2f \n',table)
```

If you put some time into your planning, you probably won't need to change the pseudocode much, once you start programming.

Flowcharts alone or flowcharts combined with pseudocode are especially appropriate for more complicated programming tasks. You can create a "big picture" of your program graphically and then convert your project to pseudocode suitable to

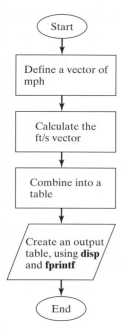

Figure 8.2
Flowcharts make it easy to visualize the structure of a program.

Table 8.3 Flowcharting for Designing Computer Programs

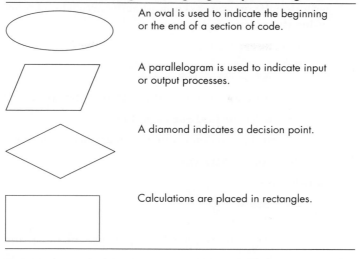

An oval is used to indicate the beginning or the end of a section of code.

A parallelogram is used to indicate input or output processes.

A diamond indicates a decision point.

Calculations are placed in rectangles.

enter into the program as comments. Before you can start flowcharting, you'll need to be introduced to some standard flowcharting symbols (see Table 8.3).

Figure 8.2 is an example of a flowchart for the mph-to-ft/s problem. For a problem this simple, you would probably never actually create a flowchart. However, as problems become more complicated, flowcharts become an invaluable tool, allowing you to organize your thoughts.

Once you've created a flowchart, you should transfer the ideas into comment lines in an M-file and then add the appropriate code between the comments.

Remember, both flowcharts and pseudocode are tools intended to help you create better computer programs. They can also be used effectively to illustrate the structure of a program to nonprogrammers, since they emphasize the logical progression of ideas over programming details.

8.3 LOGICAL FUNCTIONS

MATLAB® offers both traditional selection structures, such as the family of **if** functions, and a series of logical functions that perform much the same task. The primary logical function is find, which can often be used in place of both traditional selection structures and loops.

8.3.1 Find

The find command searches a matrix and identifies which elements in that matrix meet a given criterion. For example, the U.S. Naval Academy requires applicants to be at least 5′6″(66″) tall. Consider this list of applicant heights:

```
height = [63,67,65,72,69,78,75]
```

You can find the index numbers of the elements that meet our criterion by using the find command:

```
accept = find(height>=66 )
```

This command returns

```
accept =
    2    4    5    6    7
```

PSEUDOCODE

A list of programming tasks necessary to create a program

KEY IDEA

Logical functions are often more efficient programming tools than traditional selection structures

The find function returns the index numbers from the matrix that meet the criterion. If you want to know what the actual heights are, you can call each element, using the index number:

```
height(accept)
ans =
    67    72    69    78    75
```

An alternative approach would be to nest the commands

```
height(find(height(>=66)))
```

You could also determine which applicants do *not* meet the criterion. Use

```
decline = find(height<66)
```

which gives

```
decline =
    1    3
```

To create a more readable report use the disp and fprintf functions:

```
disp('The following candidates meet the height requirement');
    fprintf('Candidate # %4.0f is %4.0f
    inches tall \n', [accept;height(accept)])
```

These commands return the following table in the command window:

```
The following candidates meet the height requirement
Candidate #    2 is    67 inches tall
Candidate #    4 is    72 inches tall
Candidate #    5 is    69 inches tall
Candidate #    6 is    78 inches tall
Candidate #    7 is    75 inches tall
```

Clearly, you could also create a table of those who do not meet the requirement:

```
disp('The following candidates do not meet the height
    requirement')
fprintf('Candidate # %4.0f is %4.0f inches tall \n',
    [decline;height(decline)])
```

Similar to the previous code, the following table is returned in the command window:

```
The following candidates do not meet the height requirement
Candidate #    1 is    63 inches tall
Candidate #    3 is    65 inches tall
```

You can create fairly complicated search criteria that use the logical operators. For example, suppose the applicants must be at least 18 years old and less than 35 years old. Then your data might look like this:

Height, Inches	Age, Years
63	18
67	19
65	18
72	20
69	36
78	34
75	12

Now we define the matrix and find the index numbers of the elements in column 1 that are greater than 66. Then we find which of those elements in column 2 are also greater than or equal to 18 and less than or equal to 35. We use the commands

```
applicants = [ 63, 18; 67, 19; 65, 18; 72, 20; 69, 36; 78,
        34; 75, 12]
pass = find(applicants(:,1)>=66 & applicants(:,2)>=18
        & applicants(:,2) < 35)
```

which return

```
pass =
    2
    4
    6
```

the list of applicants that meet all the criteria. We could use `fprintf` to create a nicer output. First create a table of the data to be displayed:

```
results = [pass,applicants(pass,1),applicants(pass,2)]';
```

Then use `fprintf` to send the results to the command window:

```
fprintf('Applicant # %4.0f is %4.0f inches tall and
  %4.0f years old\n',results)
```

The resulting list is

```
Applicant #    2 is    67 inches tall and 19 years old
Applicant #    4 is    72 inches tall and 20 years old
Applicant #    6 is    78 inches tall and 34 years old
```

So far, we've used `find` only to return a single index number. If we define two outputs from `find`, as in

```
[row, col] = find( criteria)
```

it will return the appropriate row and column numbers (also called the row and column index numbers or subscripts).

Now, imagine that you have a matrix of patient temperature values measured in a clinic. The column represents the number of the station where the temperature was taken. Thus, the command

```
temp = [95.3, 100.2, 98.6; 97.4,99.2, 98.9; 100.1,99.3, 97]
```

gives

```
temp =
    95.3000   100.2000    98.6000
    97.4000    99.2000    98.9000
   100.1000    99.3000    97.0000
```

and

```
element = find(temp>98.6)
```

gives us the element number for the single-index representation:

```
element =
    3
    4
    5
    6
    8
```

Figure 8.3
Element-numbering sequence for a matrix.

When the `find` command is used with a two-dimensional matrix, it uses an element-numbering scheme that works down each column one at a time. For example, consider our 3×3 matrix. The element index numbers are shown in Figure 8.3. The elements that contain values greater than 98.6 are shown in bold.

In order to determine the row and column numbers, we need the syntax

```
[row, col] = find(temp>98.6)
```

which gives us the following row and column numbers:

KEY IDEA
MATLAB® is column dominant

1, 1	**1, 2**	1, 3
2, 1	**2, 2**	**2, 3**
3, 1	**3, 2**	3, 3

Figure 8.4
Row, element designation for a 3×3 matrix. The elements that meet the criterion are shown in bold.

```
row =
    3
    1
    2
    3
    2

col =
    1
    2
    2
    2
    3
```

Together, these numbers identify the elements shown in Figure 8.4.

Using `fprintf`, we can create a more readable report. For example,

```
fprintf('Patient%3.0f at station%3.0f had a temp of%6.1f
    \n', [row,col,temp(element)]')
```

returns

```
Patient 3 at station 1 had a temp of 100.1
Patient 1 at station 2 had a temp of 100.2
Patient 2 at station 2 had a temp of 99.2
Patient 3 at station 2 had a temp of 99.3
Patient 2 at station 3 had a temp of 98.9
```

8.3.2 Flowcharting and Pseudocode for Find Commands

The `find` command returns only one answer: a vector of the element numbers requested. For example, you might flowchart a sequence of commands as shown in Figure 8.5. If you use `find` multiple times to separate a matrix into categories, you may choose to employ a diamond shape, indicating the use of `find` as a selection structure.

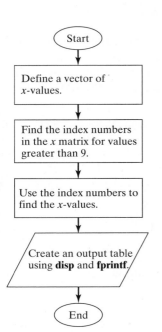

Figure 8.5
Flowchart illustrating the `find` command.

```
%Define a vector of x-values
 x = [1,2,3; 10, 5,1; 12,3,2;8, 3,1]
%Find the index numbers of the values in x >9
 element = find(x>9)
%Use the index numbers to find the x-values
```

```
%greater than 9 by plugging them into x
 values = x(element)
% Create an output table
 disp('Elements greater than 9')
 disp('Element # Value')
 fprintf('%8.0f %3.0f \n', [element';values'])
```

EXAMPLE 8.1

SIGNAL PROCESSING USING THE SINC FUNCTION

Figure 8.6
Oscilloscopes are widely used in signal-processing applications. (Courtesy of Agilent Technologies, Inc.)

The sinc function is used in many engineering applications, but especially in signal processing (Figure 8.6). Unfortunately, this function has two widely accepted definitions:

$$f_1(x) = \frac{\sin(\pi x)}{\pi x} \text{ and } f_2(x) = \frac{\sin x}{x}$$

Both of these functions have an indeterminate form of $0/0$ when $x = 0$. In this case, l'Hôpital's theorem from calculus can be used to prove that both functions are equal to 1 when $x = $ zero. For values of x not equal to zero, the two functions have a similar form. The first function, $f_1(x)$, crosses the x-axis when x is an integer; the second function crosses the x-axis when x is a multiple of π.

Suppose you would like to define a function called `sinc_x` that uses the second definition. Test your function by calculating values of `sinc_x` for x from -5π to $+5\pi$ and plotting the results.

1. State the Problem
   Create and test a function called `sinc_x`, using the second definition:

   $$f_2(x) = \frac{\sin x}{x}$$

2. Describe the Input and Output

   Input   Let x vary from -5π to $+5\pi$.

   Output   Create a plot of **sinc_x** versus **x**.

3. Develop a Hand Example
4. Develop a MATLAB® Solution
   Outline your function in a flowchart, as shown in Figure 8.7. Then convert the flowchart to pseudocode comments, and insert the appropriate MATLAB® code. Once we've created the function, we should test it in the command window:

```
sinc_x(0)
ans =
        1
sinc_x(pi/2)
ans =
        0.6366
sinc_x(pi)
ans =
        3.8982e-017
sinc_x(-pi/2)
ans =
        0.6366
```

(*continued*)

Figure 8.7
Flowchart of the sinc function.

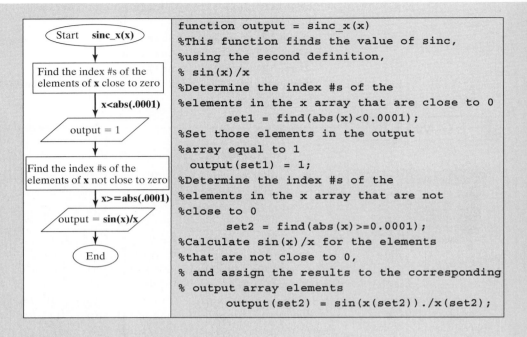

```
function output = sinc_x(x)
%This function finds the value of sinc,
%using the second definition,
% sin(x)/x
%Determine the index #s of the
%elements in the x array that are close to 0
        set1 = find(abs(x)<0.0001);
%Set those elements in the output
%array equal to 1
  output(set1) = 1;
%Determine the index #s of the
%elements in the x array that are not
%close to 0
        set2 = find(abs(x)>=0.0001);
%Calculate sin(x)/x for the elements
%that are not close to 0,
% and assign the results to the corresponding
% output array elements
        output(set2) = sin(x(set2))./x(set2);
```

Notice that sinc_x(pi/2) equals a very small number, but not zero. That is because MATLAB® treats π as a floating-point number and uses an approximation of its real value (Table 8.4).

5. Test the Solution
   When we compare the results with those of the hand example, we see that the answers match. Now we can use the function confidently in our problem. We have

```
%Example 8.1
       clear, clc
%Define an array of angles
       x = -5*pi:pi/100:5*pi;
%Calculate sinc_x
       y = sinc_x(x);
%Create the plot
       plot(x,y)
       title('Sinc Function'), xlabel('angle,
       radians'),ylabel('sinc')
```

Table 8.4 Calculating the Sinc Function

x	sin(x)	sinc_x(x) = sin(x)/x
0	0	$0/0 = 1$
$\pi/2$	1	$1/(\pi/2) = 0.637$
π	0	0
$-\pi/2$	-1	$-1/(\pi/2) = -0.637$

Figure 8.8
The sinc function.

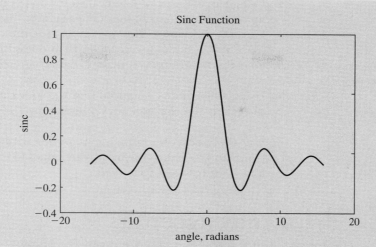

which generates the plot in Figure 8.8.

The plot also supports our belief that the function is working properly. Testing sinc_x with one value at a time validated its answers for a scalar input; however, the program that generated the plot sent a vector argument to the function. The plot confirms that it also performs properly with vector input.

If you have trouble understanding how this function works, remove the semicolons that are suppressing the output, then run the program. Understanding the output from each line will help you understand the program logic better.

In addition to find, MATLAB® offers two other logical functions: all and any. The all function checks to see if a logical condition is true for *every* member of an array, and the any function checks to see if a logical condition is true for *any* member of an array. Consult MATLAB®'s built-in help function for more information.

PRACTICE EXERCISES 8.1

Consider the following matrices:

$$x = \begin{bmatrix} 1 & 10 & 42 & 6 \\ 5 & 8 & 78 & 23 \\ 56 & 45 & 9 & 13 \\ 23 & 22 & 8 & 9 \end{bmatrix} \quad y = \begin{bmatrix} 1 & 2 & 3 \\ 4 & 10 & 12 \\ 7 & 21 & 27 \end{bmatrix} \quad z = \begin{bmatrix} 10 & 22 & 5 & 13 \end{bmatrix}$$

1. Using single-index notation, find the index numbers of the elements in each matrix that contain values greater than 10.
2. Find the row and column numbers (sometimes called subscripts) of the elements in each matrix that contain values greater than 10.
3. Find the values in each matrix that are greater than 10.

4. Using single-index notation, find the index numbers of the elements in each matrix that contain values greater than 10 and less than 40.
5. Find the row and column numbers for the elements in each matrix that contain values greater than 10 and less than 40.
6. Find the values in each matrix that are greater than 10 and less than 40.
7. Using single-index notation, find the index numbers of the elements in each matrix that contain values between 0 and 10 or between 70 and 80.
8. Use the `length` command together with results from the `find` command to determine how many values in each matrix are between 0 and 10 or between 70 and 80.

8.4 SELECTION STRUCTURES

Most of the time, the `find` command can and should be used instead of an `if` statement. In some situations, however, the `if` statement is required. This section describes the syntax used in `if` statements.

8.4.1 The Simple If

A simple `if` statement has the following form:

```
if  comparison
    statements
end
```

If the comparison (a logical expression) is true, the statements between the `if` statement and the `end` statement are executed. If the comparison is false, the program jumps immediately to the statement following `end`. It is good programming practice to indent the statements within an `if` structure for readability. However, recall that MATLAB® ignores white space. Your programs will run regardless of whether you do or do not indent any of your lines of code.

Here's a really simple example of an `if` statement:

```
if G<50
    disp('G is a small value equal to:')
    disp(G);
end
```

KEY IDEA

if statements usually work best with scalars

This statement (from `if` to `end`) is easy to interpret if **G** is a scalar. If **G** is less than 50, then the statements between the `if` and the `end` lines are executed. For example, if **G** has a value of 25, then

```
G is a small value equal to:
    25
```

is displayed on the screen. However, if **G** is not a scalar, then the **if** statement considers the comparison true **only if it is true for every element**! Thus, if G is defined from 0 to 80,

```
G = 0:10:80;
```

the comparison is false, and the statements inside the `if` statement are not executed! In general, `if` statements work best when dealing with scalars.

8.4.2 The If/Else Structure

The simple `if` allows us to execute a series of statements if a condition is true and to skip those steps if the condition is false. The `else` clause allows us to execute one set of statements if the comparison is true and a different set if the comparison is false. Suppose you would like to take the logarithm of a variable *x*. You know from basic algebra classes that the input to the `log` function must be greater than 0. Here's a set of `if/else` statements that calculates the logarithm if the input is positive and sends an error message if the input to the function is 0 or negative:

```
if x >0
    y = log(x)
else
    disp('The input to the log function must be positive')
end
```

When **x** is a scalar, this is easy to interpret. However, when **x** is a matrix, the comparison is true only if it is true for every element in the matrix. So, if

```
x = 0:0.5:2;
```

then the elements in the matrix are not all greater than 0. Therefore, MATLAB® skips to the `else` portion of the statement and displays the error message. The `if/else` statement is probably best confined to use with scalars, although you may find it to be of limited use with vectors.

HINT

MATLAB® includes a function called `beep` that causes the computer to "beep" at the user. You can use this function to alert the user to an error. For example, in the `if/else` clause, you could add a beep to the portion of the code that includes an error statement:

```
x = input('Enter a value of x greater than 0: ');
if x >0
    y = log(x)
else
    beep
    disp('The input to the log function must be positive')
end
```

8.4.3 The Elseif Structure

When we nest several levels of `if/else` statements, it may be difficult to determine which logical expressions must be true (or false) in order to execute each set of statements. The `elseif` function allows you to check multiple criteria while keeping the code easy to read. Consider the following lines of code that evaluate whether to issue a driver's license, based on the applicant's age:

```
if age<16
    disp('Sorry - You'll have to wait')
elseif age<18
    disp('You may have a youth license')
```

```
elseif age<70
    disp('You may have a standard license')
else
    disp('Drivers over 70 require a special license')
end
```

In this example, MATLAB® first checks to see if **age** < **16.** If the comparison is true, the program executes the next line or set of lines, displays the message `Sorry–You'll have to wait`, and then exits the `if` structure. If the comparison is false, MATLAB® moves on to the next `elseif` comparison, checking to see if `age < 18` this time. The program continues through the `if` structure until it finally finds a true comparison or until it encounters the `else`. Notice that the `else` line does not include a comparison, since it executes if the `elseif` immediately before it is false.

The flowchart for this sequence of commands (Figure 8.9) uses the diamond shape to indicate a selection structure.

This structure is easy to interpret if `age` is a scalar. If it is a matrix, the comparison must be true for every element in the matrix. Consider this age matrix

```
age = [15,17,25,55,75]
```

The first comparison, `if age<16`, is false, because it is not true for every element in the array. The second comparison, `elseif age<18`, is also false. The third comparison, `elseif age<70`, is false as well, since not all of the ages are below 70. The result is `Drivers over 70 require a special license`—a result that won't please the other drivers.

Figure 8.9
Flowchart using multiple
`if` statements.

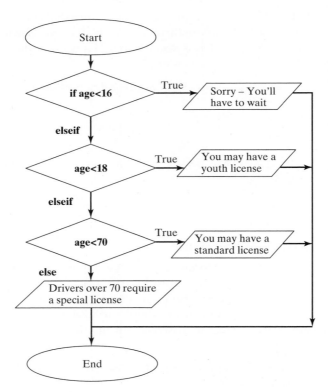

HINT

One common mistake new programmers make when using `if` statements is to overspecify the criteria. In the preceding example, it is enough to state that age < 18 in the second `if` clause, because age cannot be less than 16 and still reach this statement. You don't need to specify age < 18 and age $> = 16$. If you overspecify the criteria, you risk defining a calculational path for which there is no correct answer. For example, in the code

```
if age<16
    disp('Sorry - You''ll have to wait')
elseif age<18 & age>16
    disp('You may have a youth license')
elseif age<70 & age>18
    disp('You may have a standard license')
elseif age>70
    disp('Drivers over 70 require a special license')
end
```

there is no correct choice for age = 16, 18, or 70.

In general, `elseif` structures work well for scalars, but `find` is probably a better choice for matrices. Here's an example that uses `find` with an array of ages and generates a table of results in each category:

```
age = [15,17,25,55,75];
set1 = find(age<16);
set2 = find(age>=16 & age<18);
set3 = find(age>=18 & age<70);
set4 = find(age>=70);

fprintf('Sorry - You''ll have to wait - you"re only %3.0f
  \n',age(set1))
fprintf('You may have a youth license because you"re %3.0f
  \n',age(set2))
fprintf('You may have a standard license because you"re
  %3.0f \n',age(set3))
fprintf('Drivers over 70 require a special license. You"re
  %3.0f \n',age(set4))
```

These commands return

```
Sorry - You'll have to wait - you're only 15
You may have a youth license because you're 17
You may have a standard license because you're 25
You may have a standard license because you're 55
Drivers over 70 require a special license. You're 75
```

Since every `find` in this sequence is evaluated, it is necessary to specify the range completely (for example, `age>=16 & age<18`).

EXAMPLE 8.2

ASSIGNING GRADES

The if family of statements is used most effectively when the input is a scalar. Create a function to determine test grades based on the score and assuming a single input into the function. The grades should be based on the following criteria:

Grade	Score
A	90 to 100
B	80 to 90
C	70 to 80
D	60 to 70
E	<60

1. State the Problem
   Determine the grade earned on a test.
2. Describe the Input and Output

   Input   Single score, not an array

   Output   Letter grade

3. Develop a Hand Example
   85 should be a B
   But should 90 be an A or a B? We need to create more exact criteria.

Grade	Score
A	$\geq$90 to 100
B	$\geq$80 and <90
C	$\geq$70 and <80
D	$\geq$60 and <70
E	<60

4. Develop a MATLAB® Solution
   Outline the function, using the flowchart shown in Figure 8.10.
5. Test the Solution
   Now test the function in the command window:

```
grade(25)
ans =
E
grade(80)
ans =
B
grade(-52)
ans =
E
grade(108)
ans =
A
```

Figure 8.10
Flowchart for a grading scheme.

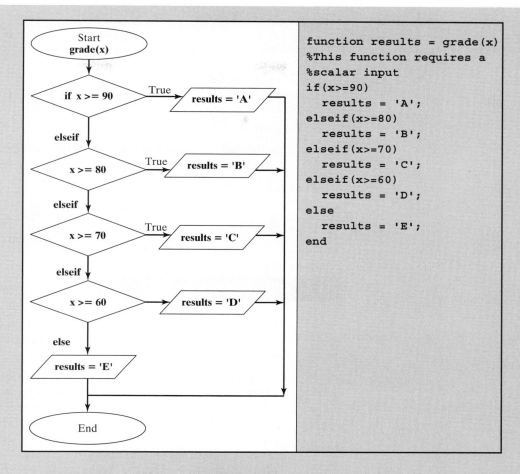

```
function results = grade(x)
%This function requires a
%scalar input
if(x>=90)
   results = 'A';
elseif(x>=80)
   results = 'B';
elseif(x>=70)
   results = 'C';
elseif(x>=60)
   results = 'D';
else
   results = 'E';
end
```

Notice that although the function seems to work properly, it returns grades for values over 100 and values less than 0. If you'd like, you can now go back and add the logic to exclude those values:

```
function results = grade(x)
%This function requires a scalar input
   if(x>=0 & x<=100)
      if(x>=90)
         results = 'A';
      elseif(x>=80)
         results = 'B';
      elseif(x>=70)
         results = 'C';
      elseif(x>=60)
         results = 'D';
      else
         results = 'E';
      end
   else
      results = 'Illegal Input';
end
```

(*continued*)

We can test the function again in the command window:

```
grade(-10)
ans =
Illegal Input
grade(108)
ans =
Illegal Input
```

This function will work great for scalars, but if you send a vector to the function, you may get some unexpected results, such as

```
score = [95,42,83,77];
grade(score)
ans =
E
```

PRACTICE EXERCISES 8.2

The `if` family of functions is particularly useful in functions. Write and test a function for each of these problems, assuming that the input to the function is a scalar:

1. Suppose the legal drinking age is 21 in your state. Write and test a function to determine whether a person is old enough to drink.
2. Many rides at amusement parks require riders to be a certain minimum height. Assume that the minimum height is 48″ for a certain ride. Write and test a function to determine whether the rider is tall enough.
3. When a part is manufactured, the dimensions are usually specified with a tolerance. Assume that a certain part needs to be 5.4 cm long, plus or minus 0.1 cm (5.4 ± 0.1 cm). Write a function to determine whether a part is within these specifications.
4. Unfortunately, the United States currently uses both metric and English units. Suppose the part in Exercise 3 was inspected by measuring the length in inches instead of centimeters. Write and test a function that determines whether the part is within specifications and that accepts input into the function in inches.
5. Many solid-fuel rocket motors consist of three stages. Once the first stage burns out, it separates from the missile and the second stage lights. Then the second stage burns out and separates, and the third stage lights. Finally, once the third stage burns out, it also separates from the missile. Assume that the following data approximately represent the times during which each stage burns:

Stage 1	0–100 seconds
Stage 2	100–170 seconds
Stage 3	170–260 seconds

Write and test a function to determine whether the missile is in Stage 1 flight, Stage 2 flight, Stage 3 flight, or free flight (unpowered).

8.4.4 Switch and Case

The `switch/case` structure is often used when a series of programming path options exists for a given variable, depending on its value. The `switch/case` is similar to the `if/else/elseif`. As a matter of fact, anything you can do with `switch/case` could be done with `if/else/elseif`. However, the code is a bit easier to read with `switch/case`, a structure that allows you to choose between multiple outcomes, based on some criterion. This is an important distinction between `switch/case` and `elseif`. The criterion can be either a scalar (a number) or a string. In practice, it is used more with strings than with numbers. The structure of `switch/case` is

```
switch variable
case option1
  code to be executed if variable is equal to option 1
case option2
  code to be executed if variable is equal to option 2
                            ⋮
case option_n
  code to be executed if variable is equal to option n
otherwise
  code to be executed if variable is not equal to any of
    the options
end
```

Here's an example: Suppose you want to create a function that tells the user what the airfare is to one of three different cities:

```
city = input('Enter the name of a city in single quotes: ')
switch city
  case 'Boston'
    disp('$345')
  case 'Denver'
    disp('$150')
  case 'Honolulu'
    disp('Stay home and study')
  otherwise
    disp('Not on file')
end
```

If, when you run this script, you reply `'Boston'` at the prompt, MATLAB® responds

```
city =
Boston
$345
```

You can tell the `input` command to expect a string by adding "s" in a second field. This relieves the user of the awkward requirement of adding single quotes around any string input. With the added "s", the preceding code now reads as follows:

```
city = input('Enter the name of a city: ','s')
switch city
  case 'Boston'
    disp('$345')
  case 'Denver'
```

```
        disp('$150')
    case 'Honolulu'
        disp('Stay home and study')
    otherwise
        disp('Not on file')
end
```

The `otherwise` portion of the **switch**/**case** structure is not required for the structure to work. However, you should include it if there is any way that the user could input a value not equal to one of the cases.

`Switch/case` structures are flowcharted exactly the same as `if/else` structures.

HINT

If you are a C programmer, you may have used `switch/case` in that language. One important difference in MATLAB® is that once a "true" case has been found, the program does not check the other cases.

EXAMPLE 8.3

BUYING GASOLINE

Figure 8.11
Gasoline is sold in both liters and gallons.

Four countries in the world do not officially use the metric system: the United States, the United Kingdom, Liberia, and Myanmar. Even in the United States, the practice is that some industries are almost completely metric and others still use the English system of units. For example, any shade-tree mechanic will tell you that although older cars have a mixture of components—some metric and others English—new cars (any car built after 1989) are almost completely metric. Wine is packaged in liters, but milk is packaged in gallons. Americans measure distance in miles, but power in watts. Confusion between metric and English units is common. American travelers to Canada are regularly confused because gasoline is sold by the liter in Canada, but by the gallon in the United States.

Imagine that you want to buy gasoline (Figure 8.11). Write a program that:

- Asks the user whether he or she wants to request the gasoline in liters or in gallons.
- Prompts the user to enter how many units he or she wants to buy.
- Calculates the total cost to the user, assuming that gasoline costs $2.89 per gallon.

Use a `switch/case` structure.

1. State the Problem
   Calculate the cost of a gasoline purchase.
2. Describe the Input and Output

   Input   Specify gallons or liters
             Number of gallons or liters

   Output Cost in dollars, assuming $2.89 per gallon

3. Develop a Hand Example
   If the volume is specified in gallons, the cost is

$$\text{volume} \times \$2.89$$

so, for 10 gallons,

$$\text{cost} = 10 \text{ gallons} \times \$2.89/\text{gallon} = \$28.90$$

If the volume is specified in liters, we need to convert liters to gallons and then calculate the cost:

$$\text{volume} = \text{liters} \times 0.264 \text{ gallon/liter}$$
$$\text{cost} = \text{volume} \times \$2.89$$

So, for 10 liters,

$$\text{volume} = 10 \text{ liters} \times 0.264 \text{ gallon/liter} = 2.64 \text{ gallons}$$
$$\text{cost} = 2.64 \text{ gallons} \times 2.89 = \$7.63$$

4. Develop a MATLAB® Solution

First create a flowchart (Figure 8.12). Then convert the flowchart into pseudocode comments. Finally, add the MATLAB® code:

Figure 8.12
Flowchart to determine the cost of gasoline, using the `switch/case` structure.

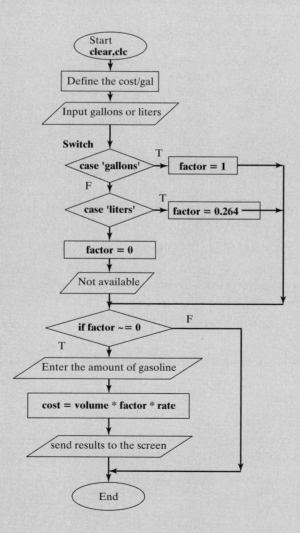

(*continued*)

```
        clear,clc
        %Define the cost per gallon
         rate = 2.89;
        %Ask the user to input gallons or liters
         unit = input('Enter gallons or liters\n ','s');
        %Use a switch/case to determine the conversion factor
         switch unit
           case 'gallons'
                factor = 1;
           case 'liters'
              factor = 0.264;
           otherwise
              disp('Not available')
              factor = 0;
        end

   %Ask the user how much gas he/she would like to buy
      volume = input( ['Enter the volume you would like to buy
      in ',unit,': \n'] );
      %Calculate the cost of the gas
      if factor ~ = 0
        cost = volume * factor*rate;
      %Send the results to the screen
      fprintf('That will be $ %5.2f for %5.1f %s
              \n',cost,volume,unit)
   end
```

There are several things to notice about this solution. First, the variable unit contains an array of character information. If you check the workspace window after you run this program, you'll notice that unit is either a 1×6 character array (if you entered liters) or a 1×7 character array (if you entered gallons). On the line

```
unit = input('Enter gallons or liters ','s');
```

the second field, **'s'**, tells MATLAB® to expect a string as input. This allows the user to enter gallons or liters without the surrounding single quotes.
On the line

```
volume = input(['Enter the volume you would like to buy in
               ',unit,': '] );
```

we created a character array out of three components:

- The string 'Enter the volume you would like to buy in'
- The character variable unit
- The string ' : '

By combining these three components, we were able to make the program prompt the user with either

```
Enter the volume you would like to buy in liters:
```

or

```
Enter the volume you would like to buy in gallons:
```

In the `fprintf` statement, we included a field for string input by using the placeholder `%s`:

```
fprintf('That will be $ %5.2f for %5.1f %s
        \n',cost,volume,unit)
```

This allowed the program to tell the users that the gasoline was measured either in gallons or in liters.

Finally, we used an **if** statement so that if the user entered something besides gallons or liters, no calculations were performed.

5. Test the Solution

We can test the solution by running the program three separate times, once for gallons, once for liters, and once for some unit not supported. The interaction in the command window for gallons is

```
Enter gallons or liters
gallons
Enter the volume you would like to buy in gallons:
10
That will be $ 28.90 for 10.0 gallons
```

For liters, the interaction is

```
Enter gallons or liters
liters
Enter the volume you would like to buy in liters:
10
That will be $ 7.63 for 10.0 liters
```

Finally, if you enter anything besides gallons or liters, the program sends an error message to the command window:

```
Enter gallons or liters
quarts
Not available
```

Since the program results are the same as the hand calculation, it appears that the program works as planned.

8.4.5 Menu

The `menu` function is often used in conjunction with a `switch/case` structure. This function causes a menu box to appear on the screen, with a series of buttons defined by the programmer.

KEY IDEA

Graphical user interfaces like the menu box reduce the opportunity for user errors, such as spelling mistakes

```
input = menu(' Message to the user',' text for button
    1',' text for button 2', etc.)
```

We can use the `menu` option in our previous airfare example to ensure that the user chooses only cities about which we have information. This also means that we don't need the `otherwise` syntax, since it is not possible to choose a city "not on file."

```
city = menu('Select a city from the menu:
    ','Boston','Denver','Honolulu')
```

Figure 8.13
The pop-up menu window.

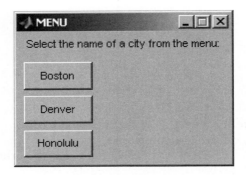

```
switch city
    case 1
        disp('$345')
    case 2
        disp('$150')
    case 3
        disp('Stay home and study')
end
```

Notice that a case number has replaced the string in each case line. When the script is executed, the menu box shown in Figure 8.13 appears and waits for the user to select one of the buttons. If you choose Honolulu, MATLAB® will respond

```
city =
    3
Stay home and study
```

Of course, you could suppress the output from the disp command, which was included here for clarity.

EXAMPLE 8.4

BUYING GASOLINE: A MENU APPROACH

In Example 8.3, we used a switch/case approach to determine whether the customer wanted to buy gasoline measured in gallons or liters. One problem with our program is that if the user can't spell, the program won't work. For example, if, when prompted for gallons or liters, the user enters

```
litters
```

The program will respond

```
Not Available
```

We can get around this problem by using a menu; then the user need only press a button to make a choice. We'll still use the switch/case structure, but will combine it with the menu.

1. State the Problem
   Calculate the cost of a gasoline purchase.

2. Describe the Input and Output

Input   Specify gallons or liters, using a menu
            Number of gallons or liters

Output   Cost in dollars, assuming $2.89 per gallon

3. Develop a Hand Example
If the volume is specified in gallons, the cost is

$$\text{volume} \times \$2.89$$

So, for 10 gallons,

$$\text{cost} = 10 \text{ gallons} \times \$2.89/\text{gallon} = \$28.90$$

If the volume is specified in liters, we need to convert liters to gallons and then calculate the cost:

$$\text{volume} = \text{liters} \times 0.264 \text{ gallon/liter}$$
$$\text{cost} = \text{volume} \times \$2.89$$

So, for 10 liters,

$$\text{volume} = 10 \text{ liters} \times 0.264 \text{ gallon/liter} = 2.64 \text{ gallons}$$
$$\text{cost} = 2.64 \text{ gallons} \times 2.89 = \$7.63$$

4. Develop a MATLAB® Solution
First create a flowchart (Figure 8.14). Then convert the flowchart into pseudocode comments. Finally, add the MATLAB® code:

```
%Example 8.4
clear,clc
%Define the cost per gallon
 rate = 2.89;
%Ask the user to input gallons or liters, using a menu
 disp('Use the menu box to make your selection ')
 choice = menu('Measure the gasoline in liters or
gallons?','gallons','liters');
%Use a switch/case to determine the conversion factor
switch choice
   case 1
     factor = 1;
     unit = 'gallons'
   case 2
    factor = 0.264;
    unit = 'liters'
end

%Ask the user how much gas he/she would like to buy
 volume = input(['Enter the volume you would like to
  buy in ',unit,': \n'] );
%Calculate the cost of the gas
 cost = volume * factor*rate;
%Send the results to the screen
 fprintf('That will be $ %5.2f for %5.1f %s
  \n',cost,volume,unit)
```

(continued)

Figure 8.14
Flowchart to determine the cost of gasoline, using a menu.

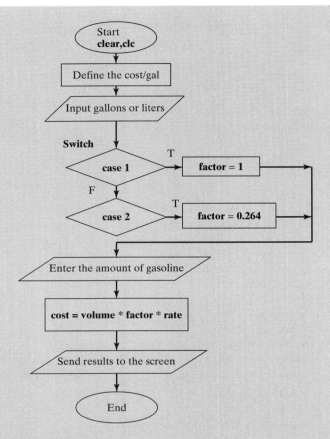

This solution is simpler than the one in Example 8.3 because there is no chance for bad input. There are a few things to notice, however.

When we define the choice by using the menu function, the result is a number, not a character array:

```
choice = menu('Measure the gasoline in liters or
gallons?','gallons','liters');
```

You can check this by consulting the workspace window, in which the choice is listed as a 1×1 double-precision number.

Because we did not use the input command to define the variable unit, which is a string (a character array), we needed to specify the value of unit as part of the case calculations:

```
case 1
    factor = 1;
    unit = 'gallons'
case 2
    factor = 0.264;
    unit = 'liters'
```

Doing this allows us to use the value of unit in the output to the command window, both in the disp command and in fprintf.

5. Test the Solution

As in Example 8.3, we can test the solution by running the program, but this time we need to try it only twice—once for gallons and once for liters. The interaction in the command window for gallons is

```
Use the menu box to make your selection
```

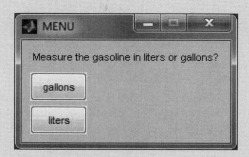

```
Enter the volume you would like to buy in gallons:
10
That will be $ 28.90 for 10.0 gallons
```

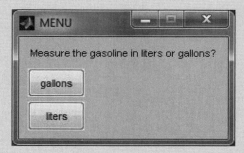

For liters, the interaction is

```
Use the menu box to make your selection

Enter the volume you would like to buy in liters:
10
That will be $ 7.63 for 10.0 liters
```

These values match those in the hand solution and have the added advantage that you can't misspell any of the input.

PRACTICE EXERCISES 8.3

Use the `switch/case` structure to solve these problems:

1. Create a program that prompts the user to enter his or her year in school—freshman, sophomore, junior, or senior. The input will be a string. Use the `switch/case` structure to determine which day finals will be given for each group—Monday for freshmen, Tuesday for sophomores, Wednesday for juniors, and Thursday for seniors.

2. Repeat Exercise 1, but this time with a menu.

3. Create a program to prompt the user to enter the number of candy bars he or she would like to buy. The input will be a number. Use the `switch/case` structure to determine the bill, where

$$1 \text{ bar} = \$0.75$$
$$2 \text{ bars} = \$1.25$$
$$3 \text{ bars} = \$1.65$$

more than 3 bars $= \$1.65 + \$0.30 \text{ (number ordered} - 3)$

8.5 DEBUGGING

As the code we are writing becomes more complicated, the debugging tools available in MATLAB® become more valuable. Consider the simple program shown in Figure 8.15 that demonstrates the use of the if/else structure. A breakpoint has been added on line two. When the code is executed by selecting the save and run icon, it will first pause on line 1 waiting for the user to enter a number. Once the number has been entered, the program moves to line two and stops because the breakpoint has been encountered. Selecting the step icon will progress the execution through the code one line at a time, allowing the programmer to observe the effect of each line of code.

Also notice that the folding capability available in MATLAB® has been activated for `if/else` structures. This was accomplish by selecting

File -> Preferences -> Editor/Debugger -> Code Folding

from the menu bar. By activating code folding for `if/else` blocks a visual cue is created, making it easier to keep track of which lines of code are included in the structure.

Figure 8.15
Using debugging tools is an effective way to evaluate how MATLAB® moves through the code as it executes.

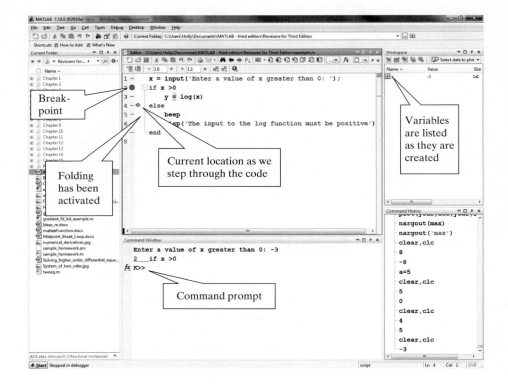

SUMMARY

Sections of computer code can be categorized as sequences, selection structures, and repetition structures. Sequences are lists of instructions that are executed in order. Selection structures allow the programmer to define criteria (conditional statements) that the program uses to choose execution paths. Repetition structures define loops in which a sequence of instructions is repeated until some criterion is met (also defined by conditional statements).

MATLAB® uses the standard mathematical relational operators, such as greater than ($>$) and less than ($<$). The not-equal-to ($\sim =$) operator's form is not usually seen in mathematics texts. MATLAB® also includes logical operators such as *and* (&) and *or* ($|$). These operators are used in conditional statements, allowing MATLAB® to make decisions regarding which portions of the code to execute.

The find command is unique to MATLAB® and should be the primary conditional function used in your programming. This command allows the user to specify a condition by using both logical and relational operators. The command is then used to identify elements of a matrix that meet the condition.

Although the if, else, and elseif commands can be used for both scalars and matrix variables, they are useful primarily for scalars. These commands allow the programmer to identify alternative computing paths on the basis of the results of conditional statements.

The following MATLAB® summary lists and briefly describes all the special characters, commands, and functions that were defined in this chapter:

MATLAB® SUMMARY

Special Characters	
<	less than
<=	less than or equal to
>	greater than
>=	greater than or equal to
==	equal to
~=	not equal to
&	and
|	or
~	not

Commands and Functions	
all	checks to see if a criterion is met by all the elements in an array
any	checks to see if a criterion is met by any of the elements in an array
case	selection structure
else	defines the path if the result of an if statement is false
elseif	defines the path if the result of an if statement is false, and specifies a new logical test
end	identifies the end of a control structure
find	determines which elements in a matrix meet the input criterion
if	checks a condition, resulting in either true or false
menu	creates a menu to use as an input vehicle
otherwise	part of the case selection structure
switch	part of the case selection structure

KEY TERMS

control structure

index

local variable

logical condition

logical operator

loop

relational operator

repetition

selection

sequence

subscript

PROBLEMS

LOGICAL OPERATORS: FIND

8.1 A sensor that monitors the temperature of a backyard hot tub records the data shown in Table 8.5.

Table 8.5 Hot-Tub Temperature Data

Time of Day	Temperature, °F
00:00	100
01:00	101
02:00	102
03:00	103
04:00	103
05:00	104
06:00	104
07:00	105
08:00	106
09:00	106
10:00	106
11:00	105
12:00	104
13:00	103
14:00	101
15:00	100
16:00	99
17:00	100
18:00	102
19:00	104
20:00	106
21:00	107
22:00	105
23:00	104
24:00	104

(a) The temperature should never exceed 105°F. Use the `find` function to find the index numbers of the temperatures that exceed the maximum allowable temperature.

(**b**) Use the `length` function with the results from part (a) to determine how many times the maximum allowable temperature was exceeded.

(**c**) Determine at what times the temperature exceeded the maximum allowable temperature, using the index numbers found in part (a).

(**d**) The temperature should never be lower than 102°F. Use the `find` function together with the `length` function to determine how many times the temperature was less than the minimum allowable temperature.

(**e**) Determine at what times the temperature was less than the minimum allowable temperature.

(**f**) Determine at what times the temperature was within the allowable limits (i.e., between 102°F and 105°F, inclusive).

(**g**) Use the `max` function to determine the maximum temperature reached and the time at which it occurred.

8.2 The height of a rocket (in meters) can be represented by the following equation:

$$\text{height} = 2.13t^2 - 0.0013t^4 + 0.000034t^{4.751}$$

Create a vector of time (t) values from 0 to 100 at 2-second intervals.

(**a**) Use the `find` function to determine when the rocket hits the ground to within 2 seconds. (*Hint*: The value of `height` will be positive for all values until the rocket hits the ground.)

(**b**) Use the `max` function to determine the maximum height of the rocket and the corresponding time.

(**c**) Create a plot with t on the horizontal axis and height on the vertical axis for times until the rocket hits the ground. Be sure to add a title and axis labels.*

8.3 Solid-fuel rocket motors are used as boosters for the space shuttle, in satellite launch vehicles, and in weapons systems (see Figure P8.3). The propellant is a solid combination of fuel and oxidizer, about the consistency of an eraser. For the space shuttle, the fuel component is aluminum and the oxidizer is ammonium perchlorate, held together with an epoxy resin "glue." The propellant mixture is poured into a motor case, and the resin is allowed to cure under controlled conditions. Because the motors are extremely large, they are cast in segments, each requiring several "batches" of propellant to fill. (Each motor contains over 1.1 million pounds of propellant!) This casting–curing process is sensitive to temperature, humidity, and pressure. If the conditions aren't just right, the fuel could ignite or the properties of the propellant grain (which means its shape; the term *grain* is borrowed from artillery) might be degraded. Solid-fuel rocket motors are extremely expensive as well as dangerous and clearly must work right every time, or the results will be disastrous. Failures can cause loss of human life and irreplaceable scientific data and equipment. Highly public failures can destroy a company. Actual processes are tightly monitored and controlled. However, for our purposes, consider these general criteria:

Figure P8.3

Solid-fuel rocket booster to a titan missile. (Courtesy of NASA.)

The temperature should remain between 115°F and 125°F.
The humidity should remain between 40% and 60%.
The pressure should remain between 100 and 200 torr.

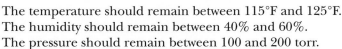

* From Etter, Kancicky, and Moore, *Introduction to Matlab* 7 (Upper Saddle River, NJ: Pearson/Prentice Hall, 2005).

Imagine that the data in Table 8.6 were collected during a casting–curing process.

Table 8.6 Casting–Curing Data

Batch Number	Temperature, °F	Humidity, %	Pressure, torr
1	116	45	110
2	114	42	115
3	118	41	120
4	124	38	95
5	126	61	118

(a) Use the `find` command to determine which batches did and did not meet the criterion for temperature.

(b) Use the `find` command to determine which batches did and did not meet the criterion for humidity.

(c) Use the `find` command to determine which batches did and did not meet the criterion for pressure.

(d) Use the `find` command to determine which batches failed for any reason and which passed.

(e) Use your results from the previous questions, along with the `length` command, to determine what percentage of motors passed or failed on the basis of each criterion and to determine the total passing rate.

8.4 Two gymnasts are competing with each other. Their scores are shown in Table 8.7.

Table 8.7 Gymnastics Scores

Event	Gymnast 1	Gymnast 2
Pommel horse	9.821	9.700
Vault	9.923	9.925
Floor	9.624	9.83
Rings	9.432	9.987
High bar	9.534	9.354
Parallel bars	9.203	9.879

(a) Write a program that uses `find` to determine how many events each gymnast won.

(b) Use the `mean` function to determine each gymnast's average score.

8.5 Create a function called **f** that satisfies the following criteria:

$$\text{For values of } x > 2, f(x) = x^2$$
$$\text{For values of } x \leq 2, f(x) = 2x$$

Plot your results for values of x from -3 to 5. Choose your spacing to create a smooth curve. You should notice a break in the curve at $x = 2$.

8.6 Create a function called *g* that satisfies the following criteria:

$$\begin{array}{lll}
\text{For } x < -\pi, & g(x) = -1 \\
\text{For } x \geq -\pi \text{ and } x \leq \pi, & g(x) = \cos(x) \\
\text{For } x > \pi, & g(x) = -1
\end{array}$$

Plot your results for values of *x* from -2π to $+2\pi$. Choose your spacing to create a smooth curve.

8.7 A file named **temp.dat** contains information collected from a set of thermocouples. The data in the file are shown in Table 8.8. The first column consists of time measurements (one for each hour of the day), and the remaining columns correspond to temperature measurements at different points in a process.

(a) Write a program that prints the index numbers (rows and columns) of temperature data values greater than 85.0. (*Hint:* You'll need to use the find command.)

(b) Find the index numbers (rows and columns) of temperature data values less than 65.0.

(c) Find the maximum temperature in the file and the corresponding hour value and thermocouple number.

Table 8.8 Temperature Data

Hour	Temp1	Temp2	Temp3
1	68.70	58.11	87.81
2	65.00	58.52	85.69
3	70.38	52.62	71.78
4	70.86	58.83	77.34
5	66.56	60.59	68.12
6	73.57	61.57	57.98
7	73.57	67.22	89.86
8	69.89	58.25	74.81
9	70.98	63.12	83.27
10	70.52	64.00	82.34
11	69.44	64.70	80.21
12	72.18	55.04	69.96
13	68.24	61.06	70.53
14	76.55	61.19	76.26
15	69.59	54.96	68.14
16	70.34	56.29	69.44
17	73.20	65.41	94.72
18	70.18	59.34	80.56
19	69.71	61.95	67.83
20	67.50	60.44	79.59
21	70.88	56.82	68.72
22	65.99	57.20	66.51
23	72.14	62.22	77.39
24	74.87	55.25	89.53

8.8 The Colorado River Drainage Basin covers parts of seven western states. A series of dams has been constructed on the Colorado River and its tributaries to store runoff water and to generate low-cost hydroelectric power (see Figure P8.8). The ability to regulate the flow of water has made the growth of agriculture and population in these arid desert states possible. Even during periods of extended drought, a steady, reliable source of water and electricity

Figure P8.8
Glen Canyon dam at Lake Powell. (Courtesy of Getty images, Inc.)

has been available to the basin states. Lake Powell is one of these reservoirs. The file **lake_powell.dat** contains data on the water level in the reservoir for the 8 years from 2000 to 2007. These data are shown in Table 8.9. Use the data in the file to answer the following questions:

(a) Determine the average elevation of the water level for each year and for the 8-year period over which the data were collected.

(b) Determine how many months each year exceed the overall average for the 8-year period.

(c) Create a report that lists the month (number) and the year for each of the months that exceed the overall average. For example, June is month 6.

(d) Determine the average elevation of the water for each month for the 8-year period.

Table 8.9 Water-Level Data for Lake Powell, Measured in Feet above Sea Level

	2000	2001	2002	2003	2004	2005	2006	2007
January	3680.12	3668.05	3654.25	3617.61	3594.38	3563.41	3596.26	3601.41
February	3678.48	3665.02	3651.01	3613	3589.11	3560.35	3591.94	3598.63
March	3677.23	3663.35	3648.63	3608.95	3584.49	3557.42	3589.22	3597.85
April	3676.44	3662.56	3646.79	3605.92	3583.02	3557.52	3589.94	3599.75
May	3676.76	3665.27	3644.88	3606.11	3584.7	3571.60	3598.27	3604.68
June	3682.19	3672.19	3642.98	3615.39	3587.01	3598.06	3609.36	3610.94
July	3682.86	3671.37	3637.53	3613.64	3583.07	3607.73	3608.79	3609.47
August	3681.12	3667.81	3630.83	3607.32	3575.85	3604.96	3604.93	3605.56
September	3678.7	3665.45	3627.1	3604.11	3571.07	3602.20	3602.08	3602.27
October	3676.96	3663.47	3625.59	3602.92	3570.7	3602.31	3606.12	3601.27
November	3674.93	3661.25	3623.98	3601.24	3569.69	3602.65	3607.46	3599.71
December	3671.59	3658.07	3621.65	3598.82	3565.73	3600.14	3604.96	3596.79

Note: This problem should be solved using the `find` function, the `mean` function, and the `length` function. Programmers with previous experience may be tempted to use a loop structure, which is not required.

IF STRUCTURES

8.9 Create a program that prompts the user to enter a scalar value of temperature. If the temperature is greater than 98.6°F, send a message to the command window telling the user that he or she has a fever.

8.10 Create a program that first prompts the user to enter a value for **x** and then prompts the user to enter a value for y. If the value of **x** is greater than the value of y, send a message to the command window telling the user that **x** > **y**. If **x** is less than or equal to y, send a message to the command window telling the user that **y** >= **x**.

8.11 The inverse sine (`asin`) and inverse cosine (`acos`) functions are valid only for inputs between −1 and +1, because both the sine and the cosine have values only between −1 and +1 (Figure P8.11). MATLAB® interprets the result of `asin` or `acos` for a value outside the range as a complex number. For example, we might have

```
acos(-2)
ans =
   3.1416 - 1.3170i
```

which is a questionable mathematical result. Create a function called my_asin that accepts a single value of **x** and checks to see if it is between −1 and +1 ($-1 <= x <= 1$). If x is outside the range, send an error message to the screen. If it is inside the allowable range, return the value of asin.

Figure P8.11
The sine function varies between −1 and +1. Thus, the inverse sine (`asin`) is not defined for values greater than 1 and values less than −1.

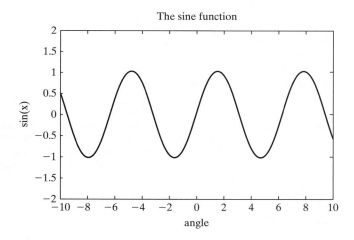

The sine function

8.12 Create a program that prompts the user to enter a scalar value for the outside air temperature. If the temperature is equal to or above 80°F, send a message to the command window telling the user to wear shorts. If the temperature is between 60°F and 80°F send a message to the command window telling the user that it is a beautiful day. If the temperature is equal to or below 60°F, send a message to the command window telling the user to wear a jacket or coat.

8.13 Suppose the following matrix represents the number of saws ordered from your company each month over the last year.

```
saws = [1,4,5,3,7,5,3,10,12,8, 7, 4]
```

All the numbers should be zero or positive.

(a) Use an `if` statement to check whether any of the values in the matrix are invalid. (Evaluate the whole matrix at once in a single `if` statement.) Send the message "All valid" or else "Invalid number found" to the screen, depending on the results of your analysis.

(b) Change the `saws` matrix to include at least one negative number, and check your program to make sure that it works for both cases.

8.14 Most large companies encourage employees to save by matching their contributions to a 401(k) plan. The government limits how much you can save in these plans, because they shelter income from taxes until the money is withdrawn during your retirement. The amount you can save is tied to your income, as is the amount your employer can contribute. The government will allow you to save additional amounts without the tax benefit. These plans change from year to year, so this example is just a made-up "what if."

Suppose the Quality Widget Company has the savings plan described in Table 8.10. Create a function that finds the total yearly contribution to your savings plan, based on your salary and the percentage you contribute. Remember, the total contribution consists of the employee contribution and the company contribution.

Table 8.10 Quality Widget Company Savings Plan

Income	Maximum You Can Save Tax Free	Maximum the Company Will Match
Up to $30,000	10%	10%
Between $30,000 and $60,000	10%	10% of the first $30,000 and 5% of the amount above $30,000
Between $60,000 and $100,000	10% of the first $60,000 and 8% of the amount above $60,000	10% of the first $30,000 and 5% of the amount between $30,000 and $60,000; nothing for the remainder above $60,000
Above $100,000	10% of the first $60,000 and 8% of the amount between $60,000 and $100,000; nothing on the amount above $100,000	Nothing—highly compensated employees are exempt from this plan and participate in stock options instead

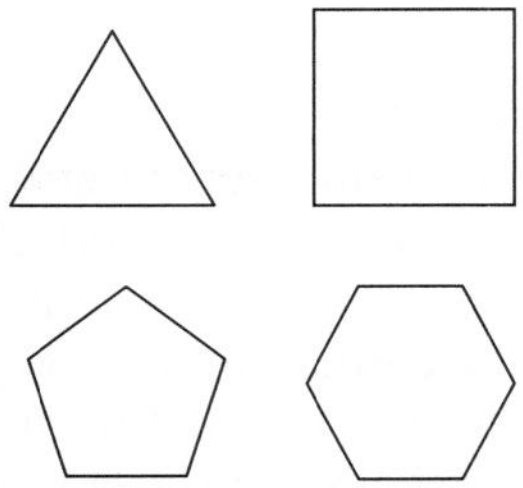

Figure P8.15
Regular Polygons.

SWITCH/CASE

8.15 In order to have a closed geometric figure composed of straight lines (Figure P8.15), the angles in the figure must add to

$$(n - 2)(180 \text{ degrees})$$

where n is the number of sides.

(a) Prove this statement to yourself by creating a vector called n from 3 to 6 and calculating the angle sum from the formula. Compare what you know about geometry with your answer.

(b) Write a program that prompts the user to enter one of the following:

triangle
square
pentagon
hexagon

Use the input to define the value of n via a `switch/case` structure; then use n to calculate the sum of the interior angles in the figure.

(c) Reformulate your program from part (b) so that it uses a menu.

8.16 At a local university, each engineering major requires a different number of credits for graduation. For example, recently the requirements were as follows:

Civil Engineering	130
Chemical Engineering	130
Computer Engineering	122
Electrical Engineering	126.5
Mechanical Engineering	129

Prompt the user to select an engineering program from a menu. Use a `switch/case` structure to send the minimum number of credits required for graduation back to the command window.

8.17 The easiest way to draw a star in MATLAB® is to use polar coordinates. You simply need to identify points on the circumference of a circle and draw lines between those points. For example, to draw a five-pointed star, start at the top of the circle ($\theta = \pi/2$, $r = 1$) and work counterclockwise (Figure P8.17).

Prompt the user to specify either a five-pointed or a six-pointed star, using a menu. Then create the star in a MATLAB® figure window. Note that a six-pointed star is made of three triangles and requires a strategy different from that used to create a five-pointed star.

CHALLENGE PROBLEMS

8.18 Most major airports have separate lots for long-term and short-term parking. The cost to park depends on the lot you select, and how long you stay. Consider this rate structure from the Salt Lake International Airport during the summer of 2008.

- Long-Term (Economy) Parking
  - The first hour is $1.00, and each additional hour or fraction thereof is $1.00
  - Daily maximum $6.00
  - Weekly maximum $42.00
- Short-Term Parking
  - The first 30 minutes are free and each additional 20 minutes or fraction thereof is $1.00
  - Daily maximum $25.00

Write a program that asks the user the following:

- Which lot are you using?
- How many weeks, hours, days, and minutes did you park? Your program should then calculate the parking bill.

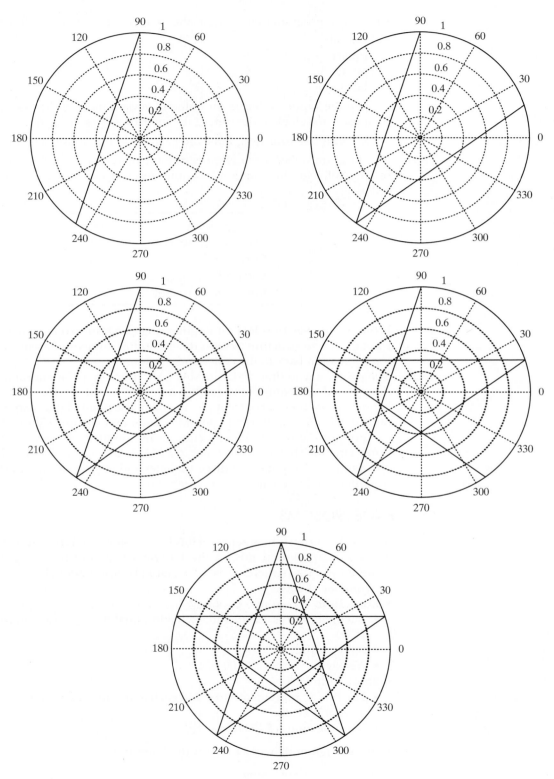

Figure P8.17
Steps required to draw a five-pointed star in polar coordinates.

Repetition Structures

Objectives

After reading this chapter, you should be able to:

- Write and use `for` loops
- Write and use `while` loops
- Create midpoint break structures

- Measure the time required to execute program components
- Understand how to improve program execution times

INTRODUCTION

As discussed in the previous chapter, one way to think of a computer program (not just MATLAB®) is to consider how the statements that compose it are organized. Usually, sections of computer code can be categorized as *sequences, selection structures,* and *repetition structures.* The previous chapter described selection structures; in this chapter we will focus on repetition structures. As a rule of thumb, if a section of code is repeated more than three times, it is a good candidate for a repetition structure.

Repetition structures are often called loops. All loops consist of five basic parts.

- A parameter to be used in determining whether or not to end the loop.
- Initialization of this parameter.
- A way to change the parameter each time through the loop. (If you don't change it, the loop will never stop executing.)
- A comparison, using the parameter, to a criterion used to decide when to end the loop.
- Calculations to do inside the loop.

MATLAB® supports two different types of loops: the for loop and the while loop. Two additional commands, break and continue, can be used to create a third type of loop, called a midpoint break loop. The for loop is the easiest choice when you know how many times you need to repeat the loop. While loops are the easiest choice when you need to keep repeating the instructions until a criterion is met. Midpoint break loops are useful for situations where the commands in the loop must be executed at least once, but where the decision to exit the loop is based on some criterion. If you have previous programming experience, you may be tempted to use loops extensively. However, in many cases you can compose MATLAB® programs that avoid loops, either by using the find command or by vectorizing the code. (In vectorization, we operate on entire vectors at a time instead of one element at a time.) It's a good idea to avoid loops whenever possible, because vectorized programs run faster and often require fewer programming steps.

9.1 FOR LOOPS

The structure of the for loop is simple. The first line identifies the loop and defines an index, which is a number that changes on each pass through the loop and is used to determine when to end the repetitions. After the identification line comes the group of commands we want to execute. Finally, the end of the loop is identified by the command end. Thus, the structure of a for loop can be summarized as

```
for index = [matrix]
      commands to be executed
end
```

The loop is executed once for each element of the index matrix identified in the first line. Here's a really simple example:

```
for k = [1,3,7]
   k
End
```

During the first pass through the loop k is assigned a value of 1, the first value in the k matrix. During the next pass the value of k is modified to 3, the second value in the k matrix. Each time through the loop k is modified and assigned to subsequent values from the index matrix. This example code returns

```
k =
    1
k =
    3
k =
    7
```

The index in this case is k. Programmers often use k as an index variable as a matter of style. The index matrix can also be defined with the colon operator or, indeed, in a number of other ways as well. Here's an example of code that finds the value of 5 raised to powers between 1 and 3:

```
for
k = 1:3
   a = 5^k
end
```

On the first line, the index, k, is defined as the matrix [1, 2, 3]. The first time through the loop, k is assigned a value of 1, and 5^1 is calculated. Then the loop repeats, but now k is equal to 2 and 5^2 is calculated. The last time through the loop, k is equal to 3 and 5^3 is calculated. Because the statements in the loop are repeated three times, the value of a is displayed three times in the command window:

```
a =
   5
a =
   25
a =
   125
```

Although we defined k as a matrix in the first line of the for loop, because k is an index number when it is used in the loop, it can equal only one value at a time. After we finish executing the loop, if we call for k, it has only one value: the value of the index the final time through the loop. For the preceding example,

```
k
```

returns

```
k =
   3
```

Notice that k is listed as a 1×1 matrix in the workspace window.

A common way to use a for loop is in defining a new matrix. Consider, for example, the code

```
for
k = 1:5
   a(k) = k^2
end
```

This loop defines a new matrix, a, one element at a time. Since the program repeats its set of instructions five times, a new element is added to the a matrix each time through the loop, with the following output in the command window:

```
a =
   1
a =
   1    4
a =
   1    4    9
a =
   1    4    9    16
a =
   1    4    9    16    25
```

HINT

Most computer programs do not have MATLAB®'s ability to handle matrices so easily; therefore, they rely on loops similar to the one just presented to define arrays. It would be easier to create the vector **a** in MATLAB® with the code

```
k = 1:5
a = k.^2
```

which returns

```
k =
    1    2    3    4    5
a =
    1    4    9   16   25
```

This is an example of *vectorizing* the code.

Another common use for a `for` loop is to combine it with an `if` statement and determine how many times something is true. For example, in the list of test scores shown in the first line, how many are above 90?

```
scores = [76,45,98,97];
count = 0;
for k=1:length(scores)
   if scores(k)>90
      count = count + 1;
   end
end
disp(count)
```

The variable `count` is initialized as zero, then each time through the loop, if the score is greater than 90, the count is incremented by 1. Notice that the `length` command was used to determine how many times the `for` loop should repeat. In this case

```
length(scores)
```

is equal to four, the number of values in the `scores` array.

Most of the time, `for` loops are created which use an index matrix that is a single row. However, if a two-dimensional matrix is defined in the index specification, MATLAB® uses an entire column as the index each time through the loop. For example, suppose we define the index matrix as

$$k = \begin{bmatrix} 1 & 2 & 3 \\ 1 & 4 & 9 \\ 1 & 8 & 27 \end{bmatrix}$$

Then

```
for k = [1,2,3; 1,4,9; 1,8,27]
   a = k'
end
```

returns

```
a =
   1     1     1
a =
   2     4     8
a =
   3     9    27
```

Notice that k was transposed when it was set equal to a, so our results are rows instead of columns. We did this to make the output easier to read.

We can summarize the use of for loops with the following rules:

- The loop starts with a for statement and ends with the word end.
- The first line in the loop defines the number of times the loop will repeat, using an index matrix.
- The index of a for loop must be a variable. (The index is the number that changes each and every time through the loop.) Although k is often used as the symbol for the index, any variable name may be employed. The use of k is a matter of style.
- Any of the techniques learned to define a matrix can be used to define the index matrix. One common approach is to use the colon operator, as in

```
for index = start:inc:final
```

- If the expression is a row vector, the elements are used one at a time—once for each time through the loop.
- If the expression is a two-dimensional matrix (this alternative is not common), each time through the loop the index will contain the next *column* in the matrix. This means that the index will be a column vector!
- Once you've completed a for loop, the index variable retains the last value used.
- For loops can often be avoided by vectorizing the code.

The basic flowchart for a for loop includes a diamond, which reflects the fact that a for loop starts each pass with a check to see if there is a new value in the index matrix (Figure 9.1). If there isn't, the loop is terminated and the program continues with the statements after the loop.

Figure 9.1
Flowchart for a for loop.

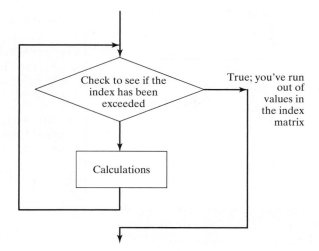

EXAMPLE 9.1

CREATING A DEGREES-TO-RADIANS TABLE

Although it would be much easier to use MATLAB®'s vector capability to create a degrees-to-radians table, we can demonstrate the use of `for` loops with this example.

1. State the Problem
   Create a table that converts angle values from degrees to radians, from 0 to 360 degrees, in increments of 10 degrees.

2. Describe the Input and Output

   Input   An array of angle values in degrees

   Output   A table of angle values in both degrees and radians

3. Develop a Hand Example
   For 10 degrees,

$$\text{Radians} = (10)\frac{\pi}{180} = 0.1745$$

4. Develop a MATLAB® Solution
   First develop a flowchart (Figure 9.2) to help you plan your code.

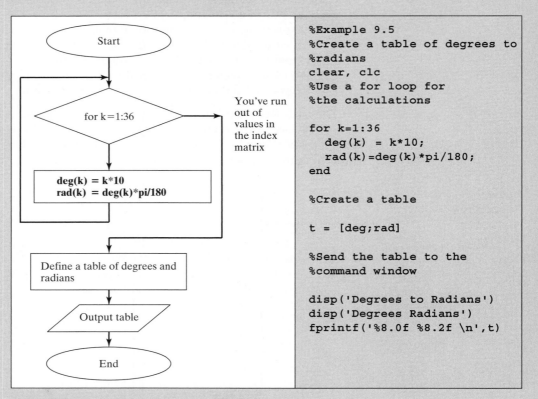

Figure 9.2
Flowchart for changing degrees to radians.

The command window displays the following results:

```
Degrees to Radians
Degrees Radians
    10 0.17
    20 0.35
    30 0.52 etc.
```

5. Test the Solution

The value for 10 degrees calculated by MATLAB® is the same as the hand calculation.

Clearly, it is much easier to use MATLAB®'s vector capabilities for this calculation. You get exactly the same answer, and it takes significantly less computing time. This approach is called vectorization of your code and is one of the strengths of MATLAB®. The vectorized code is

```
deg = 0:10:360;
rad = deg * pi/180;
t = [deg;rad]
disp('Degrees to Radians')
disp('Degrees Radians')
fprintf('%8.0f %8.2f \n',t)
```

EXAMPLE 9.2

CALCULATING FACTORIALS WITH A FOR LOOP

A factorial is the product of all the integers from 1 to N. For example, 5 factorial is

$$1 \cdot 2 \cdot 3 \cdot 4 \cdot 5$$

In mathematics texts, factorial is usually indicated with an exclamation point:

$$5! \text{ is five factorial.}$$

MATLAB® contains a built-in function for calculating factorials, called `factorial`. However, suppose you would like to program your own factorial function called `fact`.

1. State the Problem

Create a function called `fact` to calculate the factorial of any number. Assume scalar input.

2. Describe the Input and Output

Input   A scalar value N

Output   The value of $N!$

3. Develop a Hand Example

$$5! = 1 \cdot 2 \cdot 3 \cdot 4 \cdot 5 = 120$$

4. Develop a MATLAB® Solution
   First develop a flowchart (Figure 9.3) to help you plan your code.

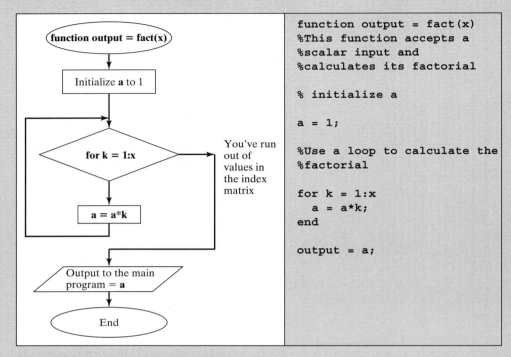

```
function output = fact(x)
%This function accepts a
%scalar input and
%calculates its factorial

% initialize a

a = 1;

%Use a loop to calculate the
%factorial

for k = 1:x
    a = a*k;
end

output = a;
```

Figure 9.3
Flowchart for finding a factorial, using a `for` loop.

5. Test the Solution
   Test the function in the command window:

```
fact(5)
ans =
   120
```

This function works only if the input is a scalar. If an array is entered, the `for` loop does not execute, and the function returns a value of 1:

```
x=1:10;
fact(x)
ans =
   1
```

You can add an `if` statement to confirm that the input is a positive integer and not an array, as shown in the flowchart in Figure 9.4 and the accompanying code.
Check the new function in the command window:

```
fact(-4)
ans =
Input must be a positive integer
```

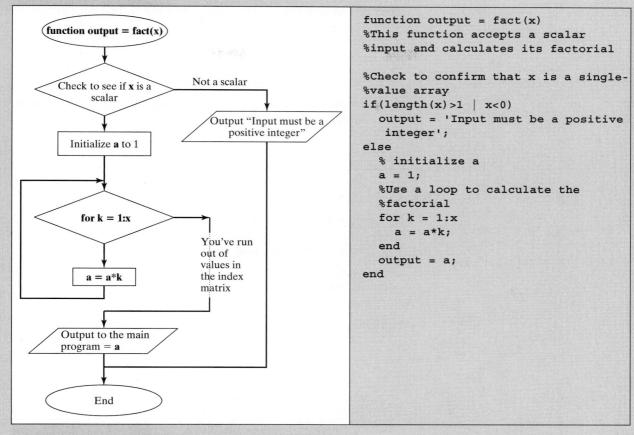

The figure contains a flowchart and the following code:

```
function output = fact(x)
%This function accepts a scalar
%input and calculates its factorial

%Check to confirm that x is a single-
%value array
if(length(x)>1 | x<0)
    output = 'Input must be a positive
    integer';
else
    % initialize a
    a = 1;
    %Use a loop to calculate the
    %factorial
    for k = 1:x
        a = a*k;
    end
    output = a;
end
```

Figure 9.4
Flowchart for finding a factorial, including error checking.

```
fact(x)
ans =
Input must be a positive integer
```

PRACTICE EXERCISES 9.1

Use a `for` loop to solve the following problems:

1. Create a table that converts inches to feet.
2. Consider the following matrix of values:

$$x = [45, 23, 17, 34, 85, 33]$$

How many values are greater than 30? (Use a counter.)
3. Repeat Exercise 2, this time using the `find` command.

4. Use a `for` loop to sum the elements of the matrix in Problem 2. Check your results with the `sum` function. (Use the `help` feature if you don't know or remember how to use `sum`.)

5. Use a `for` loop to create a vector containing the first 10 elements in the harmonic series, i.e.,

$$1/1 \quad 1/2 \quad 1/3 \quad 1/4 \quad 1/5 ... 1/10$$

6. Use a **for** loop to create a vector containing the first 10 elements in the alternating harmonic series, i.e.,

$$1/1 \quad -1/2 \quad 1/3 \quad -1/4 \quad 1/5 ... -1/10$$

9.2 WHILE LOOPS

`While` loops are similar to `for` loops. The big difference is the way MATLAB® decides how many times to repeat the loop. `While` loops continue until some criterion is met. The format for a `while` loop is

```
while criterion
    commands to be executed
end
```

Here's an example:

```
k = 0;
while k<3
   k = k+1
end
```

In this case, we initialized a counter, k, before the loop. Then the loop repeated as long as k was less than 3. We incremented k by 1 every time through the loop, so that the loop repeated three times, giving

```
k =
    1
k =
    2
k =
    3
```

Notice that when k=3 the criterion in the `while` statement

```
k<3
```

is false. Thus, when MATLAB® checks to see if it should make another pass through the loop the program makes the decision (based on the criterion) to skip to the end of the structure.

We could use k as an index number to define a matrix or just as a counter. Most `for` loops can also be coded as `while` loops. Recall the `for` loop in the previous section used to calculate the first three powers of 5. The following `while` loop accomplishes the same task:

KEY IDEA:

Any problem that can be solved using a `while` loop could also be solved using a `for` loop

```
k = 0;
while k<3
   k = k+1;
    a(k) = 5^k
end
```

The code returns

```
a =
    5
a =
    5    25
a =
    5    25    125
```

Each time through the loop, another element is added to the matrix a.
As another example, first initialize a:

```
a = 0;
```

Then find the first multiple of 3 that is greater than 10:

```
While(a<10)
    a = a + 3
end;
```

The first time through the loop, a is equal to 0, so the comparison is true. The next statement (a = a + 3) is executed, and the loop is repeated. This time a is equal to 3 and the condition is still true, so execution continues. In succession, we have

```
a =
    3
a =
    6
a =
    9
a =
    12
```

The last time through the loop, a starts out as 9 and then becomes 12 when 3 is added to 9. The comparison is made one final time, but since a is now equal to 12—which is greater than 10—the program skips to the end of the while loop and no longer repeats.

While loops can also be used to count how many times a condition is true by incorporating an if statement. Recall the test scores we counted in a for loop earlier. We can also count them with a while loop:

```
scores = [76,45,98,97];
count = 0;
k = 0;
while k<length(scores)
    k = k+1;
    if scores(k)>90
        count = count + 1;
    end
end
disp(count)
```

The variable count is used to count how many values are greater than 90. The variable k is used to count how many times the loop is executed.

The basic flow chart for a while loop (Figure 9.5) is the same as that for a for loop (Figure 9.4).

Figure 9.5
Flowchart for a while loop.

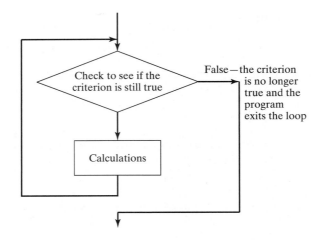

One common use for a while loop is error checking of user input. Consider a program where we prompt the user to input a positive number, and then we calculate the log base 10 of that value. We can use a while loop to confirm that the number is positive, and if it is not, to prompt the user to enter an allowed value. The program keeps on prompting for a positive value until the user finally enters a valid number.

```
x = input('Enter a positive value of x')
while (x<=0)
   disp('log(x) is not defined for negative numbers')
   x = input('Enter a positive value of x')
end
   y = log10(x);
fprintf('The log base 10 of %4.2f is %5.2f \n',x,y)
```

If, when the code is executed, a positive value of **x** is entered, the while loop does not execute (since **x** is not less than 0). If, instead, a zero or negative value is entered, the while loop is executed, an error message is sent to the command window, and the user is prompted to reenter the value of **x**. The while loop continues to execute until a positive value of **x** is finally entered.

KEY IDEA:

It is easy to create an infinite loop with a while structure

HINT

The variable used to control the while loop must be updated every time through the loop. If not, you'll generate an endless loop. When a calculation is taking a long time to complete, you can confirm that the computer is really working on it by checking the lower left-hand corner for the "busy" indicator. If you want to exit the calculation manually, type Ctrl c. (Depress the Ctrl and **c** key at the same time.) Make sure that the command window is the active window when you execute this command.

HINT

Many computer texts and manuals indicate the control key with the ^ symbol. This is confusing at best. The command ^C usually means to strike the Ctrl key and the **c** key at the same time.

EXAMPLE 9.3

CREATING A TABLE FOR CONVERTING DEGREES TO RADIANS WITH A WHILE LOOP

Just as we used a `for` loop to create a table for converting degrees to radians in Example 9.2, we can use a `while` loop for the same purpose.

1. **State the Problem**
   Create a table that converts degrees to radians, from 0 to 360 degrees, in increments of 10 degrees.

2. **Describe the Input and Output**

   Input An array of angle values in degrees

   Output A table of angle values in both degrees and radians

3. **Develop a Hand Example**
   For 10 degrees,

   $$\text{radians} = (10)\frac{\pi}{180} = 0.1745$$

4. **Develop a MATLAB® Solution**
   First develop a flowchart (Figure 9.6) to help you plan your code.

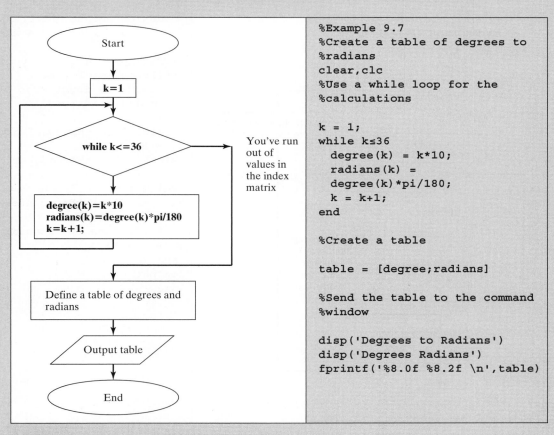

```
%Example 9.7
%Create a table of degrees to
%radians
clear,clc
%Use a while loop for the
%calculations

k = 1;
while k≤36
  degree(k) = k*10;
  radians(k) =
  degree(k)*pi/180;
  k = k+1;
end

%Create a table

table = [degree;radians]

%Send the table to the command
%window

disp('Degrees to Radians')
disp('Degrees Radians')
fprintf('%8.0f %8.2f \n',table)
```

Figure 9.6
Flowchart for converting degrees to radians with a `while` loop.

The command window displays the following results:

```
Degrees to Radians
Degrees    Radians
     10       0.17
     20       0.35
     30       0.52       etc.
```

5. Test the Solution
   The value for 10 degrees calculated by MATLAB® is the same as the hand calculation.

EXAMPLE 9.4

CALCULATING FACTORIALS WITH A WHILE LOOP

Create a new function called `fact2` that uses a `while` loop to find *N!*. Include an `if` statement to check for negative numbers and to confirm that the input is a scalar.

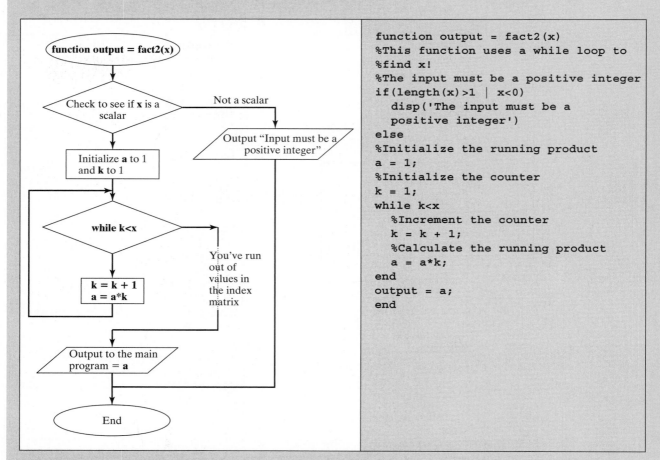

```
function output = fact2(x)
%This function uses a while loop to
%find x!
%The input must be a positive integer
if(length(x)>1 | x<0)
    disp('The input must be a
    positive integer')
else
%Initialize the running product
a = 1;
%Initialize the counter
k = 1;
while k<x
    %Increment the counter
    k = k + 1;
    %Calculate the running product
    a = a*k;
end
output = a;
end
```

Figure 9.7
Flowchart for finding a factorial with a `while` loop.

1. **State the Problem**
   Create a function called `fact2` to calculate the factorial of any number.
2. **Describe the Input and Output**

   Input   A scalar value N

   Output The value of $N!$

3. **Develop a Hand Example**
$$5! = 1 \cdot 2 \cdot 3 \cdot 4 \cdot 5 = 120$$

4. **Develop a MATLAB® Solution**
   First develop a flowchart (Figure 9.7) to help you plan your code.
5. **Test the Solution**
   Test the function in the command window:

```
fact2(5)
ans =
   120
fact2(-10)
ans =
   The input must be a positive integer
fact2([1:10])
ans =
   The input must be a positive integer
```

EXAMPLE 9.5

THE ALTERNATING HARMONIC SERIES

The ***alternating harmonic series*** converges to the natural log of 2:

$$\sum_{k=1}^{\infty} \frac{(-1)^{k+1}}{k} = 1 - \frac{1}{2} + \frac{1}{3} - \frac{1}{4} + \frac{1}{5} - \cdots = \ln(2) = 0.6931471806$$

Because of this, we can use the alternating harmonic series to approximate the $\ln(2)$. But how far out do you have to take the series to get a good approximation of the final answer? We can use a `while` loop to solve this problem.

1. **State the Problem**
   Use a `while` loop to calculate the members of the alternating harmonic sequence and the value of the series until it converges to values that vary by less than .001. Compare the result to the natural log of 2.
2. **Describe the Input and Output**

   Input   The description of the alternating harmonic series

   $$\sum_{k=1}^{\infty} \frac{(-1)^{k+1}}{k} = 1 - \frac{1}{2} + \frac{1}{3} - \frac{1}{4} + \frac{1}{5} - \cdots \frac{1}{\infty}$$

   Output The value of the truncated series, once the convergence criterion is met.
   Plot the cumulative sum of the series elements, up to the point where the convergence criterion is met.

(*continued*)

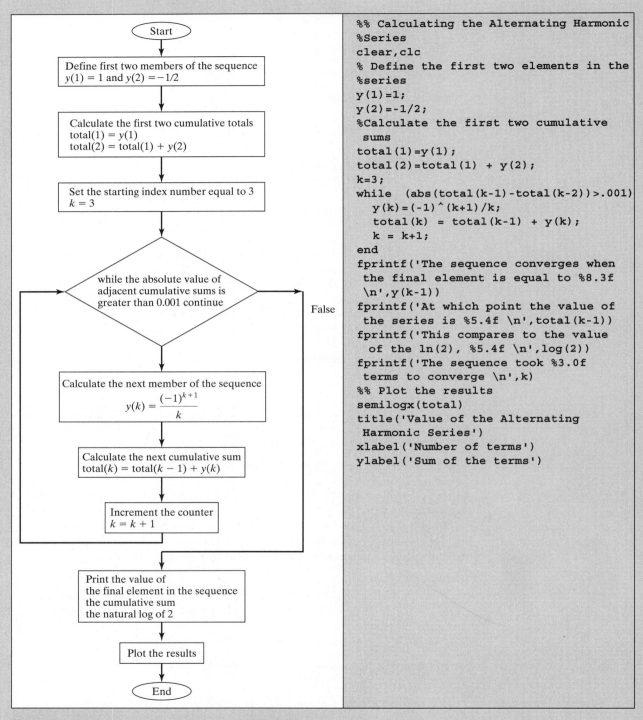

The code shown in the flowchart figure:

```
%% Calculating the Alternating Harmonic
%Series
clear,clc
% Define the first two elements in the
%series
y(1)=1;
y(2)=-1/2;
%Calculate the first two cumulative
 sums
total(1)=y(1);
total(2)=total(1) + y(2);
k=3;
while  (abs(total(k-1)-total(k-2))>.001)
    y(k)=(-1)^(k+1)/k;
    total(k)  =  total(k-1) + y(k);
    k = k+1;
end
fprintf('The sequence converges when
 the final element is equal to %8.3f
 \n',y(k-1))
fprintf('At which point the value of
 the series is %5.4f \n',total(k-1))
fprintf('This compares to the value
 of the ln(2), %5.4f \n',log(2))
fprintf('The sequence took %3.0f
 terms to converge \n',k)
%% Plot the results
semilogx(total)
title('Value of the Alternating
 Harmonic Series')
xlabel('Number of terms')
ylabel('Sum of the terms')
```

Figure 9.8
Flowchart to evaluate the alternating harmonic series until it converges.

3. Develop a Hand Example

Let's calculate the value of the alternating harmonic series for 1 to 5 terms. First find the value for each of the first five terms in the sequence

| 1.0000 | −0.5000 | 0.3333 | −0.2500 | 0.2000 |

Now calculate the sum of the series assuming 1 to 5 terms

| 1.0000 | 0.5000 | 0.8333 | 0.5833 | 0.7833 |

The calculated sums are getting closer together, as we can see if we find the difference between adjacent pairs

| −0.5000 | 0.3333 | −0.2500 | 0.2000 |

4. Develop a MATLAB® Solution

First develop a flowchart (Figure 9.8) to help you plan your code, then convert it to a MATLAB® program. When we run the program, the following results are displayed in the command window.

```
The sequence converges when the final element is equal to 0.001
At which point the value of the series is 0.6936
This compares to the value of the ln(2), 0.6931
The sequence took 1002 terms to converge
```

The series is pretty close to the ln(2), but perhaps we could get closer with more terms. If we change the convergence criterion to 0.0001 and run the program, we get the following results

```
The sequence converges when the final element is equal to
 -0.000
At which point the value of the series is 0.6931
This compares to the value of the ln(2), 0.6931
The sequence took 10001 terms to converge
```

5. Test the Solution

Compare the result of the hand solution to the MATLAB® solution, by examining the graph (Figure 9.9). The first five values for the series match those displayed in the graph. We can also see that the series seems to be converging to approximately 0.69, which is approximately the natural log of 2.

Figure 9.9

The alternating harmonic series converges to the ln(2).

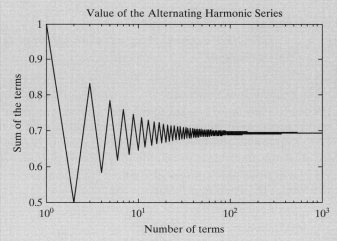

PRACTICE EXERCISES 9.2

Use a `while` loop to solve the following problems:
1. Create a conversion table of inches to feet.
2. Consider the following matrix of values:

$$x = [45, 23, 17, 34, 85, 33]$$

   How many values are greater than 30? (Use a counter.)
3. Repeat Exercise 2, this time using the `find` command.
4. Use a `while` loop to sum the elements of the matrix in Exercise 2. Check your results with the `sum` function. (Use the `help` feature if you don't know or remember how to use `sum`.)
5. Use a `while` loop to create a vector containing the first 10 elements in the harmonic series, i.e.,

$$1/1 \quad 1/2 \quad 1/3 \quad 1/4 \quad 1/5... 1/10$$

6. Use a `while` loop to create a vector containing the first 10 elements in the alternating harmonic series, i.e.,

$$1/1 \quad -1/2 \quad 1/3 \quad -1/4 \quad 1/5...-1/10$$

9.3 BREAK AND CONTINUE

The `break` command can be used to terminate a loop prematurely (while the comparison in the first line is still true). A `break` statement will cause termination of the smallest enclosing `while` or `for` loop. Here's an example:

```
n = 0;
while(n<10)
    n = n+1;
    a = input('Enter a value greater than 0:');
    if(a<=0)
        disp('You must enter a positive number')
        disp('This program will terminate')
        break
    end
    disp('The natural log of that number is')
    disp(log(a))
end
```

INITIALIZE

Define a starting value for a variable that will be changed later

In this program, the value of n is initialized outside the loop. Each time through, the input command is used to ask for a positive number. The number is checked, and if it is zero or negative, an error message is sent to the command window and the program jumps out of the loop. If the value of a is positive, the program continues and another pass through the loop occurs, until n is finally greater than 10.

The `continue` command is similar to `break`; however, instead of terminating the loop, the program just skips to the next pass:

```
n=0;
while(n<10)
```

```
      n=n+1;
      a=input('Enter a value greater than 0:');
      if(a<=0)
         disp('You must enter a positive number')
         disp('Try again')
         continue
      end
      disp('The natural log of that number is')
      disp(log(a))
   end
```

In this example, if you enter a negative number, the program lets you try again—until the value of n is finally greater than 10.

9.4 MIDPOINT BREAK LOOPS

The loops described in the previous sections are examples of *midpoint break loops.* In these constructs the loop is entered, calculations are processed, and a decision is made at some arbitrary point in the loop whether or not to exit. Then additional calculations are processed and the loop repeats. This strategy can be used either with a for loop or a while loop.

In a while structure the loop continues to repeat until the criterion specified in the first line of the loop is false. For example

```
while (x>.001)
   . . . do some calculations that result in updating x
end
```

When the comparison between **x** and 0.001 is evaluated, either a 1 (for true) or a zero (for false) is returned. If the result is 0 the loop is terminated. One potential problem with this structure is that if the original value of **x** is very small, for example, in this case 0.0005, the loop will never execute. A way around this is to force the result to true, and add an if statement and corresponding break structure

```
while(1)
   . . . do some calculations
   if (x<=.001)
      break
   end
   . . . do any additional calculations or information
        processing
end
```

The while(1) implementation allows the loop to continue executing for an infinite number of iterations. The decision to exit the loop is then controlled by the if/break structure. When would this be useful? One example is error checking, similar to the example in the previous problem. Consider another MATLAB® program that prompts the user to enter the number of candy bars purchased, and then finds the cost to the user. If the user enters a negative number the program should prompt the user to try again. If a positive number is entered the program completes the calculations and exits the loop.

```
while(1)
   num_candy_bars = input('Enter the number of candy bars ');
```

```
        if num_candy_bars<0
            disp('Must be a positive number')
        else
            total = num_candy_bars *.75;
            fprintf('The total cost is %5.2f dollars \n',total)
            break
        end
    end
end
```

Here's the command window interaction when the program is executed.

```
Enter the number of candy bars -3
Must be a positive number
Enter the number of candy bars 5
The total cost is 3.75 dollars
```

One issue with this strategy is that the loop need never end. In this program, if the user keeps replying with a negative number, the program will continue to prompt for a positive value. One way to get around this is to use a `for` loop, which has a preset number of iterations. In this example it is three.

```
for k=1:3
  num_candy_bars = input('Enter the number of candy bars');
  if num_candy_bars<0
  disp('Must be a positive number')
else
  total = num_candy_bars *.75;
  fprintf('The total cost is %5.2f dollars \n',total)
  break
  end
end
```

Here's the command window interaction.

```
Enter the number of candy bars -3
Must be a positive number
Enter the number of candy bars -2
Must be a positive number
Enter the number of candy bars -5
Must be a positive number
```

After three iterations the loop ends.

These may seem like trivial examples. A more complicated case is described in Example 9.6.

EXAMPLE 9.6

Calculating the value of the alternating harmonic series in order to approximate the value of $\ln(2)$ (as illustrated in Example 9.5) is an example of a numerical method. Many functions that you use routinely, such as sine and cosine, are approximated using similar series, called Taylor series or Maclaurin series. The alternating harmonic series is an example of a series that converges, but not every series does. For example, simply changing the alternating negative signs in the alternating harmonic series to positive numbers (the harmonic series versus the alternating harmonic series) results in a series that diverges—it just keeps getting bigger and bigger with every new term. In cases such as these, we would want to specify a maximum number of iterations in our problem before giving up and exiting the loop.

A less obvious example of a series that diverges is

$$1-2+3-4+5-6 \ldots +(n-1)-n$$

which can be expressed mathematically as

$$\sum_{k=1}^{n}(-1)^\wedge(k + 1)*k$$

Write a program that calculates the value of the summation. Assume that we don't know that it diverges, and specify an exit from the loop if two adjacent values of the cumulative sum are less than 0.001. Also specify a maximum of 50 iterations.

1. State the Problem
   Calculate the sum of the alternating series, assuming it converges, to within 0.001.
2. Describe the Input and Output

   Input $\sum_{1}^{n}(-1)^\wedge(k+1)*k$

   Output   Find the cumulative sum of the series for each iteration

   Create a plot of the cumulative sums versus the number of terms
3. Develop a Hand Example
   The first six terms in the series are

   $$1 - 2 + 3 - 4 + 5 - 6$$

   Thus, the first six cumulative sums are

$n = 1$	total = 1
$n = 2$	total = $1 - 2 = -1$
$n = 3$	total = $1 - 2 + 3 = 2$
$n = 4$	total = $1 - 2 + 3 - 4 = -2$
$n = 5$	total = $1 - 2 + 3 - 4 + 5 = 3$
$n = 6$	total = $1 - 2 + 3 - 4 + 5 - 6 = -3$

4. Develop a MATLAB® Solution
   Outline your M-file program in a flowchart, as shown in Figure 9.10. Next, convert the flowchart to pseudocode comments, and insert the appropriate MATLAB® code.
   When this program is executed the result in the command window is:

   ```
   The sequence did not converge in 50 iterations
   At which point the value of the series is -25.000
   ```

   The resulting plot is shown in Figure 9.11

5. Test the Solution
The MATLAB® solution matches the hand calculations for the first six terms of the series. If we had not programmed in a maximum number of iterations, in the form of a `for` loop structure, the program would have entered an infinite loop.

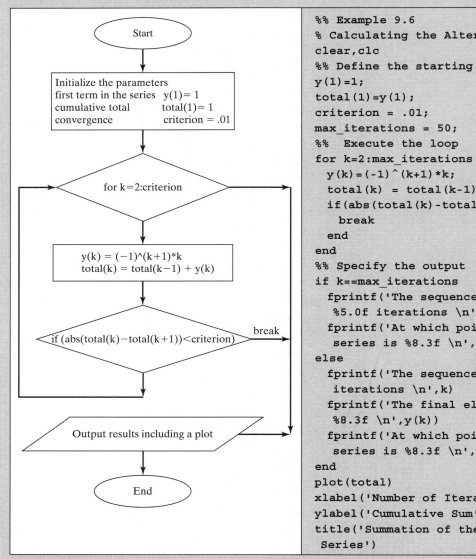

```matlab
%% Example 9.6
% Calculating the Alternating Numeric Series
clear,clc
%% Define the starting parameters
y(1)=1;
total(1)=y(1);
criterion = .01;
max_iterations = 50;
%%  Execute the loop
for k=2:max_iterations
  y(k)=(-1)^(k+1)*k;
  total(k) = total(k-1) + y(k);
  if(abs(total(k)-total(k-1))<criterion)
    break
  end
end
%% Specify the output
if k==max_iterations
  fprintf('The sequence did not converge in
    %5.0f iterations \n',max_iterations)
  fprintf('At which point the value of the
    series is %8.3f \n', total(k))
else
  fprintf('The sequence converged in %5.0f
    iterations \n',k)
  fprintf('The final element is equal to
    %8.3f \n',y(k))
  fprintf('At which point the value of the
    series is %8.3f \n', total(k))
end
plot(total)
xlabel('Number of Iterations')
ylabel('Cumulative Sum')
title('Summation of the Alternating Numeric
 Series')
```

Figure 9.10
Flowchart for calculating the cumulative sums of the alternating numeric series.

Figure 9.11
The cumulative sum of the alternating numeric series does not converge, but rather oscillates around zero.

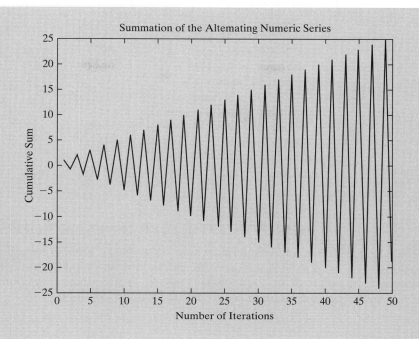

9.5 NESTED LOOPS

It is often useful to nest loops inside other loops. This is actually how many of the MATLAB® built-in functions operate. For example, consider the max function. This function looks for the maximum value for each column in a matrix. We can develop a program to find the maximum, using a simple 4 × 4 array, **x**.

```
x = [1    2    6    3;
     4    8    2    1;
     12   18   3    5;
     6    4    2    13]
```

If we use the max function

```
max(x)
```

MATLAB® returns the maximum value in each column

```
ans =
   12 18  6 13
```

KEY IDEA
Nested loops are used to evaluate multidimensional data

We can achieve the same result with nested for loops. First, we'll need to determine the dimensions of the x array, using the size function.

```
[rows,cols]=size(x);
```

Now, we can use that information to create the external for loop, which we program to execute once for each column in the array. Then, we define a provisional value for the maximum, based on the first value in each column. Finally, we can use an internal `for` loop, which will execute once for each row in the array.

```
for k=1:cols ·············· External for loop
  maximum(k)=x(1,k) ········· Internal loop
    for j=1:rows ············· If structure
      if x(j,k)>maximum(k)
      maximum(k)=x(j,k);
    end ····················
  end ······················
end ·······················
maximum % Sends the results to the screen
```

9.6 IMPROVING THE EFFICIENCY OF LOOPS

In general, using a `for` loop (or a `while` loop) is less efficient in MATLAB® than using array operations. We can test this assertion by timing the multiplication of the elements in a long array. First, we create a matrix A containing 40,000 ones. The `ones` command creates an $n \times n$ matrix of ones:

```
ones(200);
```

The result is a 200×200 matrix of ones (40,000 total values). Now, we can compare the results of multiplying each element by π, using array multiplication first and then a `for` loop. You can time the results by using the `clock` function and the function `etime`, which measures elapsed time. If you have a fast computer, you may need to use a larger array. The structure of the clocking code is

```
t0 = clock;
. . . code to be timed
etime (clock, t0)
```

The clock function polls the computer clock for the current time. The `etime` function compares the current time with the initial time and subtracts the two values to give the elapsed time.

For our problem,

```
clear, clc
A = ones(200); %Creates a 200 x 200 matrix of ones
t0 = clock;
    B = A*pi;
time = etime(clock, t0)
```

gives a result of

```
time =
    0
```

The array calculation took 0 seconds, simply meaning that it happened very quickly. Every time you run these lines of code, you should get a different answer. The `clock` and `etime` functions used here measure how long the CPU worked between receiving the original and final timing requests. However, the CPU is doing other things besides our problem: At a minimum, it is performing system tasks, and it may be running other programs in the background.

To measure the time required to perform the same calculation with a loop, we need to clear the memory and re-create the array of ones:

```
clear
A = ones(200);
```

This ensures that we are comparing calculations from the same starting point. Now, we code

```
t0 = clock;
   for k = 1:length(A(:))
B(k) = A(k)*pi;
      end
time = etime(clock, t0)
```

which gives the result

```
time =
      69.6200
```

It took almost 70 seconds to perform the same calculation! (This was on an older computer—your result will depend on the machine you use.) The number of iterations through the `for` loop was determined by finding how many elements are in A. This was accomplished with the `length` command. Recall that `length` returns the largest array dimension, which is 200 for our array and isn't what we want. To find the total number of elements, we used the colon operator (:) to represent **A** as a single list, 40,000 elements long, and then used `length`, which returned 40,000. Each time through the `for` loop, a new element was added to the B matrix. This is the step that took all the time, since the computer must allocate additional memory 40,000 times. We can reduce the time required for this calculation by creating the B matrix first (so that the memory allocation process takes place only once) and then replacing the values one at a time. The code is

```
clear
A = ones(200);
t0 = clock;
%Create a B matrix of ones
  B = A;
  for k = 1:length(A(:))
  B(k) = A(k)*pi;
  end
time = etime(clock, t0)
```

which gives the result

```
time =
       0.0200
```

This is obviously a huge improvement. You could see an even bigger difference between the first example, a simple multiplication of the elements of an array, and the last example if you created a bigger matrix. By contrast, the intermediate example, in which we did not initialize B, would take a prohibitive amount of time to execute.

MATLAB® also includes a set of commands called `tic` and `toc` that can be used in a manner similar to the `clock` and `etime` functions to time a piece of code. Thus, the code

```
clear
A = ones(200);
tic
  B = A;
  for k = 1:length(A(:))
    B(k) = A(k)*pi;
  end
toc
```

returns

Elapsed time is 0.140000 seconds.

The difference in execution time is expected, since the computer is busy doing different background tasks each time the program is executed. As with `clock/etime`, the `tic/toc` commands measure elapsed time, not the time devoted to just this program's execution.

HINT

Be sure to suppress intermediate calculations when you use a loop. Printing those values to the screen will greatly increase the amount of execution time. If you are brave, repeat the preceding example, but delete the semicolons inside the loop just to check out this claim. Don't forget that you can stop the execution of the program with `Ctrl c`. Be sure the command window is the active window when you execute `Ctrl c`.

SUMMARY

Repetition structures (loops) are used when a section of code needs to be repeated several times. As a rule of thumb, if you find yourself repeating a section of code more than three times, it probably should be in a repetition structure. MATLAB® supports two types of repetition structures: the `for` loop and the `while` loop. In addition, the `break` and `continue` commands can be used to modify either type of loop to create a midpoint break loop.

For loops are used mainly when the programmer knows how many times a sequence of commands should be executed. `While` loops are used when the commands should be executed until a condition is met. Most problems can be structured so that either `for` or `while` loops are appropriate.

The `break` and `continue` statements are used to exit a loop prematurely. They are usually used in conjunction with `if` statements. The `break` command causes a jump out of a loop and execution of the remainder of the program. The `continue` command skips execution of the current pass through a loop, but allows the loop to continue until the completion criterion is met. This type of structure is called *a midpoint break loop*, and is commonly used in many applications, especially in numerical analysis.

Vectorization of MATLAB® code allows it to execute much more efficiently and therefore more quickly. Loops, in particular, should be avoided in MATLAB® if the code can be formulated into a vectorized format. When loops are unavoidable, they can be improved by defining "dummy" variables with placeholder values, such as ones or zeros. These placeholders can then be replaced in the loop. Doing this will result in significant improvements in execution time, a fact that can be confirmed with timing experiments.

The `clock` and `etime` functions are used to poll the computer clock and then determine the time required to execute pieces of code. The time calculated is the "elapsed" time. During this time, the computer not only has been running MATLAB® code, but also has been executing background jobs and housekeeping functions. The `tic` and `toc` functions perform a similar task. Either `tic/toc` or `clock/etime` functions can be used to compare execution time for different code options.

Commands and Functions	
break	causes the execution of a loop to be terminated
case	sorts responses
clock	determines the current time on the CPU clock
continue	terminates the current pass through a loop, but proceeds to the next pass
end	identifies the end of a control structure
etime	finds elapsed time
for	generates a loop structure
ones	creates a matrix of ones
tic	starts a timing sequence
toc	stops a timing sequence
while	generates a loop structure

KEY TERMS

converge	loop	while loop
diverge	repetition	series
for loop	midpoint break loop	vectorization
infinite loop	nested loops	

PROBLEMS

9.1 Use a `for` loop to sum the elements in the following vector:

$$x = [1, 23, 43, 72, 87, 56, 98, 33]$$

Check your answer with the `sum` function.

9.2 Repeat the previous problem, this time using a `while` loop.

9.3 Use a `for` loop to create a vector of the squares of the numbers 1 through 5.

9.4 Use a while loop to create a vector of the squares of the numbers 1 through 5.

9.5 Use the `primes` function to create a list of all the primes below 100. Now use a `for` loop to multiply adjacent values together. For example, the first four prime numbers are

2   3   5   7

Your calculation would be

$$2*3 \quad 3*5 \quad 5*7$$

which gives

$$6 \quad 15 \quad 35$$

Figure P9.6
Chambered nautilus.
(Colin Keates © Dorling
Kindersley, Courtesy of
the Natural History
Museum, London.)

9.6 A Fibonacci sequence is composed of elements created by adding the two previous elements. The simplest Fibonacci sequence starts with 1, 1 and proceeds as follows:

$$1, 1, 2, 3, 5, 8, 13, \ldots$$

However, a Fibonacci sequence can be created with any two starting numbers. Fibonacci sequences appear regularly in nature. For example, the shell of the chambered nautilus (Figure P9.6) grows in accordance with a Fibonacci sequence.

Prompt the user to enter the first two numbers in a Fibonacci sequence and the total number of elements requested for the sequence. Find the sequence and store it in an array by using a `for` loop. Now plot your results on a `polar` graph. Use the element number for the angle and the value of the element in the sequence for the radius.

9.7 Repeat the preceding problem, this time using a `while` loop.

9.8 One interesting property of a Fibonacci sequence is that the ratio of the values of adjacent members of the sequence approaches a number called "the golden ratio" or Φ (phi). Create a program that accepts the first two numbers of a Fibonacci sequence as user input and then calculates additional values in the sequence until the ratio of adjacent values converges to within 0.001. You can do this in a `while` loop by comparing the ratio of element k to element k − 1 and the ratio of element k − 1 to element k − 2. If you call your sequence **x**, then the code for the `while` statement is

while abs(x(k)/x(k-1) - x(k-1)/x(k-2))>0.001

9.9 Recall from trigonometry that the tangent of both $\pi/2$ and $-\pi/2$ is infinity. This may be seen from the fact that

$$\tan(\theta) = \sin(\theta)/\cos(\theta)$$

and since

$$\sin(\pi/2) = 1$$

and

$$\cos(\pi/2) = 0$$

it follows that

$$\tan(\pi/2) = \text{infinity}$$

Because MATLAB® uses a floating-point approximation of π, it calculates the tangent of $\pi/2$ as a very large number, but not infinity.

Prompt the user to enter an angle θ between $\pi/2$ and $-\pi/2$, inclusive. If it is between $\pi/2$ and $-\pi/2$, but not equal to either of those values, calculate $\tan(\theta)$ and display the result in the command window. If it is equal to $\pi/2$ or $-\pi/2$, set the result equal to `Inf` and display the result in the command window. If it is outside the specified range, send the user an error message in the command window and prompt the user to enter another value. Continue prompting the user for a new value of theta until he or she enters a valid number.

9.10 Imagine that you are a proud new parent. You decide to start a college savings plan now for your child, hoping to have enough in 18 years to pay the sharply rising cost of education. Suppose that your folks give you $1000 to get started and that each month you can contribute $100. Suppose also that the interest rate is 6% per year compounded monthly, which is equivalent to 0.5% each month.

Because of interest payments and your contribution, each month your balance will increase in accordance with the formula

$$\text{New balance} = \text{old balance} + \text{interest} + \text{your contribution}$$

Use a `for` loop to find the amount in the savings account each month for the next 18 years. (Create a vector of values.) Plot the amount in the account as a function of time. (Plot time on the horizontal axis and dollars on the vertical axis.)

9.11 Imagine that you have a crystal ball and can predict the percentage increases in tuition for the next 22 years. The following vector `increase` shows your predictions, in percent, for each year:

```
increase = [10, 8, 10, 16, 15, 4, 6, 7, 8, 10, 8, 12,
            14, 15, 8, 7, 6, 5, 7, 8, 9, 8]
```

Use a `for` loop to determine the cost of a 4-year education, assuming that the current cost for 1 year at a state school is $5000.

9.12 Use an `if` statement to compare your results from the previous two problems. Are you saving enough? Send an appropriate message to the command window.

9.13 Edmond Halley (the astronomer famous for discovering Halley's comet) invented a fast algorithm for computing the square root of a number, A. Halley's algorithm approximates $\sqrt{A}$ as follows:

Start with an initial guess x_1. The new approximation is then given by

$$Y_n = \frac{1}{A}x_n^2$$

$$x_{n+1} = \frac{x_n}{8}(15 - y_n(10 - 3y_n))$$

These two calculations are repeated until some convergence criterion, ε, is met.

$$|x_{n+1} - x_n| \le \varepsilon$$

Write a MATLAB® function called `my_sqrt` that approximates the square root of a number. It should have two inputs, the initial guess and the convergence criterion.

Test your function by approximating the square root of 5 and comparing it to the value calculated with the built-in MATLAB® function, `sqrt`.

9.14 The value of $\cos(x)$ can be approximated using a Maclaurin series

$$\cos(x) = 1 - \frac{x^2}{2!} + \frac{x^4}{4!} - \frac{x^6}{6!} + \dots$$

which can be expressed more compactly as

$$\sum_{k=1}^{\infty}(-1)^{k-1}\frac{x^{(k-1)*2}}{((k-1)*2)!}$$

(recall that the symbol ! stands for factorial).

Use a midpoint break loop to determine how many terms must be included in the summation, in order to find the correct value of $\cos(2)$ within an error of .001. Limit the number of iterations to a maximum of 10.

9.15 The value of $\sin(x)$ can be approximated as

$$\sin(x) = x - \frac{x^3}{3!} + \frac{x^5}{5!} - \frac{x^7}{7!} + ...$$

Create a function called my_sin, using a midpoint break loop to approximate the value of $\sin(x)$. Determine convergence by comparing successive values of the summation as you add additional terms. These successive sums should be within an absolute value of 0.001 of each other. Test your function by evaluating the my_sin(2) and comparing it to the built-in MATLAB® sine function.

9.16 A store owner asks you to write a program for use in the checkout process. The program should:

- Prompt the user to enter the cost of the first item.
- Continue to prompt for additional items, until the user enters 0.
- Display the total.
- Prompt for the dollar amount the customer submits as payment.
- Display the change due.

Nested Loops

9.17 In the previous chapter, the water elevation data for Lake Powell were evaluated using the `find` function. Repeat the calculations, using a nested loop structure.

(a) Determine the average elevation of the water level for each year and for the eight-year period over which the data were collected.

(b) Determine how many months each year exceed the overall average for the eight-year period.

(c) Create a report that lists the month (number) and the year for each of the months that exceed the overall average. For example, June is month 6.

(d) Determine the average elevation of the water for each month for the eight-year period.

Faster Loops

9.18 Whenever possible, it is better to avoid using `for` loops, because they are slow to execute.

(a) Generate a 100,000-item vector of random digits called **x**; square each element in this vector and name the result y; use the commands `tic` and `toc` to time the operation.

(b) Next, perform the same operation element by element in a `for` loop. Before you start, clear the values in your variables with

clear x y

Use `tic` and `toc` to time the operation.

Depending on how fast your computer runs, you may need to stop the calculations by issuing the `Ctrl c` command in the command window.

(c) Now convince yourself that suppressing the printing of intermediate answers will speed up execution of the code by allowing these same operations to run and print the answers as they are calculated. You will almost undoubtedly need to cancel the execution of this loop because of the large amount of time it takes. ***Recall that Ctrl c terminates the program.***

(d) If you are going to use a constant value several times in a `for` loop, calculate it once and store it, rather than calculating it each time through the loop. Demonstrate the increase in speed of this process by adding `(sin(0.3) + cos(pi/3))*5!` to every value in the long vector in a `for` loop. (Recall that ! means factorial, which can be calculated with the MATLAB® function `factorial`.)

(e) As discussed in this chapter, if MATLAB® must increase the size of a vector every time through a loop, the process will take more time than if the vector were already the appropriate size. Demonstrate this fact by repeating part (b) of this problem. Create the following vector of **y**-values, in which every element is equal to zero before you enter the `for` loop:

```
y = zeros(1,100000);
```

You will be replacing the zeros one at a time as you repeat the calculations in the loop.

Challenge Problems

9.19 (a) Create a function called `polygon` that draws a polygon in a polar plot. Your function should have a single input parameter—the number of sides.

(b) Use a `for` loop to create a figure with four subplots, showing a triangle in the first subplot, a square in the second subplot, a pentagon in the third subplot, and a hexagon in the fourth subplot. You should use the function you created in part (a) to draw each polygon. Use the index parameter from the `for` loop to specify the subplot in which each polygon is drawn, and in an expression to determine the number of sides used as input to the `polygon` function.

9.20 Consider the following method to approximate the mathematical constant, e. Start by generating K uniform random integers between 1 and K. Compute J, the number of integers between 1 and K, which were never generated. We then approximate e by the ratio

$$\frac{K}{J}$$

Consider the following example for $K = 5$. Assume that the following five integers are randomly generated between 1 and 5.

$$1 \quad 1 \quad 2 \quad 3 \quad 2$$

The number of times the integers are generated is given by

Integers	1	2	3	4	5
Number of instances	2	2	1	0	0

In this example, there are two integers, namely 4 and 5, which were never generated. This means that $J = 2$. Consequently, e is approximated by

$$\frac{5}{2} = 2.5$$

Write a function called `eapprox` that takes the value of K as input, and which then approximates e using the method described above. Test your function several times with different values of K, and compare the result to the value of e calculated using the built-in MATLAB® function.

exp(1)

HINT

Use a rounding function to transform the array of random numbers to random integers.

9.21   Vectorize (replace loops with a single statement) the calculations in the function created in the previous problem, by using the built-in MATLAB® functions `hist` and `sum`.

10

Matrix Algebra

Objectives

After reading this chapter, you should be able to:

- Perform the basic operations of matrix algebra
- Solve simultaneous equations by using

MATLAB® matrix operations
- Use some of MATLAB®'s special matrices

INTRODUCTION

The terms *array* and *matrix* are often used interchangeably in engineering. However, technically, an array is an orderly grouping of information, whereas a matrix is a two-dimensional numeric array used in linear algebra. Arrays can contain numeric information, but they can also contain character data, symbolic data, and so on. Thus, not all arrays are matrices. Only those upon which you intend to perform linear transformations meet the strict definition of a matrix.

Matrix algebra is used extensively in engineering applications. The mathematics of matrix algebra is first introduced in college algebra courses and is extended in linear algebra courses and courses in differential equations. Students start using matrix algebra regularly in statics and dynamics classes.

10.1 MATRIX OPERATIONS AND FUNCTIONS

In this chapter, we introduce MATLAB® functions and operators that are intended specifically for use in matrix algebra. These functions and operators are contrasted with MATLAB®'s array functions and operators, from which they differ significantly. Much of this material may be a review, but is included for completeness.

10.1.1 Transpose

The `transpose` operator changes the rows of a matrix into columns and the columns into rows. In mathematics texts, you will often see the transpose indicated with superscript T (as in A^T). Don't confuse this notation with MATLAB® syntax, however: In MATLAB®, the transpose operator is a single quote (`'`), so that the transpose of matrix A is A`'`.

Consider the following matrix and its transpose:

$$A = \begin{bmatrix} 1 & 2 & 3 \\ 4 & 5 & 6 \\ \boxed{7} & 8 & 9 \\ 10 & 11 & 12 \end{bmatrix} \quad A^T = \begin{bmatrix} 1 & 4 & \boxed{7} & 10 \\ 2 & 3 & 8 & 11 \\ 3 & 6 & 9 & 12 \end{bmatrix}$$

The rows and columns have been switched. Notice that the value in position $(3, 1)$ of A has now moved to position $(1, 3)$ of A^T, and the value in position $(4, 2)$ of A has now moved to position $(2, 4)$ of A^T. In general, the row and column subscripts (also called index numbers) are interchanged to form the transpose.

In MATLAB®, one of the most common uses of the transpose operation is to change row vectors into column vectors. For example:

```
A = [1 2 3];
A'
```

returns

```
A = 1
    2
    3
```

When used with complex numbers, the transpose operation returns the complex conjugate. For example, we may define a vector of negative numbers, take the square root, and then transpose the resulting matrix of complex numbers. Thus, the code

```
x = [-1:-1:-3]
```

returns

```
x =
   -1    -2    -3
```

Then, taking the square root with the code

```
y = sqrt(x)
y =
   0 + 1.0000i    0 + 1.4142i    0 + 1.7321i
```

and finally transposing y

```
y'
```

gives

```
ans =
   0 - 1.0000i
   0 - 1.4142i
   0 - 1.7321i
```

Notice that the results (y`'`) are the complex conjugates of the elements in y.

ARRAY
An orderly grouping of information

MATRIX
A two-dimensional numeric array used in linear algebra

KEY IDEA
The terms array and matrix are often used interchangeably

TRANSPOSE
Switch the positions of the rows and columns

DOT PRODUCT
The sum of the results of the array multiplications of two vectors

10.1.2 Dot Product

The dot product (sometimes called the scalar product) is the sum of the results you obtain when you multiply two vectors together, element by element. Consider the following two vectors:

```
A = [ 1 2 3];
B = [ 4 5 6];
```

The result of the array multiplication of these two vectors is

```
y = A.*B
y =
      4      10      18
```

If you add the elements up, you get the dot product:

```
sum(y)
ans =
      32
```

A mathematics text would represent the dot product as

$$\sum_{i=1}^{n} A_i \cdot B_i$$

which we could write in MATLAB® as

```
sum(A.*B)
```

MATLAB® includes a function called dot to compute the dot product:

```
dot(A,B)
ans =
      32
```

It doesn't matter whether A and B are row or column vectors, just as long as they have the same number of elements.

The dot product finds wide use in engineering applications, such as in calculating the center of gravity (Example 10.1) and in carrying out vector algebra (Example 10.2).

HINT

With dot products, it doesn't matter if both the vectors are rows, both are columns, or one is a row and the other a column. It also doesn't matter what order you use to perform the process: The result of dot (A, B) is the same as that of dot (B, A). This isn't true for most matrix operations.

EXAMPLE 10.1

CALCULATING THE CENTER OF GRAVITY

The mass of a space vehicle is an extremely important quantity. Whole groups of people in the design process keep track of the location and mass of every nut and bolt. Not only is the total mass of the vehicle important, but information about mass is also used to determine the center of gravity of the vehicle. One reason the center

(continued)

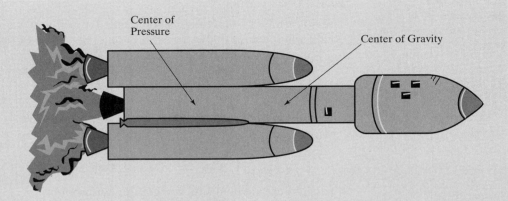

of gravity is important is that rockets tumble if the center of pressure is forward of the center of gravity (Figure 10.1). You can demonstrate the importance of the center of gravity to flight characteristics with a paper airplane. Put a paper clip on the nose of the paper airplane and observe how the flight pattern changes.

Although finding the center of gravity is a fairly straightforward calculation, it becomes more complicated when you realize that both the mass of the vehicle and the distribution of mass change as the fuel is burned.

The location of the center of gravity can be calculated by dividing the vehicle into small components. In a rectangular coordinate system,

$$\bar{x}W = x_1 W_1 + x_2 W_2 + x_3 W_3 + L$$

$$\bar{y}W = y_1 W_1 + y_2 W_2 + y_3 W_3 + L$$

$$\bar{z}W = z_1 W_1 + z_2 W_2 + z_3 W_3 + L$$

where

$\bar{x}$, $\bar{y}$, and $\bar{z}$ are the coordinates of the center of gravity,

W is the total mass of the system,

x_1, x_2, x_3, ... are the x-coordinates of system components 1, 2, 3, ..., respectively,

y_1, y_2, y_3, ... are the y-coordinates of system components 1, 2, 3, ..., respectively,

z_1, z_2, z_3, ... are the z-coordinates of system components 1, 2, 3, ..., respectively, and

W_1, W_2, W_3, ... are the weights of system components 1, 2, 3, ..., respectively.

In this example, we will find the center of gravity of a small collection of the components used in a complicated space vehicle (see Table 10.1). We can formulate this problem in terms of the dot product.

Table 10.1 Vehicle Component Locations and Mass

Item	x, m	y, m	z, m	Mass
Bolt	0.1	2.0	3.0	3.50 g
Screw	1.0	1.0	1.0	1.50 g
Nut	1.5	0.2	0.5	0.79 g
Bracket	2.0	2.0	4.0	1.75 g

1. State the Problem
   Find the center of gravity of the space vehicle.
2. Describe the Input and Output

   Input   Location of each component in an x–y–z coordinate system
             Mass of each component
   Output Location of the center of gravity of the vehicle

3. Develop a Hand Example
   The x-coordinate of the center of gravity is equal to

   $$\bar{x} = \frac{\displaystyle\sum_{i=1}^{3} x_i m_i}{m_{\text{Total}}} = \frac{\displaystyle\sum_{i=1}^{3} x_i m_i}{\displaystyle\sum_{i=1}^{3} m_i}$$

   so, from Table 10.2,

   $$\bar{x} = \frac{6.535}{7.54} = 0.8667 \text{ m}$$

   Notice that the summation of the products of the x-coordinates and the corresponding masses could be expressed as a dot product.
4. Develop a MATLAB® Solution
   The MATLAB® code

   ```
   % Example 10.1
   mass = [3.5, 1.5, 0.79, 1.75];
   x = [0.1, 1, 1.5, 2];
   x_bar = dot(x,mass)/sum(mass)
   y = [2, 1, 0.2, 2];
   y_bar = dot(y,mass)/sum(mass)
   z = [3, 1, 0.5, 4];
   z_bar = dot(z,mass)/sum(mass)
   ```

returns the following result:

   ```
   x_bar =
     0.8667
   y_bar =
     1.6125
   z_bar =
     2.5723
   ```

Table 10.2 Finding the x-Coordinate of the Center of Gravity

Item	x, m		Mass, g	x × m, gm
Bolt	0.1	×	3.50	= 0.35
Screw	1.0	×	1.50	= 1.50
Nut	1.5	×	0.79	= 1.1850
Bracket	2.0	×	1.75	= 3.50
Sum			7.54	6.535

(*continued*)

5. Test the Solution

Compare the MATLAB® solution with the hand solution. The *x*-coordinate appears to be correct, so the *y*- and *z*-coordinates are probably correct, too. Plotting the results would also help us evaluate them:

```
plot3(x,y,z,'o',x_bar,y_bar,z_bar,'s')
grid on
xlabel('x-axis')
ylabel('y-axis')
zlabel('z-axis')
title('Center of Gravity')
axis([0,2,0,2,0,4])
```

The resulting plot is shown in Figure 10.2.

Now that we know the program works, we can use it for any number of items. The program will be the same for three components as for 3000.

Figure 10.2
Center of gravity of some sample data. This plot was enhanced with the use of MATLAB®'s interactive plotting tools.

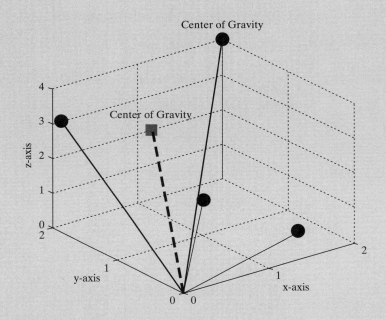

EXAMPLE 10.2

FORCE VECTORS

Statics is the study of forces in systems that don't move (and hence are *static*). These forces are usually described as vectors. If you add the vectors up, you can determine the total force on an object. Consider the two force vectors **A** and **B** shown in Figure 10.3.

Each has a magnitude and a direction. One typical notation would show these vectors as $\vec{A}$ and $\vec{B}$, but would represent the magnitude of each (their physical

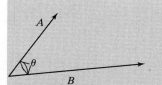

Figure 10.3
Force vectors are used in the study of both statics and dynamics.

length) as A and B. The vectors could also be represented in terms of their magnitudes along the x-, y-, and z-axes, multiplied by a unit vector ($\vec{i}$, $\vec{j}$, $\vec{k}$). Then

$$\vec{A} = A_x \vec{j} + A_y \vec{j} + A_z \vec{k}$$

and

$$\vec{B} = B_x \vec{i} + B_y \vec{j} + B_z \vec{k}$$

The dot product of $\vec{A}$ and $\vec{B}$ is equal to the magnitude of $\vec{A}$ times the magnitude of $\vec{B}$, times the cosine of the angle between them:

$$\vec{A} \cdot \vec{B} = AB \cos(\theta)$$

Finding the magnitude of a vector involves using the Pythagorean theorem. In the case of three dimensions,

$$A = \sqrt{A_x^2 + A_y^2 + A_z^2}$$

We can use MATLAB® to solve problems like this if we define the vector $\vec{A}$ as

```
A = [Ax Ay Az]
```

where `Ax`, `Ay`, and `Az` are the component magnitudes in the x-, y-, and z-directions, respectively. As our MATLAB® problem, use the dot product to find the angle between the following two force vectors:

$$\vec{A} = 5\vec{i} + 6\vec{j} + 3\vec{k}$$
$$\vec{B} = 1\vec{i} + 3\vec{j} + 2\vec{k}$$

1. State the Problem
   Find the angle between two force vectors.
2. Describe the Input and Output
   Input $\vec{A} = 5\vec{i} + 6\vec{j} + 3\vec{k}$
   $\quad\vec{B} = 1\vec{i} + 3\vec{j} + 2\vec{k}$

   Output θ, the angle between the two vectors

3. Develop a Hand Example

$$\vec{A} \cdot \vec{B} = 5 \cdot 1 + 6 \cdot 3 + 3 \cdot 2 = 29$$
$$A = \sqrt{5^2 + 6^2 + 3^2} = 8.37$$
$$B = \sqrt{1^2 + 3^2 + 2^2} = 3.74$$
$$\cos(\theta) = \vec{A} \cdot \vec{B}/AB = 0.9264$$
$$\cos^{-1}(\theta) = 0.386$$

Thus, the angle between the two vectors is 0.386 radians or 22.12 degrees.
4. Develop a MATLAB® Solution
   The MATLAB® code

```
%Example 10.2
%Find the angle between two force vectors
%Define the vectors
```

(*continued*)

```
A = [5 6 3];
B = [1 3 2];
%Calculate the magnitude of each vector
mag_A = sqrt(sum(A.^2));
mag_B = sqrt(sum(B.^2));
%Calculate the cosine of theta
cos_theta = dot(A,B)/(mag_A*mag_B);
%Find theta
theta = acos(cos_theta);
%Send the results to the command window
fprintf('The angle between the vectors is %4.3f radians
  \n',theta)
fprintf('or %6.2f degrees \n',theta*180/pi)
```

generates the following interaction in the command window:

```
The angle between the vectors is 0.386 radians
    or 22.12 degrees
```

5. Test the Solution

In this case, we just reproduced the hand solution in MATLAB®. However, doing so gives us confidence in our solution process. We could expand our problem to allow the user to enter any pair of vectors. Consider this example:

```
%Example 10.2—expanded
%Finding the angle between two force vectors
%Define the vectors
disp('Component magnitudes should be entered')
disp('Using matrix notation, i.e.')
disp('[ A B C]')
A = input('Enter the x y z component magnitudes of vector A: ')
B = input('Enter the x y z component magnitudes of vector B: ')
%Calculate the magnitude of each vector
mag_A = sqrt(sum(A.^2));
mag_B = sqrt(sum(B.^2));
%Calculate the cosine of theta
cos_theta = dot(A,B)/(mag_A*mag_B);
%Find theta
theta = acos(cos_theta);
%Send the results to the command window
fprintf('The angle between the vectors is %4.3f radians
  \n',theta)
fprintf('or %6.2f degrees \n',theta*180/pi)
```

gives the following interaction in the command window:

```
Component magnitudes should be entered
Using matrix notation, i.e.
 [ A B C]
```

```
Enter the x y z component magnitudes of vector A: [1 2 3]
A =
        1    2    3
Enter the x y z component magnitudes of vector B: [4 5 6]
B =
        4    5    6
The angle between the vectors is 0.226 radians or 12.93 degrees
```

PRACTICE EXERCISES 10.1

1. Use the dot function to find the dot product of the following vectors:

$$\vec{A} = [1\,2\,3\,4]$$
$$\vec{B} = [12\,20\,15\,7]$$

2. Find the dot product of $\vec{A}$ and $\vec{B}$ by summing the array products of $\vec{A}$ and $\vec{B}$ (sum(A.*B)).

3. A group of friends went to a local fast-food establishment. They ordered four hamburgers at $0.99 each, three soft drinks at $1.49 each, one milk shake at $2.50, two orders of fries at $0.99 each, and two orders of onion rings at $1.29. Use the dot product to determine the bill.

10.1.3 Matrix Multiplication

Matrix multiplication is similar to the dot product. If you define

```
A = [1 2 3]
B = [ 3;
      4;
      5]
```

then

```
A*B
ans =
  26
```

gives the same result as

```
dot(A,B)
ans =
  26
```

KEY IDEA

Matrix multiplication results in an array in which each element is a dot product

Matrix multiplication results in an array in which each element is a dot product. The preceding example is just the simplest case. In general, the results are found by taking the dot product of each row in matrix A with each column in matrix B. For example, if

```
A = [ 1 2 3;
      4 5 6 ]
```

and

```
B = [ 10 20 30;
      40 50 60;
      70 80 90 ]
```

then the first element of the resulting matrix is the dot product of row 1 in matrix A and column 1 in matrix B, the second element is the dot product of row 1 in matrix A and column 2 in matrix B, and so on. Once the dot product is found for the first row in matrix A with all the columns in matrix B, we start over again with row 2 in matrix A. Thus,

```
C = A*B
```

returns

```
C =
  300 360 420
  660 810 960
```

Consider the result in row 2, column 2, of the matrix C. We can call this result C(2,2). It is the dot product of row 2 of matrix A and column 2 of matrix B:

```
dot(A(2,:), B(:,2))
ans =
    810
```

We could express this relationship in mathematical notation (instead of MATLAB® syntax) as

$$C_{i,j} = \sum_{k=1}^{N} A_{i,k} B_{k,j}$$

Because matrix multiplication is a series of dot products, the number of columns in matrix A must equal the number of rows in matrix B. If matrix A is an $m \times n$ matrix, matrix B must be $n \times p$, and the results will be an $m \times p$ matrix. In this example, A is a 2×3 matrix and B is a 3×3 matrix. The result is a 2×3 matrix.

One way to visualize this set of rules is to write the sizes of the two matrices next to each other, in the order of their operation. In this example, we have

$$2 \times \underline{3 \quad\quad 3} \times 3$$

The two inner numbers must match, and the two outer numbers determine the size of the resulting matrix.

Matrix multiplication is not in general commutative, which means that, in MATLAB®,

$$\mathbf{A} * \mathbf{B} \neq \mathbf{B} * \mathbf{A}$$

We can see this in our example: When we reverse the order of the matrices, we have

$$3 \times \underline{3 \quad\quad 2} \times 3$$

and it is no longer possible to take the dot product of the columns in the first matrix and the rows in the second matrix. If both matrices are square, we can indeed calculate an answer for A * B and an answer for B * A, but the answers are not the same. Consider this example:

```
A = [1 2 3
   4 5 6
   7 8 9];
B = [2 3 4
   5 6 7
   8 9 10];
A*B
ans =
     36     42     48
     81     96    111
    126    150    174
B*A
ans =
     42     51     60
     78     96    114
    114    141    168
```

| EXAMPLE 10.3 |

USING MATRIX MULTIPLICATION TO FIND THE CENTER OF GRAVITY

In Example 10.1, we used the dot product to find the center of gravity of a space vehicle. We could also use matrix multiplication to do the calculation in one step, instead of calculating each coordinate separately. Table 10.1 is repeated in this example for clarity.

1. State the Problem
   Find the center of gravity of the space vehicle.
2. Describe the Input and Output

   Input   Location of each component in an x–y–z coordinate system
           Mass of each component
   Output Location of the center of gravity of the vehicle

3. Develop a Hand Example
   We can create a two-dimensional matrix containing all the information about the coordinates and a corresponding one-dimensional matrix containing

Table 10.1 Vehicle Component Locations and Mass

Item	x, m	y, m	z, m	Mass
Bolt	0.1	2.0	3.0	3.50 g
Screw	1.0	1.0	1.0	1.50 g
Nut	1.5	0.2	0.5	0.79 g
Bracket	2.0	2.0	4.0	1.75 g

(continued)

information about the mass. If there are n components, the coordinate information should be in a $3 \times n$ matrix and the masses should be in an $n \times 1$ matrix. The result would then be a 3×1 matrix representing the x–y–z coordinates of the center of gravity times the total mass.

4. Develop a MATLAB® Solution
   The MATLAB® code

```
% Example 10.3
coord =    [0.1      2      3
            1        1      1
            1.5      0.2    0.5
            2        2      4 ]';
mass = [3.5, 1.5, 0.79, 1.75]';
location=coord*mass/sum(mass)
```

sends the following results to the screen:

```
location =
   0.8667
   1.6125
   2.5723
```

5. Test the Solution
   The results are the same as those in Example 10.1.

PRACTICE EXERCISES 10.2

Which of the following sets of matrices can be multiplied together?

1. $A = \begin{bmatrix} 2 & 5 \\ 2 & 9 \\ 6 & 5 \end{bmatrix}$ $B = \begin{bmatrix} 2 & 5 \\ 2 & 9 \\ 6 & 5 \end{bmatrix}$

2. $A = \begin{bmatrix} 2 & 5 \\ 2 & 9 \\ 6 & 5 \end{bmatrix}$ $B = \begin{bmatrix} 1 & 3 & 12 \\ 5 & 2 & 9 \end{bmatrix}$

3. $A = \begin{bmatrix} 5 & 1 & 9 \\ 7 & 2 & 2 \end{bmatrix}$ $B = \begin{bmatrix} 8 & 5 \\ 4 & 2 \\ 8 & 9 \end{bmatrix}$

4. $A = \begin{bmatrix} 1 & 9 & 8 \\ 8 & 4 & 7 \\ 2 & 5 & 3 \end{bmatrix}$ $B = \begin{bmatrix} 7 \\ 1 \\ 5 \end{bmatrix}$

Show that, for each case, $A \cdot B \neq B \cdot A$.

10.1.4 Matrix Powers

KEY IDEA

A matrix must be square to be raised to a power

Raising a matrix to a power is equivalent to multiplying the matrix by itself the requisite number of times. For example, A^2 is the same as $A \cdot A$, A^3 is the same as $A \cdot A \cdot A$. Recalling that the number of columns in the first matrix of a multiplication must be equal to the number of rows in the second matrix, we see that in order to raise a matrix to a power, the matrix must be square (have the same number of rows and columns). Consider the matrix

$$A = \begin{bmatrix} 1 & 2 & 3 \\ 4 & 5 & 6 \end{bmatrix}$$

If we tried to square this matrix, we would get an error statement because of the rows and columns mismatch:

$$2 \times 3 \qquad 2 \times 3$$

| rows and columns |
| must match |

KEY IDEA

Array multiplication and matrix multiplication are different operations and yield different results

However, consider another example. The code

```
A = randn(3)
```

creates a 3×3 matrix of random numbers, such as

```
A =
   -1.3362   -0.6918   -1.5937
    0.7143    0.8580   -1.4410
    1.6236    1.2540    0.5711
```

HINT

Remember that `randn` produces random numbers, so your computer may produce numbers different from those listed.

If we square this matrix, the result is also a 3×3 matrix:

```
A^2
ans =
   -1.2963   -1.6677    2.2161
   -2.6811   -1.5650   -3.1978
   -0.3463    0.6690   -4.0683
```

Raising a matrix to a noninteger power gives a complex result:

```
A^1.5
ans =
   -1.8446 - 0.0247i   -1.5333 + 0.0153i   -0.3150 - 0.0255i
   -0.7552 + 0.0283i    0.0668 - 0.0176i   -3.0472 + 0.0292i
    1.3359 + 0.0067i    1.5292 - 0.0042i   -1.5313 + 0.0069i
```

Note that raising A to the `matrix` power of two is different from raising A to the `array` power of two:

```
C = A.^2;
```

Raising A to the array power of two produces the following results:

```
C =
    1.7854    0.4786    2.5399
    0.5102    0.7362    2.0765
    2.6361    1.5725    0.3262
```

and is equivalent to squaring each term.

10.1.5 Matrix Inverse

In mathematics, what do we mean when we say "Take the inverse"? For a function, the inverse "undoes" the function, or gets us back where we started. For example, $\sin^{-1}(x)$ is the inverse function of $\sin(x)$. We can demonstrate the relationship in MATLAB®:

> **asin(sin(1.5))**   (Recall that the MATLAB® syntax for the inverse sine is asin.)
>
> **ans =**
>   1.5

HINT

Remember that $\sin^{-1}(x)$ does not mean the same thing as $1/\sin(x)$. Most current mathematics texts use the $\sin^{-1}(x)$ notation, but on your calculator and in computer programs $\sin^{-1}(x)$ is represented as $\text{asin}(x)$.

Another example of functions that are inverses is $\ln(x)$ and e^x:

> **log(exp(3))**   (Recall that the MATLAB® syntax for the natural logarithm is log, not ln.)
>
> **ans =**
>   3

But what does taking the inverse of a number mean? One way to think about it is that if you operated on the number 1 by multiplying it by a number, what could you do to undo this operation and get the number 1 back? Clearly, you'd need to divide by your number, or multiply by 1 over the number. This leads us to the conclusion that $1/x$ and x are inverses, since

KEY IDEA

A function times its inverse is equal to one

$$\frac{1}{x}x = 1$$

These are, of course, *multiplicative* inverses, as opposed to the function inverse we first discussed. (There are also additive inverses, such as $-a$ and a.) Finally, what is the inverse of a matrix? It's the matrix you need to multiply by using matrix algebra to get the identity matrix. The identity matrix consists of ones down the main diagonal and zeros in all the other locations:

$$\begin{bmatrix} 1 & 0 & 0 & 0 \\ 0 & 1 & 0 & 0 \\ 0 & 0 & 1 & 0 \\ 0 & 0 & 0 & 1 \end{bmatrix}$$

The inverse operation is one of the few matrix multiplications that is commutative; that is,

$$A^{-1}A = AA^{-1} = 1$$

In order for the preceding statement to be true, matrix A must be square, which leads us to the conclusion that, in order for a matrix to have an inverse, it must be square.

We can demonstrate these concepts in MATLAB® by first defining a matrix and then experimenting with its behavior. The "magic matrix," in which the sum of the rows equals the sum of the columns, as well as the sum of each diagonal, is easy to create, so we'll choose it for our experiment:

```
A = magic(3)
A =
   8   1   6
   3   5   7
   4   9   2
```

MATLAB® offers two approaches for finding the inverse of a matrix. We could raise A to the -1 power with the code

```
A^-1
ans =
    0.1472   -0.1444    0.0639
   -0.0611    0.0222    0.1056
   -0.0194    0.1889   -0.1028
```

or we could use the built-in function `inv`:

```
inv(A)
ans =
    0.1472   -0.1444    0.0639
   -0.0611    0.0222    0.1056
   -0.0194    0.1889   -0.1028
```

Using either approach, we can show that multiplying the inverse of A by A gives the identity matrix:

```
inv(A)*A
ans =
    1.0000        0        -0.0000
         0   1.0000             0
         0   0.0000        1.0000
```

and

```
A*inv(A)
ans =
    1.0000        0        -0.0000
   -0.0000   1.0000             0
    0.0000        0        1.0000
```

Determining the inverse of a matrix by hand can be difficult, so we'll leave that exercise to a course in matrix mathematics. There are matrices for which an inverse does not exist; these are called **singular matrices** or **ill-conditioned matrices**. When you attempt to compute the inverse of an ill-conditioned matrix in MATLAB®, an error message is sent to the command window.

The matrix inverse is widely used in matrix algebra, although from a computational point of view it is rarely the most efficient way to solve a problem. This subject is discussed at length in linear algebra courses.

10.1.6 Determinants

Determinants are used in linear algebra and are related to the matrix inverse. If the determinant of a matrix is 0, the matrix does not have an inverse, and we say that it is singular. Determinants are calculated by multiplying together the elements along the matrix's left-to-right diagonals and subtracting the product of the right-to-left diagonals. For example, for a 2×2 matrix

$$A = \begin{bmatrix} A_{11} & A_{12} \\ A_{21} & A_{22} \end{bmatrix}$$

the determinant is

$$|A| = A_{11}A_{22} - A_{12}A_{21}$$

Thus, for

$$A = \begin{bmatrix} 1 & 2 \\ 3 & 4 \end{bmatrix}$$

$$|A| = (1)(4) - (2)(3) = -2$$

MATLAB® has a built-in determinant function, det, that will find the determinant for you:

```
A = [1 2;3 4];
det(A)
ans =
    -2
```

Figuring out the diagonals for a 3×3 matrix

$$A = \begin{bmatrix} A_{11} & A_{12} & A_{13} \\ A_{21} & A_{22} & A_{23} \\ A_{31} & A_{32} & A_{33} \end{bmatrix}$$

is a bit harder. If you copy the first two columns of the matrix into columns 4 and 5, it becomes easier to see. Multiply each left-to-right diagonal and add them up:

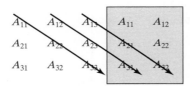

$$(A_{11}A_{22}A_{33}) + (A_{12}A_{23}A_{31}) + (A_{13}A_{21}A_{32})$$

Then multiply each right-to-left diagonal and add them up:

$$(A_{13}A_{22}A_{31}) + (A_{11}A_{23}A_{32}) + (A_{12}A_{21}A_{33})$$

Finally, subtract the second calculation from the first. For example, we might have

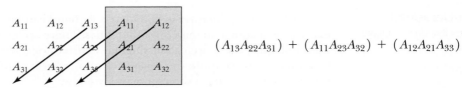

$$|A| = \begin{bmatrix} 1 & 2 & 3 \\ 4 & 5 & 6 \\ 7 & 8 & 9 \end{bmatrix} = (1 \times 5 \times 9) + (2 \times 6 \times 7) + (3 \times 4 \times 8)$$

$$-(3 \times 5 \times 7) - (1 \times 6 \times 8) - (2 \times 4 \times 9) = 225 - 225 = 0$$

Using MATLAB® for the same calculation yields

```
A = [1 2 3;4 5 6;7 8 9];
det(A)
ans =
     0
```

Since we know that matrices with a determinant of zero do not have inverses, let's see what happens when we ask MATLAB® to find the inverse of *A*:

```
inv(A)
Warning: Matrix is close to singular or badly scaled.
    Results may be inaccurate. RCOND = 1.541976e-018.
ans =
  1.0e+016 *
   -0.4504      0.9007     -0.4504
    0.9007     -1.8014      0.9007
   -0.4504      0.9007     -0.4504
```

PRACTICE EXERCISES 10.3

1. Find the inverse of the following magic matrices, both by using the inv function and by raising the matrix to the −1 power:
   (a) magic(3)
   (b) magic(4)
   (c) magic(5)
2. Find the determinant of each of the matrices in Exercise 1.
3. Consider the following matrix:

$$A = \begin{bmatrix} 1 & 2 & 3 \\ 2 & 4 & 6 \\ 3 & 6 & 9 \end{bmatrix}$$

Would you expect it to be singular or not? (Recall that singular matrices have a determinant of 0 and do not have an inverse.)

KEY IDEA
The result of a cross product is a vector

ORTHOGONAL
At right angles

10.1.7 Cross Products

Cross products are sometimes called vector products, because, unlike dot products, which return a scalar, the result of a cross product is a vector. The resulting vector is always at right angles (normal) to the plane defined by the two input vectors—a property that is called *orthogonality*.

Consider two vectors in three-space that represent both a direction and a magnitude. (Force is often represented this way.) Mathematically,

$$\overrightarrow{A} = A_x \overrightarrow{i} + A_y \overrightarrow{j} + A_z \overrightarrow{k}$$
$$\overrightarrow{B} = B_x \overrightarrow{i} + B_y \overrightarrow{j} + B_z \overrightarrow{k}$$

The values A_x, A_y, A_z and B_x, B_y, B_z represent the magnitude of the vector in the *x*, *y*, and *z* directions, respectively. The $\overrightarrow{i}$, $\overrightarrow{j}$, $\overrightarrow{k}$ symbols represent unit vectors in the *x*, *y*, and *z* directions. The *cross product* of $\overrightarrow{A}$ and $\overrightarrow{B}$, $\overrightarrow{A} \times \overrightarrow{B}$, is defined as

$$\vec{A} \times \vec{B} = (A_y B_z - A_z B_y)\vec{i} + (A_z B_x - A_x B_z)\vec{j} + (A_x B_y - A_y B_x)\vec{k}$$

You can visualize this operation by creating a table

$$
\begin{array}{ccc}
i & j & k \\
A_x & A_y & A_z \\
B_x & B_y & B_z
\end{array}
$$

and then repeating the first two columns at the end of the table:

$$
\begin{array}{ccccc}
i & j & k & i & j \\
A_x & A_y & A_z & A_x & A_y \\
B_x & B_y & B_z & B_x & B_y
\end{array}
$$

The component of the cross product in the i direction is found by obtaining the product $A_y B_z$ and subtracting the product $A_z B_y$ from it:

$$
\begin{array}{ccccc}
\textcircled{i} & j & k & i & j \\
A_x & A_y & A_z & A_x & A_y \\
B_x & B_y & B_z & B_x & B_y
\end{array}
$$

Moving across the diagram, the component of the cross product in the j direction is found by obtaining the product $A_z B_x$ and subtracting the product $A_x B_z$ from it:

$$
\begin{array}{ccccc}
i & \textcircled{j} & k & i & j \\
A_x & A_y & A_z & A_x & A_y \\
B_x & B_y & B_z & B_x & B_y
\end{array}
$$

Finally, the component of the cross product in the k direction is found by obtaining the product $A_x B_y$ and subtracting the product $A_y B_x$ from it:

$$
\begin{array}{ccccc}
i & j & \textcircled{k} & i & j \\
A_x & A_y & A_z & A_x & A_y \\
B_x & B_y & B_z & B_x & B_y
\end{array}
$$

HINT

You may have noticed that the cross product is just a special case of a determinant whose first row is composed of unit vectors.

In MATLAB®, the cross product is found using the function `cross`, which requires two inputs: the vectors A and B. Each of these MATLAB® vectors must have three elements, since they represent the vector components in three-space. For example, we might have

```
A = [1 2 3];
```
(which represents $\vec{A} = 1\vec{i} + 2\vec{j} + 3\vec{k}$)
```
B = [4 5 6];
```
(which represents $\vec{B} = 4\vec{i} + 5\vec{j} + 6\vec{k}$)
```
cross(A,B)
ans =
   -3   6   -3
```
(which represents $\vec{C} = -3\vec{i} + 6\vec{j} - 3\vec{k}$)

Consider two vectors in the x–y plane (with no z component):

```
A = [1 2 0]
B = [3 4 0]
```

The magnitude of these vectors in the z direction needs to be specified as zero in MATLAB®.

The result of the cross product must be at right angles to the plane that contains the vectors A and B, which tells us that in this case it must be straight out of the x–y plane, with only a z component.

```
cross(A,B)
ans =
      0      0     -2
```

Cross products find wide use in statics, dynamics, fluid mechanics, and electrical engineering problems.

EXAMPLE 10.4

MOMENT OF A FORCE ABOUT A POINT

The moment of a force about a point is found by computing the cross product of a vector that defines the *position* of the force with respect to a point, with the force vector:

$$M_0 = r \times F$$

Consider the force applied at the end of a lever, as shown in Figure 10.4. If you apply a force to the lever close to the pivot point, the effect is different than if you apply a force further out on the lever. That effect is called the *moment*.

Calculate the moment about the pivot point on a lever for a force described as the vector

$$\vec{F} = -100\vec{i} + 20\vec{j} + 0\vec{k}$$

Assume that the lever is 12 inches long, at an angle of 45° from the horizontal. This means that the position vector can be represented as

$$\vec{r} = \frac{12}{\sqrt{2}}\vec{i} + \frac{12}{\sqrt{2}}\vec{j} + 0\vec{k}$$

1. **State the Problem**

   Find the moment of a force vector about the pivot point of a lever.

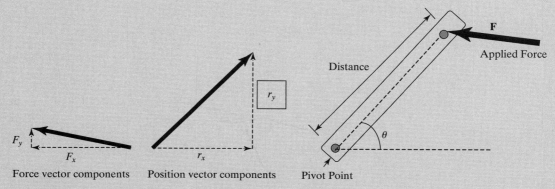

Force vector components   Position vector components   Pivot Point

Figure 10.4
The force applied to a lever creates a moment about the pivot point.

2. Describe the Input and Output

Input   position vector $\vec{r} = \dfrac{12}{\sqrt{2}}\vec{i} + \dfrac{12}{\sqrt{2}}\vec{j} + 0\vec{k}$

   force vector $\vec{F} = -100\vec{i} + 20\vec{j} + 0\vec{k}$

Output   Moment about the pivot point of the lever

3. Develop a Hand Example
Visualize the problem as the determinant of a 3 × 3 array:

$$M_0 = \begin{bmatrix} \vec{i} & \vec{j} & \vec{k} \\ \dfrac{12}{\sqrt{2}} & \dfrac{12}{\sqrt{2}} & 0 \\ -100 & 20 & 0 \end{bmatrix}$$

Clearly, there can be no $\vec{i}$ or $\vec{j}$ component in the answer. The moment must be

$$M_0 = \left(\frac{12}{\sqrt{2}} \times 20 - \frac{12}{\sqrt{2}} \times (-100) \times \vec{k}\right) = 1018.23\vec{k}$$

4. Develop a MATLAB® Solution
The MATLAB® code

```
%Example 10.4
%Moment about a pivot point
%Define the position vector
r = [12/sqrt(2), 12/sqrt(2), 0];
%Define the force vector
F = [-100, 20, 0];
%Calculate the moment
moment=cross(r,F)
```

returns the following result:

```
moment =
    0    0    1018.23
```

This corresponds to a moment vector

$$M_0 = 0\vec{i} + 0\vec{j} + 1018.23\,\vec{k}$$

Notice that the moment is at right angles to the plane defined by the position and force vectors.

5. Test the Solution
Clearly, the hand and MATLAB® solutions match, which means that we can now expand our program to a more general solution. For example, the following program prompts the user for the *x*, *y*, and *z* components of the position and force vectors and then calculates the moment:

```
%Example 10.4
%Moment about a pivot point
```

```
%Define the position vector
  clear,clc
rx = input('Enter the x component of the position vector: ');
ry = input('Enter the y component of the position vector: ');
rz = input('Enter the z component of the position vector: ');
r  = [rx, ry, rz];
  disp('The position vector is')
  fprintf('%8.2f i + %8.2f j + %8.2f k ft\n',r)
%Define the force vector
Fx = input('Enter the x component of the force vector: ');
Fy = input('Enter the y component of the force vector: ');
Fz = input('Enter the z component of the force vector: ');
F  = [Fx, Fy, Fz];
  disp('The force vector is')
  fprintf('%8.2f i + %8.2f j + %8.2f k lbf\n',F)
%Calculate the moment
  moment = cross(r,F);
  fprintf('The moment vector about the pivot point is \n')
  fprintf('%8.2f i + %8.2f j + %8.2f k ft-lbf\n',moment)
```

A sample interaction in the command window is

```
Enter the x component of the position vector: 2
Enter the y component of the position vector: 3
Enter the z component of the position vector: 4
The position vector is
   2.00 i +    3.00 j +    4.00 k ft
Enter the x component of the force vector: 20
Enter the y component of the force vector: 10
Enter the z component of the force vector: 30
The force vector is
  20.00 i +   10.00 j +   30.00 k lbf
The moment vector about the pivot point is
  50.00 i +   20.00 j +   -40.00 k ft-lbf
```

10.2 SOLUTIONS OF SYSTEMS OF LINEAR EQUATIONS

Consider the following system of three equations with three unknowns:

$$
\begin{aligned}
3x &+ 2y &- z &= 10 \\
-x &+ 3y &+ 2z &= 5 \\
x &- y &- z &= -1
\end{aligned}
$$

We can rewrite this system of equations by using the following matrices:

$$
A = \begin{bmatrix} 3 & 2 & 1 \\ -1 & 3 & 2 \\ 1 & -1 & -1 \end{bmatrix} \quad X = \begin{bmatrix} x \\ y \\ z \end{bmatrix} \quad B = \begin{bmatrix} 10 \\ 5 \\ -1 \end{bmatrix}
$$

Using matrix multiplication, we can then write the **system of equations**

$$AX = B.$$

10.2.1 Solution Using the Matrix Inverse

Probably the most straightforward way of solving this system of equations is to use the matrix inverse. Since we know that

$$A^{-1}A = 1$$

we can multiply both sides of the matrix equation $AX = B$ by A^{-1} to get

$$A^{-1}AX = A^{-1}B$$

giving

$$X = A^{-1}B$$

As in all matrix mathematics, the order of multiplication is important. Since A is a 3×3 matrix, its inverse A^{-1} is also a 3×3 matrix. The multiplication $A^{-1}B$

$$3 \times \underbrace{3 \quad 3} \times 1$$

works because the dimensions match up. The result is the 3×1 matrix X. If we change the order to BA^{-1} the dimensions would no longer match, and the operation would be impossible.

Since, in MATLAB®, the matrix inverse is computed with the `inv` function, we can use the following set of commands to solve this problem:

```
A = [3 2 -1; -1 3 2; 1 -1 -1];
B = [10; 5; -1];
X = inv(A)*B
```

This code returns

```
X =
   -2.0000
    5.0000
   -6.0000
```

Alternatively, you could represent the matrix inverse as `A^-1`, so that

```
X = A^-1*B
```

which gives the same result.

```
X =
   -2.0000
    5.0000
   -6.0000
```

Although this technique corresponds well with the approach taught in college algebra classes when matrices are introduced, it is not very efficient and can result in excessive round-off errors. In general, using the matrix inverse to solve linear systems of equations should be avoided.

KEY IDEA

Gaussian elimination is more efficient and less susceptible to round-off error than the matrix inverse method

10.2.2 Solution Using Matrix Left Division

A better way of solving a system of linear equations is to use a technique called *Gaussian elimination*. This is actually the way you probably learned to solve systems of

EXAMPLE 10.5

SOLVING SIMULTANEOUS EQUATIONS: AN ELECTRICAL CIRCUIT*

In solving an electrical circuit problem, one quickly finds oneself mired in a large number of simultaneous equations. For example, consider the electrical circuit shown in Figure 10.5.

Figure 10.5
An electrical circuit.

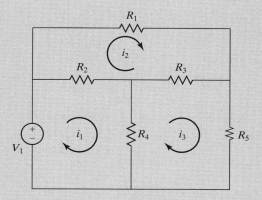

It contains a single voltage source and five resistors. You can analyze this circuit by dividing it up into smaller pieces and using two basic facts about electricity:

$$\sum voltage \text{ around a circuit must be zero (Kirchhoff's second law—see Figure 10.6)}$$

$$voltage = current \times resistance \ (V = iR)$$

Following the lower left-hand loop results in our first equation:

$$-V_1 + R_2(i_1 - i_2) + R_4(i_1 - i_3) = 0$$

Figure 10.6
Gustav Kirchhoff was a German physicist, who formulated many of the basic laws of circuit theory.

---

*From *Introduction to MATLAB® 7*, by Etter, Kuncicky, and Moore (Upper Saddle River, NJ: Pearson Prentice Hall, 2005).

Following the upper loop results in our second equation:

$$R_1 i_2 + R_3 (i_2 - i_3) + R_2 (i_2 - i_1) = 0$$

Finally, following the lower right-hand loop results in the last equation:

$$R_3 (i_3 - i_2) + R_5 i_3 + R_4 (i_3 - i_1) = 0$$

Since we know all the resistances (the R values) and the voltage, we have three equations and three unknowns. Now we need to rearrange the equations so that they are in a form to which we can apply a matrix solution. In other words, we need to isolate the i's as follows:

$$(R_2 + R_4) i_1 + (-R_2) i_2 + (-R_4) i_3 = V_1$$
$$(-R_2) i_1 + (R_1 + R_2 + R_3) i_2 + (-R_3) i_3 = 0$$
$$(-R_4) i_1 + (-R_3) i_2 + (R_3 + R_4 + R_5) i_3 = 0$$

Create a MATLAB® program to solve these equations, using the matrix inverse method. Allow the user to enter the five values of R and the voltage from the keyboard.

1. State the Problem
   Find the three currents for the circuit shown.
2. Describe the Input and Output

   Input   Five resistances R_1, R_2, R_3, R_4, R_5, and the voltage V, provided from the keyboard

   Output   Three current values i_1, i_2, i_3

3. Develop a Hand Example
   If there is no applied voltage in a circuit, there can be no current, so if we enter any value for the resistances and enter zero for the voltage, the answer should be zero.
4. Develop a MATLAB® Solution
   The MATLAB® code

```
%Example 10.5
%Finding Currents
clear,clc
R1 = input('Input the value of R1: ');
R2 = input('Input the value of R2: ');
R3 = input('Input the value of R3: ');
R4 = input('Input the value of R4: ');
R5 = input('Input the value of R5: ');
V = input('Input the value of voltage, V: ');
coef = [(R2+R4), -R2, -R4;
        -R2, (R1 + R2 + R3), (-R3);
        -R4, - R3,(R3 + R4 + R5)];
result = [V; 0; 0];
I = inv(coef)*result
```

generates the following interaction in the command window:

```
Input the value of R1: 5
Input the value of R2: 5
Input the value of R3: 5
Input the value of R4: 5
Input the value of R5: 5
Input the value of voltage, V: 0
I =

        0

        0

        0
```

5. **Test the Solution**

We purposely chose to enter a voltage of zero in order to check our solution. Circuits without a driving force (voltage) cannot have a current flowing through them. Now try the program with other values:

```
Input the value of R1: 2
Input the value of R2: 4
Input the value of R3: 6
Input the value of R4: 8
Input the value of R5: 10
Input the value of voltage, V: 10
```

Together, these values give

```
I =
      1.69
      0.97
      0.81
```

equations in college algebra. Gaussian elimination was developed by Carl Friedrich Gauss, a German mathematician and scientist (see Figure 10.7).

Consider our problem of three equations in x, y, and z:

$$
\begin{array}{rrrcr}
3x & +2y & -z & = & 10 \\
-x & +3y & +2z & = & 5 \\
x & -y & -z & = & -1
\end{array}
$$

To solve this problem by hand, we would first consider the first two equations in the set and eliminate one of the variables—for example, x. To do this, we'll need to multiply the second equation by 3 and then add the resulting equation to the first one:

$$
\begin{array}{rrrcr}
3x & +2y & -z & = & 10 \\
-3x & +9y & +6z & = & 15 \\
\hline
0 & +11y & -5z & = & 25
\end{array}
$$

Now, we need to repeat the process for the second and third equations:

$$
\begin{array}{rrrrr}
-x & +3y & -2z & = & 5 \\
x & -y & -z & = & -1 \\
\hline
0 & +2y & +z & = & 4
\end{array}
$$

At this point, we've eliminated one variable and reduced our problem to two equations and two unknowns:

$$
\begin{array}{rrrr}
11y & +5z & = & 25 \\
2y & +z & = & 4
\end{array}
$$

Now, we can repeat the elimination process by multiplying row 3 by $-11/2$:

$$
\begin{array}{rrrr}
11y & +5z & = & 25 \\
-\dfrac{11}{2} * 2y & -\dfrac{11}{2}z & = & -\dfrac{11}{2} * 4 \\
\hline
0 & -\dfrac{1}{2}z & = & 3
\end{array}
$$

Finally, we can solve for z:

$$z = -6$$

Once we know the value of z, we can substitute back into either of the two equations in just z and y—namely,

$$
\begin{array}{rr}
11y +5z = 25 \\
2y +z = 4
\end{array}
$$

to find that

$$y = 5$$

The last step is to substitute back into one of our original equations,

$$
\begin{array}{rrrrr}
3x & +2y & -z & = & 10 \\
-x & +3y & +2z & = & 5 \\
x & -y & -z & = & -1
\end{array}
$$

to find that

$$x = -2$$

GAUSSIAN ELIMINATION

An organized approach to eliminating variables and solving a set of simultaneous equations

The technique of Gaussian elimination is an organized approach to eliminating variables until only one unknown exists and then substituting back until all the unknowns are determined. In MATLAB®, we can use left division to solve the problem by Gaussian elimination. Thus,

```
X = A\B
```

returns

```
X =
   -2.0000
    5.0000
   -6.0000
```

Clearly, this is the same result we obtained with the hand solution and the matrix inverse approach.

MATLAB® is also capable of solving problems which are either overdefined or underdefined using left division. Consider, for example, the following problem:

$$
\begin{aligned}
3*x \quad +2*y \quad +5*z &= 22 \\
4*x \quad +5*y \quad -2*z &= 8 \\
x \quad +y \quad +z &= 6
\end{aligned}
$$

This problem is appropriately defined with three equations and three unknowns. When formulated as

```
A = [3    2    5
     4    5   -2
     1    1    1]
```

and

```
B = [22;  8;  6]
```

the left division operator can be used to solve for x, y, and z

```
X = A\B
```

which results in the solution

```
X =
   1
   2
   3
```

Suppose, however, that we knew four equations relating x, y, and z, such as

```
3*x +2*y +5*z =  22
4*x +5*y -2*z =   8
  x   +y   +z =   6
2*x -4*y -7*z = -27
```

Now, we have four equations and three unknowns and the problem is overdefined. We can still solve it using the left division operator. The coefficient matrix is defined as

```
A = [3   2   5
     4   5  -2
     1   1   1
     2  -4  -7]
```

and the result matrix as

```
B = [22; 8; 6; -27]
```

When we execute the statement

```
X = A\B
```

we get the same result, because the equations were consistent.

```
X =
   1
   2
   3
```

However, it is possible when gathering data that there might be small errors that result in different numbers in the result matrix. Assume that instead the fourth equation tells us that the result is -28, instead of -27. This means that we'll need to adjust the B vector

```
B = [22; 8; 6; -28]
```

Now, when we execute

```
X = A\B
```

the result is

```
X =
   0.8618
   2.1234
   3.0328
```

MATLAB® uses a least squared approach to find the set of X values (which correspond to x, y, z in our equations), which is the best match to the equations. If we use these values to find B

```
A*X
```

The result is

```
ans =
    21.9962
     7.9982
     6.0180
   -27.9997
```

The least squared approach minimizes the absolute value of the difference between the calculated B values and the actual B values. This approach is described in a later chapter on numerical methods.

What if your system of equations is underdefined? For example, what if we only had two equations for three unknowns?

```
3*x +2*y +5*z = 22
4*x +5*y -2*z =  8
```

In this case we'd define the coefficient matrix as

```
A = [3 2   5
     4 5 -2]
```

and the result matrix as

```
B = [22; 8]
```

MATLAB® solves the problem by setting the first variable equal to 0, which effectively reduces the problem to two equations and two unknowns.

```
X = A\B
```

which results in

```
X =
   0
   2.8966
   3.2414
```

This is only one of an infinite number of possible solutions, but it does give the correct answer if we substitute back into our equation

```
A*X
ans =
    22.0000
    8.0000
```

10.2.3 Solution Using the Reverse Row Echelon Function

In a manner similar to left division we could solve the system of linear equations

$$
\begin{array}{rrrrr}
3x & +2y & -z & = & 10 \\
-x & +3y & +2z & = & 5 \\
x & -y & -z & = & -1
\end{array}
$$

using the reduced row echelon function, rref. Recall that we can rewrite this system of equations by using the following matrices:

$$
A = \begin{bmatrix} 3 & 2 & 1 \\ -1 & 3 & 2 \\ 1 & -1 & -1 \end{bmatrix} \quad X = \begin{bmatrix} x \\ y \\ z \end{bmatrix} \quad B = \begin{bmatrix} 10 \\ 5 \\ -1 \end{bmatrix}
$$

The rref function requires an expanded matrix as input, representing the coefficients and results. For our example system of equations the input would be

```
C = [A,B]
C =
   3   2  -1   10
  -1   3   2    5
   1  -1  -1   -1
rref(C)
ans =
   1   0   0   -2
   0   1   0    5
   0   0   1   -6
```

The solution to our problem is represented by the last column in the output array, and corresponds to the results achieved with the other methods.

In a simple problem like this, no matter which technique we use, round-off error and execution time are not big factors. However, some numerical techniques require the solution of matrices with thousands or even millions of elements. Execution times are measured in hours or days for these problems, and round-off error and execution time become critical considerations. For such problems the matrix inverse technique is not appropriate.

Not all systems of linear equations have a unique solution. If there are fewer equations than variables, the problem is underspecified. If there are more equations than variables, the problem is overspecified. MATLAB® includes functions that will allow you to solve each of these systems of equations, by using numerical best-fit approaches or adding constraints. Consult the MATLAB® help function for more information on these techniques.

EXAMPLE 10.6

MATERIAL BALANCES ON A DESALINATION UNIT: SOLVING SIMULTANEOUS EQUATIONS

Freshwater is a scarce resource in many parts of the world. For example, Israel supports a modern industrial society in the middle of a desert. To supplement local water sources, Israel depends on water desalination plants along the Mediterranean coast. Current estimates predict that the demand for freshwater in Israel will increase to 60% by the year 2020, and most of that new water will have to come from desalination. Modern desalination plants use reverse osmosis, the process used in kidney dialysis! Chemical engineers make wide use of material-balance calculations to design and analyze plants such as the water desalination plants in Israel.

Consider the hypothetical desalination unit shown in Figure 10.8. The salty water flowing into the unit contains 4 wt% salt and 96 wt% water. Inside the unit, the water is separated into two streams by a series of reverse-osmosis operations. The stream flowing out the top is almost pure water. The remaining concentrated solution of salty water is 10 wt% salt and 90 wt% water. Calculate the mass flow rates coming out of the top and bottom of the desalination unit.

Figure 10.8
Water desalination is an important source of freshwater for desert nations such as Israel.

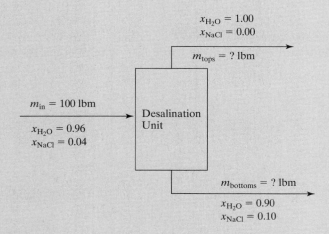

$x_{H_2O} = 1.00$
$x_{NaCl} = 0.00$
$m_{tops} = ?\ \text{lbm}$

$m_{in} = 100\ \text{lbm}$
$x_{H_2O} = 0.96$
$x_{NaCl} = 0.04$

Desalination Unit

$m_{bottoms} = ?\ \text{lbm}$
$x_{H_2O} = 0.90$
$x_{NaCl} = 0.10$

This problem requires us to perform a material balance on the reactor for both the salt and the water. The amount of any component flowing into the reactor must be the same as the amount of that component flowing out in the two exit streams. That is,

$$m_{\mathrm{inA}} = m_{\mathrm{topsA}} + m_{\mathrm{bottomsA}}$$

which could be rewritten as

$$x_A m_{\mathrm{in\ total}} = x_{\mathrm{Atops}} m_{\mathrm{tops}} + x_{\mathrm{Abottoms}} m_{\mathrm{bottoms}}$$

Thus, we can formulate this problem as a system of two equations in two unknowns:

$$0.96 \times 100 = 1.00 m_{\mathrm{tops}} + 0.90 m_{\mathrm{bottoms}} \ (\text{for water})$$
$$0.04 \times 100 = 0.00 m_{\mathrm{tops}} + 0.10 m_{\mathrm{bottoms}} \ (\text{for salt})$$

1. **State the Problem**
   Find the mass of freshwater produced and the mass of brine rejected from the desalination unit.
2. **Describe the Input and Output**

   Input Mass of 100 lb into the system
   Concentrations (mass fractions) of the input stream:

   $$x_{\mathrm{H2O}} = 0.96$$
   $$x_{\mathrm{NaCl}} = 0.04$$

   Concentrations (mass fractions) in the output streams:
   water-rich stream (tops)

   $$x_{\mathrm{H2O}} = 1.00$$

   brine (bottoms)

   $$x_{\mathrm{H2O}} = 0.90$$
   $$x_{\mathrm{NaCl}} = 0.10$$

   Output Mass out of the water-rich stream (tops)
   Mass out of the brine (bottoms)

3. **Develop a Hand Example**
   Since salt (NaCl) is present only in one of the outlet streams, it is easy to solve the following system of equations:

   $$(0.96)(100) = 1.00 m_{\mathrm{tops}} + 0.90 m_{\mathrm{bottoms}} \ (\text{for water})$$
   $$(0.04)(100) = 0.00 m_{\mathrm{tops}} + 0.10 m_{\mathrm{bottoms}} \ (\text{for salt})$$

   Starting with the salt material balance, we find that

   $$4 = 0.1 m_{\mathrm{bottoms}}$$
   $$m_{\mathrm{bottoms}} = 40 \ \mathrm{lbm}$$

   Once we know the value of m_{bottoms} we can substitute back into the water balance:

   $$96 = 1 m_{\mathrm{tops}} + (0.90)(40)$$
   $$m_{\mathrm{tops}} = 60 \ \mathrm{lb}$$

4. **Develop a MATLAB® Solution**

We can use matrix mathematics to solve this problem, once we realize it is of the form

$$AX = B$$

where A is the coefficient matrix and thus the mass fractions of the water and salt. The result matrix, B, consists of the mass flow rate into the system of water and salt:

$$A = \begin{bmatrix} 1 & 0.9 \\ 0 & 0.1 \end{bmatrix} \quad B = \begin{bmatrix} 96 \\ 4 \end{bmatrix}$$

The matrix of unknowns, X, consists of the total mass flow rates out of the top and bottom of the desalination unit. Using MATLAB® to solve this system of equations requires only three lines of code:

```
A = [1, 0.9; 0, 0.1];
B = [96; 4];
X = A\B
```

This code returns

```
X =
    60
    40
```

5. **Test the Solution**

Notice that in this example we chose to use matrix left division. Using the matrix inverse approach gives the same result:

```
X = inv(A)*B
X =
    60
    40
```

The results from both approaches match that from the hand example, but one additional check can be made to verify the results. We performed material balances based on water and on salt, but an additional balance can be performed on the *total* mass in and out of the system:

$$m_{in} = m_{tops} + m_{bottoms}$$
$$m_{in} = 40 + 60 = 100$$

Verifying that 100 lbm actually exits the system serves as one more confirmation that we performed the calculations correctly.

Although it was easy to solve the system of equations in this problem by hand, most real material-balance calculations include more process streams and more components. Matrix solutions such as the one we created are an important tool for chemical-process engineers.

EXAMPLE 10.7

A FORCE BALANCE ON A STATICALLY DETERMINATE TRUSS

A statically determinate truss is one of the early problems addressed in sophomore Statics classes. A typical problem is shown in Figure 10.9.

Figure 10.9
A simple statically determinate truss.

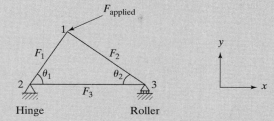

At the hinge (point 2) the truss cannot move in either the x or the y direction. At the roller (point 3) movement is allowed in the x direction, but not in the y direction. This results in reactive forces at point 2 in both the x and the y directions, and at point 3 in just the y direction. If we also separate the applied force (at point 1) into x and y components, the freebody diagram can be draw as shown in Figure 10.10.

Figure 10.10
Freebody diagram for a statically determinate truss.

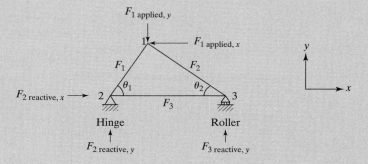

Because we assume that the truss is not moving, the sum of the forces at each of the nodes (1, 2, and 3) must be zero in both the x and the y directions. This gives us a total of six equations.

$$\sum F_{\text{at node 1, } x \text{ direction}} = 0 = -F_1 \cos(\theta_1) + F_2 \cos(\theta_2) + F_{1 \text{ applied, } x}$$
$$\sum F_{\text{at node 1, } y \text{ direction}} = 0 = -F_1 \sin(\theta_1) - F_2 \sin(\theta_2) + F_{1 \text{ applied, } y}$$
$$\sum F_{\text{at node 2, } x \text{ direction}} = 0 = +F_{2\text{reactive, } x} + F_1 \cos(\theta_1) + F_3$$
$$\sum F_{\text{at node 2, } y \text{ direction}} = 0 = +F_{2\text{reactive, } y} + F_1 \sin(\theta)$$
$$\sum F_{\text{at node 3, } x \text{ direction}} = 0 = -F_2 \cos(\theta_2) - F_3$$
$$\sum F_{\text{at node 3, } y \text{ direction}} = 0 = -F_2 \sin(\theta_2) - F_{3 \text{ reactive, } y}$$

If the applied force is known, as well as the angles, this results in six equations and six unknowns ($F_1, F_2, F_3, F_{2 \text{ reactive, } x}, F_{2 \text{ reactive, } y}$, and $F_{3 \text{ reactive, } y}$). It turns out that with a little rearranging, we can see that this is a linear system of equations.

$$-\cos(\theta_1)*F_1 + \cos(\theta_2)*F_2 + 0*F_3 + 0*F_{2\ reactive,\ x} + 0*F_{2\ reactive,\ y} + 0*F_{3\ reactive,\ y} = -F_{1\ applied,\ x}$$

$$-\sin(\theta_1)*F_1 - \sin(\theta_2)*F_2 + 0*F_3 + 0*F_{2\ reactive,\ x} + 0*F_{2\ reactive,\ y} + 0*F_{3\ reactive,\ y} = -F_{1\ applied,\ y}$$

$$\cos(\theta_1)*F_1 + 0*F_2 + 1*F_3 + 1*F_{2\ reactive,\ x} + 0*F_{2\ reactive,\ y} + 0*F_{3\ reactive,\ y} = 0$$

$$\sin(\theta_1)*F_1 + 0*F_2 + 0*F_3 + 0*F_{2\ reactive,\ x} + 1*F_{2\ reactive,\ y} + 0*F_{3\ reactive,\ y} = 0$$

$$+0*F_1 - \cos(\theta_2)*F_2 - 1*F_3 + 0*F_{2\ reactive,\ x} + 0*F_{2\ reactive,\ y} + 0*F_{3\ reactive,\ y} = 0$$

$$+0*F_1 + \sin(\theta_2)*F_2\ 1*F_3 + 0*F_{2\ reactive,\ x} + 0*F_{2\ reactive,\ y} + 1*F_{3\ reactive,\ y} = 0$$

This system can be expressed, using matrix notation as:

$$
\begin{vmatrix}
-\cos(\theta_1) & \cos(\theta_2) & 0 & 0 & 0 & 0 \\
-\sin(\theta_1) & -\sin(\theta_2) & 0 & 0 & 0 & 0 \\
\cos(\theta_1) & 0 & 1 & 1 & 0 & 0 \\
\sin(\theta_1) & 0 & 0 & 0 & 1 & 0 \\
0 & -\cos(\theta_2) & -1 & 0 & 0 & 0 \\
0 & \sin(\theta_2) & 0 & 0 & 0 & 1
\end{vmatrix}
*
\begin{vmatrix}
F_1 \\
F_2 \\
F_3 \\
F_{2\ reactive,\ x} \\
F_{2\ reactive,\ y} \\
F_{3\ reactive,\ y}
\end{vmatrix}
=
\begin{vmatrix}
-F_{1\ applied,\ x} \\
-F_{1\ applied,\ y} \\
0 \\
0 \\
0 \\
0
\end{vmatrix}
$$

Now that we've derived the appropriate equations, solve this system for the case where:

$$\theta_1 = 45°,$$
$$\theta_2 = 45°$$

and the applied load at node 1 is 1000 lbf in the negative vertical direction.

1. State the Problem
   Find the loads experienced on the truss, shown in Figure 10.10.
2. Describe the Input and Output

   Input Negative vertical load at node 1 of 1000 lbf

   $$\theta_1 = 45°$$
   $$\theta_2 = 45°$$

   Output Force experienced in each beam of the truss, F_1, F_2, and F_3, the reactive forces at the hinge, $F_{2\ reactive,\ x}$ and $F_{2\ reactive,\ y}$, and the reactive force at the roller, $F_{3\ reactive,\ y}$.

3. Develop a Hand Example
   Substituting into the matrix previously derived gives

$$
\begin{vmatrix}
-0.7071 & +0.7071 & 0 & 0 & 0 & 0 \\
-0.7071 & -0.7071 & 0 & 0 & 0 & 0 \\
+0.7071 & 0 & 1 & 1 & 0 & 0 \\
+0.7071 & 0 & 0 & 0 & 1 & 0 \\
0 & -0.7071 & -1 & 0 & 0 & 0 \\
0 & +0.7071 & 0 & 0 & 0 & 1
\end{vmatrix}
*
\begin{vmatrix}
F_1 \\
F_2 \\
F_3 \\
F_{2\ reactive,\ x} \\
F_{2\ reactive,\ y} \\
F_{3\ reactive,\ y}
\end{vmatrix}
=
\begin{vmatrix}
0 \\
1000 \\
0 \\
0 \\
0 \\
0
\end{vmatrix}
$$

Figure 10.11
Freebody diagram for
a balanced truss.

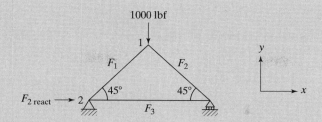

We could solve this equation using matrix algebra, however, an examination of
the truss in Figure 10.11 leads to a more simple solution. Notice that there is no
horizontal applied force. The reactive force resulting at node 2 must therefore
be zero. Because the geometry of the truss is symmetrical that also leads to the
conclusion that nodes 2 and 3 must also experience the same load—hence, in
order for the net vertical force to equal zero $F_{2\ reactive,\ y}$ and $F_{3\ reactive,\ y}$, must
both be 500 lbf. We've now determined three of the unknowns,

$$F_{2\ reactive,\ x} = 0$$
$$F_{2\ reactive,\ y} = 0$$
$$F_{3\ reactive,\ y} = 0$$

Examining the set of equations we notice that the force balance in the vertical
direction at node 2 can now be solved

$$\sum F_{\text{at node 2, } y\text{ direction}} = 0 = +F_{2\ reactive,\ y} + F_1 \sin(\theta_1)$$
$$\sum F_{\text{at node 2, } y\text{ direction}} = 0 = +500 + F_1 \sin(45°)$$
$$F_1 = \frac{-500}{\sin(45°)} = -707.1 \text{ lbf similarly...}$$
$$F_2 = -707.1 \text{ lbf}$$

Finally, we can use the balance at node 3 in the horizontal direction to give:

$$\sum F_{\text{at node 3, } x\text{ direction}} = 0 = -F_2\cos(\theta_2) - F_3$$
$$F_3 = -F_2\cos(\theta_2) = 707.1*\cos(45°) = 500$$

4. Develop a MATLAB® Solution
   We can develop a general solution to this problem, and use the given data to
   check it.

```
theta1=45 % angle in degrees
theta2=45 % angle in degrees
F1x=0 % horizontal load
F1y=-1000 % vertical load
A=[-cosd(theta1),cosd(theta2),0,0,0,0
-sind(theta1),-sind(theta2),0,0,0,0
cosd(theta1),0,1,1,0,0
sind(theta1),0,0,0,1,0
0,-cosd(theta2),-1,0,0,0
0,sind(theta2),0,0,0,1]
B=[F1x,-F1y,0,0,0,0]'
```

```
x=(A\B )' % use left division
```

This code returns the result

```
x =
   -707.11   -707.11   500.00   0   500.00   500.00
```

which corresponds to the hand solution.

5. Test the Solution

Notice that in this example we chose to use matrix left division. Using the matrix inverse approach gives the same result:

```
x =( inv(A)*B)'
```

returns the following to the command window

```
x =
   -707.11 -707.11 500.00 0 500.00 500.00
```

The results from both approaches match that from the hand example, which did not depend on matrix algebra. Now, we can use the same program to analyze the truss at different conditions. For example, assume the following...

$$\theta_1 = 30°$$
$$\theta_2 = 60°$$

and an applied load of 1000 lbf in the horizontal direction at node 1. The MATLAB® code would be modified to read...

```
theta1=30 % angle in degrees
theta2=60 % angle in degrees
F1x=1000 % horizontal load
F1y=0 % vertical load
A=[-cosd(theta1),cosd(theta2),0,0,0,0
-sind(theta1),-sind(theta2),0,0,0,0
cosd(theta1),0,1,1,0,0
sind(theta1),0,0,0,1,0
0,-cosd(theta2),-1,0,0,0
0,sind(theta2),0,0,0,1]
B=[F1x,-F1y,0,0,0,0]'
x=inv(A)*B
x=A\B
```

giving a result of

```
x =
   -866.03 500.00 -250.00 1000.00 433.01 -433.01
```

Notice that the fourth value in the array, which corresponds to the reactive force in the *x* direction at node 2 is 1000, just what we would expect.

10.3 SPECIAL MATRICES

MATLAB® contains a group of functions that generate special matrices, some of which we review in this section.

10.3.1 Ones and Zeros

The `ones` and `zeros` functions create matrices consisting entirely of ones and zeros, respectively. When a single input is used, the result is a square matrix. When two inputs are used, they specify the number of rows and columns. For example,

```
ones(3)
```

returns

```
ans =
     1    1    1
     1    1    1
     1    1    1
```

and

```
zeros(2,3)
```

returns

```
ans =
     0    0    0
     0    0    0
```

If more than two inputs are specified in either function, MATLAB® creates a multi-dimensional matrix. For instance,

```
ones(2,3,2)
ans(:,:,1) =
            1.00    1.00    1.00
            1.00    1.00    1.00
ans(:,:,2) =
            1.00    1.00    1.00
            1.00    1.00    1.00
```

creates a three-dimensional matrix with two rows, three columns, and two pages.

10.3.2 Identity Matrix

An identity matrix is a matrix with ones on the main diagonal and zeros everywhere else. For example, here is an identity matrix with four rows and four columns:

$$\begin{bmatrix} 1 & 0 & 0 & 0 \\ 0 & 1 & 0 & 0 \\ 0 & 0 & 1 & 0 \\ 0 & 0 & 0 & 1 \end{bmatrix}$$

Note that the main diagonal contains elements in which the row number is the same as the column number. The subscripts for elements on the main diagonal are (1, 1), (2, 2), (3, 3), and so on.

In MATLAB®, identity matrices can be generated with the eye function. The arguments of the eye function are similar to those of the zeros and the ones functions. If the argument of the function is a scalar, as in eye(6), the function will generate a square matrix, using the argument as both the number of rows and the number of columns. If the function has two scalar arguments, as in `eye(m,n)`, the function will generate a matrix with *m* rows and *n* columns. To generate an identity matrix that is the same size as another matrix, use the `size` function to determine the correct number of rows and columns. Although most applications use a square identity matrix, the definition can be extended to nonsquare matrices. The following statements illustrate these various cases:

```
A = eye(3)
A =
     1    0    0
     0    1    0
     0    0    1
B = eye(3,2)
B =
     1    0
     0    1
     0    0
C = [1, 2, 3 ; 4, 2, 5]
C =
     1    2    3
     4    2    5
D = eye(size(C))
D =
     1    0    0
     0    1    0
```

HINT

We recommend that you do not name an identity matrix *i*, because *i* will no longer represent $\sqrt{-1}$ in any statements that follow.

Recall that `A * inv(A)` equals the identity matrix. We can illustrate this with the following statements:

```
A = [1,0,2; -1, 4, -2; 5,2,1]
A =
     1    0    2
    -1    4   -2
     5    2    1
inv(A)
ans =
    -0.2222   -0.1111    0.2222
     0.2500    0.2500    0.0000
     0.6111    0.0556   -0.1111
A*inv(A)
ans =
     1.0000        0    0.0000
    -0.0000   1.0000    0.0000
    -0.0000  -0.0000    1.0000
```

As we discussed earlier, matrix multiplication is not in general commutative—that is,

$$AB \neq b$$

However, for identity matrices,
$$AI = IA$$
which we can show with the following MATLAB® code:

```
I = eye(3)
I =
      1   0   0
      0   1   0
      0   0   1
A*I
ans =
      1   0   2
     -1   4  -2
      5   2   1
I*A
ans =
      1   0   2
     -1   4  -2
      5   2   1
```

10.3.3 Other Matrices

MATLAB® includes a number of matrices that are useful for testing numerical techniques, that serve in computational algorithms, or that are just interesting.

Pascal	Creates a Pascal matrix, using Pascal's triangle.	**pascal(4)** **ans =** 1.00 1.00 1.00 1.00 1.00 2.00 3.00 4.00 1.00 3.00 6.00 10.00 1.00 4.00 10.00 20.00
Magic	Creates a Magic Matrix, in which all the rows, all the columns, and all the diagonals add up to the same value.	**Magic(3)** **ans =** 8.00 1.00 6.00 3.00 5.00 7.00 4.00 9.00 2.00
rosser	The Rosser Matrix is used as an eigenvalue test matrix. It requires no input.	**rosser** **ans = [8 × 8]**
Gallery	The gallery contains over 50 different test matrices.	The syntax for the gallery functions is different for each function. Use **help** to determine which is right for your needs.

SUMMARY

One of the most common matrix operations is the transpose, which changes rows into columns and columns into rows. In mathematics texts, the transpose is indicated with a superscript T, as in A^T. In MATLAB®, the single quote is used as the transpose operator. Thus,

```
A'
```

is the transpose of A.

Another common matrix operation is the dot product, which is the sum of the array multiplications of two equal-size vectors:

$$C = \sum_{i=1}^{N} A_i * B_i$$

The MATLAB® function for dot products is

> `dot(A,B)`

Similar to the dot product is matrix multiplication. Each element in the result of a matrix multiplication is a dot product:

$$C_{i,j} = \sum_{k=1}^{N} A_{i,k} B_{k,j}$$

Matrix multiplication uses the asterisk operator in MATLAB®, so that

> `C = A*B`

indicates that the matrix A is multiplied by the matrix B in accordance with the rules of matrix algebra. Matrix multiplication is not commutative—that is,

$$AB \neq BA$$

Raising a matrix to a power is similar to multiple multiplication steps:

$$A^3 = AAA$$

Since a matrix must be square in order to be multiplied by itself, only square matrices can be raised to a power. When matrices are raised to noninteger powers, the result is a matrix of complex numbers.

A matrix times its inverse is the identity matrix:

$$AA^{-1} = I$$

MATLAB® provides two techniques for determining a matrix inverse: the `inv` function,

> `inv_of_A = inv(A)`

and raising the matrix to the -1 power, given by

> `inv_of_A = A^-1`

If the determinant of a matrix is zero, the matrix is singular and does not have an inverse. The MATLAB® function used to find the determinant is

> `det(A)`

In addition to computing dot products, MATLAB® contains a function that calculates the cross product of two vectors in three-space. The cross product is often called the vector product because it returns a vector:

$$C = A \times B$$

The cross product produces a vector that is at right angles (normal) to the two input vectors, a property called orthogonality. Cross products can be thought of as the determinant of a matrix composed of the unit vectors in the x, y, and z directions and the two input vectors:

$$C = \begin{vmatrix} \vec{i} & \vec{j} & \vec{k} \\ A_x & A_y & A_z \\ B_x & B_y & B_z \end{vmatrix}$$

The MATLAB® syntax for calculating a cross product uses the **cross** function:

```
C = cross(A,B)
```

One common use of the matrix inverse is to solve systems of linear equations. For example, the system

$$
\begin{array}{rrrr}
3x & +2y & -z & = 10 \\
-x & +3y & +2z & = 5 \\
x & -y & -z & = -1
\end{array}
$$

can be expressed with matrices as

$$
\mathbf{AX = B}
$$

To solve this system of equations with MATLAB®, you could multiply B by the inverse of A:

```
X = inv(A)*B
```

However, this technique is less efficient than Gaussian elimination, which is accomplished in MATLAB® by using left division:

```
X = A\B
```

The left division technique can also be used to solve both overdefined and underdefined systems of equations. When the system is overdefined a least squared approach is used to find the best fit result. When the system is underdefined one or more of the variables is set equal to 0, and the remaining variables calculated.

MATLAB® includes a number of special matrices that can be used to make calculations easier or to test numerical techniques. For example, the ones and zeros functions can be used to create matrices of ones and zeros, respectively. The pascal and magic functions are used to create Pascal matrices and magic matrices, respectively, which have no particular computational use but are interesting mathematically. The gallery function contains over 50 matrices especially formulated to test numerical techniques.

MATLAB® SUMMARY

The following MATLAB® summary lists and briefly describes all the special characters, commands, and functions that are defined in this chapter:

Special Characters	
'	indicates a matrix transpose
*	matrix multiplication
\	matrix left division
^	matrix exponentiation

Commands and Functions	
cross	computes the cross product
det	computes the determinant of a matrix
dot	computes the dot product
eye	generates an identity matrix
gallery	contains sample matrices
inv	computes the inverse of a matrix
magic	creates a "magic" matrix
ones	creates a matrix containing all ones
pascal	creates a pascal matrix
rref	uses the reduced row echelon format scheme for solving a series of linear equations
size	determines the number of rows and columns in a matrix
zeros	creates a matrix containing all zeros

KEY TERMS

cross product	inverse	system of equations
determinant	matrix multiplication	transpose
dot product	normal	unit vector
Gaussian elimination	orthogonal	
identity matrix	singular	

PROBLEMS

Dot Products

10.1 Compute the dot product of the following pairs of vectors, and then show that

$$A \cdot B = B \cdot A$$

(a) $\mathbf{A} = \begin{bmatrix} 1 & 3 & 5 \end{bmatrix}$, $\mathbf{B} = \begin{bmatrix} -3 & -2 & 4 \end{bmatrix}$

(b) $\mathbf{A} = \begin{bmatrix} 0 & -1 & -4 & -8 \end{bmatrix}$, $\mathbf{B} = \begin{bmatrix} 4 & -2 & -3 & 24 \end{bmatrix}$

10.2 Compute the total mass of the components shown in Table 10.3, using a dot product.

Table 10.3 Component Mass Properties

Component	Density, g/cm^3	Volume, cm^3
Propellant	1.2	700
Steel	7.8	200
Aluminum	2.7	300

10.3 Use a dot product and the shopping list in Table 10.4 to determine your total bill at the grocery store.

Table 10.4 Shopping List

Item	Number Needed	Cost
Milk	2 gallons	$3.50 per gallon
Eggs	1 dozen	$1.25 per dozen
Cereal	2 boxes	$4.25 per box
Soup	5 cans	$1.55 per can
Cookies	1 package	$3.15 per package

10.4 Bomb calorimeters are used to determine the energy released during chemical reactions. The total heat capacity of a bomb calorimeter is defined as the sum of the products of the mass of each component and the specific heat capacity of each component, or

$$CP = \sum_{i=1}^{n} m_i C_i$$

where

m_i = mass of component i, g

C_i = heat capacity of component, i, J/g K

CP = total heat capacity, J/K

Find the total heat capacity of a bomb calorimeter, using the thermal data in Table 10.5.

Table 10.5 Thermal Data

Component	Mass, g	Heat Capacity, J/gK
Steel	250	0.45
Water	100	4.2
Aluminum	10	0.90

10.5 Organic compounds are composed primarily of carbon, hydrogen, and oxygen and for that reason are often called hydrocarbons. The molecular weight (MW) of any compound is the sum of the products of the number of atoms of each element (Z) and the atomic weight (AW) of each element present in the compound.

$$MW = \sum_{i=1}^{n} AW_i \cdot Z_i$$

The atomic weights of carbon, hydrogen, and oxygen are approximately 12, 1, and 16, respectively. Use a dot product to determine the molecular weight of ethanol (C_2H_5OH), which has two carbon, one oxygen, and six hydrogen atoms.

10.6 It is often useful to think of air as a single substance with a molecular weight (molar mass) determined by a weighted average of the molecular weights of the different gases present. With little error, we can estimate the molecular weight of air using in our calculation only nitrogen, oxygen, and argon. Use a dot product and Table 10.6 to approximate the molecular weight of air.

Table 10.6 Composition of Air

Compound	Fraction in Air	Molecular Weight, g/mol
Nitrogen, N_2	0.78	28
Oxygen, O_2	0.21	32
Argon, Ar	0.01	40

Matrix Multiplication

10.7 Compute the matrix product A*B of the following pairs of matrices:

(a) $A = \begin{bmatrix} 12 & 4 \\ 3 & -5 \end{bmatrix}$ $\quad B = \begin{bmatrix} 2 & 12 \\ 0 & 0 \end{bmatrix}$

(b) $A = \begin{bmatrix} 1 & 3 & 5 \\ 2 & 4 & 6 \end{bmatrix}$ $\quad B = \begin{bmatrix} -2 & 4 \\ 3 & 8 \\ 12 & -2 \end{bmatrix}$

Show that A*B is not the same as B*A.

10.8 You and a friend are both going to a grocery store. Your lists are shown in Table 10.7.

Table 10.7 Ann and Fred's Shopping List

Item	Number Needed by Ann	Number Needed by Fred
Milk	2 gallons	3 gallons
Eggs	1 dozen	2 dozen
Cereal	2 boxes	1 box
Soup	5 cans	4 cans
Cookies	1 package	3 packages

The items cost as follows:

Item	Cost
Milk	$3.50 per gallon
Eggs	$1.25 per dozen
Cereal	$4.25 per box
Soup	$1.55 per can
Cookies	$3.15 per package

Find the total bill for each shopper.

10.9 A series of experiments was performed with a bomb calorimeter. In each experiment, a different amount of water was used. Calculate the total heat capacity for the calorimeter for each of the experiments, using matrix multiplication, the data in Table 10.8, and the information on heat capacity that follows the table.

Table 10.8 Thermal Properties of a Bomb Calorimeter

Experiment No.	Mass of Water, g	Mass of Steel, g	Mass of Aluminum, g
1	110	250	10
2	100	250	10
3	101	250	10
4	98.6	250	10
5	99.4	250	10

Component	Heat Capacity, J/gK
Steel	0.45
Water	4.2
Aluminum	0.90

10.10 The molecular weight (MW) of any compound is the sum of the products of the number of atoms of each element (Z) and the atomic weight (AW) of each element present in the compound, or

$$MW = \sum_{i=1}^{n} AW_i \cdot Z_i$$

The compositions of the first five straight-chain alcohols are listed in Table 10.9. Use the atomic weights of carbon, hydrogen, and oxygen (12, 1, and 16, respectively) and matrix multiplication to determine the molecular weight (more correctly called the molar mass) of each alcohol.

Table 10.9 Composition of Alcohols

Name	Carbon	Hydrogen	Oxygen
Methanol	1	4	1
Ethanol	2	6	1
Propanol	3	8	1
Butanol	4	10	1
Pentanol	5	12	1

Matrix Exponentiation

10.11   Given the array

$$A = \begin{bmatrix} -1 & 3 \\ 4 & 2 \end{bmatrix}$$

(a) Raise **A** to the second power by array exponentiation. (Consult `help` if necessary.)
(b) Raise **A** to the second power by matrix exponentiation.
(c) Explain why the answers are different.

10.12   Create a 3×3 array called A by using the `pascal` function:

```
pascal(3)
```

(a) Raise A to the third power by array exponentiation.
(b) Raise A to the third power by matrix exponentiation.
(c) Explain why the answers are different.

Determinants and Inverses

10.13   Given the array $\mathbf{A} = \begin{bmatrix} -1 & 3; & 4 & 2 \end{bmatrix}$, compute the determinant of **A** both by hand and by using MATLAB®.

10.14   Recall that not all matrices have an inverse. A matrix is singular (i.e., it doesn't have an inverse) if its determinant equals 0 (i.e., $|A| = 0$). Use the determinant function to test whether each of the following matrices has an inverse:

$$A = \begin{bmatrix} 2 & -1 \\ 4 & 5 \end{bmatrix}, \quad B = \begin{bmatrix} 4 & 2 \\ 2 & 1 \end{bmatrix}, \quad C = \begin{bmatrix} 2 & 0 & 0 \\ 1 & 2 & 2 \\ 5 & -4 & 0 \end{bmatrix}$$

If an inverse exists, compute it.

Cross Products

10.15   Compute the moment of force around the pivot point for the lever shown in Figure P10.15. You'll need to use trigonometry to determine the x and y components of both the position vector and the force vector. Recall that the moment of force can be calculated as the cross product

$$M_0 = r \times F$$

A force of 200 lbf is applied vertically at a position 20 feet along the lever. The lever is positioned at an angle of 60° from the horizontal.

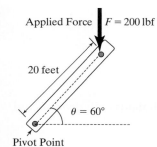

Applied Force $F = 200$ lbf

20 feet

$\theta = 60°$

Pivot Point

Figure P10.15
Moment of force acting on a lever about the origin.

10.16 Determine the moment of force about the point where a bracket is attached to a wall. The bracket is shown in Figure P10.16. It extends 10 inches out from the wall and 5 inches up. A force of 35 lbf is applied to the bracket at an angle of 55° from the vertical. Your answer should be in ft-lbf, so you'll need to do some conversions of units.

Figure P10.16
A bracket attached to a wall.

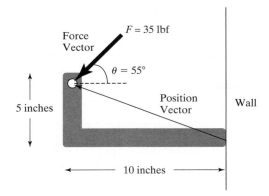

10.17 A rectangular shelf is attached to a wall by two brackets 12 inches apart at points *A* and *B*, as shown in Figure P10.17. A wire with a 10-lbf weight attached to it is hanging from the edge of the shelf at point *C*. Determine the moment of force about point *A* and about point *B* caused by the weight at point *C*.

You can formulate this problem by solving it twice, once for each bracket, or by creating a 2×3 matrix for the position vector and another 2×3 matrix for the force vector. Each row should correspond to a different bracket. The **cross** function will return a 2×3 result, each row corresponding to the moment about a separate bracket.

Figure P10.17
Calculation of moment of force in three dimensions.

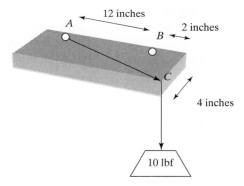

Solving Linear Systems of Equations

10.18 Solve the following systems of equations, using both matrix left division and the inverse matrix method:

(a) $-2x + y = 3 \quad x + y = 10$

(b) $5x + 3y - z = 10$
$3x + 2y + z = 4$
$4x - y + 3z = 12$

(c) $3x + y + z + w = 24$

$\quad\quad x - 3y + 7z + w = 12$

$\quad\quad 2x + 2y - 3z + 4w = 17$

$\quad\quad x + y + z + w = 0$

10.19 In general, matrix left division is faster and more accurate than the matrix inverse. Using both techniques, solve the following system of equations and time the execution with the `tic` and `toc` functions:

$$3x_1 + 4x_2 + 2x_3 - x_4 + x_5 + 7x_6 + x_7 = 42$$
$$2x_1 - 2x_2 + 3x_3 - 4x_4 + 5x_5 + 2x_6 + 8x_7 = 32$$
$$x_1 + 2x_2 + 3x_3 + x_4 + 2x_5 + 4x_6 + 6x_7 = 12$$
$$5x_1 + 10x_2 + 4x_3 + 3x_4 + 9x_5 - 2x_6 + x_7 = -5$$
$$3x_1 + 2x_2 - 2x_3 - 4x_4 - 5x_5 - 6x_6 + 7x_7 = 10$$
$$-2x_1 + 9x_2 + x_3 + 3x_4 - 3x_5 + 5x_6 + x_7 = 18$$
$$x_1 - 2x_2 - 8x_3 + 4x_4 + 2x_5 + 4x_6 + 5x_7 = 17$$

If you have a new computer, you may find that this problem executes so quickly that you won't be able to detect a difference between the two techniques. If so, see if you can formulate a larger problem to solve.

10.20 In Example 10.5, we demonstrated that the circuit shown in Figure 10.5 could be described by the following set of linear equations:

$$(R_2 + R_4)i_1 + (-R_2)i_2 + (-R_4)i_3 = V_1$$

$$(-R_2)i_1 + (R_1 + R_2 + R_3)i_2 + (-R_3)i_3 = 0$$

$$(-R_4)i_1 + (-R_3)i_2 + (R_3 + R_4 + R_5)i_3 = 0$$

We solved this set of equations by the matrix inverse approach. Redo the problem, but this time use the left-division approach.

10.21 Consider a separation process in which a stream of water, ethanol, and methanol enters a process unit. Two streams leave the unit, each with varying amounts of the three components (see Figure P10.21).

Determine the mass flow rates into the system and out of the top and bottom of the separation unit.

Figure P10.21

Separation process with three components.

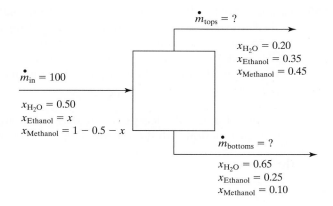

$\dot{m}_{tops} = ?$

$x_{H_2O} = 0.20$
$x_{Ethanol} = 0.35$
$x_{Methanol} = 0.45$

$\dot{m}_{in} = 100$

$x_{H_2O} = 0.50$
$x_{Ethanol} = x$
$x_{Methanol} = 1 - 0.5 - x$

$\dot{m}_{bottoms} = ?$

$x_{H_2O} = 0.65$
$x_{Ethanol} = 0.25$
$x_{Methanol} = 0.10$

(a) First set up material-balance equations for each of the three components:

Water

$$(0.5)(100) = 0.2m_{\text{tops}} + 0.65m_{\text{bottoms}}$$

$$50 = 0.2m_{\text{tops}} + 0.65m_{\text{bottoms}}$$

Ethanol

$$100x = 0.35m_{\text{tops}} + 0.25m_{\text{bottoms}}$$

$$0 = -100x + 0.35m_{\text{tops}} + 0.25m_{\text{bottoms}}$$

Methanol

$$100(1 - 0.5 - x) = 0.45m_{\text{tops}} + 0.1m_{\text{bottoms}}$$

$$50 = 100x + 0.45m_{\text{tops}} + 0.1m_{\text{bottoms}}$$

(b) Arrange the equations you found in part (a) into a matrix representation:

$$A = \begin{bmatrix} 0 & 0.2 & 0.65 \\ -100 & 0.35 & 0.25 \\ 100 & 0.45 & 0.1 \end{bmatrix} \qquad B = \begin{bmatrix} 50 \\ 0 \\ 50 \end{bmatrix}$$

(c) Use MATLAB® to solve the linear system of three equations.

10.22 Consider the statically determinate truss shown in Figure P10.22. The applied force has a magnitude of 1000 lbf at an angle of 30° from the horizontal, as shown in the figure. The inner angles, θ_1 and θ_2 are 45° and 65° respectively. Determine the values of the forces in each member of the truss, and the reactive forces experienced at the hinge and the roller (nodes 2 and 3).

Figure P10.22
A statically determinate truss.

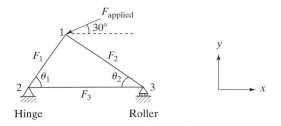

Challenge Problem

10.23 Create a MATLAB® function M-file called **my_matrix_solver** to solve a system of linear equations, using nested `for` loops instead of MATLAB®'s built-in operators or functions. Your function should accept a coefficient matrix and a result matrix, and should return the values of the variables. For example, if you wish to solve the following matrix equation for X

$$AX = B$$

your function should accept A and B as input, and return X as the result. Test your function with the system of equations from the previous problem.

11
Other Kinds of Arrays

Objectives

After reading this chapter, you should be able to:

- Understand the different kinds of data used in MATLAB®
- Create and use both numeric and character arrays

- Create multidimensional arrays and access data in those arrays
- Create and use cell and structure arrays

INTRODUCTION

In MATLAB®, scalars, vectors, and two-dimensional matrices are used to store data. In reality, all these are two dimensional. Thus, even though

```
A = 1;
```

creates a scalar,

```
B = 1:10;
```

creates a vector, and

```
C = [1,2,3;4,5,6];
```

creates a two-dimensional matrix, they are all still two-dimensional arrays. Notice in Figure 11.1 that the size of each of these variables is listed as a *two-dimensional* matrix 1×1 for A, 1×10 for B, and 2×3 for C. The class listed for each is also the same: Each is a "double," which is short for double-precision floating-point number. (To ensure that you see all the columns shown in Figure 11.1 right click on the title bar and select the appropriate parameters. You can also access this menu by selecting View from the menu bar.)

Figure 11.1
MATLAB® supports a
variety of array types.

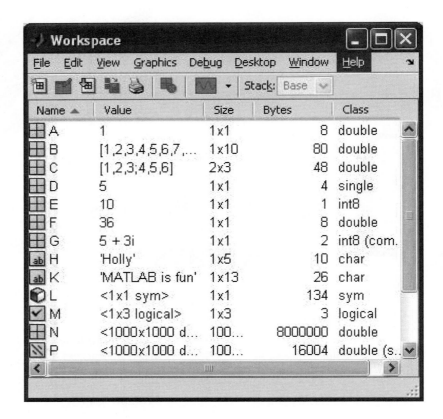

MATLAB® includes the capability to create multidimensional matrices and to store data that are not doubles, such as characters. In this chapter, we'll introduce the data types supported by MATLAB® and explore how they can be stored and used by a program.

11.1 DATA TYPES

The primary data type (also called a class) in MATLAB® is the *array* or *matrix*. Within the array, MATLAB® supports a number of different secondary data types. Because MATLAB® was written in C, many of those data types parallel the data types supported in C. In general, all the data within an array must be the same type. However, MATLAB® also includes functions to convert between data types, and array types to store different kinds of data in the same array (cell and structure arrays).

The kinds of data that can be stored in MATLAB® are listed in Figure 11.2. They include numerical data, character data, logical data, and symbolic data types. Each can be stored either in arrays specifically designed for that data type or in arrays that can store a variety of data. Cell arrays and structure arrays fall into the latter category (Figure 11.3).

11.1.1 Numeric Data Types

Double-Precision Floating-Point Numbers
The default numeric data type in MATLAB® is the double-precision floating-point number, as defined by IEEE Standard 754. (IEEE, the Institute of Electrical and

IEEE
Institute of Electrical and
Electronics Engineers

Figure 11.2
Many different kinds of data can be stored in MATLAB®.

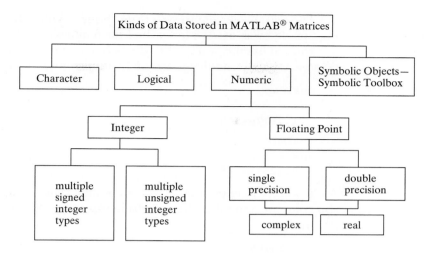

Figure 11.3
MATLAB® supports multiple data types, all of which are arrays.

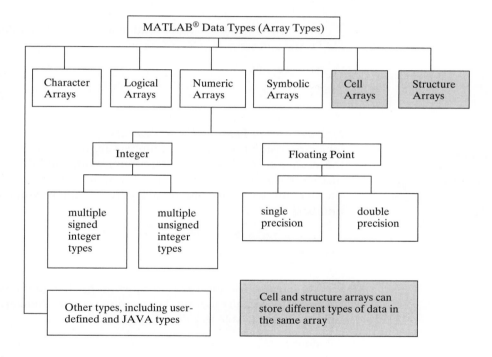

Electronics Engineers, is the professional organization for electrical engineers.) Recall that when we create a variable such as A, as in

```
A = 1;
```

the variable is listed in the workspace window and the class is "double," as shown in Figure 11.1. Notice that the array requires 8 bytes of storage space. Each byte is equal to 8 bits, so the number 1 requires 64 bits of storage space. Also in Figure 11.1, notice how much storage space is required for variables B and C:

```
B = 1:10;                    C=[1,2,3; 4,5,6];
```

The variable B requires 80 bytes, 8 for each of the 10 values stored, and C requires 48 bytes, again 8 for each of the 6 values stored.

You can use the `realmax` and `realmin` functions to determine the maximum possible value of a double-precision floating-point number:

```
realmax
ans =
   1.7977e+308
realmin
ans =
   2.2251e-308
```

If you try to enter a value whose absolute value is greater than `realmax`, or if you compute a number that is outside this range, MATLAB® will assign a value of ± infinity:

```
x = 5e400
x =
    Inf
```

Similarly, if you try to enter a value whose absolute value is less than `realmin`, MATLAB® will assign a value of zero:

```
x = 1e-400
x =
    0
```

Single-Precision Floating-Point Numbers

Single-precision floating-point numbers are new to MATLAB® 7. They use only half the storage space of a double-precision number and thus store only half the information. Each value requires only 4 bytes, or $4 \times 8 = 32$ bits, of storage space, as shown in the workspace window in Figure 11.1 when we define D as a single-precision number:

```
D = single(5)
D =
     5
```

We need to use the `single` function to change the value 5 (which is double precision by default) to a single-precision number. Similarly, the `double` function will convert a variable to a double, as in

```
double(D)
```

which changes the variable D into a double.

Since single-precision numbers are allocated only half as much storage space, they cannot cover as large a range of values as double-precision numbers. We can use the `realmax` and `realmin` functions to show this:

```
realmax('single')
ans =
   3.4028e+038
realmin('single')
ans =
   1.1755e-038
```

Engineers will rarely need to convert to single-precision numbers, because today's computers have plenty of storage space for most applications and will execute most of the problems we pose in extremely short amounts of time. However, in some numerical analysis applications, you may be able to improve the run time of a long problem by changing from double to single precision. Note, though, that this has the disadvantage of making round-off error more of a problem.

We can demonstrate the effect of round-off error in single-precision versus double-precision problems with an example. Consider the series

$$\Sigma\left(\frac{1}{1} + \frac{1}{2} + \frac{1}{3} + \frac{1}{4} + \frac{1}{5} + \frac{1}{6} + \cdots + \frac{1}{n} + \cdots\right)$$

A series is the sum of a sequence of numbers, and this particular series is called the *harmonic series*, represented with the following shorthand notation:

$$\sum_{n=1}^{\infty} \frac{1}{n}$$

The harmonic series diverges; that is, it just keeps getting bigger as you add more terms together. You can represent the first 10 terms of the harmonic sequence with the following commands:

```
n = 1:10;
harmonic = 1./n
```

You can view the results as fractions if you change the format to rational:

```
format rat
harmonic =
   1    1/2    1/3    1/4    1/5    1/6    1/7    1/8    1/9    1/10
```

Or you can use the short format, which shows decimal representations of the numbers:

```
format short
harmonic =
  1.0000    0.5000    0.3333    0.2500    0.2000    0.1667    0.1429
            0.1250    0.1111    0.1000
```

No matter how the values are displayed on the screen, they are stored as double-precision floating-point numbers inside the computer. By calculating the partial sums (also called cumulative sums), we can see how the value of the sum of these numbers changes as we add more terms:

```
partial_sum = cumsum(harmonic)
partial_sum =
  Columns 1 through 6
    1.0000    1.5000    1.8333    2.0833    2.2833    2.4500
  Columns 7 through 10
    2.5929    2.7179    2.8290    2.9290
```

The cumulative sum (cumsum) function calculates the sum of the values in the array up to the element number displayed. Thus, in the preceding calculation, the value in column 3 is the partial sum of the values in columns 1 through 3 of the input array (in this case, the array named harmonic). No matter how big we make the harmonic array, the partial sums continue to increase.

The only problem with this process is that the values in `harmonic` keep getting smaller and smaller. Eventually, when n is big enough, `1./n` is so small that the computer can't distinguish it from zero. This happens much more quickly with single-precision than with double-precision representations of numbers. We can demonstrate this property with a large array of *n*-values:

```
n = 1:1e7;
harmonic = 1./n;
partial_sum = cumsum(harmonic);
```

(This may take your computer a while to calculate, especially if you have an older machine.) All these calculations are performed with double-precision numbers, because double precision is the default data type in MATLAB®. Now we'd like to plot the results, but there are really too many numbers (10 million, in fact). We can select every thousandth value with the following code:

```
m = 1000:1000:1e7;
partial_sums_selected = partial_sum(m);
plot(partial_sums_selected)
```

Now we can repeat the calculations, but change to single-precision values. You may need to clear your computer memory before this step, depending on how much memory is available on your system. The code is

```
n = single(1:1e7);
harmonic = 1./n;
partial_sum = cumsum(harmonic);
m = 1000:1000:1e7;
partial_sums_selected = partial_sum(m);
hold on
plot(partial_sums_selected,':')
```

The results are presented in Figure 11.4. The solid line represents the partial sums calculated with double precision. The dashed line represents the partial sums calculated with single precision. The single-precision calculation levels off, because we reach the point where each successive term is so small that the computer sets it equal to zero. We haven't reached that point yet for the double-precision values.

KEY IDEA
Round-off error is a bigger problem in single-precision than in double-precision calculations

Figure 11.4
Round-off error degrades the harmonic series calculation for single-precision faster than for double-precision numbers.

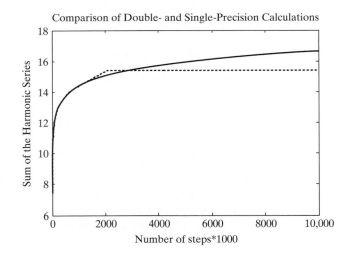

Comparison of Double- and Single-Precision Calculations

Integers

New to MATLAB® are several integer-number types. Traditionally, integers are used as counting numbers. For example, there can't be 2.5 people in a room, and you can't specify element number 1.5 in an array. Eight different types of integers are supported by MATLAB®. They differ in how much storage space is allocated for the type and in whether the values are signed or unsigned. The more storage space, the larger the value of an integer number you can use. The eight types are shown in Table 11.1.

Since 8 bits is 1 byte, when we assign E as an `int8` with the code

```
E = int8(10)
E =
   10
```

it requires only 1 byte of storage, as shown in Figure 11.1.

You can determine the maximum value of any of the integer types by using the `intmax` function. For example, the code

```
intmax('int8')
ans =
   127
```

indicates that the maximum value of an 8-bit signed integer is 127.

The four signed-integer types allocate storage space to specify whether the number is plus or minus. The four unsigned-integer types assume that the number is positive and thus do not need to store that information, leaving more room to store numerical values.

The code

```
intmax('uint8')
ans =
   255
```

reveals that the maximum value of an 8-bit unsigned integer is 255.

One place where integer arrays find use is to store image information. These arrays are often very large, but a limited number of colors are used to create the picture. Storing the information as unsigned-integer arrays reduces the storage requirement dramatically.

Complex Numbers

The default storage type for complex numbers is double; however, twice as much storage room is needed, because both the real and imaginary components must be stored:

```
F = 5+3i;
```

Table 11.1 MATLAB® Integer Types

8-bit signed integer	**int8**	8-bit unsigned integer	**uint8**
16-bit signed integer	**int16**	16-bit unsigned integer	**uint16**
32-bit signed integer	**int32**	32-bit unsigned integer	**uint32**
64-bit signed integer	**int64**	64-bit unsigned integer	**uint64**

Thus, 16 bytes ($= 128$ bits) are required to store a double complex number. Complex numbers can also be stored as singles or integers (see Figure 11.1), as the following code illustrates:

```
G = int8(5+3i);
```

PRACTICE EXERCISES 11.1

1. Enter the following list of numbers into arrays of each of the numeric data types [1, 4, 6; 3, 15, 24; 2, 3, 4]:
   (a) Double-precision floating point—name this array A
   (b) Single-precision floating point—name this array B
   (c) Signed integer (pick a type)—name this array C
   (d) Unsigned integer (pick a type)—name this array D

2. Create a new matrix E by adding A to B:

$$\mathbf{E = A + B}$$

   What data type is the result?

3. Define x as an integer data type equal to 1 and y as an integer data type equal to 3.
   (a) What is the result of the calculation x/y?
   (b) What is the data type of the result?
   (c) What happens when you perform the division when x is defined as the integer 2 and y as the integer 3?

4. Use intmax to determine the largest number you can define for each of the numeric data types. (Be sure to include all eight integer data types.)

5. Use MATLAB® to determine the smallest number you can define for each of the numeric data types. (Be sure to include all eight integer data types.)

KEY IDEA
Each character, including spaces, is a separate element in a character array

11.1.2 Character and String Data

In addition to storing numbers, MATLAB® can store character information. Single quotes are used to identify a string and to differentiate it from a variable name. When we type the string

```
H ='Holly';
```

a 1×5 character array is created. Each letter is a separate element of the array, as is indicated by the code

```
H(5)
ans =
   y
```

Any string represents a character array in MATLAB®. Thus,

```
K = 'MATLAB is fun'
```

becomes a 1×13 character array. Notice that the spaces between the words are counted as characters. Notice also that the name column in Figure 11.1 displays a

ASCII
American Standard Code for Information—a standard code for exchanging information between computers

symbol containing the letters "ab," which indicates that H and K are character arrays. Each character in a character array requires 2 bytes of storage space.

All information in computers is stored as a series of zeros and ones. There are two major coding schemes to do this: ASCII and EBCDIC. Most small computers use the ASCII coding scheme, whereas many mainframes and supercomputers use EBCDIC. You can think of the series of zeros and ones as a binary, or base-2, number. In this sense, all computer information is stored numerically. Every base-2 number has a decimal equivalent. The first several numbers in each base are shown in Table 11.2.

EBCDIC
Extended Binary Coded Decimal Interchange Code—a standard code for exchanging information between computers

Every ASCII (or EBCDIC) character stored has both a binary representation and a decimal equivalent. When we ask MATLAB® to change a character to a double, the number we get is the decimal equivalent in the ASCII coding system. Thus, we may have

BINARY
A coding scheme using only zeros and ones

```
double('a')
ans =
   97
```

Conversely, when we use the char function on a double, we get the character represented by that decimal number in ASCII—for example,

```
char (98)
ans =
   b
```

If we try to create a matrix containing both numeric and character information, MATLAB® converts all the data to character information:

```
['a',98]
ans =
   ab
```

(The character b is equivalent to the number 98.) Not all numbers have a character equivalent. If this is the case they are represented as a blank in the resulting character array

```
['a',3]
ans =
   a
```

Although this result looks like it has only one character in the array, check the workspace window. You'll find that the size is a 1 × 3 character array.

Table 11.2 Binary-to-Decimal Conversions

Base 2 (binary)	Base 10 (decimal)
1	1
10	2
11	3
100	4
101	5
110	6
111	7
1000	8

If we try to perform mathematical calculations with both numeric and character information, MATLAB® converts the character to its decimal equivalent:

```
'a' + 3
ans =
   100
```

Since the decimal equivalent of **'a'** is 97, the problem is converted to

$$97 + 3 = 100$$

PRACTICE EXERCISES 11.2

1. Create a character array consisting of the letters in your name.
2. What is the decimal equivalent of the letter *g*?
3. Upper- and lowercase letters are 32 apart in decimal equivalent. (Uppercase comes first.) Using nested functions, convert the string "matlab" to the uppercase equivalent, "MATLAB."

11.1.3 Symbolic Data

The symbolic toolbox uses symbolic data to perform symbolic algebraic calculations. One way to create a symbolic variable is to use the `sym` function:

```
L = sym('x^2-2')
L =
x^2-2
```

The storage requirements of a symbolic object depend on how large the object is. Notice, however, in Figure 11.1, that L is a 1×1 array. Subsequent symbolic objects could be grouped together into an array of mathematical expressions. The symbolic-variable icon shown in the left-hand column of Figure 11.1 is a cube.

KEY IDEA

Computer programs use the number 0 to mean false and the number 1 to mean true

11.1.4 Logical Data

Logical arrays may look like arrays of ones and zeros because MATLAB® (as well as other computer languages) uses these numbers to denote true and false:

```
M = [true,false,true]
M =
   1   0   1
```

We don't often create logical arrays this way. Usually, they are the result of logical operations. For example,

```
x = 1:5;
y = [2,0,1,9,4];
z = x>y
```

returns

```
z =
   0   1   1   0   1
```

We can interpret this to mean that $x > y$ is false for elements 1 and 3, and true for elements 2, 3, and 5. These arrays are used in logical functions and usually are not even seen by the user. For example,

```
find(x>y)

ans =
    2    3    5
```

tells us that elements 2, 3, and 5 of the **x** array are greater than the corresponding elements of the **y** array. Thus, we don't have to analyze the results of the logical operation ourselves. The icon representing logical arrays is a check mark (Figure 11.1).

11.1.5 Sparse Arrays

Both double-precision and logical arrays can be stored either in full matrices or as sparse matrices. Sparse matrices are "sparsely populated," which means that many or most of the values in the array are zero. (Identity matrices are examples of sparse matrices.) If we store double-precision sparse arrays in the full-matrix format, every data value takes 8 bytes of storage, be it a zero or not. The sparse-matrix format stores only the nonzero values and remembers where they are—a strategy that saves a lot of computer memory.

For example, define a 1000×1000 identity matrix, which is a one-million-element matrix:

```
N = eye(1000);
```

At 8 bytes per element, storing this matrix takes 8 MB. If we convert it to a sparse matrix, we can save some space. The code to do this is

```
P = sparse(N);
```

Notice in the workspace window that array P requires only 16,004 bytes! Sparse matrices can be used in calculations just like full matrices. The icon representing a sparse array is a group of diagonal lines (Figure 11.1).

11.2 MULTIDIMENSIONAL ARRAYS

When the need arises to store data in multidimensional (more than two-dimensional) arrays, MATLAB® represents the data with additional pages. Suppose you would like to combine the following four two-dimensional arrays into a three-dimensional array:

```
x = [1,2,3;4,5,6];
y = 10*x;
z = 10*y;
w = 10*z;
```

You need to define each page separately:

```
my_3D_array(:,:,1) = x;
my_3D_array(:,:,2) = y;
my_3D_array(:,:,3) = z;
my_3D_array(:,:,4) = w;
```

Read each of the previous statements as all the rows, all the columns, page 1, and so on.

When you call up my_3D_array, using the code

```
my_3D_array
```

the result is

```
my_3D_array
my_3D_array(:,:,1) =
   1    2    3
   4    5    6
my_3D_array(:,:,2) =
   10    20    30
   40    50    60
my_3D_array(:,:,3) =
   100    200    300
   400    500    600
my_3D_array(:,:,4) =
   1000    2000    3000
   4000    5000    6000
```

An alternative approach is to use the cat function. When you concatenate a list you group the members together in order, which is what the cat function does. The first field in the function specifies which dimension to use to concatenate the arrays, which follow in order. For example, to create the array we used in the previous example the syntax is

```
cat(3,x,y,z,w)
```

A multidimensional array can be visualized as shown in Figure 11.5. Even higher-dimensional arrays can be created in a similar fashion.

HINT

The squeeze function can be used to eliminate singleton dimensions in multidimensional arrays. For example, consider the three-dimensional array with the following dimensions

$$3 \times 1 \times 4$$

This represents an array with three rows, one column, and four pages. It could be stored more efficiently as a two-dimensional array by *squeezing* out the singleton column dimension

```
b = squeeze(a)
```

to give a new array with the dimensions

$$3 \times 4$$

Figure 11.5
Multidimensional arrays are grouped into pages.

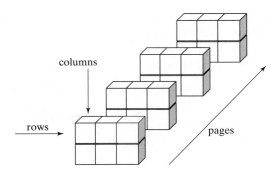

1. Create a three-dimensional array consisting of a 3 × 3 magic square, a 3 × 3 matrix of zeros, and a 3 × 3 matrix of ones.
2. Use triple indexing such as A(m,n,p) to determine what number is in row 3, column 2, page 1 of the matrix you created in Exercise 1.
3. Find all the values in row 2, column 3 (on all the pages) of the matrix.
4. Find all the values in all the rows and pages of column 3 of the matrix.

11.3 CHARACTER ARRAYS

We can create two-dimensional character arrays only if the number of elements in each row is the same. Thus, a list of names such as the following won't work, because each name has a different number of characters:

```
Q = ['Holly';'Steven';'Meagan';'David';'Michael';'Heidi']
??? Error using ==> vertcat
All rows in the bracketed expression must have the same number
of columns.
```

The char function "pads" a character array with spaces, so that every row has the same number of elements:

```
Q = char('Holly','Steven','Meagan','David','Michael','Heidi')
Q =
Holly
Steven
Meagan
David
Michael
Heidi
```

Q is a 6 × 7 character array. Notice that commas are used between each string in the char function.

Not only alphabetic characters can be stored in a MATLAB® character array. Any of the symbols or numbers found on the keyboard can be stored as characters. We can take advantage of this feature to create tables that appear to include both character and numeric information, but really are composed of just characters.

For example, let's assume that the array R contains test scores for the students in the character array Q:

```
R = [98;84;73;88;95;100]

R =
   98
   84
   73
   88
   95
  100
```

If we try to combine these two arrays, we'll get a strange result, because they are two different data types:

```
table = [Q,R]

table =
```

```
Holly b
Steven T
Meagan I
David X
Michael_
Heidi d
```

The double-precision values in R were used to define characters on the basis of their ASCII equivalent. When doubles and chars are used in the same array, MATLAB® converts all the information to chars. This is confusing, since, when we combine characters and numeric data in mathematical computations, MATLAB® converts the character information to numeric information.

The num2str (number to string) function allows us to convert the double R matrix to a matrix composed of character data:

```
S = num2str(R)
S =
  98
  84
  73
  88
  95
 100
```

R and S look alike, but if you check the workspace window (Figure 11.6), you'll see that R is a 6 × 1 double array and S is the 6 × 3 char array shown below.

space	9	8
space	8	4
space	7	3
space	8	8
space	9	5
1	0	0

Now we can combine Q, the character array of names, with S, the character array of scores:

Figure 11.6
Character and numeric data can be combined in a single array by changing the numeric values to characters with the num2str function.

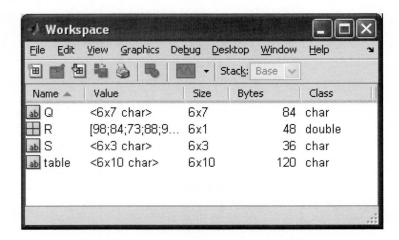

```
table = [Q,S]

table =
Holly     98
Steven    84
Meagan    73
David     88
Michael   95
Heidi    100
```

We show the results in the monospace font, which is evenly spaced. You can control the font that MATLAB® uses; if you choose a proportional font, such as Times New Roman, your columns won't line up.

We could also use the `disp` function to display the results:

```
disp([Q,S])
Holly     98
Steven    84
Meagan    73
David     88
Michael   95
Heidi    100
```

HINT

Put a space after your longest string, so that when you create a padded character array, there will be a space between the character information and the numeric information you've converted to character data.

KEY IDEA

Combine character and numeric arrays using the `num2str` function to create data file names

A useful application of character arrays and the `num2str` function is the creation of file names. On occasion you may want to save data into .dat or .mat files, without knowing ahead of time how many files will be required. One solution would be to name your files using the following pattern:

```
my_data1.dat
my_data2.dat
my_data3.dat  etc.
```

Imagine that you load a file of unknown size, called `some_data`, into MATLAB® and want to create new files, each composed of a single column from `some_data`:

```
load some_data
```

You can determine how big the file is by using the `size` function:

```
[rows,cols] = size(some_data)
```

If you want to store each column of the data into its own file, you'll need a file name for each column. You can do this in a `for` loop, using the function form of the `save` command:

```
for k = 1:cols
  file_name = ['my_data',num2str(k)]
```

```
        data = some_data(:,k) '
        save(file_name,'data')
end
```

The loop will execute once for each column. You construct the file name by creating an array that combines characters and numbers with the statement

```
file_name = ['my_data',num2str(k)];
```

This statement sets the variable file_name equal to a character array, such as my_data1 or my_data2, depending on the current pass through the loop. The **save** function accepts character input. In the line

```
save(file_name,'data')
```

file_name is a character variable, and 'data' is recognized as character information because it is inside single quotes. If you run the preceding for loop on a file that contains a 5 × 3 matrix of random numbers, you get the following result:

```
rows =
    5
cols =
    3
file_name =
my_data1
data =
   -0.4326  -1.6656   0.1253   0.2877  -1.1465
file_name =
my_data2
data =
    1.1909   1.1892  -0.0376   0.3273   0.1746
file_name =
my_data3
data =
   -0.1867   0.7258  -0.5883   2.1832  -0.1364
```

The current folder now includes three new files.

PRACTICE EXERCISES 11.4

1. Create a character matrix called names of the names of all the planets. Your matrix should have nine rows.
2. Some of the planets can be classified as rocky midgets and others as gas giants. Create a character matrix called type, with the appropriate designation on each line.
3. Create a character matrix of nine spaces, one space per row.
4. Combine your matrices to form a table listing the names of the planets and their designations, separated by a space.
5. Use the Internet to find the mass of each of the planets, and store the information in a matrix called mass. (Or use the data from Example 11.2.) Use the num2str function to convert the numeric array into a character array, and add it to your table.

EXAMPLE 11.1

CREATING A SIMPLE SECRET CODING SCHEME

Keeping information private in an electronic age is becoming more and more difficult. One approach is to encode information, so that even if an unauthorized person sees the information, he or she won't be able to understand it. Modern coding techniques are extremely complicated, but we can create a simple code by taking advantage of the way character information is stored in MATLAB®. If we add a constant integer value to character information, we can transform the string into something that is difficult to interpret.

1. State the Problem
   Encode and decode a string of character information.
2. Describe the Input and Output

   Input Character information entered from the command window
   Output Encoded information

3. Develop a Hand Example
   The lowercase letter *a* is equivalent to the decimal number 97. If we add 5 to *a* and convert it back to a character, it becomes the letter *f.*
4. Develop a MATLAB® Solution

```
%Example 11.1
%Prompt the user to enter a string of character information.
A=input('Enter a string of information to be encoded: ')
encoded=char(A+5);
disp('Your input has been transformed!');
disp(encoded);
disp('Would you like to decode this message?');
response=menu('yes or no?','YES','NO');
switch response
  case 1
    disp(char(encoded-5));
  case 2
    disp('OK - Goodbye');
end
```

5. Test the Solution
   Run the program and observe what happens. The program prompts you for input, which must be entered as a string (inside single quotes):

```
Enter a string of information to be encoded:
'I love rock and roll'
```

   Once you hit the return key, the program responds

```
Your input has been transformed!
N%qt{j%wthp%fsi%wtqq
Would you like to decode this message?
```

Because we chose to use a menu option for the response, the menu window pops up. When we choose YES, the program responds with

```
I love rock and roll
```

If we choose NO, it responds with

```
OK - Goodbye
```

11.4 CELL ARRAYS

Unlike the numeric, character, and symbolic arrays, the cell array can store different types of data inside the same array. Each element in the array is also an array. For example, consider these three different arrays:

```
A = 1:3;
B = ['abcdefg'];
C = single([1,2,3;4,5,6]);
```

We have created three separate arrays, all of a different data type and size. A is a double, B is a char, and C is a single. We can combine them into one cell array by using curly brackets as our cell-array constructor (the standard array constructors are square brackets):

```
my_cellarray = {A,B,C}
```

returns

```
my_cellarray =
   [1x3 double] 'abcdefg' [2x3 single]
```

To save space, large arrays are listed just with size information. You can show the entire array by using the celldisp function:

```
celldisp(my_cellarray)
my_cellarray{1} =
   1   2   3
my_cellarray{2} =
abcdefg
my_cellarray{3} =
   1   2   3
```

The indexing system used for cell arrays is the same as that used in other arrays. You may use either a single index or a row-and-column indexing scheme. There are two approaches to retrieving information from cell arrays: You can use parentheses, as in

```
my_cellarray(1)
ans =
   [1x3 double]
```

Figure 11.7
The `Cellplot` function provides a graphical representation of the structure of a cell array.

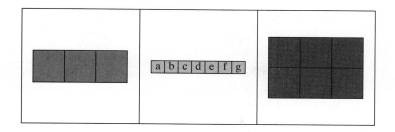

which returns a result as new cell array. An alternative is to use curly brackets, as in

```
my_cellarray{1}
ans =
   1    2    3
```

In this case the answer is a double. To access a particular element inside an array stored in a cell array, you must use a combination of curly brackets and parentheses:

```
my_cellarray{3}(1,2)
ans =
   2
```

Cell arrays can become quite complicated. The `cellplot` function is a useful way to view the structure of the array graphically, as shown in Figure 11.7.

```
cellplot(my_cellarray)
```

Cell arrays are useful for complicated programming projects or for database applications. A use in common engineering applications would be to store various kinds of data from a project in one variable name that can be disassembled and used later.

11.5 STRUCTURE ARRAYS

Structure arrays are similar to cell arrays. Multiple arrays of differing data types can be stored in structure arrays, just as they can in cell arrays. Instead of using content indexing, however, each matrix stored in a structure array is assigned a location called a *field*. For example, using the three arrays from the previous section on cell arrays,

KEY IDEA
Structure arrays can store information using various data types

```
A = 1:3;
B = ['abcdefg'];
C = single([1,2,3;4,5,6]);
```

we can create a simple structure array called my_structure:

```
my_structure.some_numbers = A
```

which returns

```
my_structure =
  some_numbers: [1 2 3]
```

The name of the structure array is `my_structure`. It has one field, called `some_numbers`. We can now add the content in the character matrix B to a second field called `some_letters`:

```
my_structure.some_letters = B
my_structure =
  some_numbers: [1 2 3]
  some_letters: 'abcdefg'
```

Finally, we add the single-precision numbers in matrix C to a third field called `some_more_numbers`:

```
my_structure.some_more_numbers = C
my_structure =
  some_numbers: [1 2 3]
  some_letters: 'abcdefg'
  some_more_numbers: [2x3 single]
```

Notice in the workspace window (Figure 11.8) that the structure matrix (called a `struct`) is a 1 × 1 array that contains all the information from all three dissimilar matrices. The structure has three fields, each containing a different data type:

some_numbers	double-precision numeric data
some_letters	character data
some_more_numbers	single-precision numeric data

We can add more content to the structure, and expand its size, by adding more matrices to the fields we've defined:

```
my_structure(2).some_numbers = [2 4 6 8]
my_structure =
1x2 struct array with fields:
  some_numbers
  some_letters
  some_more_numbers
```

Figure 11.8
Structure arrays can contain many different types of data.

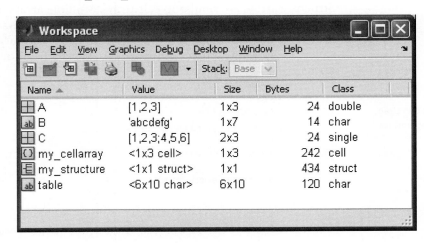

You can access the information in structure arrays by using the matrix name, field name, and index numbers. The syntax is similar to what we have used for other types of matrices. An example is

```
my_structure(2)
ans =
        some_numbers: [2 4 6 8]
        some_letters: []
   some_more_numbers: []
```

Notice that `some_letters` and `some_more_numbers` are empty matrices, because we didn't add information to those fields.

To access just a single field, add the field name:

```
my_structure(2).some_numbers
ans =
   2      4      6      8
```

Finally, if you want to know the content of a particular element in a field, you must specify the element index number after the field name:

```
my_structure(2).some_numbers(2)
ans =
   4
```

The `disp` function displays the contents of structure arrays. For example,

```
disp(my_structure(2).some_numbers(2))
```

returns

```
4
```

You can also use the array editor to access the content of a structure array (and any other array, for that matter). When you double-click the structure array in the workspace window, the array editor opens (Figure 11.9). If you double-click one of the elements of the structure in the array editor, the editor expands to show you the contents of that element (Figure 11.10).

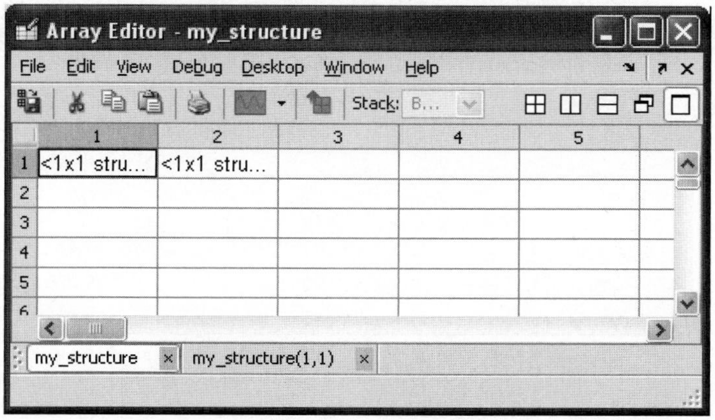

Figure 11.9
The array editor reports the size of an array in order to save space.

Figure 11.10
Double-clicking on a component in the array editor allows us to see the data stored in the array.

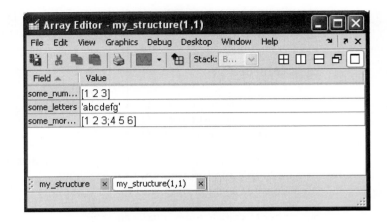

Structure arrays are of limited use in engineering **calculations**, but they are widely used in large computer programs to pass information between functions. The GUIDE program in MATLAB®, which is used to design graphical user interfaces, uses this approach. They are also extremely useful in applications such as **database management**. Since large amounts of engineering data are often stored in a database, the structure array is extremely useful for data analysis. The examples that follow will give you a better idea of how to manipulate and use structure arrays.

EXAMPLE 11.2

STORING PLANETARY DATA WITH STRUCTURE ARRAYS

Structure arrays can be used much like a database. You can store numeric information, as well as character data or any of the other data types supported by MATLAB®. Create a structure array to store information about the planets. Prompt the user to enter the data.

1. **State the Problem**
   Create a structure array to store planetary data and input the information from Table 11.3.

Table 11.3 Planetary Data

Planet Name	Mass, in Earth Multiples	Length of Year, in Earth Years	Mean Orbital Velocity, km/s
Mercury	0.055	0.24	47.89
Venus	0.815	0.62	35.03
Earth	1	1	29.79
Mars	0.107	1.88	24.13
Jupiter	318	11.86	13.06
Saturn	95	29.46	9.64
Uranus	15	84.01	6.81
Neptune	17	164.8	5.43
Pluto	0.002	247.7	4.74

2. Describe the Input and Output

Input

Output A structure array storing the data

3. Develop a Hand Example

Developing a hand example for this problem would be difficult. Instead, a flow-chart would be useful.

4. Develop a MATLAB® Solution

```matlab
%% Example 11.2
clear,clc
k = 1;
response = menu('Would you like to enter planetary
data?','yes','no');
while response==1
  disp('Remember to enter strings in single quotes')
  planetary(k).name = input('Enter a planet name in single
    quotes: ');
  planetary(k).mass = input('Enter the mass in multiples of
    earth''s mass: ');
  planetary(k).year = input('Enter the length of the
    planetary year in Earth years: ');
  planetary(k).velocity = input('Enter the mean orbital
    velocity in km/sec: ');
  %Review the input
  planetary(k)
  increment = menu('Was the data correct?','Yes','No');
  switch increment
    case 1
      increment = 1;
    case 2
      increment = 0;
end
k = k+increment;
response = menu('Would you like to enter more planetary
  data?','yes','no');
end
%%
planetary %output the information stored in planetary
```

Here's a sample interaction in the command window when we run the program and start to enter data:

```
Remember to enter strings in single quotes
Enter a planet name in single quotes: 'Mercury'
Enter the planetary mass in multiples of Earth's mass: 0.055
Enter the length of the planetary year in Earth years: 0.24
Enter the mean orbital velocity in km/sec: 47.89
ans =
    name: 'Mercury'
    mass: 0.0550
    year: 0.2400
    velocity: 47.8900
```

5. Test the Solution

Enter the data, and compare your array with the input table. As part of the program, we reported the input values back to the screen so that the user could check for accuracy. If the user responds that the data are not correct, the information is overwritten the next time through the loop. We also used menus instead of free responses to some questions, so that there would be no ambiguity regarding the answers. Notice that the structure array we built, called `planetary`, is listed in the workspace window. If you double-click on `planetary`, the array editor pops up and allows you to view any of the data in the array (Figure 11.11). You can also update any of the values in the array editor.

Figure 11.11
The array editor allows you to view (and change) data in the structure array.

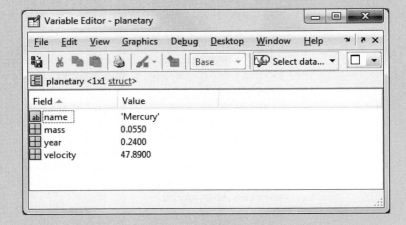

We'll be using this structure array in Example 11.3 to perform some calculations. You'll need to save your results as

```
save    planetary_information    planetary
```

This command sequence saves the structure array `planetary` into the file `planetary_information.mat`.

EXAMPLE 11.3

EXTRACTING AND USING DATA FROM STRUCTURE ARRAYS

Structure arrays have some advantages for storing information. First, they use field names to identify array components. Second, information can be added to the array easily and is always associated with a group. Finally, it's hard to accidentally scramble information in structure arrays. To demonstrate these advantages, use the data you stored in the `planetary_information` file to complete the following tasks:

- Identify the field names in the array, and list them.
- Create a list of the planet names.
- Create a table representing the data in the structure array. Include the field names as column headings in the table.
- Calculate and report the average of the mean orbital velocity values.

- Find the biggest planet and report its size and name.
- Find and report the orbital period of Jupiter.

1. State the Problem
   Create a program to perform the tasks listed.
2. Describe the Input and Output

   Input   `planetary_information.mat`, stored in the current folder

   Output   Create a report in the command window

3. Develop a Hand Example
   You can complete most of the designated tasks by accessing the information in
   the planetary structural array through the array editor
4. Develop a MATLAB® Solution

```
%Example 11.3
clear,clc
load planetary_information
%Identify the field names in the structure array
planetary          %recalls the contents of the structure
                   %array named planetary
pause(2)
%Create a list of planets in the file
disp('These names are OK, but they''re not in an array');
planetary.name
pause(4)
fprintf('\n')   %Creates an empty line in the output
%Using square brackets puts the results into an array
disp('This array isn''t too great');
disp('Everything runs together');
names = [planetary.name]
pause(4)
fprintf('\n')   %Creates an empty line in the output
%Using char creates a padded list, which is more useful
disp('By using a padded character array we get what we
    want');
names = [char(planetary.name)]
pause(4)
%Create a table by first creating character arrays of all
%the data
disp('These arrays are character arrays too');
mass = num2str([planetary.mass]')
fprintf('\n')   %Creates an empty line in the output
pause(4)
year = num2str([planetary.year]')
fprintf('\n')   %Creates an empty line in the output
pause(2)
velocity = num2str([planetary(:).velocity]')
fprintf('\n')   %Creates an empty line in the output
```

```
pause(4)
fprintf('\n')   %Creates an empty line in the output
%Create an array of spaces to separate the data
spaces = ['          ']';
%Use disp to display the field names
disp('The overall result is a big character array');
fprintf('\n')   %Creates an empty line in the output
disp('Planet  mass    year velocity');
table = [names,spaces,mass,spaces,year,spaces,velocity];
disp(table);
fprintf('\n')   %Creates an empty line in the output
pause(2)
%Find the average planet mean orbital velocity
MOV = mean([planetary.velocity]);
fprintf('The mean orbital velocity is %8.2f km/sec\n',MOV)
pause(1)
%Find the planet with the maximum mass
max_mass = max([planetary.mass]);
fprintf('The maximum mass is %8.2f times the earth''s
    \n',max_mass)
pause(1)
%Jupiter is planet #5
%Find the orbital period of Jupiter
planet_name = planetary(5).name;
planet_year = planetary(5).year;
fprintf(' %s has a year %6.2f times the earth''s
    \n',planet_name,planet_year)
```

Most of this program consists of formatting commands. Before you try to analyze the code, run the program in MATLAB® and observe the results.

5. Test the Solution

Compare the information extracted from the array with that available from the array editor. Using the array editor becomes unwieldy as the data stored in planetary increases. It is easy to add new fields and new information as they become available. For example, we could add the number of moons to the existing structure:

```
planetary(1).moons = 0;
planetary(2).moons = 0;
planetary(3).moons = 1;
planetary(4).moons = 2;
planetary(5).moons = 60;
planetary(6).moons = 31;
planetary(7).moons = 27;
planetary(8).moons = 13;
planetary(9).moons = 1;
```

This code adds a new field called moons to the structure. We can report the number of moons for each planet to the command window with the command

```
disp([planetary.moons]);
```

SUMMARY

MATLAB®'s primary data structure is the array. Within the array, MATLAB® allows the user to store a number of different types of data. The default numeric data type is the double-precision floating-point number, usually referred to as a double. MATLAB® also supports single-precision floating-point numbers, as well as eight different types of integers. Character information, too, is stored in arrays. Characters can be grouped together into a string, although the string represents a one-dimensional array in which each character is stored in its own element. The `char` function allows the user to create two-dimensional character arrays from strings of different sizes by "padding" the array with an appropriate number of blank spaces. In addition to numeric and character data, MATLAB® includes a symbolic data type.

All these kinds of data can be stored as two-dimensional arrays. Scalar and vector data are actually stored as two-dimensional arrays—they just have a single row or column. MATLAB® also allows the user to store data in multidimensional arrays. Each two-dimensional slice of a three-dimensional or higher array is called a page.

In general, all data stored in a MATLAB® array must be of the same type. If character and numeric data are mixed, the numeric data are changed to character data on the basis of their ASCII-equivalent decimal values. When calculations are attempted on combined character and numeric data, the character data are converted to their ASCII equivalents.

MATLAB® offers two array types that can store multiple types of data at the same time: the cell array and the structure array. Cell arrays use curly brackets, {and} as array constructors. Structure arrays depend on named fields. Both cell and structure arrays are particularly useful in database applications.

MATLAB® SUMMARY

The following MATLAB® summary lists and briefly describes all the special characters, commands, and functions that are defined in this chapter:

Special Characters	
{ }	cell-array constructor
' '	string data (character information)
abc	character array
⊞	numeric array
▣	symbolic array
✓	logical array
⧄	sparse array
{}	cell array
⫟	structure array

Commands and Functions	
celldisp	displays the contents of a cell array
char	creates a padded character array

(Continued)

Commands and Functions

cumsum	finds the cumulative sum of the members of an array
double	changes an array to a double-precision array
eye	creates an identity matrix
format rat	converts the display format to rational numbers (fractions)
int16	16-bit signed integer
int32	32-bit signed integer
int64	64-bit signed integer
int8	8-bit signed integer
intmax	determines the largest integer that can be stored in MATLAB®
intmin	determines the smallest integer that can be stored in MATLAB®
num2str	converts a numeric array to a character array
realmax	determines the largest real number that can be expressed in MATLAB®
realmin	determines the smallest real number that can be expressed in MATLAB®
single	changes an array to a single-precision array
sparse	converts a full-format matrix to a sparse-format matrix
squeeze	removes singleton dimensions from multidimensional arrays
str2num	converts a character array to a numeric array
uint16	16-bit unsigned integer
uint32	32-bit unsigned integer
uint64	64-bit unsigned integer
uint8	8-bit unsigned integer

KEY TERMS

ASCII	double precision	rational numbers
base 2	drawers	single precision
cell	EBCDIC	string
character	floating-point numbers	structure
class	integer	symbolic data
complex numbers	logical data	
data type	pages	

PROBLEMS

Numeric Data Types

11.1 Calculate the sum (not the partial sums) of the first 10 million terms in the harmonic series

$$\frac{1}{1} + \frac{1}{2} + \frac{1}{3} + \frac{1}{4} + \frac{1}{5} + \frac{1}{6} + \cdots + \frac{1}{n} + \cdots$$

using both double-precision and single-precision numbers. Compare the results. Explain why they are different.

11.2 Define an array of the first 10 integers, using the `int8` type designation. Use these integers to calculate the first 10 terms in the harmonic series. Explain your results.

11.3 Explain why it is better to allow MATLAB® to default to double-precision floating-point number representations for most engineering calculations than to specify single and integer types.

11.4 Complex numbers are automatically created in MATLAB® as a result of calculations. They can also be entered directly, as the addition of a real and an imaginary number, and can be stored as any of the numeric data types. Define two variables: a single- and a double-precision complex number, as

$$\textbf{doublea} = \textbf{5} + \textbf{3}\boldsymbol{i}$$
$$\textbf{singlea} = \textbf{single}(\textbf{5} + \textbf{3}\boldsymbol{i})$$

Raise each of these numbers to the 100th power. Explain the difference in your answers.

Character Data

11.5 Use an Internet search engine to find a list showing the binary equivalents of characters in both ASCII and EBCDIC. Briefly outline the differences in the two coding schemes.

11.6 Sometimes it is confusing to realize that numbers can be represented as both numeric data and character data. Use MATLAB® to express the number 85 as a character array.

(a) How many elements are in this array?
(b) What is the numeric equivalent of the character 8?
(c) What is the numeric equivalent of the character 5?

Multidimensional Arrays

11.7 Create each of the following arrays:

$$A = \begin{bmatrix} 1 & 2 \\ 3 & 4 \end{bmatrix}, \quad B = \begin{bmatrix} 10 & 20 \\ 30 & 40 \end{bmatrix}, \quad C = \begin{bmatrix} 3 & 16 \\ 9 & 12 \end{bmatrix}$$

(a) Combine them into one large $2 \times 2 \times 3$ multidimensional array called `ABC`.
(b) Extract each column 1 into a 2×3 array called `Column_A1B1C1`.
(c) Extract each row 2 into a 3×2 array called `Row_A2B2C2`.
(d) Extract the value in row 1, column 2, page 3.

11.8 A college professor would like to compare how students perform on a test she gives every year. Each year, she stores the data in a two-dimensional array. The first and second year's data are as follows:

Year 1	Question 1	Question 2	Question 3	Question 4
Student 1	3	6	4	10
Student 2	5	8	6	10
Student 3	4	9	5	10
Student 4	6	4	7	9
Student 5	3	5	8	10

Year 2	Question 1	Question 2	Question 3	Question 4
Student 1	2	7	3	10
Student 2	3	7	5	10
Student 3	4	5	5	10
Student 4	3	3	8	10
Student 5	3	5	2	10

(a) Create a two-dimensional array called `year1` for the first year's data, and another called `year2` for the second year's data.

(b) Combine the two arrays into a three-dimensional array with two pages, called `testdata`.

(c) Use your three-dimensional array to perform the following calculations:

- Calculate the average score for each question, for each year, and store the results in a two-dimensional array. (Your answer should be either a 2×4 array or a 4×2 array.)
- Calculate the average score for each question, using `all` the data.
- Extract the data for Question 3 for each year, and create an array with the following format:

	Question 3, Year 1	Question 3, Year 2
Student 1		
Student 2		
and so on		

11.9 If the teacher described in the preceding problem wants to include the results from a second and third test in the array, she would have to create a four-dimensional array. (The fourth dimension is sometimes called a *drawer*.) All the data are included in a file called `test_results.mat` consisting of six two-dimensional arrays similar to those described in Problem 11.8. The array names are

```
test1year1
test2year1
test3year1
test1year2
test2year2
test3year2
```

Organize these data into a four-dimensional array that looks like the following:

dimension 1	(row)	student
dimension 2	(column)	question
dimension 3	(page)	year
dimension 4	(drawer)	test

(a) Extract the score for Student 1, on Question 2, from the first year, on Test 3.

(b) Create a one-dimensional array representing the scores from the first student, on Question 1, on the second test, for all the years.

(c) Create a one-dimensional array representing the scores from the second student, on all the questions, on the first test, for Year 2.

(d) Create a two-dimensional array representing the scores from all the students, on Question 3, from the second test, for all the years.

Character Arrays

11.10 (a) Create a padded character array with five different names.

(b) Create a two-dimensional array called `birthdays` to represent the birthday of each person. For example, your array might look something like this:

```
birthdays=
       6    11    1983
       3    11    1985
       6    29    1986
      12    12    1984
      12    11    1987
```

(c) Use the `num2str` function to convert `birthdays` to a character array.

(d) Use the `disp` function to display a table of names and birthdays.

11.11 Imagine that you have the following character array, which represents the dimensions of some shipping boxes:

```
box_dimensions =

box1    1    3    5
box2    2    4    6
box3    6    7    3
box4    1    4    3
```

You need to find the volumes of the boxes to use in a calculation to determine how many packing "peanuts" to order for your shipping department. Since the array is a 4×12 character array, the character representation of the numeric information is stored in columns 6 to 12. Use the `str2num` function to convert the information into a numeric array, and use the data to calculate the volume of each box. (You'll need to enter the `box_dimensions` array as string data, using the `char` function.)

11.12 Consider the file called `thermocouple.dat` as shown in the table on the next page:

(a) Create a program that:

- Loads `thermocouple.dat` into MATLAB®.
- Determines the size (number of rows and columns) of the file.
- Extracts each set of thermocouple data and stores it into a separate file. Name the various files `thermocouple1.mat`, `thermocouple2. mat`, etc.

(b) Your program should be able to accept a two-dimensional file of any size. Do not assume that there are only three columns; let the program determine the array size and assign appropriate file names.

Thermocouple 1	Thermocouple 2	Thermocouple 3
84.3	90.0	86.7
86.4	89.5	87.6

Thermocouple 1	Thermocouple 2	Thermocouple 3
85.2	88.6	88.3
87.1	88.9	85.3
83.5	88.9	80.3
84.8	90.4	82.4
85.0	89.3	83.4
85.3	89.5	85.4
85.3	88.9	86.3
85.2	89.1	85.3
82.3	89.5	89.0
84.7	89.4	87.3
83.6	89.8	87.2

11.13 Create a program that encodes text entered by the user and saves it into a file. Your code should add 10 to the decimal equivalent value of each character entered.

11.14 Create a program to decode a message stored in a data file by subtracting 10 from the decimal equivalent value of each character.

Cell Arrays

11.15 Create a cell array called `sample_cell` to store the following individual arrays:

$$A = \begin{bmatrix} 1 & 3 & 5 \\ 3 & 9 & 2 \\ 11 & 8 & 2 \end{bmatrix} \text{ (a double-precision floating-point array)}$$

$$B = \begin{bmatrix} fred & ralph \\ ken & susan \end{bmatrix} \text{ (a padded character array)}$$

$$C = \begin{bmatrix} 4 \\ 6 \\ 3 \\ 1 \end{bmatrix} \text{ (an \textbf{int8} integer array)}$$

(a) Extract array A from `sample_cell`.
(b) Extract the information in array C, row 3, from `sample_cell`.
(c) Extract the name *fred* from `sample_cell`. Remember that the name `fred` is a 1×4 array, not a single entity.

11.16 Cell arrays can be used to store character information without padding the character arrays. Create a separate character array for each of the strings

aluminum
copper
iron
molybdenum
cobalt

and store them in a cell array.

11.17 Consider the following information about metals:

Metal	Symbol	Atomic Number	Atomic Weight	Density, g/cm³	Crystal Structure
Aluminum	Al	13	26.98	2.71	FCC
Copper	Cu	29	63.55	8.94	FCC
Iron	Fe	26	55.85	7.87	BCC
Molybdenum	Mo	42	95.94	10.22	BCC
Cobalt	Co	27	58.93	8.9	HCP
Vanadium	V	23	50.94	6.0	BCC

(a) Create the following arrays:
- Store the name of each metal into an individual character array, and store all these character arrays into a cell array.
- Store the symbol for all these metals into a single padded character array.
- Store the atomic number into an `int8` integer array.
- Store the atomic weight into a double-precision numeric array.
- Store the density into a single-precision numeric array.
- Store the structure into a single padded character array.

(b) Group the arrays you created in part (a) into a single cell array.

(c) Extract the following information from your cell array:
- Find the name, atomic weight, and structure of the fourth element in the list.
- Find the names of all the elements stored in the array.
- Find the average atomic weight of the elements in the table. (Remember, you need to extract the information to use in your calculation from the cell array.)

Structure Arrays

11.18 Store the information presented in Problem 11.17 in a structure array. Use your structure array to determine the element with the maximum density.

11.19 Create a program that allows the user to enter additional information into the structure array you created in Problem 11.18. Use your program to add the following data to the array:

Metal	Symbol	Atomic Number	Atomic Weight	Density, g/cm³	Crystal Structure
Lithium	Li	3	6.94	0.534	BCC
Germanium	Ge	32	72.59	5.32	Diamond cubic
Gold	Au	79	196.97	19.32	FCC

11.20 Use the structure array you created in Problem 11.19 to find the element with the maximum atomic weight.

12

Symbolic Mathematics

Objectives

After reading this chapter, you should be able to:

- Create and manipulate symbolic variables
- Factor and simplify mathematical expressions
- Solve symbolic expressions
- Solve systems of equations
- Determine the symbolic derivative of an expression
- Integrate an expression

INTRODUCTION

MATLAB® has a number of different data types, including both double-precision and single-precision numeric data, character data, logical data, and symbolic data, all of which are stored in a variety of different arrays. In this chapter, we will explore how symbolic arrays allow MATLAB® users to manipulate and use symbolic data.

MATLAB®'s symbolic capability is based on the MuPad software, originally produced by SciFace Software (based on research done at the University of Paderborn, Germany). SciFace was purchased by the Mathworks (publishers of MATLAB®) in 2008. The MuPad engine is part of the symbolic toolbox, which is included with the Student Edition of MATLAB®. It is available for purchase separately for the Professional Edition of MATLAB®. There are two ways to use MuPad inside the MATLAB® software. You can access it directly and create a MuPad notebook by typing

```
mupad
```

at the command prompt. The MuPad notebook interface opens as a MATLAB® figure window, as shown in Figure 12.1. If you are familiar with other symbolic algebra programs such as MAPLE the syntax will probably look familiar.

Figure 12.1
The MuPad interface in MATLAB®.

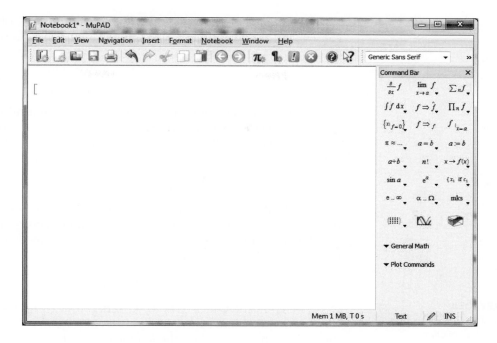

Mupad can also be used to create symbolic objects inside MATLAB® itself. This offers the advantage of a familiar interface, and the ability to interact with MATLAB®'s other functions. Earlier versions of MATLAB® (before 2007b) used the MAPLE symbolic algebra program as the engine for the symbolic math toolbox. Most of the symbolic manipulations performed in this chapter should work with these earlier versions of MATLAB®; however, some of the results will be represented in the command window in a different order. If your version of MATLAB® is 2007b or later, the Mupad interface should be functional; however, problems can occur if Maple is also installed on your computer. The standard installation of Maple adds a Maple toolbox to MATLAB®, which supersedes the Symbolic Toolbox. You can determine if this has occurred on your system, by checking the help feature, which lists the installed toolboxes. If the Maple toolbox is installed, you won't be able to use the MuPad interface.

MATLAB®'s symbolic toolbox allows us to manipulate symbolic expressions to simplify them, to solve them symbolically, and to evaluate them numerically. It also allows us to take derivatives, to integrate, and to perform linear algebraic manipulations. More advanced features include LaPlace transforms, Fourier transforms, and variable-precision arithmetic.

12.1 SYMBOLIC ALGEBRA

Symbolic mathematics is used regularly in math, engineering, and science classes. It is often preferable to manipulate equations symbolically before you substitute values for variables. For example, consider the equation

$$y = \frac{2(x + 3)^2}{x^2 + 6x + 9}$$

At first glance, y appears to be a fairly complicated function of x. However, if you expand the quantity $(x + 3)^2$, it becomes apparent that you can simplify the equation to

$$y = \frac{2*(x + 3)^2}{x^2 + 6x + 9} = \frac{2*(x^2 + 6x + 9)}{(x^2 + 6x + 9)} = 2$$

You may or may not want to perform this simplification, because, in doing so, you lose some information. For example, for values of x equal to -3, y is undefined, since $x + 3$ becomes 0, as does $x^2 + 6x + 9$. Thus,

$$y = \frac{2(-3 + 3)^2}{9 - 18 + 9} = 2\frac{0}{0} = \text{undefined}$$

MATLAB®'s symbolic algebra capabilities allow you to perform this simplification or to manipulate the numerator and denominator separately.

Relationships are not always constituted in forms that are so easy to solve. For instance, consider the equation

$$k = k_0 e^{-Q/RT}$$

If we know the values of k_0, Q, R, and T, it's easy to solve for k. It's not so easy if we want to find T and we know the values of k, k_0, R, and Q. We have to manipulate the relationship to get T on the left-hand side of the equation:

$$\ln(k) = \ln(k_0) - \frac{Q}{RT}$$

$$\ln\left(\frac{k}{k_0}\right) = -\frac{Q}{RT}$$

$$\ln\left(\frac{k_0}{k}\right) = \frac{Q}{RT}$$

$$T = \frac{Q}{R\ln(k_0/k)}$$

Although solving for T was awkward manually, it's easy with MATLAB®'s symbolic capabilities.

12.1.1 Creating Symbolic Variables

Before we can solve any equations, we need to create some symbolic variables. Simple symbolic variables can be created in two ways. For example, to create the symbolic variable x, type either

```
x = sym('x')
```

or

```
syms x
```

Both techniques set the character 'x' equal to the symbolic variable x. More complicated variables can be created by using existing symbolic variables, as in the expression

```
y = 2*(x + 3)^2/(x^2 + 6*x + 9)
```

Notice in the workspace window (Figure 12.2) that both x and y are listed as symbolic variables and that the array size for each is 1×1.

Figure 12.2
Symbolic variables are identified in the workspace window. They require a variable amount of storage.

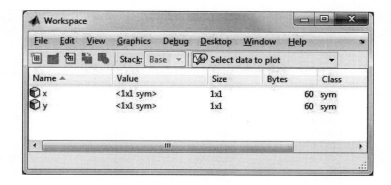

The `syms` command is particularly convenient, because it can be used to create multiple symbolic variables at the same time, as with the command

```
syms Q R T k0
```

These variables could be combined mathematically to create another symbolic variable, `k`:

```
k = k0*exp(-Q/(R*T))
```

EXPRESSION
A set of mathematical operations

Notice that in both examples we used the standard algebraic operators, not the array operators, such as `.*` or `.^`. This makes sense when we observe that array operators specify that corresponding elements in arrays are used in the associated calculations—a situation that does not apply here.

EQUATION
An expression set equal to a value or another expression

The `sym` function can also be used to create either an entire expression or an entire equation. For example,

```
E = sym('m*c^2')
```

creates a symbolic variable named `E`. Notice that m and **c** are not listed in the workspace window (Figure 12.3); they have not been specifically defined as symbolic variables. Instead, the input to sym was a character string, identified by the single quotes inside the function.

Figure 12.3
Unless a variable is explicitly defined, it is not listed in the workspace window.

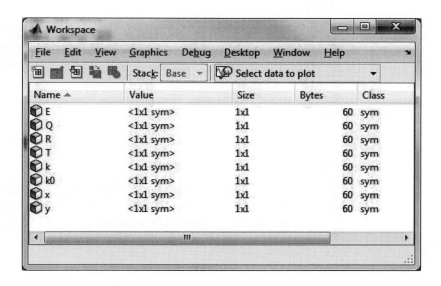

Figure 12.4
The variable `ideal_gas_law` is an equation, not an expression.

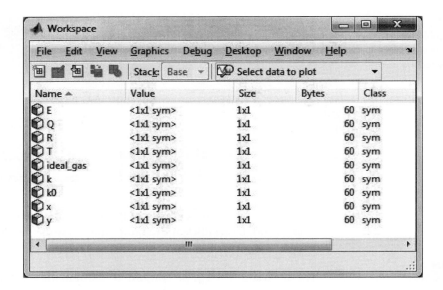

KEY IDEA
The symbolic toolbox uses standard algebraic operators

In this example, we set the ***expression*** m * c^2 equal to the variable E. We can also create an entire ***equation*** and give it a name. For example, we can define the ideal gas law

```
ideal_gas_law = sym('P*V = n*R*Temp')
```

At this point, if you've been typing in the examples as you read along, your workspace window should look like Figure 12.4. Notice that only `ideal_gas_law` is listed as a symbolic variable, since P, V, n, R, and Temp have not been explicitly defined, but were part of the character string input to the `sym` function.

HINT

One ideosyncracy of the implementation of MuPad inside MATLAB® is that a number of commonly used variables are reserved. They can be overwritten, however, if you try to use them inside expressions or equations you may run into problems. Try to avoid these names:

```
D, E, I, O, beta, zeta, theta, psi, gamma, Ci, Si, Ei
```

HINT

Notice that when you use symbolic variables, MATLAB® does not indent the result, unlike the format used for numeric results. This can help you keep track of variable types without referring to the workspace window.

PRACTICE EXERCISES 12.1

1. Create the following symbolic variables, using either the `sym` or `syms` command:

   ```
   x, a, b, c, d
   ```

2. Verify that the variables you created in Exercise 1 are listed in the workspace window as symbolic variables. Use them to create the following symbolic *expressions*:

   ```
   ex1 = x^2-1
   ex2 = (x+1)^2
   ex3 = a*x^2-1
   ex4 = a*x^2 + b*x + c
   ex5 = a*x^3 + b*x^2 + c*x + d
   ex6 = sin(x)
   ```

3. Create the following symbolic *expressions*, using the `sym` function:

   ```
   EX1 = sym('X^2 - 1 ')
   EX2 = sym(' (X + 1)^2 ')
   EX3 = sym('A*X ^2 - 1 ')
   EX4 = sym('A*X ^2 + B*X + C ')
   EX5 = sym('A*X ^3 + B*X ^2 + C*X + F ')
   EX6 = sym('sin(X) ')
   ```

4. Create the following symbolic *equations*, using the `sym` function:

   ```
   eq1 = sym(' x^2=1 ')
   eq2 = sym(' (x+1)^2=0 ')
   eq3 = sym(' a*x^2=1 ')
   eq4 = sym('a*x^2 + b*x + c=0 ')
   eq5 = sym('a*x^3 + b*x^2 + c*x + d=0 ')
   eq6 = sym('sin(x)=0 ')
   ```

5. Create the following symbolic *equations*, using the `sym` function:

   ```
   EQ1 = sym('X^2 = 1 ')
   EQ2 = sym('(X +1)^2=0 ')
   EQ3 = sym('A*X ^2 =1 ')
   EQ4 = sym('A*X ^2 + B*X + C = 0 ')
   EQ5 = sym('A*X ^3 + B*X ^2 + C*X + F = 0 ')
   EQ6 = sym(' sin(X) = 0 ')
   ```

Notice that only the explicitly defined variables, expressions, and equations are listed in the workspace window. Also notice that instead of D in the places where it should logically occur, we've used F. The reason is that D (and E for that matter) is a reserved name, and can cause problems if used in expressions or equations. Save the variables, expressions, and equations you created in this practice to use in later practice exercises in the chapter.

12.1.2 Manipulating Symbolic Expressions and Symbolic Equations

First, we need to remind ourselves how expressions and equations differ. Equations are set equal to something; expressions are not. The variable `ideal_gas_law` has been set equal to an equation. If you type in

```
ideal_gas_law
```

MATLAB® will respond

```
ideal_gas_law =
P*V = n*R*Temp
```

However, if you type in

```
E
```

MATLAB® responds

```
E=
m*c^2
```

or if you type in

```
y
```

MATLAB® responds

```
y =
2*(x+3)^2/(x^2+6*x+9)
```

The variables E and y are *expressions*, but the variable `ideal_gas_law` is an ***equation***. Most of the time you will be working with symbolic ***expressions***.

MATLAB® has a number of functions designed to manipulate symbolic variables, including functions to separate an expression into its numerator and denominator, to expand or factor expressions, and to simplify expressions in a number of ways.

Extracting Numerators and Denominators

The `numden` function extracts the numerator and denominator from an expression. For example, if you've defined **y** as

```
y = 2*(x+3)^2/(x^2+6*x+9)
```

then you can extract the numerator and denominator with

```
[num,den] = numden(y)
```

MATLAB® creates two new variables, num and den (of course, you could name them whatever you please):

```
num =
2*(x+3)^2
den =
x^2+6*x+9
```

We can recombine these expressions or any symbolic expressions by using standard algebraic operators:

```
num*den
ans =
2*(x+3)^2*(x^2+6*x+9)
```

```
num/den
ans =
2*(x+3)^2/(x^2+6*x+9)
num+den
ans =
2*(x+3)^2+x^2+6*x+9
```

Expanding Expressions, Factoring Expressions, and Collecting Terms

We can use the expressions we have defined to demonstrate the use of the expand, factor, and collect functions. Thus,

```
expand(num)
```

returns

```
ans =
2*x^2+12*x+18
```

and

```
factor(den)
```

returns

```
ans =
(x+3)^2
```

The collect function collects like terms and is similar to the expand function:

```
collect(num)
ans =
2*x^2 + 12*x + 18
```

This works regardless of whether each individual variable in an expression has or has not been defined as a symbolic variable. Define a new variable z:

```
z = sym('3*a-(a+3)*(a-3)^2')
```

In this case, both expand and factor give the same result:

```
factor(z)
ans =
-a^3 + 3*a^2 + 12*a - 27
expand(z)
ans =
-a^3 + 3*a^2 + 12*a - 27
```

The result obtained by using collect is also the same:

```
collect(z)
ans =
-27-a^3+3*a^2+12*a
```

You can use both factor and expand with equations as well as with expressions. The collect function requires an expression. With equations, each side of the equation is treated as a separate expression. To illustrate, we can define an equation w:

```
w = sym('x^3-1 = (x-3)*(x+3)')
expand(w)
ans =
x^3-1 = x^2-9
```

```
factor(w)
ans =
(x-1)*(x^2+x+1) = (x-3)*(x+3)
collect(w)
??? Error using ==> mupadmex
```

Note that an error was generated when we tried to use the `collect` function with w, because w is an equation, not an expression.

Simplification Functions

We can think of the `expand`, `factor`, and `collect` functions as ways to simplify an expression or equation. However, what constitutes a "simple" equation is not always obvious. The `simplify` function simplifies each part of an expression or equation, using MuPad's built-in simplification rules. For example, assume again that z has been defined as

```
z = sym('3*a-(a+3)*(a-3)^2')
```

Then, the command

```
simplify(z)
```

returns

```
ans =
3*a-(a-3)^2*(a+3)
```

If the equation w has been defined as

```
w = sym('x^3-1 = (x-3)*(x+3)')
```

then

```
simplify(w)
```

returns

```
ans =
x^3 + 8 = x^2
```

Notice again that this works regardless of whether each individual variable in an expression has or has not been defined as a symbolic variable: The expression z contains the variable a, which has not been explicitly defined and is not listed in the workspace window.

The `simple` function is slightly different. It tries a number of different simplification techniques and reports the result that is *shortest*. All the attempts are reported to the screen. For example,

```
simple(w)
```

gives the following results:

```
simplify:
x^3-1 = x^2 - 9
radsimp:
x^3-1 = (x-3)*(x+3)
simplify(100):
x in RootOf(X90^3 - X90^2 + 8, X90)
combine(sincos):
x^3-1 = (x-3)*(x+3)
```

```
combine(sinhcosh):
x^3-1 = (x-3)*(x+3)
combine(ln):
x^3-1 = (x-3)*(x+3)
factor:
x^3-1 = x^2-9
expand:
x^3-1 = x^2-9
combine:
x^3-1 = (x-3)*(x+3)
rewrite(exp):
x^3-1 = (x-3)*(x+3)
rewrite(sincos):
x^3-1 = (x-3)*(x+3)
rewrite(sinhcosh):
x^3-1 = (x-3)*(x+3)
rewrite(tan):
x^3-1 = (x-3)*(x+3)
mwcos2sin:
x^3-1 = (x-3)*(x+3)
ans =
x^3-1 = x^2-9
```

KEY IDEA
Many, but not all, symbolic functions work for both expressions and equations

Notice that although a large number of results are displayed, there is only one answer:

```
ans =
x^2-1 = x^2-9
```

Both `simple` and `simplify` work on expressions as well as equations.

Table 12.1 lists some of the MATLAB® functions used to manipulate expressions and equations.

HINT

A shortcut to create a symbolic polynomial is the `poly2sym` function. This function requires a vector as input and creates a polynomial, using the vector for the coefficients of each term of the polynomial.

```
a = [1,3,2]
a =
   1   3   2
b = poly2sym(a)
b =
x^2+3*x+2
```

Similarly, the `sym2poly` function converts a polynomial into a vector of coefficient values:

```
c = sym2poly(b)
c =
   1   3   2
```

Table 12.1 Functions Used to Manipulate Expressions and Equations

expand(S)	Multiplies out all the portions of the expression or equation	**syms x** **expand((x-5)*(x+5))** **ans =** **x^2-25**
factor(S)	Factors the expression or equation	**syms x** **factor(x^3-1)** **ans =** **(x-1)*(x^2+x+1)**
collect(S)	Collects like terms	**S=2*(x+3)^2+x^2+6*x+9** **collect(S)** **S =** **27+3*x^2+18*x**
simplify(S)	Simplifies in accordance with MuPad's simplification rules	**syms a** **simplify(exp(log(a)))** **ans =** **a**
simple(S)	Simplifies to the shortest representation of the expression or equation	**syms x** **simple(sin(x)^2+** **cos(x)^2)** **ans =** **1**
numden(S)	Finds the numerator of an expression; this function is not valid for equations	**syms x** **numden((x-5)/(x+5))** **ans =** **x-5**
[num,den]=numden(S)	Finds both the numerator and the denominator of an expression; this function is not valid for equations	**syms x** **[num,den] = numden((x-5)/** **(x+5))** **num =** **x-5** **den =** **x+5**

PRACTICE EXERCISES 12.2

Use the variables defined in Practice Exercises 12.1 in these exercises.

1. Multiply ex1 by ex2, and name the result y1.
2. Divide ex1 by ex2, and name the result y2.
3. Use the numden function to extract the numerator and denominator from y1 and y2.
4. Multiply EX1 by EX2, and name the result Y1.
5. Divide EX1 by EX2, and name the result Y2.
6. Use the numden function to extract the numerator and denominator from Y1 and Y2.
7. Try using the numden function on one of the equations you've defined. Does it work?
8. Use the factor, expand, collect, and simplify functions on y1, y2, Y1, and Y2.
9. Use the factor, expand, collect, and simplify functions on the expressions ex1 and ex2 and on the corresponding equations eq1 and eq2. Explain any differences you observe.

12.2 SOLVING EXPRESSIONS AND EQUATIONS

A highly useful function in the symbolic toolbox is `solve`. It can be used to determine the roots of expressions, to find numerical answers when there is a single variable, and to solve for an unknown symbolically. The `solve` function can also solve systems of equations, both linear and nonlinear. When paired with the substitution function (`subs`), the `solve` function allows the user to find analytical solutions to a variety of problems.

12.2.1 The Solve Function

When used with an expression, the `solve` function sets the expression equal to zero and solves for the roots. For example (assuming that x has already been defined as a symbolic variable), if

```
E1 = x-3
```

then

```
solve(E1)
```

returns

```
ans =
3
```

`Solve` can be used either with an expression name or by creating a symbolic expression directly in the `solve` function. Thus,

```
solve('x^2-9')
```

returns

```
ans =
-3
3
```

Notice that `ans` is a 2×1 symbolic array. If **x** has been previously defined as a symbolic variable, then *single quotes are not necessary*. If not, the entire expression must be enclosed within single quotes.

You can readily solve symbolic expressions with more than one variable. For example, for the quadratic expression $ax^2 + bx + c$,

```
solve('a*x^2+b*x +c')
```

returns

```
ans =
-(b + (b^2-4*a*c)^(1/2))/(2*a)
-(b - (b^2-4*a*c)^(1/2))/(2*a)
```

KEY IDEA
MATLAB® solves preferentially for x

MATLAB® preferentially solves for x. If there is no x in the expression, MATLAB® finds the variable closest to x. If you want to specify the variable to solve for, just include it in the second field. For instance, to solve the quadratic expression for a, the command

```
solve('a*x^2+b*x +c', 'a')
```

returns

```
ans =
-(c+b*x)/x^2
```

Again, if a has been specifically defined as a symbolic variable, it is not necessary to enclose it in single quotes:

```
syms a b c x
solve(a*x*x^2+b*x+c, b)
ans =
-(a*x^3+c)/x
```

To solve an expression set equal to something besides zero, you must use one of two approaches. If the equation is simple, you can transform it into an expression by subtracting the right-hand side from the left-hand side. For example,

$$5x^2 + 6x + 3 = 10$$

could be reformulated as

$$5x^2 + 6x - 7 = 0$$

```
solve('5*x^2+6*x-7')
ans =
-(2*11^(1/2))/5-3/5
(2*11^(1/2))/5-3/5
```

If the equation is more complicated, you may prefer to define a new equation, as in

```
E2 = sym('5*x^2 + 6*x +3 = 10')
solve(E2)
```

which returns

```
ans =
-(2*11^(1/2))/5-3/5
(2*11^(1/2))/5-3/5
```

Notice that in both cases the results are expressed as simply as possible, using fractions (i.e., rational numbers). In the workspace, ans is listed as a 2×1 symbolic matrix. You can use the `double` function to convert a symbolic representation to a double-precision floating-point number:

```
double(ans)
ans =
0.7266
-1.9266
```

HINT

Because MATLAB®'s symbolic capability is based on MuPad, we need to understand how MuPad handles calculations. MuPad recognizes two types of numeric data: integers and floating point. Floating-point numbers are approximations and use decimal points, whereas integers are exact and are represented without decimal points. In calculations using integers, MuPad forces an exact answer resulting in fractions. If there are decimal points (floating-point numbers) in MuPad calculations, the result will also be an approximation and will contain decimal points. MuPad defaults to 32 significant figures, so 32 digits are shown in the results. Consider an example using `solve`. If the expression uses floating-point numbers, we get the following result:

```
solve('5.0*x^2.0+6.0*x-7.0')
ans =
.72664991614215993964597309466828
-1.9266499161421599396459730946683
```

If the expression uses integers, the results are fractions:

```
solve('5*x^2+6*x-7')
ans =
-(2*11^(1/2))/5-3/5
 (2*11^(1/2))/5-3/5
```

The `solve` function is particularly useful with symbolic expressions having multiple variables:

```
E3 = sym('P = P0*exp(r*t)')
solve(E3,'t')
ans =
log(P/P0)/r
```

If you have previously defined t as a symbolic variable, it does not need to be in single quotes. (Recall that the log function is a natural log.)

It is often useful to redefine a variable, such as t, in terms of the other variables:

```
t = solve(E3,'t')
t =
log(P/P0)/r
```

PRACTICE EXERCISES 12.3

Use the variables and expressions you defined in Practice Exercises 12.1 to solve these exercises:

1. Use the `solve` function to solve all four versions of expression/equation 1: ex1, EX1, eq1, and EQ1.
2. Use the `solve` function to solve all four versions of expression/equation 2: ex2, EX2, eq2, and EQ2.
3. Use the `solve` function to solve ex3, and eq3 for both x and a.
4. Use the `solve` function to solve EX3, and EQ3 for both X and A. Recall that neither X nor A has been explicitly defined as a symbolic variable.
5. Use the `solve` function to solve ex4, and eq4 for both x and a.
6. Use the `solve` function to solve EX4, and EQ4 for both X and A. Recall that neither X nor A has been explicitly defined as a symbolic variable.
7. All four versions of expression/equation 4 represent the quadratic equation—the general form of a second-order polynomial. The solution for *x* is usually memorized by students in early algebra classes. Expression/equation 5 in these exercises is the general form of a third-order polynomial. Use the `solve` function to solve these expressions/equations, and comment on why students do not memorize the general solution of a third-order polynomial.
8. Use the `solve` function to solve ex6, EX6, eq6, and EQ6. On the basis of your knowledge of trigonometry, comment on this solution.

EXAMPLE 12.1

USING SYMBOLIC MATH

MATLAB®'s symbolic capability allows us to let the computer do the math. Consider the equation for reaction rate constants:

$$k = k_0 \exp\left(\frac{-Q}{RT}\right)$$

Solve this equation for Q, using MATLAB®.

1. **State the Problem**

   Find the equation for Q.

2. **Describe the Input and Output**

   Input Equation for the reaction rate constant, k

   Output Equation for Q

3. **Develop a Hand Example**

$$k = k_0 \exp\left(\frac{-Q}{RT}\right)$$

$$\frac{k}{k_0} = \exp\left(\frac{-Q}{RT}\right)$$

$$\ln\left(\frac{k}{k_0}\right) = \frac{-Q}{RT}$$

$$Q = RT \ln\left(\frac{k_0}{k}\right)$$

Notice that the minus sign caused the values inside the natural logarithm to be inverted.

4. **Develop a MATLAB® Solution**

   First, define a symbolic equation and give it a name (recall that it's OK to use an equation as the function input argument):

   ```
   X = sym('k = k0*exp(-Q/(R*T))')
   X =
   k = k0/exp(Q/(R*T))
   ```

   Now, we can ask MATLAB® to solve our equation. We need to specify that MATLAB® is to solve for Q, and Q needs to be in single quotes, because it has not been separately defined as a symbolic variable:

   ```
   solve(X,'Q')
   ans =
   -R*T*log(k/k0)
   ```

Alternatively, we could define our answer as **Q**:

```
Q = solve(X,'Q')
Q =
-R*T*log(k/k0)
```

5. Test the Solution
   Compare the MATLAB® solution with the hand solution. The only difference is that we pulled the minus sign outside the logarithm instead of inverting the ratio of k/k_0. Notice that MATLAB® (as well as most computer programs) represents ln as `log` ($\log_{10}$ is represented as `log10`).
   Now that we know this strategy works, we can solve for any of the variables. For example, we could have

```
T = solve(X,'T')
T =
-Q/(R*log(k/k0))
```

HINT

The `findsym` command is useful in determining which variables exist in a symbolic expression or equation. In the previous example, the variable *X* was defined as

```
X = sym('k = k0*exp(-Q/(R*T))')
```

The `findsym` function identifies all the variables, whether explicitly defined or not:

```
findsym(X)
ans =
k, k0, Q, R, T
```

12.2.2 Solving Systems of Equations

Not only can the `solve` function solve single equations or expressions for any of the included variables, it can also solve systems of equations. Take, for example, these three symbolic equations:

```
one = sym('3*x + 2*y -z = 10');
two = sym('-x + 3*y + 2*z = 5');
three = sym('x - y - z = -1');
```

To solve for the three embedded variables x, y, and z, simply list all three equations in the `solve` function:

```
answer = solve(one,two,three)
answer =
  x: [1x1 sym]
  y: [1x1 sym]
  z: [1x1 sym]
```

These results are puzzling. Each answer is listed as a 1 × 1 symbolic variable, but the program doesn't reveal the values of those variables. In addition, `answer` is listed in the workspace window as a 1 × 1 structure array. To access the actual values, you'll need to use the structure array syntax:

```
answer.x
ans =
-2
answer.y
ans =
5
answer.z
ans =
-6
```

To force the results to be displayed without using a structure array and the associated syntax, we must assign names to the individual variables. Thus, for our example, we have

```
[x,y,z] = solve(one,two,three)
x =
-2
y =
5
z =
-6
```

KEY IDEA

The results of the symbolic `solve` function are listed alphabetically

The results are assigned alphabetically. For instance, if the variables used in your symbolic expressions are q, x, and p, the results will be returned in the order p, q, and x, *independently* of the names you have assigned for the results.

Notice in our example that x, y, and z are still listed as symbolic variables, even though the results are numbers. The result of the `solve` function is a symbolic variable, either `ans` or a user-defined name. If you want to use that result in a MATLAB® expression requiring a double-precision floating-point input, you can change the variable type with the `double` function. For example,

```
double(x)
```

changes x from a symbolic variable to a corresponding numeric variable.

HINT

Using the `solve` function for multiple equations has both advantages and disadvantages over using linear algebra techniques. In general, if a problem can be solved by means of matrices, the matrix solution will take less computer time. However, linear algebra is limited to first-order equations. The `solve` function may take longer, but it can solve nonlinear problems and problems with symbolic variables. Table 12.2 lists some uses of the `solve` function.

Table 12.2 Using the Solve Function

solve(S)	Solves an expression with a single variable	**solve('x-5')** **ans =** **5**
solve(S)	Solves an equation with a single variable	**solve('x^2-2 = 5')** **ans =** **7^(1/2)** **-7^(1/2)**
solve(S)	Solves an equation whose solutions are complex numbers	**solve('x^2 = -5')** **ans =** **i*5^(1/2)** **-i*5^(1/2)**
solve(S)	Solves an equation with more than one variable for x or the closest variable to x	**solve('y = x^2+2')** **ans =** **(y-2)^(1/2)** **-(y-2)^(1/2)**
solve(S,y)	Solves an equation with more than one variable for a specified variable	**solve('y+6*x',x)** **ans =** **-1/6*y**

solve(S1,S2,S3)	Solves a system of equations and presents the solutions as a structure array	**one = sym('3*x+2*y-z =10');**
		two = sym('-x+3*y+2*z =5');
		three = sym('x - y- z = - 1');
		solve(one,two,three)
		ans =
		   x: [1x1 sym]
		   y: [1x1 sym]
		   z: [1x1 sym]

[A,B,C]= solve (S1,S2,S3)	Solves a system of equations and assigns the solutions to user-defined variable names; displays the results alphabetically	**one = sym('3*x+2*y -z =10');** **two = sym('-x+3*y+2*z =5');** **three = sym('x - y- z = -1');** **[x,y,z] = solve(one,two, three)** **x =-2** **y = 5** **z = -6**

PRACTICE EXERCISES 12.4

Consider the following system of linear equations to use in Exercises 12.1 through 12.5:

$$5x + 6y - 3z = 10$$
$$3x - 3y + 2z = 14$$
$$2x - 4y - 12z = 24$$

1. Solve this system of equations by means of the linear algebra techniques discussed in Chapter 10.
2. Define a symbolic equation representing each equation in the given system of equations. Use the `solve` function to solve for x, y, and z.
3. Display the results from Exercise 2 by using the structure array syntax.
4. Display the results from Exercise 2 by specifying the output names.
5. Add decimal points to the numbers in your equation definitions and `solve` them again. How do your answers change?
6. Consider the following nonlinear system of equations:

$$x^2 + 5y - 3z^3 = 15$$
$$4x + y^2 - z = 10$$
$$x + y + z = 15$$

Solve the nonlinear system with the `solve` function. Use the `double` function on your results to simplify the answer.

KEY IDEA

If a variable is not listed as a symbolic variable in the workspace window, it must be enclosed in single quotes when used in the subs function

12.2.3 Substitution

Particularly for engineers or scientists, once we have a symbolic expression, we often want to substitute values into it. Consider the quadratic equation again:

```
E4 = sym('a*x^2+b*x+c')
```

There are a number of substitutions we might want to make. For example, we might want to change the variable x into the variable y. To accomplish this, the subs function requires three inputs: the expression to be modified, the variable to be modified, and the new variable to be inserted. To substitute y for all the x's, we would use the command

```
subs(E4,'x','y')
```

which returns

```
ans =
a*(y)^2+b*(y)+c
```

The variable E4 has not been changed; rather, the new information is stored in ans, or it could be given a new name, such as E5:

```
E5 = subs(E4,'x','y')
E5 =
a*(y)^2+b*(y)+c
```

Recalling E4, we see that it remains unchanged:

```
E4
E4 =
a*x^2+b*x+c
```

To substitute numbers, we use the same procedure:

```
subs(E4,'x',3)
ans =
9*a+3*b+c
```

As with other symbolic operations, if the variables have been previously explicitly defined as symbolic, the single quotes are not required. For example,

```
syms a b c x
subs(E4,x,4)
```

returns

```
ans =
16*a+4*b+c
```

We can make multiple substitutions by listing the variables inside curly brackets, defining a cell array:

```
subs(E4,{a,b,c,x},{1,2,3,4})
ans =
     27
```

We can even substitute in numeric arrays. For example, first we create a new expression containing only x:

```
E6 = subs(E4,{a,b,c},{1,2,3})
```

This gives us

```
E6 =
x^2+2*x+3
```

Now we define an array of numbers and substitute them into E6:

```
numbers = 1:5;
subs(E6,x,numbers)
ans =
     6    11    18    27    38
```

PRACTICE EXERCISES 12.5

1. Using the subs function, substitute 4 into each expression/equation defined in Practice Exercises 12.1 for x (or X). Comment on your results.
2. Define a vector v of the even numbers from 0 to 10. Substitute this vector into all four versions of expression/equation 1: ex1, EX1, eq1, and EQ1. Does this work for all four versions? Comment on your results.
3. Substitute the following values into all four versions of expression/equation 4—ex4, EX4, eq4, and EQ4 (this is a two-step process because x is a vector):

a = 3		A = 3
b = 4	or	B = 4
c = 5		C = 5
x = 1:0.5:5		X = 1:0.5:5

4. Check your results for Exercise 3 in the workspace window. What kind of a variable is your result—double or symbolic?

EXAMPLE 12.2

USING SYMBOLIC MATH TO SOLVE A BALLISTICS PROBLEM

We can use the symbolic math capabilities of MATLAB® to explore the equations representing the trajectory of an unpowered projectile, such as the cannonball shown in Figure 12.5.

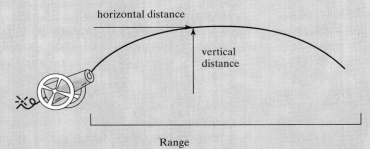

Figure 12.5
The range of a projectile depends on the initial velocity and the launch angle.

We know from elementary physics that the distance a projectile travels horizontally is

$$d_x = v_0 t \cos(\theta)$$

and the distance traveled vertically is

$$d_y = v_0 t \sin(\theta) - \frac{1}{2}gt^2$$

where

v_0 = velocity at launch,
t = time,
θ = launch angle, and
g = acceleration due to gravity.

Use these equations and MATLAB®'s symbolic capability to derive an equation for the distance the projectile has traveled horizontally when it hits the ground (the range).

1. State the Problem
   Find the range equation.

2. Describe the Input and Output

   Input Equations for horizontal and vertical distances

   Output Equation for range

3. Develop a Hand Example

$$d_y = v_0 t \sin(\theta) - \frac{1}{2}gt^2 = 0$$

Rearrange to give

$$v_0 t \sin(\theta) = \frac{1}{2}gt^2$$

Divide by t and solve:

$$t = \frac{2v_0 \sin(\theta)}{g}$$

Now substitute this expression for t into the horizontal-distance formula to obtain

$$d_x = v_0 t \cos(\theta)$$

$$\text{range} = v_0 \left(\frac{2v_0 \sin(\theta)}{g}\right) \cos(\theta)$$

We know from trigonometry that $2 \sin \theta \cos \theta$ is the same as $\sin(2\theta)$, which would allow a further simplification if desired.

4. Develop a MATLAB® Solution

First define the symbolic variables:

```
syms v0 t theta g
```

Next define the symbolic expression for the vertical distance traveled:

```
Distancey = v0 * t *sin(theta) - 1/2*g*t^2;
```

Now define the symbolic expression for the horizontal distance traveled:

```
Distancex = v0 * t *cos(theta);
```

Solve the vertical-distance expression for the time of impact, since the vertical distance = 0 at impact:

```
impact_time = solve(Distancey,t)
```

This returns two answers:

```
impact_time =
[                 0]
[ 2*v0*sin(theta)/g]
```

This result makes sense, since the vertical distance is zero at launch and again at impact. Substitute the impact time into the horizontal-distance expression. Since we are interested only in the second time, we'll need to use `impact_time(2)`:

```
impact_distance = subs(Distancex,t,impact_time(2))
```

The substitution results in an equation for the distance the projectile has traveled when it hits the ground:

```
impact_distance =
2*v0^2*sin(theta)/g*cos(theta)
```

5. Test the Solution

Compare the MATLAB® solution with the hand solution. Both approaches give the same result.

MATLAB® can simplify the result, although it is already pretty simple. We chose to use the `simple` command to demonstrate all the possibilities. The command

```
simple(impact_distance)
```

(continued)

gives the following results:

```
simplify:               (v0^2*sin(2*theta))/g
radsimp:                (2*v0^2*cos(theta)*sin(theta))/g
simplify(100):          (v0^2*sin(2*theta))/g
combine(sincos):        (v0^2*sin(2*theta))/g
combine(sinhcosh):      (2*v0^2*cos(theta)*sin(theta))/g
combine(ln):            (2*v0^2*cos(theta)*sin(theta))/g
factor:                 (2*v0^2*cos(theta)*sin(theta))/g
expand:                 (2*v0^2*cos(theta)*sin(theta))/g
combine:                (2*v0^2*cos(theta)*sin(theta))/g
rewrite(exp):           (2*v0^2*((1/exp(theta*i))/2
                        +exp(theta*i)/2)*(((1/exp(theta*i))*i)/
                        2-(exp(theta*i)*i)/2))/g
rewrite(sincos):        (2*v0^2*cos(theta)*sin(theta))/g
rewrite(sinhcosh):      (2*v0^2*cosh(-theta*i)*sinh
                        (-theta*i)*i)/g
rewrite(tan):           -(4*v0^2*tan(theta/2)*(tan(theta/2)^
                        2-1))/(g*(tan(theta/2)^2 + 1)^2)
mwcos2sin:              -(2*v0^2*sin(theta)*(2*sin(theta/2)^
                        2-1))/g
collect(v0):            ((2*cos(theta)*sin(theta))/g)*v0^2
                        ans =
                        (v0^2*sin(2*theta))/g
```

12.3 SYMBOLIC PLOTTING

The symbolic toolbox includes a group of functions that allow you to plot symbolic functions. The most basic is `ezplot`.

12.3.1 The Ezplot Function

Consider a simple function of **x**, such as

```
y = sym('x^2-2')
```

To plot this function, use

```
ezplot(y)
```

The resulting graph is shown in Figure 12.6. The `ezplot` function defaults to an **x** range from -2π to $+2\pi$. MATLAB®created this plot by choosing values of x and calculating corresponding values of y, so that a smooth curve is produced. Notice that the expression plotted is automatically displayed as the title of an `ezplot`.

The user who does not want to accept the default values can specify the minimum and maximum values of **x** in the second field of the `ezplot` function:

```
ezplot(y,[-10,10])
```

The values are enclosed in square brackets, indicating that they are elements in the array that defines the plot extremes. You can also specify titles, axis labels, and annotations, just as you do for other MATLAB® plots. For example, to add a title and labels to the plot, use

Figure 12.6
Symbolic expressions can be plotted with `ezplot`. In the left-hand graph, the default title is the plotted expression and the default range is -2π to $+2\pi$. In the right-hand graph, titles, labels, and other annotations are added to `ezplot` with the use of standard MATLAB® annotation functions.

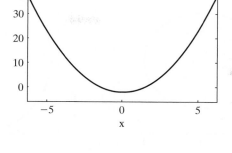

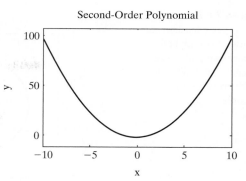

```
title('Second Order Polynomial')
xlabel('x')
ylabel('y')
```

The `ezplot` function also allows you to plot implicit functions of x and y, as well as parametric functions. For instance, consider the implicit equation

$$x^2 + y^2 = 1$$

which you may recognize as the equation for a circle of radius 1. You could solve for y, but it's not necessary with `ezplot`. Any of the commands

```
ezplot('x^2 + y^2 = 1',[-1.5,1.5])
ezplot('x^2 + y^2 -1',[-1.5,1.5])
```

and

```
z = sym('x^2 + y^2 -1')
ezplot(z,[-1.5,1.5])
```

can be used to create the graph of the circle shown on the left-hand side in Figure 12.7.

Another way to define an equation is parametrically; that is, define separate equations for x and for y in terms of a third variable. A circle can be defined parametrically as

$$x = \sin(t)$$
$$y = \cos(t)$$

To plot the circle parametrically with `ezplot`, list first the symbolic expression for x and then that for y:

```
ezplot('sin(t)','cos(t)')
```

The results are shown on the right-hand side of Figure 12.7.

Although annotation is done the same way for symbolic plots as for standard numeric plots, in order to plot multiple lines on the same graph, you'll need to use the `hold on` command. To adjust colors, line styles, and marker styles, use the interactive tools available in the plotting window. For example, to plot $\sin(x)$, $\sin(2x)$, and $\sin(3x)$ on the same graph, first define some symbolic expressions:

```
y1 = sym('sin(x)')
y2 = sym('sin(2*x)')
y3 = sym('sin(3*x)')
```

PARAMETRIC EQUATIONS
Equations that define x and y in terms of another variable, typically t

Figure 12.7

The `ezplot` function can be used to graph both implicit and parametric functions, in addition to functions of a single variable.

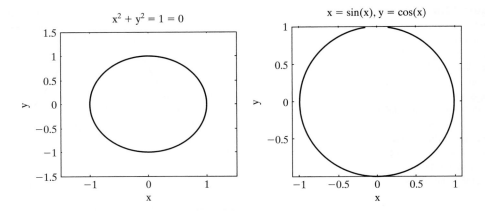

Then plot each expression:

```
ezplot(y1)
hold on
ezplot(y2)
ezplot(y3)
```

The results are shown in Figure 12.8. To change the line colors, line styles, or marker styles, you'll need to select the arrow on the menu bar (circled in the figure) and then select the line you'd like to edit. Once you've selected the line, right-click to activate the editing menu. When you've done plotting, don't forget to issue the

```
hold off
```

command.

Figure 12.8

Use the interactive plotting tools to adjust line style, color, and markers.

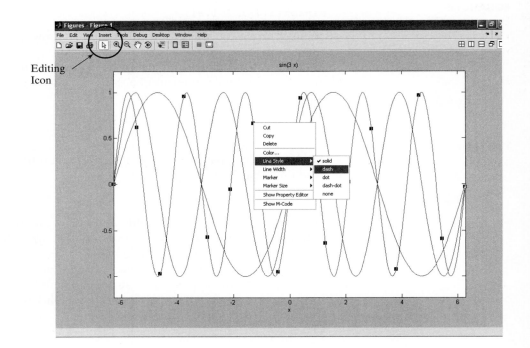

HINT

Most symbolic functions will allow you to enter either a symbolic variable that represents a function or the function itself enclosed in single quotes. For example,

```
y = sym('x^2-1')
ezplot(y)
```

is equivalent to

```
ezplot('x^2-1')
```

PRACTICE EXERCISES 12.6

Be sure to add titles and axis labels to all your plots.
1. Use ezplot to plot ex1 from -2π to $+2\pi$.
2. Use ezplot to plot EX1 from -2π to $+2\pi$.
3. Use ezplot to plot ex2 from -10 to $+10$.
4. Use ezplot to plot EX2 from -10 to $+10$.
5. Why can't we plot equations with only one variable?
6. Use ezplot to plot ex6 from -2π to $+2\pi$.
7. Use ezplot to plot $\cos(x)$ from -2π to $+2\pi$. Don't define an expression for $\cos(x)$; just enter it into ezplot as a character string:

```
ezplot('cos(x)')
```

8. Use ezplot to create an implicit plot of $x^2 - y^4 = 5$.
9. Use ezplot to plot $\sin(x)$ and $\cos(x)$ on the same graph. Use the interactive plotting tools to change the color of the sine graph.
10. Use ezplot to create a parametric plot of $x = \sin(t)$ and $y = 3\cos(t)$.

12.3.2 Additional Symbolic Plots

Additional symbolic plotting functions that mirror the functions used in numeric MATLAB® plotting options are listed in Table 12.3.

To demonstrate how the three-dimensional surface plotting functions (ezmesh, ezmeshc, ezsurf, and ezsurfc) work, first define a symbolic version of the peaks function:

```
z1 = sym('3*(1-x)^2*exp(-(x^2) - (y+1)^2)')
z2 = sym('- 10*(x/5 - x^3 - y^5)*exp(-x^2-y^2)')
z3 = sym('- 1/3*exp(-(x+1)^2 - y^2)')
z = z1+z2+z3
```

We broke this function into three parts to make it easier to enter into the computer. Notice that no "dot" operators are used in these expressions, since

Table 12.3 Symbolic Plotting Functions

ezplot	Function plotter	If z is a function of x: `ezplot(z)`
ezmesh	Mesh plotter	If z is a function of x and y: `ezmesh(z)`
ezmeshc	Combined mesh and contour plotter	If z is a function of x and y: `ezmeshc(z)`
ezsurf	Surface plotter	If z is a function of x and y: `ezsurf(z)`
ezsurfc	Combined surface and contour plotter	If z is a function of x and y: `ezsurfc(z)`
ezcontour	Contour plotter	If z is a function of x and y: `ezcontour(z)`
ezcontourf	Filled contour plotter	If z is a function of x and y: `ezcontourf(z)`
ezplot3	Three-dimensional parametric curve plotter	If x is a function of t, if y is a function of t, and if z is a function of t: `ezplot3(x,y,z)`
ezpolar	Polar coordinate plotter	If r is a function of θ: `ezpolar(r)`

they are all symbolic. The `ezplot` functions work similarly to their numeric counterparts:

subplot(2,2,1)
ezmesh(z)
title('ezmesh')

subplot(2,2,2)
ezmeshc(z)
title('ezmeshc')

The plots resulting from these commands are shown in Figure 12.9. When we created the same plots via a standard MATLAB® approach, it was necessary to define an array of both x- and y-values, mesh them together, and calculate the values of z on the basis of the two-dimensional arrays.
The symbolic plotting capability contained in the symbolic toolbox makes creating these graphs much easier.

KEY IDEA
Most of the MATLAB® plotting functions for arrays have corresponding functions for symbolic applications

subplot(2,2,3)
ezsurf(z)
title('ezsurf')
subplot(2,2,4)
ezsurfc(z)
title('ezsurfc')

All these graphs can be annotated by using the standard MATLAB® functions, such as `title`, `xlabel`, `text`, etc.

Figure 12.9
Examples of three-dimensional symbolic surface plots.

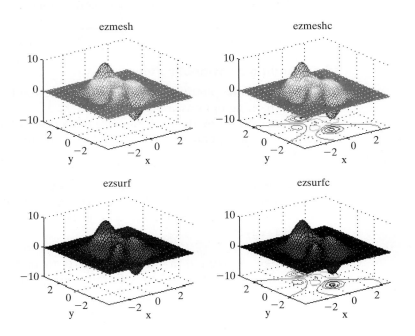

Figure 12.10
A variety of symbolic plots.

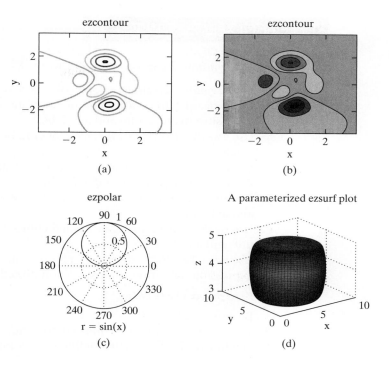

The two-dimensional plots and contour plots are also similar to their numeric counterparts. For example, these contour plots are a two-dimensional representation of the three-dimensional peaks function and are shown in Figure 12.10a and b.

```
subplot(2,2,1)
ezcontour(z)
title ('ezcontour')
subplot(2,2,2)
ezcontourf(z)
title('ezcontourf')
```

To demonstrate the use of ezpolar we need a new function to graph. For example, when $\sin(x)$ is plotted in polar coordinates the result is a circle, as shown in Figure 12.10c.

```
subplot(2,2,3)
z = sym('sin(x)')
ezpolar(z)
title('ezpolar')
```

Any of these functions (ezmesh, ezsurf, ezmeshc, ezsurfc, and ezcontour) can also handle parameterized functions (one function for x, one for y, and one for z). For example, the following code produces the torus shown in Figure 12.10d.

```
subplot(2,2,4)
x=sym('4+(3+cos(v))*sin(u)')
y=sym('4 + (3 + cos(v))*cos(u)')
z=sym('4+sin(v)')
ezsurf(x,y,z)
title('A Parameterized ezsurf Plot')
```

PRACTICE EXERCISES 12.7

Create a symbolic expression for $Z = \sin(\sqrt{X^2 + Y^2})$.

1. Use `ezmesh` to create a mesh plot of Z. Be sure to add a title and axis labels.
2. Use `ezmeshc` to create a combination mesh plot and contour plot of Z. Be sure to add a title and axis labels.
3. Use `ezsurf` to create a surface plot of Z. Be sure to add a title and axis labels.
4. Use `ezsurfc` to create a combination surface plot and contour plot of Z. Be sure to add a title and axis labels.
5. Use `ezcontour` to create a contour plot of Z. Be sure to add a title and axis labels.
6. Use `ezcontourf` to create a filled contour plot of Z. Be sure to add a title and axis labels.
7. Use `ezpolar` to create a polar plot of $x\sin(x)$. Don't define a symbolic expression, but enter this expression directly into `ezpolar`:

   ```
   ezpolar('x*sin(x)')
   ```

   Be sure to add a title.
8. The `ezplot3` function requires us to define three variables as a function of a fourth. To do this, first define t as a symbolic variable, and then let

$$x = t$$
$$y = \sin(t)$$
$$z = \cos(t)$$

Use `ezplot3` to plot this parametric function from 0 to 30.
You may have problems creating `ezplot3` graphs inside subplot windows, because of a MATLAB® program idiosyncrasy. Later versions may fix this problem.

EXAMPLE 12.3

USING SYMBOLIC PLOTTING TO ILLUSTRATE A BALLISTICS PROBLEM

In Example 12.2, we used MATLAB®'s symbolic capabilities to derive an equation for the distance a projectile travels before it hits the ground. The horizontal-distance formula

$$d_x = v_0 t \cos(\theta)$$

and the vertical-distance formula

$$d_y = v_0 t \sin(\theta) - \frac{1}{2}gt^2$$

where

v_0 = the velocity at launch,
t = time,
θ = launch angle, and
g = acceleration due to gravity,

were combined to give

$$\text{range} = v_0 \left(\frac{2v_0 \sin(\theta)}{g} \right) \cos(\theta)$$

Using MATLAB®'s symbolic plotting capability, create a plot showing the range traveled for angles from 0 to $\pi/2$. Assume an initial velocity of 100 m/s and an acceleration due to gravity of 9.8 m/s².

1. **State the Problem**
   Plot the range as a function of launch angle.

2. **Describe the Input and Output**

   Input   Symbolic equation for range

   $$v_0 = 100 \text{ m/s}$$
   $$g = 9.8 \text{ m/s}^2$$

   Output   Plot of range versus angle

3. **Develop a Hand Example**

   $$\text{range} = v_0 \left(\frac{2v_0 \sin(\theta)}{g} \right) \cos(\theta)$$

   We know from trigonometry that $2 \sin \theta \cos \theta$ equals $\sin(2\theta)$. Thus, we can simplify the result to

   $$\text{range} = \frac{v_0^2}{g} \sin(2\theta)$$

   With this equation, it is easy to calculate a few data points:

Angle	Range, m
0	0
$\pi/6$	884
$\pi/4$	1020
$\pi/3$	884
$\pi/2$	0

   The range appears to increase with increasing angle and then decrease back to zero when the cannon is pointed straight up.

4. **Develop a MATLAB® Solution**
   First, we need to modify the equation from Example 12.2 to include the launch velocity and the acceleration due to gravity. Recall that

   ```
   impact_distance =
   2*v0^2*sin(theta)/g*cos(theta)
   ```

   Use the subs function to substitute the numerical values into the equation:

   ```
   impact_100 = subs(impact_distance,{v0,g},{100, 9.8})
   ```

   (*continued*)

This returns

```
impact_100 =
100000/49*sin(theta)*cos(theta)
```

Finally, plot the results and add a title and labels:

```
ezplot(impact_100,[0, pi/2])
title('Maximum Projectile Distance Traveled')
xlabel('angle, radians')
ylabel('range, m')
```

This generates Figure 12.11.

5. Test the Solution
The MATLAB® solution agrees with the hand solution. The range is zero when the cannon is pointed straight up and zero when it is pointed horizontally. The range appears to peak at an angle of about 0.8 radian, which corresponds roughly to 45°.

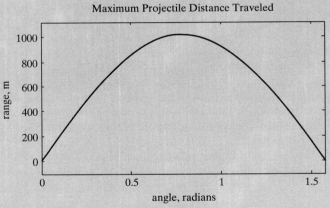

Figure 12.11
Projectile range.

12.4 CALCULUS

MATLAB®'s symbolic toolbox allows the user to differentiate symbolically and to perform integrations. This makes it possible to find analytical solutions, instead of numeric approximations, for many problems.

12.4.1 Differentiation

Differential calculus is studied extensively in first-semester calculus. The derivative can be thought of as the slope of a function or as the rate of change of the function. For example, consider a race car. The velocity of the car can be approximated by the change in distance divided by the change in time. Suppose that, during a race, the car starts slowly and reaches its fastest speed at the finish line. Of course, to avoid running

Figure 12.12
Position of a race car. The car speeds up until it reaches the finish line. Then it slows to a stop. (The dotted line indicating the finish line was added after the graph was created.)

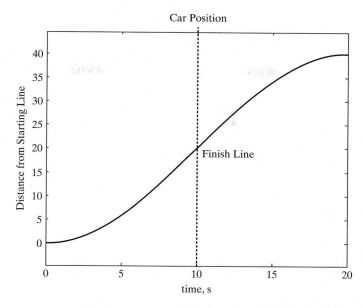

into the stands, the car must then slow down until it finally stops. We might model the position of the car with a sine wave, as shown in Figure 12.12. The relevant equation is

$$d = 20 + 20 \sin\left(\frac{\pi (t - 10)}{20}\right)$$

The graph in Figure 12.12 was created with `ezplot` and symbolic mathematics. First, we define a symbolic expression for distance:

```
dist = sym('20+20*sin(pi*(t-10)/20)')
```

Once we have the symbolic expression, we can substitute it into the `ezplot` function and annotate the resulting graph:

```
ezplot(dist,[0,20])
title('Car Position')
xlabel('time, s')
ylabel('Distance from Starting Line')
text(10,20,'Finish Line')
```

MATLAB® includes a function called `diff` to find the derivative of a symbolic expression. (The word *differential* is another term for the derivative.) The velocity is the derivative of the position, so to find the equation of the velocity of the car, we'll use the `diff` function:

```
velocity = diff(dist)
velocity =
pi*cos((pi*(t-10))/20)
```

We can use the `ezplot` function to plot the velocity:

```
ezplot(velocity,[0,20])
title('Race Car Velocity')
xlabel('time, s')
ylabel('velocity, distance/time')
text(10,3,'Finish Line')
```

Figure 12.13
The maximum velocity is
reached at the finish line.

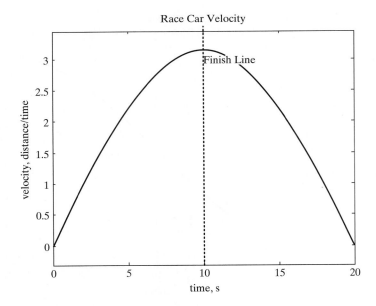

Figure 12.13
The maximum velocity is
reached at the finish line.

The results are shown in Figure 12.13.

The acceleration of the race car is the change in the velocity divided by the change in time, so the acceleration is the derivative of the velocity function:

```
acceleration = diff(velocity)
acceleration =
-(pi^2*sin((pi*(t-10))/20))/20
```

The plot of acceleration (Figure 12.14) was also created with the use of the symbolic plotting function:

```
ezplot(acceleration,[0,20])
title('Race Car Acceleration')
xlabel('time, s')
```

Figure 12.14
The race car is accelerating
up to the finish line and
then is decelerating. The
acceleration at the finish
line is zero.

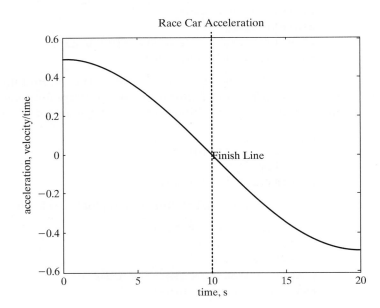

Table 12.4 Symbolic Differentiation

diff(f)	Returns the derivative of the expression f with respect to the default independent variable	`y=sym('x^3+z^2')`
		`diff(y)`
		`ans =`
		`3*x^2`

diff(f,'t')	Returns the derivative of the expression f with respect to the variable t	`y=sym('x^3+z^2')`
		`diff(y,'z')`
		`ans =`
		`2*z`

diff(f,n)	Returns the nth derivative of the expression f with respect to the default independent variable	`y=sym('x^3+z^2')`
		`diff(y,2)`
		`ans =`
		`6*x`

diff(f,'t',n)	Returns the nth derivative of the expression f with respect to the variable t	`y=sym('x^3+z^2')`
		`diff(y,'z',2)`
		`ans =`
		`2`

DERIVATIVE

The instantaneous rate of change of one variable with respect to a second variable

```
ylabel('acceleration, velocity/time')
text(10,0,'Finish Line')
```

The acceleration is the first derivative of the velocity and the second derivative of the position. MATLAB® offers several slightly different techniques to find both first derivatives and *n*th derivatives (see Table 12.4).

If we have a more complicated equation with multiple variables, such as

```
y = sym('x^2+t-3*z^3')
```

MATLAB® will calculate the derivative with respect to **x**, the default variable:

```
diff(y)
ans =
2*x
```

Our result is the rate of change of y as x changes (if we keep all the other variables constant). This is usually depicted as $\partial y/\partial x$ and is called a *partial derivative*. If we want to see how y changes with respect to another variable, such as t, we must specify it in the diff function (remember that if **t** has been previously defined as a symbolic variable, we don't need to enclose it in single quotes):

```
diff(y,'t')
ans =
1
```

Similarly, to see how y changes with z when everything else is kept constant, we use

```
diff(y,'z')
ans =
-9*z^2
```

KEY IDEA

Integration is the opposite of taking the derivative

To find higher-order derivatives, we can either nest the `diff` function or specify the order of the derivative in the `diff` function. Either of the statements

```
diff(y,2)
```

and

```
diff(diff(y))
```

returns the same result:

```
ans =
2
```

Notice that although the result appears to be a number, it is a symbolic variable. In order to use it in a MATLAB® calculation, you'll need to convert it to a double-precision floating-point number.

If we want to take a higher derivative of y with respect to a variable that is not the default, we need to specify both the degree of the derivative and the variable. For example, to find the second derivative of y with respect to z, we type

```
diff(y,'z',2)
ans =
-18*z
```

PRACTICE EXERCISES 12.8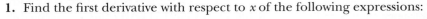

1. Find the first derivative with respect to x of the following expressions:

$$x^2 + x + 1$$
$$\sin(x)$$
$$\tan(x)$$
$$\ln(x)$$

2. Find the first partial derivative with respect to x of the following expressions:

$$ax^2 + bx + c$$
$$x^{0.5} - 3y$$
$$\tan(x + y)$$
$$3x + 4y - 3xy$$

3. Find the second derivative with respect to x for each of the expressions in Exercises 12.1 and 12.2.

4. Find the first derivative with respect to y for the following expressions:

$$y^2 - 1$$
$$2y + 3x^2$$
$$ay + bx + cz$$

5. Find the second derivative with respect to y for each of the expressions in Problem 12.4.

EXAMPLE 12.4

USING SYMBOLIC MATH TO FIND THE OPTIMUM LAUNCH ANGLE

In Example 12.3, we used the symbolic plotting capability of MATLAB® to create a graph of range versus launch angle, based on the range formula derived in Example 12.2, namely

$$\text{range} = v_0 \left(\frac{2 v_0 \sin(\theta)}{g} \right) \cos(\theta)$$

where

v_0 = velocity at launch, which we chose to be 100 m/s,
θ = launch angle, and
g = acceleration due to gravity, which we chose to be 9.8 m/s².

Use MATLAB®'s symbolic capability to find the angle at which the maximum range occurs and to find the maximum range.

1. **State the Problem**
   Find the angle at which the maximum range occurs.
   Find the maximum range.

2. **Describe the Input and Output**

   Input Symbolic equation for range

   $$v_0 = 100 \text{ m/s}$$

   $$g = 9.8 \text{ m/s}^2$$

   Output The angle at which the maximum range occurs
   The maximum range

3. **Develop a Hand Example**
   From the graph in Figure 12.15, the maximum range appears to occur at a launch angle of approximately 0.7 or 0.8 radian, and the maximum height appears to be approximately 1000 m.

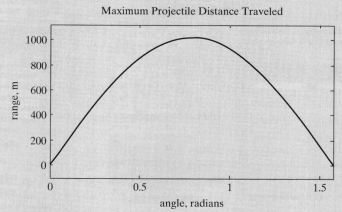

Figure 12.15
The projectile range as a function of launch angle.

4. Develop a MATLAB® Solution
   Recall that the symbolic expression for the impact distance with v_0 and g defined as 100 m/s and 9.8 m/s², respectively, is

```
impact_100 =
100000/49*sin(theta)*cos(theta)
```

   From the graph, we can see that the maximum distance occurs when the slope is equal to zero. The slope is the derivative of `impact_100`, so we need to set the derivative equal to zero and solve. Since MATLAB® automatically assumes that an expression is equal to zero, we have

```
max_angle = solve(diff(impact_100))
```

   which returns the angle at which the maximum height occurs:

```
max_angle =
[ 1/4*pi]
```

   Now the result can be substituted into the expression for the range:

```
max_distance = subs(impact_100,theta,max_angle)
```

   Finally, the result should be changed to a double precision number

```
double(max_distance)
ans =
1.0204e+003
```

12.4.2 Integration

Integration can be thought of as the opposite of differentiation (finding a derivative) and is even sometimes called the antiderivative. It is commonly visualized as the area under a curve. For example, work done by a piston–cylinder device as it moves up or down can be calculated by taking the integral of P with respect to V—that is,

$$W = \int_1^2 P dV$$

In order to do the calculation, we need to know how P changes with V. If, for example, P is a constant, we could create the plot shown in Figure 12.16.

The work consumed or produced as we move the piston is the area under the curve from the initial volume to the final volume. For example, if we moved the piston from 1 cm³ to 4 cm³, the work would correspond to the area shown in Figure 12.17

As you may know from a course in integral calculus (usually Calculus II), the integration is quite simple:

$$W = \int_1^2 P dV = P \int_1^2 dV = PV \big|_1^2 = PV_2 - PV_1 = P \, \Delta V$$

If

$$P = 100 \text{ psia, and } \Delta V = 3 \text{ cm}^3$$

Figure 12.16
Pressure profile in a
piston–cylinder device. In
this example, the pressure
is constant.

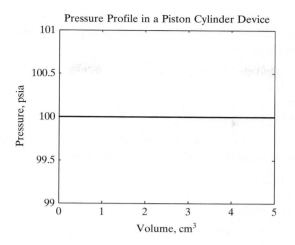

Figure 12.17
The work produced in a
piston–cylinder device is
the area under the curve.

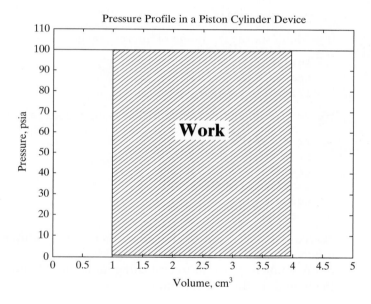

then

$$W = 3 \text{ cm}^3 \times 100 \text{ psia}$$

The symbolic toolbox allows us to easily take integrals of some very complicated functions. For example, if we want to find an indefinite integral (an integral for which we don't specify the boundary values of the variable), we can use the int function. First, we need to specify a function:

```
y = sym('x^3 + sin(x)')
```

To find the indefinite integral, we type

```
int(y)
ans =
 1/4*x^4-cos(x)
```

The int function uses x as the default variable. For example, if we define a function with two variables, the int function will find the integral with respect to **x** or the variable closest to x:

```
y = sym('x^3 +sin(t)')
int(y)
ans =
1/4*x^4+sin(t)*x
```

If we want to take the integral with respect to a user-defined variable, that variable needs to be specified in the second field of the int function:

```
int(y,'t')
ans =
x^3*t-cos(t)
```

To find the definite integral, we need to specify the range of interest. Consider this expression:

```
y = sym('x^2')
```

If we don't specify the range of interest, we get

```
int(y)
ans =
1/3*x^3
```

We could evaluate this from 2 to 3 by using the subs function:

```
yy = int(y)
yy =
1/3*x^3
subs(yy,3)-subs(yy,2)
ans =
   6.3333
```

Notice that the result of the subs function is a double-precision floating-point number.

A simpler approach to evaluating an integral between two points is to specify the bounds in the int function:

```
int(y,2,3)
ans =
19/3
```

Notice, however, that the result is a symbolic number. To change it to a double we can use the double function.

```
double(ans)
ans =
   6.3333
```

If we want to specify both the variable and the bounds, we need to list them all:

```
y = sym('sin(x)+cos(z)')
int(y,'z',2,3)
ans =
sin(x)+sin(3)-sin(2)
```

Table 12.5 Symbolic Integration

int(f)	Returns the integral of the expression f with respect to the default independent variable	`y = sym('x^3+z^2')`
		`int(y)`
		`ans =`
		`1/4*x^4+z^2*x`

int(f,'t')	Returns the integral of the expression f with respect to the variable t	`y = sym('x^3+z^2')`
		`int(y,'z')`
		`ans =`
		`x^3*z+1/3*z^3`

int(f,a,b)	Returns the integral, with respect to the default variable, of the expression f between the numeric bounds a and b	`y = sym('x^3+z^2')`
		`int(y,2,3)`
		`ans =`
		`65/4+z^2`

int(f,'t',a,b)	Returns the integral, with respect to the variable t, of the expression f between the numeric bounds a and b	`y = sym('x^3+z^2')`
		`int(y,'z',2,3)`
		`ans =`
		`x^3+19/3`

int(f,'t',a,b)	Returns the integral, with respect to the variable t, of the expression f between the symbolic bounds a and b	`y = sym('x^3+z^2')`
		`int(y,'z','a','b')`
		`ans =`
		`x^3*(b-a)+1/3*b^3-1/3*a^3`

Bounds can be numeric, or they can be symbolic variables:

```
int(y,'z','b','c')
ans =
sin(x)*c+sin(c)-sin(x)*b-sin(b)
```

Table 12.5 lists the MATLAB® functions having to do with integration.

PRACTICE EXERCISES 12.9

1. Integrate the following expressions with respect to x:

$$x^2 + x + 1$$
$$\sin(x)$$
$$\tan(x)$$
$$\ln(x)$$

2. Integrate the following expressions with respect to x:

$$ax^2 + bx + c$$
$$x^{0.5} - 3y$$
$$\tan(x + y)$$
$$3x + 4y - 3xy$$

3. Perform a double integration with respect to x for each of the expressions in Exercises 1 and 2.
4. Integrate the following expressions with respect to y:

$$y^2 - 1$$
$$2y + 3x^2$$
$$ay + bx + cz$$

5. Perform a double integration with respect to y for each of the expressions in Exercise 12.4.
6. Integrate each of the expressions in Exercise 1 with respect to x from 0 to 5.

EXAMPLE 12.5

USING SYMBOLIC MATH TO FIND WORK PRODUCED IN A PISTON–CYLINDER DEVICE

Piston–cylinder devices are used in a wide range of scientific instrumentation and engineering devices. Probably, the most pervasive is the internal combustion engine (Figure 12.18), which typically uses four to eight cylinders.

Figure 12.18
Internal combustion engine.

The work produced by a piston–cylinder device depends on the pressure inside the cylinder and the amount the piston moves, resulting in a change in volume inside the cylinder. Mathematically,

$$W = \int P \, dV$$

In order to integrate this equation, we need to understand how the pressure changes with the volume. We can model most combustion gases as air and assume that they follow the ideal gas law

$$PV = nRT$$

where

P = pressure, kPa,
V = volume, m^3,
n = number of moles, kmol,
R = universal gas constant, 8.314 kPa m^3/kmol K, and
T = temperature, K.

If we assume that there is 1 mole of gas at 300 K and that the temperature stays constant during the process, we can use these equations to calculate the work either done on the gas or produced by the gas as it expands or contracts between two known volumes.

1. State the Problem

   Calculate the work done per mole in an isothermal (constant-temperature) piston–cylinder device as the gas expands or contracts between two known volumes.

2. Describe the Input and Output

   Input

   Temperature = 300 K

   Universal gas constant = 8.314 kPa m^3/kmol K = 8.314 kJ/kmol K

   Arbitrary values of initial and final volume; for this example, we'll use

   $$\text{initial volume} = 1 \text{ m}^3$$

   $$\text{final volume} = 5 \text{ m}^3$$

   Output

   Work produced by the piston–cylinder device, in kJ.

3. Develop a Hand Example

   First, we'll need to solve the ideal gas law for P:

   $$PV = nRT$$

   $$P = nRT/V$$

   Since n, R, and T are constant during the process, we can now perform the integration:

   $$W = \int \frac{nRT}{V}\, dV = nRT \int \frac{dV}{V} = nRT \ln\left(\frac{V_2}{V_1}\right)$$

   Substituting the values, we find that

   $$W = 1 \text{ kmol} \times 8.314 \text{ kJ/kmol K} \times 300 \text{ K} \times \ln\left(\frac{V_2}{V_1}\right)$$

   If we use the arbitrary values $V_1 = 1$ m^3 and $V_2 = 5$ m^3, then the work becomes

   $$W = 4014 \text{ kJ}$$

   Because the work is positive, it is produced *by* (not *on*) the system.

4. Develop a MATLAB® Solution

   First, we'll need to solve the ideal gas law for pressure. The code

   ```
   syms P V n R T V1 V2                    %Define variables
   ideal_gas_law = sym('P*V = n*R*T')      %Define ideal gas law
   P = solve(ideal_gas_law,'P')            %Solve for P
   ```

   returns

   ```
   P =
   n*R*T/V
   ```

Once we have the equation for *P*, we can integrate. The command

```
W = int(P,V,V1,V2)          %Integrate P with respect
                            %to V from V1 to V2
```

returns

```
W =
n*R*T*log(V2)-n*R*T*log(V1)
```

Finally, we can substitute the values into the equation. We type

```
work = subs(W,{n,R,V1,V2,T},{1,8.314,1,5,300.0})
```

giving us

```
work =
    4.0143e+003
```

5. **Test the Solution**
The most obvious test is to compare the hand and computer solutions. However, the same answer with both techniques just means that we did the calculations the same way. One way to check reasonability would be to create a *PV* plot and estimate the area under the curve.

To create the plot, we'll need to return to the equation for *P* and substitute values for *n*, *R*, and *T*:

```
p = subs(P,{n,R,T},{1,8.314, 300})
```

This returns the following equation for *P*:

```
p =
12471/5/V
```

Now, we can use `ezplot` to create a graph of *P* versus *V* (see Figure 12.19):

Figure 12.19
For an isothermal system, as the volume increases the pressure decreases.

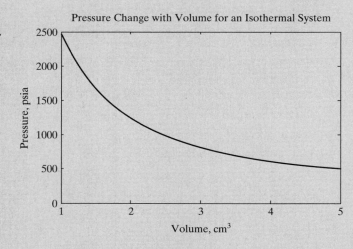

Pressure Change with Volume for an Isothermal System

Figure 12.20
We can estimate the area under the curve with a triangle.

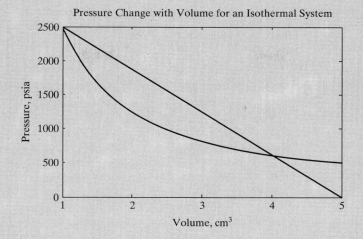

```
ezplot(p,[1,5]) %Plot the pressure versus V
title('Pressure Change with Volume for an Isothermal System')
xlabel('Volume')
ylabel('Pressure, psia')
xlabel('Volume, cm^3')
axis([1,5,0,2500])
```

To estimate the work, we could find the area of a triangle that approximates the shape shown in Figure 12.20. We have

$$\text{area} = \frac{1}{2}\text{base} * \text{height}$$

$$\text{area} = 0.5 * (5 - 1) * 2400 = 4800$$

which corresponds to 4800 kJ. This matches quite nicely with the calculated value of 4014 kJ.

Now that we have a process that works, we could create an M-file that prompts the user to enter values for any change in volume:

```
clear,clc
syms P V n R T V1 V2                  %Define variables
ideal_gas_law = sym('P*V = n*R*T')    %Define ideal gas law
P = solve(ideal_gas_law,'P')          %Solve for P
W = int(P,V,V1,V2)                    %Integrate to find work

%Now let the user input the data

temp = input('Enter a temperature: ')
v1 = input('Enter the initial volume: ')
v2 = input('Enter the final volume: ')
work = subs(W,{n,R,V1,V2,T},{1,8.314,v1,v2,temp})
```

This M-file generates the following user interaction:

```
Enter a temperature: 300
temp =
    300
Enter the initial volume: 1
v1 =
    1
Enter the final volume: 5
v2 =
    5
work =
    4.0143e+003
```

12.5 DIFFERENTIAL EQUATIONS

Differential equations contain both the dependent variable and its derivative with respect to the independent variable. For example,

$$\frac{dy}{dt} = y$$

is a differential equation.

Although any symbol can be used for either the independent or the dependent variable, the default independent variable in MATLAB® is t (and it is the usual choice for most ordinary differential equation formulations). Consider this simple equation:

$$y = e^t$$

The derivative of y with respect to t is

$$\frac{dy}{dt} = e^t$$

We could also express this as a differential equation, since $y = e^t$:

$$\frac{dy}{dt} = y$$

When we solve a differential equation, we are looking for an expression for y in terms of t. Differential equations typically have more than one solution. The following family of functions of t could be expressed by the same differential equation $(dy/dt = y)$:

$$y = C_1 e^t$$

We can specify the particular equation of interest by specifying an initial condition. For example, if

$$y(0) = 1,$$

then

$$C_1 = 1$$

A slightly more complicated function of y might be

$$y = t^2$$

The derivative of y with respect to t is

$$\frac{dy}{dt} = 2t$$

If we wanted to, we could rewrite this equation as

$$\frac{dy}{dt} = \frac{2t^2}{t} = \frac{2y}{t}$$

The symbolic toolbox includes a function called `dsolve` that solves differential equations, that is, it solves for y in terms of t. This function requires the user to enter the differential equation, using the symbol D to specify derivatives with respect to the independent variable, as in

```
dsolve('Dy = y')
ans =
C1*exp(t)
```

Using a single input results in a family of results. If you also include a second field specifying an initial condition (or a boundary condition), the exact answer is returned:

```
dsolve('Dy = y','y(0) = 1')
ans =
exp(t)
```

Similarly,

```
dsolve('Dy = 2*y/t','y(-1) = 1')
ans =
t^2
```

If t is not the independent variable in your differential equation, you can specify the independent variable in a third field:

```
dsolve('Dy = 2*y/t','y(-1) = 1', 't')
ans =
t^2
```

If a differential equation includes only a first derivative, it's called a first-order differential equation. Second-order differential equations include a second derivative, third-order equations a third derivative, and so on. To specify a higher-order derivative in the `dsolve` function, put the order immediately after the D. For example,

```
dsolve('D2y = -y')
ans =
C1*sin(t)+C2*cos(t)
```

solves a second-order differential equation.

HINT

■
■ Don't use the letter D in your variable names in differential equations. The
■ function will interpret the D as specifying a derivative.
■

The `dsolve` function can also be used to solve systems of differential equations. First, list the equations to be solved, then list the conditions. The `dsolve`
function will accept up to 12 inputs. For example:

```
dsolve('eq1,eq2, . . .', 'cond1,cond2, . . .', 'v')
```

or

```
dsolve('eq1','eq2',. . .,'cond1','cond2',. . .,'v')
```

(The variable **v** is the independent variable.) Now consider the following example:

```
a = dsolve('Dx = y','Dy = x')
a =
  x: [1x1 sym]
  y: [1x1 sym]
```

The results are reported as symbolic elements in a structure array, just as the results
were reported with the `solve` command. To access these elements, use the structure array syntax:

```
a.x
ans =
C1*exp(t)-C2*exp(-t)
```

and

```
a.y
ans =
C1*exp(t)+C2*exp(-t)
```

You could also specify multiple outputs from the function:

```
[x,y] = dsolve('Dx = y','Dy = x')
x =
C1*exp(t)-C2*exp(-t)
y =
C1*exp(t)+C2*exp(-t)
```

KEY IDEA

Not every differential
equation can be solved
analytically

MATLAB® cannot solve every differential equation symbolically. For complicated (or ill-behaved) systems of equations, you may find it easier to use MuPad.
(Remember that MATLAB®'s symbolic capability is based on the MuPad engine.)
Many differential equations can't be solved analytically at all, no matter how sophisticated the tool. For those equations, numerical techniques often suffice.

12.6 CONVERTING SYMBOLIC EXPRESSIONS
TO MATLAB® FUNCTIONS

It is often useful to evaluate mathematical expressions symbolically before using the
results in more traditional MATLAB® functions. To accomplish this the `matlab-
Function` function converts a symbolic expression into an anonymous function.
Here's a really simple example.

```
syms x
y=cos(x)
dy=diff(y)
```

which returns the derivative of $\cos(x)$

```
dy=-sin(x)
```

To convert this symbolic variable, dy, into an anonymous function use the following approach.

```
f=matlabFunction(dy)
```

which returns

```
f =
    @x -sin(x)
```

Now f can be used to evaluate $-\sin(x)$. For example to evaluate $-\sin(x)$ at $x = 2$

```
f(2)
ans =
    -0.9093
```

Here's a more complicated example, which also involves symbolically finding a derivative.

```
syms x
y=(exp(-x)-1)/x
dy=diff(y)
g=matlabFunction(dy)
```

which results in a new anonymous function called g.

```
g=
   @(x) -1./(x.*exp(x))-(1./exp(x)-1)./x.^2
```

Anonymous functions can be used like any other MATLAB® function.

HINT

If you have a version of MATLAB® before 2007b, or if the Maple toolbox is installed on your computer, the matlabFunction will not work.

SUMMARY

MATLAB®'s symbolic mathematics toolbox uses the MuPad software engine. The symbolic toolbox is an optional component of the professional version of MATLAB®, but is included with the Student Version. The syntax used by the symbolic toolbox is similar to that used by MuPad. However, because the underlying structure of each program is different, MuPad users will recognize some differences in syntax.

Symbolic variables are created in MATLAB® with either the `sym` or the `syms` command:

```
x = sym('x')  or
syms x
```

The `syms` command has the advantage of making it easy to create multiple symbolic variables in one statement:

```
syms a b c
```

The `sym` command can be used to create complete expressions or equations in a single step:

```
y = sym('z^2-3')
```

Although z is included in this symbolic expression, it has not been explicitly defined as a symbolic variable.

Once symbolic variables have been defined, they can be used to create more complicated expressions. Since x, a, b, and c were defined as symbolic variables, they can be combined to create the quadratic equation:

```
EQ = a*x^2 + b*x + c
```

MATLAB® allows users to manipulate either symbolic expressions or symbolic equations. Equations are set equal to something; expressions are not. All the statements in this summary so far have created expressions. By contrast, the statement

```
EQ = sym('n = m/MW')
```

defines a symbolic equation.

Both symbolic expressions and equations can be manipulated by using built-in MATLAB® functions from the symbolic toolbox. The `numden` function extracts the numerator and denominator from an expression but is not valid for equations. The `expand`, `factor`, and `collect` functions can be used to modify either an expression or an equation. The `simplify` function simplifies an expression or an equation on the basis of built-in MuPad rules, and the `simple` function tries each member of the family of simplification functions and reports the shortest answer.

A highly useful symbolic function is `solve`, which allows the user to solve equations symbolically. If the input to the function is an expression, MATLAB® sets the expression equal to zero. The `solve` function can solve not only a single equation for the specified variable, but also systems of equations. Unlike the techniques used in matrix algebra to solve systems of equations, the input to `solve` need not be linear.

The substitution function, `subs`, allows the user to replace variables with either numeric values or new variables. It is important to remember that if a variable has not been explicitly defined as symbolic, it must be enclosed in single quotes when it is used in the `subs` function. When **y** is defined as

```
y = sym('m +2*n + p')
```

the variables m, n, and p are not explicitly defined as symbolic and must therefore be enclosed in single quotes. Notice that when multiple variables are replaced, they are listed inside curly brackets. If a single variable is replaced, the brackets are not required. Given the preceding definition of y, the command

```
subs(y,{'m','n','p'}, {1,2,3})
```

returns

```
ans =
    8
```

The `subs` command can be used to substitute both numeric values and symbolic variables.

MATLAB®'s symbolic plotting capability roughly mirrors the standard plotting options. The most useful of these plots for engineers and scientists is probably the *x–y* plot, `ezplot`. This function accepts a symbolic expression and plots it for values of x from -2π to $+2\pi$. The user can also assign the minimum and maximum values of x. Symbolic plots are annotated with the use of the same syntax as standard MATLAB® plots.

The symbolic toolbox includes a number of calculus functions, the most basic being `diff` (differentiation) and `int` (integration). The `diff` function allows the user to take the derivative with respect to a default variable (**x** or whatever is closest to x in the expression) or to specify the differentiation variable. Higher-order derivatives can also be specified. The `int` function also allows the user to integrate with respect to the default variable (x) or to specify the integration variable. Both definite and indefinite integrals can be evaluated. Additional calculus functions not discussed in this chapter are available. Use the `help` function for more information.

When solving a problem it is often useful to manipulate expressions symbolically before creating MATLAB® functions. The `matlabFunction` function allows you to do this easily.

MATLAB® SUMMARY

The following MATLAB® summary lists all the special characters, commands, and functions that are defined in this chapter:

Special Characters	
"	identifies a symbolic variable that has not been explicitly defined
{ }	encloses a cell array, used in the `solve` function to create lists of symbolic variables

Commands and Functions	
collect	collects like terms
diff	finds the symbolic derivative of a symbolic expression
dsolve	differential equation solver
expand	expands an expression or equation
ezcontour	creates a contour plot
ezcontourf	creates a filled contour plot
ezmesh	creates a mesh plot from a symbolic expression

(continued)

Commands and Functions	
ezmeshc	plots both a mesh and a contour plot created from a symbolic expression
ezplot	plots a symbolic expression (creates an x–y plot)
ezplot3	creates a three-dimensional line plot
ezpolar	creates a plot in polar coordinates
ezsurf	creates a surface plot from a symbolic expression
ezsurfc	plots both a mesh and a contour plot created from a symbolic expression
factor	factors an expression or equation
int	finds the symbolic integral of a symbolic expression
matlabFunction	converts a symbolic expression into an anonymous MATLAB® function
numden	extracts the numerator and denominator from an expression or an equation
simple	tries and reports all the simplification functions and selects the shortest answer
simplify	simplifies, using MuPad's built-in simplification rules
solve	solves a symbolic expression or equation
subs	substitutes into a symbolic expression or equation
sym	creates a symbolic variable, expression, or equation
syms	creates symbolic variables

PROBLEMS

Algebra

12.1   Create the symbolic variables

    a b c d x

and use them to create the following symbolic expressions:

    se1 = x^3 -3*x^2 +x
    se2 = sin(x) + tan(x)
    se3 =(2*x^2 - 3*x - 2)/(x^2 - 5*x)
    se4 = (x^2 -9)/(x+3)

12.2   **(a)** Divide se1 by se2.

   (b) Multiply se3 by se4.

   (c) Divide se1 by x.

   (d) Add se1 to se3.

12.3   Create the following symbolic equations:

   (a) $sq1 = sym('x^2 + y^2 = 4')$

   (b) $sq2 = sym('5*x^5 - 4*x^4 + 3*x^3 + 2*x^2 -x = 24 ')$

   (c) $sq3 = sym('sin(a) + cos(b) -x*c = d')$

   (d) $sq4 = sym('(x^3 - 3*x)/(3-x) = 14')$

12.4 Try to use the `numden` function to extract numerator and denominator from `se4` and `sq4`. Does this function work for both expressions and equations? Describe how your results vary. Try to explain the differences.

12.5 Use the `expand`, `factor`, `collect`, `simplify`, and `simple` functions on `se1` to `se4`, and on `sq1` to `sq4`. In your own words, describe how these functions work for the various types of equations and expressions.

Solving Symbolically and Using the Subs Command

12.6 Solve each of the expressions created in Problem 12.1 for x.

12.7 Solve each of the equations created in Problem 12.3 for x.

12.8 Solve equation `sq3`, created in Problem 12.3, for a.

12.9 A pendulum is a rigid object suspended from a frictionless pivot point (see Figure P12.9). If the pendulum is allowed to swing back and forth with a given inertia, we can find the frequency of oscillation with the equation

$$2\pi f = \sqrt{\frac{mgL}{I}}$$

where
$\quad f$ = frequency,
$\quad m$ = mass of the pendulum,
$\quad g$ = acceleration due to gravity,
$\quad L$ = distance from the pivot point to the center of gravity of the pendulum, and
$\quad I$ = inertia.

Use MATLAB®'s symbolic capability to solve for the length L.

Figure P12.9
Pendulum described in Problem 12.9.

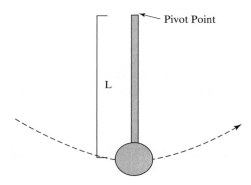

12.10 Let the mass, inertia, and frequency of the pendulum in the previous problem be, respectively,

$$m = 10 \text{ kg}$$

$$f = 0.2 \text{ s}^{-1}$$

$$I = 60 \text{ kg m/s.}$$

If the pendulum is on the earth ($g = 9.8 \text{ m/s}^2$) what is the length from the pivot point to the center of gravity? (Use the `subs` function to solve this problem.)

12.11 Kinetic energy is defined as

$$KE = \frac{1}{2}mV^2$$

where

KE = kinetic energy, measured in J
m = mass, measured in kg
V = velocity, measured in m/s.

Create a symbolic equation for kinetic energy, and solve it for velocity.

12.12 Find the kinetic energy of a car that weighs 2000 lb_m and is traveling at 60 mph (see Figure P12.12). Your units will be $\text{lb}_m \text{ mile}^2/\text{h}^2$. Once you've calculated this result, change it to Btu by using the following conversion factors:

$$1 \text{ lb}_f = 32.174 \text{ lb}_m \cdot \text{ft/s}^2$$
$$1 \text{ h} = 3600 \text{ s}$$
$$1 \text{ mile} = 5280 \text{ ft}$$
$$1 \text{ Btu} = 778.169 \text{ ft} \cdot \text{lb}_f$$

Figure P12.12
Car described in problem 12.12.

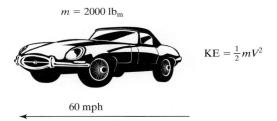

$m = 2000 \text{ lb}_m$

$KE = \frac{1}{2}mV^2$

60 mph

12.13 The heat capacity of a gas can be modeled with the following equation, composed of the empirical constants a, b, c, and d and the temperature T in kelvins:

$$C_P = a + bT + cT^2 + dT^3$$

Empirical constants do not have a physical meaning but are used to make the equation fit the data. Create a symbolic equation for heat capacity and solve it for T.

12.14 Substitute the following values for a, b, c, and d into the heat-capacity equation from the previous problem and give your result a new name [these values model the heat capacity of nitrogen gas in kJ/(kmol K) as it changes temperature between approximately 273 and 1800 K]:

$$a = 28.90$$
$$b = -0.1571 \times 10^{-2}$$
$$c = 0.8081 \times 10^{-5}$$
$$d = -2.873 \times 10^{-9}$$

Solve your new equation for T if the heat capacity (C_p) is equal to 29.15 kJ/(kmol K).

12.15 The Antoine equation uses empirical constants to model the vapor pressure of a gas as a function of temperature. The model equation is

$$\log_{10}(P) = A - \frac{B}{C + T}$$

where

P = pressure, in mmHg

A = empirical constant

B = empirical constant

C = empirical constant

T = temperature in °C.

The normal boiling point of a liquid is the temperature at which the vapor pressure (P) of the gas is equal to atmospheric pressure, 760 mmHg. Use MATLAB®'s symbolic capability to find the normal boiling point of benzene if the empirical constants are

$$A = 6.89272$$
$$B = 1203.531$$
$$C = 219.888$$

12.16 A hungry college student goes to the cafeteria and buys lunch. The next day he spends twice as much. The third day he spends $1 less than he did the second day. At the end of 3 days he has spent $35. How much did he spend each day? Use MATLAB®'s symbolic capability to help you solve this problem.

Solving Systems of Equations

12.17 Consider the following set of seven equations:

$$3x_1 + 4x_2 + 2x_3 - x_4 + x_5 + 7x_6 + x_7 = 42$$
$$2x_1 - 2x_2 + 3x_3 - 4x_4 + 5x_5 + 2x_6 + 8x_7 = 32$$
$$x_1 + 2x_2 + 3x_3 + x_4 + 2x_5 + 4x_6 + 6x_7 = 12$$
$$5x_1 + 10x_2 + 4x_3 + 3x_4 + 9x_5 - 2x_6 + x_7 = -5$$
$$3x_1 + 2x_2 - 2x_3 - 4x_4 - 5x_5 - 6x_6 + 7x_7 = 10$$
$$-2x_1 + 9x_2 + x_3 + 3x_4 - 3x_5 + 5x_6 + x_7 = 18$$
$$x_1 - 2x_2 - 8x_3 + 4x_4 + 2x_5 + 4x_6 + 5x_7 = 17$$

Define a symbolic variable for each of the equations, and use MATLAB®'s symbolic capability to solve for each unknown.

12.18 Compare the amount of time it takes to solve the preceding problem by using left division and by using symbolic math with the `tic` and `toc` functions, whose syntax is

```
tic
⋮
code to be timed
⋮
toc
```

12.19 Use MATLAB®'s symbolic capabilities to solve the following problem by means of matrix algebra:

Consider a separation process in which streams of water, ethanol, and methanol enter a process unit. Two streams leave the unit, each with varying amounts of the three components (see Figure P12.19).

Determine the mass flow rates into the system and out of the top and bottom of the separation unit.

(a) First set up the following material-balance equations for each of the three components:

Water
$$0.5(100) = 0.2m_{tops} + 0.65m_{bottoms}$$
$$50 = 0.2m_{tops} + 0.65m_{bottoms}$$

Ethanol
$$100x = 0.35m_{tops} + 0.25m_{bottoms}$$
$$0 = -100x + 0.35m_{tops} + 0.25m_{bottoms}$$

Methanol
$$100(1 - 0.5 - x) = 0.45m_{tops} + 0.1m_{bottoms}$$
$$50 = 100x + 0.45m_{tops} + 0.1m_{bottoms}$$

(b) Create symbolic equations to represent each material balance.

(c) Use the solve function to solve the system of three equations and three unknowns.

Figure P12.19
Separation process with three components: Water, ethanol, and methanol.

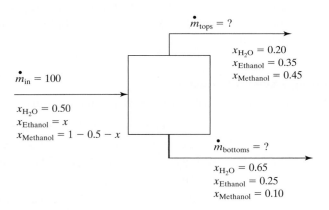

$\dot{m}_{tops} = ?$

$x_{H_2O} = 0.20$
$x_{Ethanol} = 0.35$
$x_{Methanol} = 0.45$

$\dot{m}_{in} = 100$

$x_{H_2O} = 0.50$
$x_{Ethanol} = x$
$x_{Methanol} = 1 - 0.5 - x$

$\dot{m}_{bottoms} = ?$

$x_{H_2O} = 0.65$
$x_{Ethanol} = 0.25$
$x_{Methanol} = 0.10$

12.20 Consider the following two equations:

$$x^2 + y^2 = 42$$
$$x + 3y + 2y^2 = 6$$

Define a symbolic equation for each, and solve it by using MATLAB®'s symbolic capability. Could you solve these equations by using matrices? Try this problem twice, once using only integers in your equation definitions and once using floating-point numbers (those with decimal points). How do your results vary? Check the workspace window to determine whether the results are still symbolic.

Symbolic Plotting

12.21 Create plots of the following expressions from $x = 0$ to 10:

(a) $y = e^x$
(b) $y = \sin(x)$
(c) $y = ax^2 + bx + c$, where $a = 5$, $b = 2$, and $c = 4$
(d) $y = \sqrt{x}$

Each of your plots should include a title, an x-axis label, a y-axis label, and a grid.

12.22 Use `ezplot` to graph the following expressions on the same figure for x-values from -2π to 2π (you'll need to use the `hold on` command):

$$y_1 = \sin(x)$$
$$y_2 = \sin(2x)$$
$$y_3 = \sin(3x)$$

Use the interactive plotting tools to assign each line a different color and line style.

12.23 Use `ezplot` to graph the following implicit equations:

(a) $x^2 + y^3 = 0$
(b) $x + x^2 - y = 0$
(c) $x^2 + 3y^2 = 3$
(d) $x \cdot y = 4$

12.24 Use `ezplot` to graph the following parametric functions:

(a) $f_1(t) = x = \sin(t)$
   $f_2(t) = y = \cos(t)$

(b) $f_1(t) = x = \sin(t)$
   $f_2(t) = y = 3\cos(t)$

(c) $f_1(t) = x = \sin(t)$
   $f_2(t) = y = \cos(3t)$

(d) $f_1(t) = x = 10\sin(t)$ from $t = 0$ to 30
   $f_2(t) = y = t\cos(t)$

(e) $f_1(t) = x = t\sin(t)$ from $t = 0$ to 30
   $f_2(t) = y = t\cos(t)$

12.25 The distance a projectile travels when fired at an angle θ is a function of time and can be divided into horizontal and vertical distances (see Figure P12.25), given respectively by

$$\text{horizontal}(t) = tV_0 \cos(\theta)$$

and

$$\text{vertical}(t) = tV_0 \sin(\theta) - \tfrac{1}{2}gt^2$$

where

horizontal = distance traveled in the x direction
vertical = distance traveled in the y direction
V_0 = initial velocity of the projectile
g = acceleration due to gravity, 9.8 m/s^2
t = time, s.

Suppose a projectile is fired at an initial velocity of 100 m/s and a launch angle of $\pi/4$ radians (45°). Use ezplot to graph horizontal distance on the x-axis and vertical distance on the y-axis for times from 0 to 20 seconds.

Figure P12.25
Trajectory of a projectile.

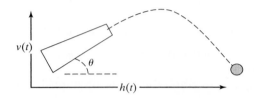

12.26 For each of the following expressions, use the ezpolar plot function to create a graph of the expression, and use the subplot function to put all four of your graphs in the same figure:

(a) $\sin^2(\theta) + \cos^2(\theta)$
(b) $\sin(\theta)$
(c) $e^{\theta/5}$ for θ from 0 to 20
(d) $\sinh(\theta)$ for θ from 0 to 20

12.27 Use ezplot3 to create a three-dimensional line plot of the following functions:

$$f_1(t) = x = t\sin(t)$$
$$f_2(t) = y = t\cos(t)$$
$$f_3(t) = z = t$$

12.28 Use the following equation to create a symbolic function Z:

$$Z = \frac{\sin(\sqrt{X^2 + Y^2})}{\sqrt{X^2 + Y^2}}$$

(a) Use the ezmesh plotting function to create a three-dimensional plot of Z.
(b) Use the ezsurf plotting function to create a three-dimensional plot of Z.
(c) Use ezcontour to create a contour map of Z.
(d) Generate a combination surface and contour plot of Z, using ezsurfc.

Use subplots to put all the graphs you create into the same figure.

Calculus

12.29 Determine the first and second derivatives of the following functions, using MATLAB®'s symbolic functions:

(a) $f_1(x) = y = x^3 - 4x^2 + 3x + 8$
(b) $f_2(x) = y = (x^2 - 2x + 1)(x - 1)$

(c) $f_3(x) = y = \cos(2x)\sin(x)$

(d) $f_4(x) = y = 3xe^{4x^2}$

12.30 Use MATLAB®'s symbolic functions to perform the following integrations:

(a) $\displaystyle\int (x^2 + x)\,dx$

(b) $\displaystyle\int_{0.3}^{1.3} (x^2 + x)\,dx$

(c) $\displaystyle\int (x^2 + y^2)\,dx$

(d) $\displaystyle\int_{3.5}^{24} (ax^2 + bx + c)\,dx$

12.31 Let the following polynomial represent the altitude in meters during the first 48 hours following the launch of a weather balloon:

$$h(t) = -0.12t^4 + 12t^3 - 380t^2 + 4100t + 220$$

Assume that the unit of t is hours.

(a) Use MATLAB® together with the fact that the velocity is the first derivative of the altitude to determine the equation for the velocity of the balloon.

(b) Use MATLAB® together with the fact that acceleration is the derivative of velocity, or the second derivative of the altitude, to determine the equation for the acceleration of the balloon.

(c) Use MATLAB® to determine when the balloon hits the ground. Because $h(t)$ is a fourth-order polynomial, there will be four answers. However, only one answer will be physically meaningful.

(d) Use MATLAB®'s symbolic plotting capability to create plots of altitude, velocity, and acceleration from time 0 until the balloon hits the ground [which was determined in part (c)]. You'll need three separate plots, since altitude, velocity, and acceleration have different units.

(e) Determine the maximum height reached by the balloon.

Use the fact that the velocity of the balloon is zero at the maximum height.

12.32 Suppose that water is being pumped into an initially empty tank (see Figure P12.32). It is known that the rate of flow of water into the tank at time t (in seconds) is $50 - t$ l/s. The amount of water Q that flows into the tank during the first x seconds can be shown to be equal to the integral of the expression $(50 - t)$ evaluated from 0 to x seconds.[*]

Figure P12.32
Tank-filling problem.

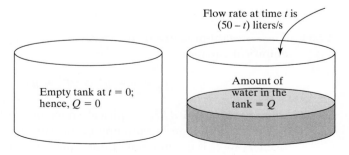

Flow rate at time t is $(50 - t)$ liters/s

Empty tank at $t = 0$; hence, $Q = 0$

Amount of water in the tank = Q

---

[*]From Etter, Kuncicky, and Moore, *Introduction to MATLAB 7* (Upper Saddle River, NJ: Pearson/Prentice Hall, 2005).

(a) Determine a symbolic equation that represents the amount of water in the tank after x seconds.
(b) Determine the amount of water in the tank after 30 seconds.
(c) Determine the amount of water that flowed into the tank between 10 and 15 seconds after the flow was initiated.

12.33 Consider a spring with the left end held fixed and the right end free to move along the x-axis (see Figure P12.33). We assume that the right end of the spring is at the origin $x = 0$ when the spring is at rest. When the spring is stretched, the right end of the spring is at some new value of x greater than zero. When the spring is compressed, the right end of the spring is at some value less than zero. Suppose that the spring has a natural length of 1 ft and that a force of 10 lb is required to compress it to a length of 0.5 ft. Then, it can be shown that the work, in ft lb$_f$ performed to stretch the spring from its natural length to a total of n ft is equal to the integral of $20x$ over the interval from 0 to $n - 1$.

(a) Use MATLAB® to determine a symbolic expression that represents the amount of work necessary to stretch the spring to a total length of n ft.
(b) What is the amount of work done to stretch the spring to a total of 2 ft?
(c) If the amount of work exerted is 25 ft lb$_f$, what is the length of the stretched spring?

Figure P12.33
Spring problem described in Problem 12.33.

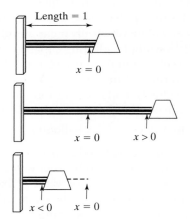

12.34 The constant-pressure heat capacity C_p of a gas can be modeled with the empirical equation

$$C_p = a + bT + cT^2 + dT^3$$

where a, b, c, and d are empirical constants and T is the temperature in Kelvin. The change in enthalpy (a measure of energy) as a gas is heated from T_1 to T_2 is the integral of this equation with respect to T:

$$\Delta h = \int_{T_1}^{T_2} C_p \, dT$$

Find the change in enthalpy of oxygen gas as it is heated from 300 to 1000 K. The values of a, b, c, and d for oxygen are

$$a = 25.48$$

$$b = 1.520 \times 10^{-2}$$

$$c = -0.7155 \times 10^{-5}$$

$$d = 1.312 \times 10^{-9}$$

Creating Anonymous Functions from Symbolic Expressions

12.35 A third-order polynomial is often represented as

$$ax^3 + bx^2 + cx^3 + d = 0$$

(a) Use the symbolic algebra capability in MATLAB® to solve this equation for x.

(b) Use the `matlabFunction` function to convert your result from part a into a MATLAB® function.

(c) Evaluate your function with the following input:

$$a = 4$$
$$b = 3$$
$$c = 1$$
$$d = 3$$

12.36 Consider the simple trigonometric function $\tan(x)$.

(a) Use the symbolic algebra capability in MATLAB® to integrate this function.

(b) Use the `matlabFunction` function to convert your result from part a into a MATLAB® function.

(c) Use `fplot` to plot your function from -5 to $+5$.

13

Numerical Techniques

Objectives

After reading this chapter, you should be able to:

- Interpolate between data points, using either linear or cubic spline models
- Model a set of data points as a polynomial
- Use the basic fitting tool

- Use the curve-fitting toolbox
- Perform numerical differentiations
- Perform numerical integrations
- Solve differential equations numerically

13.1 INTERPOLATION

Especially when we measure things, we don't gather data at every possible data point. Consider a set of *x–y* data collected during an experiment. By using an interpolation technique, we can estimate the value of *y* at values of *x* where we didn't take a measurement (see Figure 13.1). The two most common interpolation techniques are linear interpolation and cubic spline interpolation, both of which are supported by MATLAB®.

13.1.1 Linear Interpolation

The most common way to estimate a data point between two known points is *linear interpolation*. In this technique, we assume that the function between the points can be estimated by a straight line drawn between them, as shown in Figure 13.2. If we find the equation of a straight line defined by the two known points, we can find *y* for any value of *x*. The closer together the points are, the more accurate our approximation is likely to be.

Figure 13.1
Interpolation between data points.

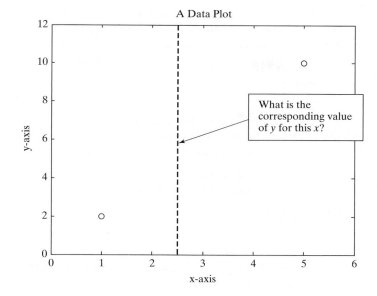

Figure 13.2
Linear interpolation: Connect the points with a straight line to find *y*.

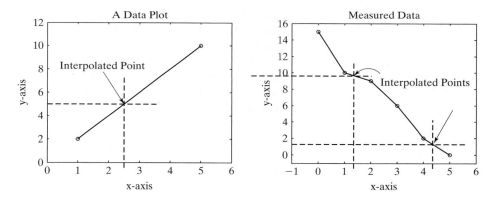

HINT

Although possible, it is rarely wise to *extrapolate* past the region where you've collected data. It may be tempting to assume that data continue to follow the same pattern, but this assumption can lead to large errors.

HINT

The last character in the function name `interp1` is the number 1. Depending on the font, it may look like the lowercase letter "ell" (l).

We can perform linear interpolation in MATLAB® with the `interp1` function. We'll first need to create a set of ordered pairs to use as input to the function. The data used to create the right-hand graph of Figure 13.2 are

```
x = 0:5;
y = [15, 10, 9, 6, 2, 0];
```

To perform a single interpolation, the input to `interp1` is the x data, the y data, and the new x value for which you'd like an estimate of y. For example, to estimate the value of y when x is equal to 3.5, type

```
interp1(x,y,3.5)
ans =
    4
```

INTERPOLATION

A technique for estimating an intermediate value based on nearby values

You can perform multiple interpolations all at the same time by putting a vector of x-values in the third field of the `interp1` function. For example, to estimate y-values for new x's spaced evenly from 0 to 5 by 0.2, type

```
new_x = 0:0.2:5;
new_y = interp1(x,y,new_x)
```

which returns

```
new_y =
  Columns 1 through 5
  15.0000 14.0000 13.0000 12.0000 11.0000
  Columns 6 through 10
  10.0000 9.8000 9.6000 9.4000 9.2000
  Columns 11 through 15
  9.0000 8.4000 7.8000 7.2000 6.6000
  Columns 16 through 20
  6.0000 5.2000 4.4000 3.6000 2.8000
  Columns 21 through 25
  2.0000 1.6000 1.2000 0.8000 0.4000
  Column 26
     0
```

We can plot the results on the same graph with the original data in Figure 13.3:

```
plot(x,y,new_x,new_y,'o')
```

Figure 13.3

Both measured data points and interpolated data were plotted on the same graph. The original points were modified in the interactive plotting function to make them solid circles.

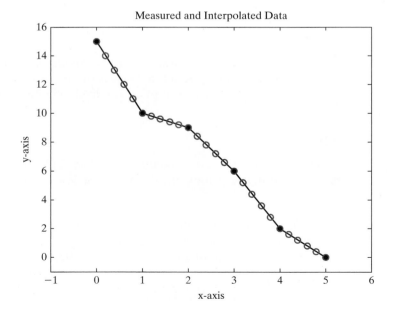

(For simplicity, the commands used to add titles and axis labels to plots in this chapter have been left out.)

The `interp1` function defaults to linear interpolation to make its estimates. However, as we will see in the next section, other approaches are possible. If we want (probably for documentation purposes) to explicitly define the approach used in `interp1` as linear interpolation, we can specify it in a fourth field:

```
interp1(x, y, 3.5, 'linear')
ans =
   4
```

13.1.2 Cubic Spline Interpolation

Connecting data points with straight lines probably isn't the best way to estimate intermediate values, although it is surely the simplest. We can create a smoother curve by using the cubic spline interpolation technique, included in the `interp1` function. This approach uses a third-order polynomial to model the behavior of the data. To call the cubic spline, we need to add a fourth field to `interp1`:

```
interp1(x,y,3.5,'spline')
```

This command returns an improved estimate of y at **x** = 3.5:

```
ans =
   3.9417
```

Of course, we could also use the cubic spline technique to create an array of new estimates for y for every member of an array of x-values:

```
new_x = 0:0.2:5;
new_y_spline = interp1(x,y,new_x,'spline');
```

A plot of these data on the same graph as the measured data (Figure 13.4) using the command

```
plot(x,y,new_x,new_y_spline,'-o')
```

results in two different lines.

Figure 13.4
Cubic spline interpolation. The data points on the smooth curve were calculated.

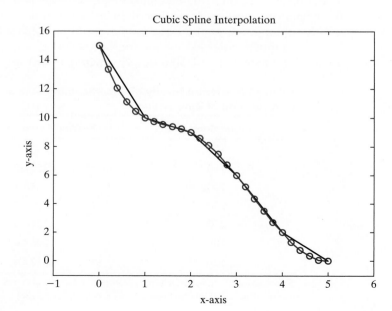

Table 13.1 Interpolation Options in the `Interp1` Function

`'linear'`	linear interpolation, which is the default	`interp1(x,y,3.5,'linear')`
		`ans =`
		   `4`

`'nearest'`	nearest-neighbor interpolation	`interp1(x,y,3.5,'nearest')`
		`ans =`
		   `2`

`'spline'`	piecewise cubic spline interpolation	`interp1(x,y,3.5,'spline')`
		`ans =`
		   `3.9417`

`'pchip'`	shape-preserving piecewise cubic interpolation	`interp1(x,y,3.5,'pchip')`
		`ans =`
		   `3.9048`

`'cubic'`	same as `'pchip'`	`interp1(x,y,3.5,'cubic')`
		`ans =`
		   `3.9048`

`'v5cubic'`	the cubic interpolation from MATLAB® 5, which does not extrapolate and uses `'spline'` if x is not equally spaced	`interp1(x,y,3.5,'v5cubic')`
		`ans =`
		   `3.9375`

The data points on the straight-line segments were measured. Note that every measured point also falls on the curved line.

The curved line in Figure 13.4 was drawn with the use of the interpolated data points. The line composed of straight-line segments was drawn through just the original data.

Although the most common ways to interpolate between data points are linear and spline approaches, MATLAB® does offer some other choices, as listed in Table 13.1.

EXAMPLE 13.1

THERMODYNAMIC PROPERTIES: USING THE STEAM TABLES

The subject of thermodynamics makes extensive use of tables. Although many thermodynamic properties can be described by fairly simple equations, others are either poorly understood, or the equations describing their behavior are very complicated. It is much easier to tabulate the values. For example, consider the values in Table 13.2 for steam at 0.1 MPa (approximately 1 atm) (Figure 13.5).

Table 13.2 Internal Energy of Superheated Steam at 0.1 MPa, as a Function of Temperature

Temperature, °C	Internal Energy u, kJ/kg
100	2506.7
150	2582.8
200	2658.1
250	2733.7
300	2810.4
400	2967.9
500	3131.6

Source: Data from Joseph H. Keenan, Frederick G. Keyes, Philip G. Hill, and Joan G. Moore, *Steam Tables, SI units* (New York: John Wiley & Sons, 1978).

Figure 13.5
Geysers spray high-temperature and high-pressure water and steam. (Rod Redfern © Dorling Kindersley.)

Use linear interpolation to determine the internal energy at 215°C. Use linear interpolation to determine the temperature if the internal energy is 2600 kJ/kg.

1. **State the Problem**
   Find the internal energy of steam, using linear interpolation.
   Find the temperature of the steam, using linear interpolation.

2. **Describe the Input and Output**

   Input   Table of temperature and internal energy
            u unknown
            T unknown

   Output Internal energy
            Temperature

3. **Develop a Hand Example**
   In the first part of the problem, we need to find the internal energy at 215°C. The table includes values at 200°C and 250°C. First we need to determine the fraction of the distance between 200 and 250 at which the value 215 falls:

   $$\frac{215 - 200}{250 - 200} = 0.30$$

   If we model the relationship between temperature and internal energy as linear, the internal energy should also be 30% of the distance between the tabulated values:

   $$0.30 = \frac{u - 2658.1}{2733.7 - 2658.1}$$

   Solving for u gives

   $$u = 2680.78 \text{ kJ/kg}$$

4. **Develop a MATLAB® Solution**
   Create the MATLAB® solution in an M-file, then run it in the command environment:

   ```
   %Example 13.1
   %Thermodynamics
   T=[100, 150, 200, 250, 300, 400, 500];
   u= [2506.7, 2582.8, 2658.1, 2733.7, 2810.4, 2967.9, 3131.6];
   newu=interp1(T,u,215)
   newT=interp1(u,T,2600)
   ```

(continued)

The code returns

```
newu =
    2680.78
newT =
    161.42
```

5. Test the Solution
   The MATLAB® result matches the hand result. This approach could be used for any of the properties tabulated in the steam tables. The JANAF tables published by the National Institute of Standards and Technology are a similar source of thermodynamic properties.

EXAMPLE 13.2

THERMODYNAMIC PROPERTIES: EXPANDING THE STEAM TABLES

As we saw in Example 13.1, thermodynamics makes extensive use of tables. Commonly, many experiments are performed at atmospheric pressure, so you may regularly need to use Table 13.3, which is just a portion of the steam tables (Figure 13.6).

Notice that the table is spaced first at 50°C intervals and then at 100°C intervals. Suppose you have a project that requires you to use this table and you prefer not to

Table 13.3 Properties of Superheated Steam at 0.1 MPa (Approximately 1 atm)

Temperature, °C	Specific Volume, v, m³/kg	Internal Energy, u, kJ/kg	Enthalpy, h, kJ/kg
100	1.6958	2506.7	2676.2
150	1.9364	2582.8	2776.4
200	2.172	2658.1	2875.3
250	2.406	2733.7	2974.3
300	2.639	2810.4	3074.3
400	3.103	2967.9	3278.2
500	3.565	3131.6	3488.1

Source: Data from Joseph H. Keenan, Frederick G. Keyes, Philip G. Hill, and Joan G. Moore, *Steam Tables, SI units* (New York: John Wiley & Sons, 1978).

Figure 13.6
Power plants use steam as a "working fluid."

perform a linear interpolation every time you use it. Use MATLAB® to create a table, employing linear interpolation, with a temperature spacing of 25°C.

1. State the Problem
   Find the specific volume, internal energy, and enthalpy every 5°C.
2. Describe the Input and Output

   Input   Table of temperature and internal energy
                New table interval of 25°C

   Output   Table

3. Develop a Hand Example
   In Example 13.1, we found the internal energy at 215°C. Since 215 is not on our output table, we'll redo the calculations at 225°C:

   $$\frac{225 - 200}{250 - 200} = 0.50$$

   and

   $$0.50 = \frac{u - 2658.1}{2733.7 - 2658.1}$$

   Solving for u gives

   $$u = 2695.9 \text{ kJ/kg}$$

   We can use this same calculation to confirm those in the table we create.

4. Develop a MATLAB® Solution
   Create the MATLAB® solution in an M-file, then run it in the command environment:

```
%Example 13.2
%Thermodynamics
clear, clc
T = [100, 150, 200, 250, 300, 400, 500]';
v = [1.6958, 1.9364, 2.172, 2.406, 2.639, 3.103, 3.565]';
u = [2506.7, 2582.8, 2658.1, 2733.7, 2810.4, 2967.9, 3131.6]';
h = [2676.2, 2776.4, 2875.3, 2974.3, 3074.3, 3278.2, 3488.1]';
props = [v,u,h];
newT = [100:25:500]';
newprop = interp1(T,props,newT);
disp('Steam Properties at 0.1 MPa')
disp('Temp Specific Volume Internal Energy Enthalpy')
disp(' C m^3/kg kJ/kg kJ/kg')
fprintf('%6.0f %10.4f %8.1f %8.1f \n',[newT,newprop]')
```

The program prints the following table to the command window:

```
Steam Properties at 0.1 MPa
Temp Specific Volume Internal Energy Enthalpy
 C      m^3/kg       kJ/kg        kJ/kg
100     1.6958      2506.7       2676.2
125     1.8161      2544.8       2726.3
150     1.9364      2582.8       2776.4
175     2.0542      2620.4       2825.9
200     2.1720      2658.1       2875.3
225     2.2890      2695.9       2924.8
```

(*continued*)

250	2.4060	2733.7	2974.3
275	2.5225	2772.1	3024.3
300	2.6390	2810.4	3074.3
325	2.7550	2849.8	3125.3
350	2.8710	2889.2	3176.3
375	2.9870	2928.5	3227.2
400	3.1030	2967.9	3278.2
425	3.2185	3008.8	3330.7
450	3.3340	3049.8	3383.1
475	3.4495	3090.7	3435.6
500	3.5650	3131.6	3488.1

5. Test the Solution
   The MATLAB result matches the hand result. Now that we know the program works, we can create more extensive tables by changing the definition of `newT` from

   ```
   newT = [100:25:500]';
   ```

   to a vector with a smaller temperature increment—for example,

   ```
   newT = [100:1:500]';
   ```

PRACTICE EXERCISES 13.1

Create x and y vectors to represent the following data:

x	y
10	23
20	45
30	60
40	82
50	111
60	140
70	167
80	198
90	200
100	220

1. Plot the data on an x–y plot.
2. Use linear interpolation to approximate the value of y when $x = 15$.
3. Use cubic spline interpolation to approximate the value of y when $x = 15$.
4. Use linear interpolation to approximate the value of x when $y = 80$.
5. Use cubic spline interpolation to approximate the value of x when $y = 80$.
6. Use cubic spline interpolation to approximate y-values for x-values evenly spaced between 10 and 100 at intervals of 2.
7. Plot the original data on an x–y plot as data points not connected by a line. Also, plot the values calculated in Exercise 6.

13.1.3 Multidimensional Interpolation

Imagine you have a set of data z that depends on two variables, x and y. For example, consider this table:

	x = 1	x = 2	x = 3	x = 4
y = 2	7	15	22	30
y = 4	54	109	164	218
y = 6	403	807	1210	1614

If you wanted to determine the value of z at $y = 3$ and $x = 1.5$, you would have to perform two interpolations. One approach would be to find the values of z at $y = 3$ and all the given x-values by using `interp1` and then do a second interpolation in your new chart. First let's define x, y, and z in MATLAB®:

```
y = 2:2:6;
x = 1:4;
z = [ 7      15      22      30
      54     109     164     218
      403    807     1210    1614];
```

Now we can use `interp1` to find the values of z at $y = 3$ for all the x-values:

```
new_z = interp1(y,z,3)  returns
new_z =
     30.50    62.00    93.00    124.00
```

Finally, since we have z-values at $y = 3$, we can use `interp1` again to find z at $y = 3$ and $x = 1.5$:

```
new_z2 = interp1(x,new_z,1.5)
new_z2 =
     46.25
```

Although this approach works, performing the calculations in two steps is awkward. MATLAB® includes a two-dimensional linear interpolation function, `interp2`, that can solve the problem in a single step:

```
interp2(x,y,z,1.5,3)
ans =
  46.2500
```

The first field in the `interp2` function must be a vector defining the value associated with each column (in this case, x), and the second field must be a vector defining the values associated with each row (in this case, y). The array z must have the same number of columns as the number of elements in x and must have the same number of rows as the number of elements in y. The fourth and fifth fields correspond to the values of x and of y for which you would like to determine new z-values.

MATLAB® also includes a function, `interp3`, for three-dimensional interpolation. Consult the `help` feature for the details on how to use this function and `interpn`, which allows you to perform n-dimensional interpolation. All these functions default to the linear interpolation technique but will accept any of the other techniques listed in Table 13.1.

PRACTICE EXERCISES 13.2

Create *x* and *y* vectors to represent the following data:

$y\downarrow/x\rightarrow$	x = 15	x = 30
y = 10	z = 23	33
20	45	55
30	60	70
40	82	92
50	111	121
60	140	150
70	167	177
80	198	198
90	200	210
100	20	230

1. Plot both sets of *y–z* data on the same plot. Add a legend identifying which value of *x* applies to each data set.
2. Use two-dimensional linear interpolation to approximate the value of *z* when *y* = 15 and *x* = 20.
3. Use two-dimensional cubic spline interpolation to approximate the value of *z* when *y* = 15 and *x* = 20.
4. Use linear interpolation to create a new subtable for *x* = 20 and *x* = 25 for all the *y*-values.

KEY IDEA

Curve fitting is a technique for modeling data with an equation

13.2 CURVE FITTING

Although we could use interpolation techniques to find values of *y* between measured *x*-values, it would be more convenient if we could model experimental data as $y = f(x)$. Then we could just calculate any value of *y* we wanted. If we know something about the underlying relationship between *x* and *y*, we may be able to determine an equation on the basis of those principles. For example, the ideal gas law is based on two underlying assumptions:

- All the molecules in a gas collide elastically.
- The molecules don't take up any room in their container.

Neither assumption is entirely accurate, so the ideal gas law works only when they are a good approximation of reality, but that is true for many situations, and the ideal gas law is extremely valuable. However, when real gases deviate from this simple relationship, we have two choices for how to model their behavior. Either we can try to understand the physics of the situation and adjust the equation accordingly or we can just take the data and model them empirically. Empirical equations are not related to any theory of why a behavior occurs; they just do a good job of predicting how a parameter changes in relationship to another parameter.

MATLAB® has built-in curve-fitting functions that allow us to model data empirically. It's important to remind ourselves that these models are good only in the region where we've collected data. If we don't understand why a parameter such as *y* changes as it does with *x*, we can't predict whether our data-fitting equation will still work outside the range where we've collected data.

13.2.1 Linear Regression

The simplest way to model a set of data is as a straight line. Let's revisit the data from Section 13.1.1:

```
x = 0:5;
y = [15, 10, 9, 6, 2, 0];
```

If we plot the data in Figure 13.7, we can try to draw a straight line through the data points to get a rough model of the data's behavior. This process is sometimes called "eyeballing it"—meaning that no calculations were done, but it looks like a good fit.

Looking at the plot, we can see that several of the points appear to fall exactly on the line, but others are off by varying amounts. In order to compare the quality of the fit of this line to other possible estimates, we find the difference between the actual *y*-value and the value calculated from the estimate. This difference is called the *residual*.

We can find the equation of the line in Figure 13.7 by noticing that at $x = 0$, $y = 0$ and at $x = 5$, $y = 0$. Thus, the slope of the line is

$$\frac{\text{rise}}{\text{run}} = \frac{\Delta y}{\Delta x} = \frac{y_2 - y_1}{x_2 - x_1} = \frac{0 - 15}{5 - 0} = -3$$

The line crosses the *y*-axis at 15, so the equation of the line is

$$y = -3x + 15$$

The differences between the actual values and the calculated values are listed in Table 13.4.

Figure 13.7
A linear model; the line was "eyeballed."

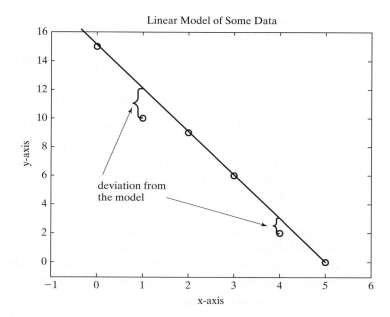

Linear Model of Some Data

Table 13.4 Difference between Actual and Calculated Values

x	y (actual)	y_calc (calculated)	difference = y − y_calc
0	15	15	0
1	10	12	−2
2	9	9	0
3	6	6	0
4	2	3	−1
5	0	0	0

LINEAR REGRESSION
A technique for modeling data as a straight line

The *linear regression* technique uses an approach called least squares fit to compare how well different equations model the behavior of the data. In this technique, the differences between the actual and calculated values are squared and added together. This has the advantage that positive and negative deviations don't cancel each other out. We could use MATLAB® to calculate this parameter for our data. We have

```
sum_of_the_squares = sum((y-y_calc).^2)
```

which gives us

```
sum_of_the_squares =
   5
```

It's beyond the scope of this chapter to explain how the linear regression technique works, except to say that it compares different models and chooses the appropriate one in which the sum of the squares is the smallest. Linear regression is accomplished in MATLAB® with the `polyfit` function. Three fields are required by `polyfit`: a vector of x-values, a vector of y-values, and an integer indicating what order polynomial should be used to fit the data. Since a straight line is a first-order polynomial, we'll enter the number 1 into the `polyfit` function:

```
polyfit(x,y,1)
ans =
 -2.9143 14.2857
```

The results are the coefficients corresponding to the best-fit first-order polynomial equation:

$$y = -2.9143x + 14.2857$$

Is this really a better fit than our "eyeballed" model? We can calculate the sum of the squares to find out:

```
best_y = -2.9143*x+14.2857;
new_sum = sum((y-best_y).^2)
new_sum =
   3.3714
```

Since the result of the sum-of-the-squares calculation is indeed less than the value found for the "eyeballed" line, we can conclude that MATLAB® found a better fit to the data. We can plot the data and the best-fit line determined by linear regression (see Figure 13.8) to try to get a visual sense of whether the line fits the data well:

```
plot(x,y,'o',x,best_y)
```

Figure 13.8
Data and best-fit line using linear regression.

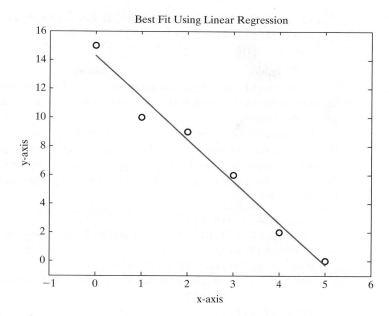

13.2.2 Polynomial Regression

Of course, straight lines are not the only equations that could be analyzed with the regression technique. For example, a common approach is to fit the data with a higher-order polynomial of the form

$$y = a_1 x^n + a_2 x^{n-1} + a_3 x^{n-2} + \cdots + a_n x + a_{n+1}$$

Polynomial regression is used to get the best fit by minimizing the sum of the squares of the deviations of the calculated values from the data. The `polyfit` function allows us to do this easily in MATLAB®. We can fit our sample data to second- and third-order equations with the commands

```
a=polyfit(x,y,2)
a =
   0.0536 -3.1821 14.4643
```

and

```
a=polyfit(x,y,3)
a =
  -0.0648 0.5397 -4.0701 14.6587
```

which correspond to the following equations

$$y_2 = 0.0536x^2 - 3.1821x + 14.4643$$

$$y_3 = -0.0648x^3 + 0.5397x^2 - 4.0701x + 14.6587$$

We can find the sum of the squares to determine whether these models fit the data better:

```
y2 = 0.0536*x.^2-3.182*x + 14.4643;
sum((y2-y).^2)
ans =
   3.2643
```

```
y3 = -0.0648*x.^3+0.5398*x.^2-4.0701*x + 14.6587
sum((y3-y).^2)
ans =
   2.9921
```

As we might expect, the more terms we add to our equation, the "better" is the fit, at least in the sense that the distance between the measured and predicted data points decreases.

In order to plot the curves defined by these new equations, we'll need more than the six data points used in the linear model. Remember that MATLAB® creates plots by connecting calculated points with straight lines, so if we want a smooth curve, we'll need more points. We can get more points and plot the curves with the following code:

```
smooth_x = 0:0.2:5;
smooth_y2 = 0.0536*smooth_x.^2-3.182*smooth_x + 14.4643;
subplot(1,2,1)
plot(x,y,'o',smooth_x,smooth_y2)
smooth_y3 = -0.0648*smooth_x.^3+0.5398*smooth_x.^2-4.0701*
smooth_x + 14.6587;
subplot(1,2,2)
plot(x,y,'o',smooth_x,smooth_y3)
```

The results are shown in Figure 13.9. Notice the slight curvature in each model. Although mathematically these models fit the data better, they may not be as good a representation of reality as the straight line. As an engineer or scientist, you'll need to evaluate any modeling you do. You'll need to consider what you know about the physics of the process you're modeling and how accurate and reproducible your measurements are.

13.2.3 The Polyval Function

The `polyfit` function returns the coefficients of a polynomial that best fits the data, at least on the basis of a regression criterion. In the previous section, we entered those coefficients into a MATLAB® expression for the corresponding polynomial and used it to calculate new values of y. The `polyval` function can perform the same job without our having to reenter the coefficients.

Figure 13.9
Second- and third-order polynomial fits.

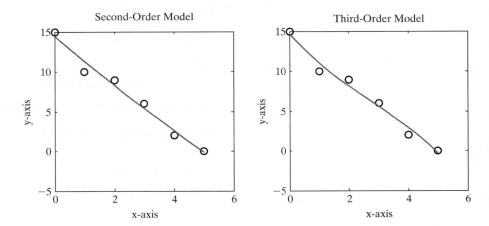

The `polyval` function requires two inputs. The first is a coefficient array, such as that created by `polyfit`. The second is an array of *x*-values for which we would like to calculate new *y*-values. For example, we might have

```
coef = polyfit(x,y,1)
y_first_order_fit = polyval(coef,x)
```

These two lines of code could be shortened to one line by nesting functions:

```
y_first_order_fit = polyval(polyfit(x,y,1),x)
```

We can use our new understanding of the `polyfit` and `polyval` functions to write a program to calculate and plot the fourth- and fifth-order fits for the data from Section 13.1.1:

```
y4 = polyval(polyfit(x,y,4),smooth_x);
y5 = polyval(polyfit(x,y,5),smooth_x);

subplot(1,2,1)
plot(x,y,'o',smooth_x,y4)
axis([0,6,-5,15])
subplot(1,2,2)
plot(x,y,'o',smooth_x,y5)
axis([0,6,-5,15])
```

Figure 13.10 gives the results of our plot.

As expected, the higher-order fits match the data better and better. The fifth-order model matches exactly because there were only six data points.

HINT

You could create all four of the graphs shown in Figures 13.9 and 13.10 by using a `for` loop that makes use of subplots and the `sprintf` function.

```
x = 0:5;
y = [15, 10, 9, 6, 2, 0];
smooth_x = 0:0.2:5;
for k = 1:4
    subplot(2,2,k)
    plot(x,y,'o',smooth_x,polyval(polyfit(x,y,k+1),smooth_x))
    axis([0,6,-5,15])
    a = sprintf('Polynomial plot of order %1.0f \n',k+1);
    title(a)
end
```

Figure 13.10
Fourth- and fifth-order model of six data points.

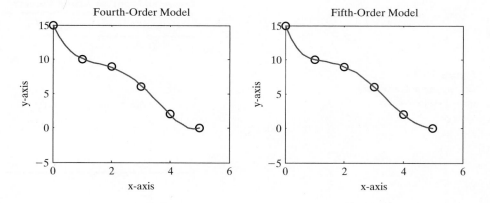

PRACTICE EXERCISES 13.3

Create *x* and *y* vectors to represent the following data:

z = 15		z = 30	
x	y	x	y
10	23	10	33
20	45	20	55
30	60	30	70
40	82	40	92
50	111	50	121
60	140	60	150
70	167	70	177
80	198	80	198
90	200	90	210
100	220	100	230

1. Use the `polyfit` function to fit the data for $z = 15$ to a first-order polynomial.
2. Create a vector of new *x* values from 10 to 100 in intervals of 2. Use your new vector in the `polyval` function together with the coefficient values found in Exercise 1 to create a new *y* vector.
3. Plot the original data as circles without a connecting line and the calculated data as a solid line on the same graph. How well do you think your model fits the data?
4. Repeat Exercises 1 through 3 for the *x* and *y* data corresponding to $z = 30$.

EXAMPLE 13.3

WATER IN A CULVERT

Determining how much water will flow through a culvert is not as easy as it might first seem. The channel could have a nonuniform shape (see Figure 13.11), obstructions might influence the flow, friction is important, and so on. A numerical approach allows us to fold all those concerns into a model of how the water actually behaves.

Figure 13.11
Culverts do not necessarily have a uniform cross section.

Consider the Following Data Collected From an Actual Culvert

Height, ft	Flow, ft³/s
0	0
1.7	2.6
1.95	3.6
2.60	4.03
2.92	6.45
4.04	11.22
5.24	30.61

Compute a best-fit linear, quadratic, and cubic equation for the data, and plot them on the same graph. Which model best represents the data? (Linear is first order, quadratic is second order, and cubic is third order.)

1. State the Problem
   Perform a polynomial regression on the data, plot the results, and determine which order best represents the data.
2. Describe the Input and Output

   Input   Height and flow data

   Output   Plot of the results

3. Develop a Hand Example
   Draw an approximation of the curve by hand. Be sure to start at zero, since, if the height of water in the culvert is zero, no water should be flowing (see Figure 13.12).
4. Develop a MATLAB® Solution
   Create the MATLAB® solution in an M-file, then run it in the command environment:

```
%13.3 Example - Water in a Culvert
height = [1.7, 1.95, 2.6, 2.92, 4.04, 5.24];
flow = [2.6, 3.6, 4.03, 6.45, 11.22, 30.61];
new_height = 0:0.5:6;
newf1 = polyval(polyfit(height,flow,1),new_height);
newf2 = polyval(polyfit(height,flow,2),new_height);
newf3 = polyval(polyfit(height,flow,3),new_height);
plot(height,flow,'o',new_height,newf1,new_height,newf2,
new_height,newf3)
title('Fit of Water Flow')
xlabel('Water Height, ft')
ylabel('Flow Rate, CFS')
legend('Data','Linear Fit','Quadratic Fit', 'Cubic Fit')
```

The MATLAB® code generates the plot shown in Figure 13.13.

Figure 13.12
Hand fit of water flow.

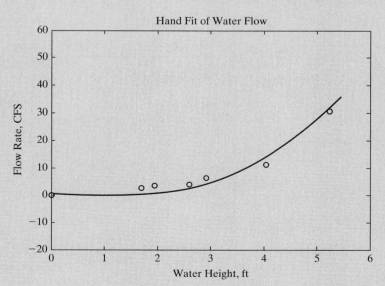

(*continued*)

Figure 13.13
Different curve-fitting approaches.

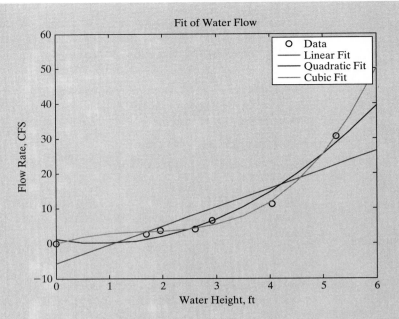

5. Test the Solution
   The question of which line best represents the data is difficult to answer. The higher-order polynomial approximation will follow the data points better, but it doesn't necessarily represent reality better.

   The linear fit predicts that the water flow rate will be approximately -5 CFS at a height of zero, which doesn't match reality. The quadratic fit goes back up after a minimum at a height of approximately 1.5 m—again a result inconsistent with reality. The cubic (third-order) fit follows the points the best and is probably the best polynomial fit. We should also compare the MATLAB® solution with the hand solution. The third-order (cubic) polynomial fit approximately matches the hand solution.

EXAMPLE 13.4

HEAT CAPACITY OF A GAS

The amount of energy necessary to warm a gas 1°C (called the *heat capacity* of the gas) depends not only on the gas, but on its temperature as well. This relationship is commonly modeled with polynomials. For example, consider the data for carbon dioxide in Table 13.5.

Use MATLAB® to model these data as a polynomial. Then compare the results with those obtained from the model published in B. G. Kyle, *Chemical and Process Thermodynamics* (Upper Saddle River, NJ: Prentice Hall PTR, 1999), namely

$$C_p = 1.698 \times 10^{-10}T^3 - 7.957 \times 10^{-7}T^2 + 1.359 \times 10^{-3}T + 5.059 \times 10^{-1}$$

1. State the Problem
   Create an empirical mathematical model that describes heat capacity as a function of temperature. Compare the results with those obtained from published models.

Table 13.5 Heat Capacity of Carbon Dioxide

Temperature, T, in K	Heat Capacity, C_p in kJ/(kg K)
250	0.791
300	0.846
350	0.895
400	0.939
450	0.978
500	1.014
550	1.046
600	1.075
650	1.102
700	1.126
750	1.148
800	1.169
900	1.204
1000	1.234
1500	1.328

Source: *Tables of Thermal Properties of Gases*, NBS Circular 564, 1955.

2. **Describe the Input and Output**

   Input   Use the table of temperature and heat-capacity data provided.

   Output   Find the coefficients of a polynomial that describes the data. Plot the results.

3. **Develop a Hand Example**
   By plotting the data (Figure 13.14) we can see that a straight-line fit (first-order polynomial) is not a good approximation of the data. We'll need to evaluate several different models—for example, from first to fourth order.

Figure 13.14
Heat capacity of
carbon dioxide as a
function of temperature.

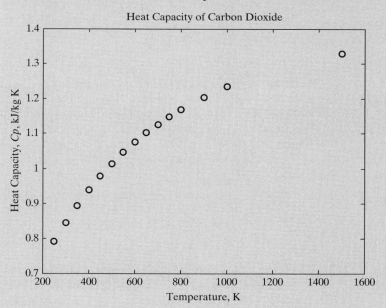

4. Develop a MATLAB® Solution

```
%Example 13.4 Heat Capacity of a Gas
%Define the measured data
T=[250:50:800,900,1000,1500];
Cp=[0.791, 0.846, 0.895, 0.939, 0.978, 1.014, 1.046, . . .
 1.075, 1.102, 1.126, 1.148, 1.169, 1.204, 1.234, 1.328];
%Define a finer array of temperatures
new_T = 250:10:1500;

%Calculate new heat capacity values, using four different
polynomial models
Cp1 = polyval(polyfit(T,Cp,1),new_T);
Cp2 = polyval(polyfit(T,Cp,2),new_T);
Cp3 = polyval(polyfit(T,Cp,3),new_T);
Cp4 = polyval(polyfit(T,Cp,4),new_T);

%Plot the results
subplot(2,2,1)
plot(T,Cp,'o',new_T,Cp1)
axis([0,1700,0.6,1.6])
subplot(2,2,2)
plot(T,Cp,'o',new_T,Cp2)
axis([0,1700,0.6,1.6])
subplot(2,2,3)
plot(T,Cp,'o',new_T,Cp3)
axis([0,1700,0.6,1.6])
subplot(2,2,4)
plot(T,Cp,'o',new_T,Cp4)
axis([0,1700,0.6,1.6])
```

By looking at the graphs shown in Figure 13.15, we can see that a second- or third-order model adequately describes the behavior in this temperature region. If we decide to use a third-order polynomial model, we can find the coefficients with `polyfit`:

```
polyfit(T,Cp,3)
ans =
2.7372e-010 -1.0631e-006 1.5521e-003 4.6837e-001
```

The results correspond to the equation

$$C_p = 2.7372 \times 10^{-10}T^3 - 1.0631 \times 10^{-6}T^2 + 1.5521 \times 10^{-3}T$$
$$+ 4.6837 \times 10^{-1}$$

5. Test the Solution
Comparing our result with that reported, we see that they are close, but not exact:

$$C_p = 2.737 \times 10^{-10}T^3 - 10.63 \times 10^{-7}T^2 + 1.552 \times 10^{-3}T + 4.683 \times 10^{-1}$$

(our fit)

$$C_p = 1.698 \times 10^{-10}T^3 - 7.957 \times 10^{-7}T^2 + 1.359 \times 10^{-3}T + 5.059 \times 10^{-1}$$

(literature)

Figure 13.15
A comparison of different polynomials used to model the heat-capacity data of carbon dioxide.

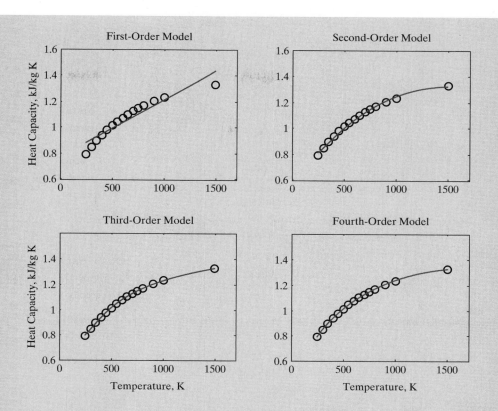

This is not too surprising, since we modeled a limited number of data points. The models reported in the literature use more data and are therefore probably more accurate.

13.3 USING THE INTERACTIVE FITTING TOOLS

MATLAB® 7 includes new interactive plotting tools that allow you to annotate your plots without using the command window. Also included are basic curve fitting, more complicated curve fitting, and statistical tools.

13.3.1 Basic Fitting Tools

To access the basic fitting tools, first create a figure:

```
x = 0:5;
y = [0,20,60,68,77,110]
plot(x,y,'o')
axis([-1,7,-20,120])
```

These commands produce a graph (Figure 13.16) with some sample data.

To activate the curve-fitting tools, select Tools → Basic Fitting from the menu bar in the figure. The basic fitting window opens on top of the plot. By checking `linear`, `cubic`, and `show equations` (see Figure 13.16), we generated the plot shown in Figure 13.17.

Figure 13.16
Interactive basic fitting window.

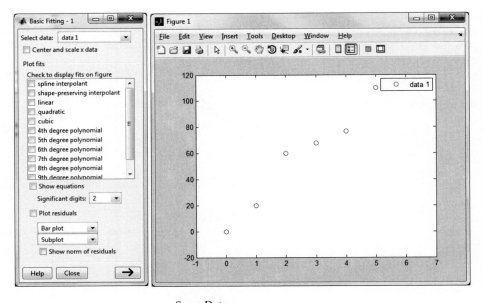

Figure 13.17
Plot generated with the basic fitting window.

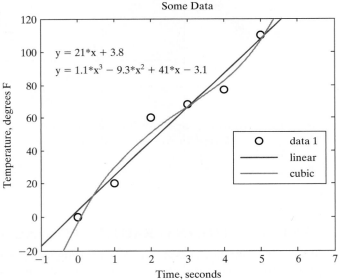

RESIDUAL
The difference between the actual and calculated value

Checking the plot residuals box generates a second plot, showing how far each data point is from the calculated line, as shown in Figure 13.18.

In the lower right-hand corner of the basic fitting window is an arrow button. Selecting that button twice opens the rest of the window (Figure 13.19).

The center panel of the window shows the results of the curve fit and offers the option of saving those results into the workspace. The right-hand panel allows you to select *x*-values and calculate *y*-values based on the equation displayed in the center panel.

In addition to the basic fitting window, you can access the data statistics window (Figure 13.20) from the figure menu bar. Select **Tools → Data Statistics** from the figure window. The data statistics window allows you to calculate statistical functions such as the mean and standard deviation interactively, based on the data in the figure, and allows you to save the results to the workspace.

Figure 13.18
Residuals are the difference between the actual and calculated data points.

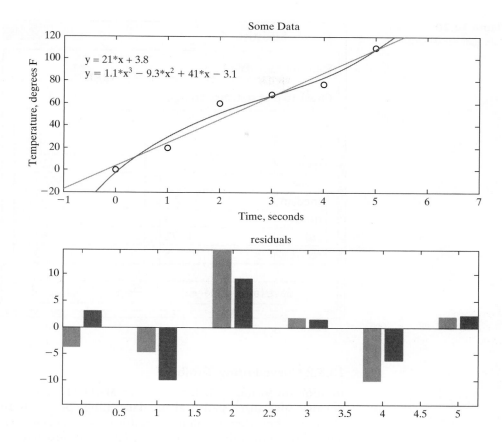

Figure 13.19
Basic fitting window.

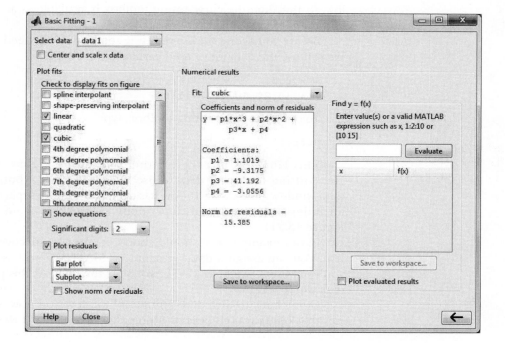

Figure 13.20
Data statistics window.

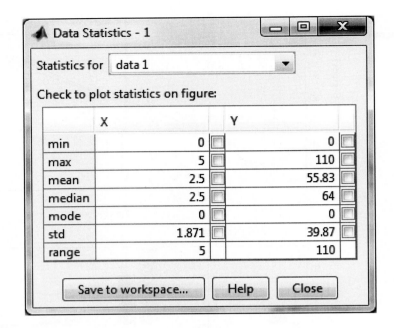

13.3.2 Curve-Fitting Toolbox

In addition to the basic fitting utility, MATLAB® contains toolboxes to help you perform specialized statistical and data-fitting operations. In particular, the curve-fitting toolbox contains a graphical user interface (GUI) that allows you to fit curves with more than just polynomials. You must have the curve-fitting toolbox installed in your copy of MATLAB® before you can execute the examples that follow. At this time the curve-fitting toolbox is available as an add-on for the student edition of MATLAB®.

Before you access the curve-fitting toolbox, you'll need a set of data to analyze. We can use the data we've used earlier in the chapter:

```
x = 0:5;
y = [0,20,60,68,77,110];
```

To open the curve-fitting toolbox, type

```
cftool
```

This launches the curve-fitting tool window. Now you'll need to tell the curve-fitting tool what data to use. Select the data button, which will open a data window. The data window has access to the workspace and will let you select an independent (x) and dependent (y) variable from a drop-down list (see Figure 13.21).

In our example, you should choose x and y, respectively, from the drop-down lists. You can assign a data-set name, or MATLAB® will assign one for you. Once you've chosen variables, MATLAB® plots the data. At this point, you can close the data window.

Going back to the curve-fitting tool window, you now select the Fitting button that offers you choices of fitting algorithms. Select New fit, and select a fit

Figure 13.21

The curve-fitting and data windows.

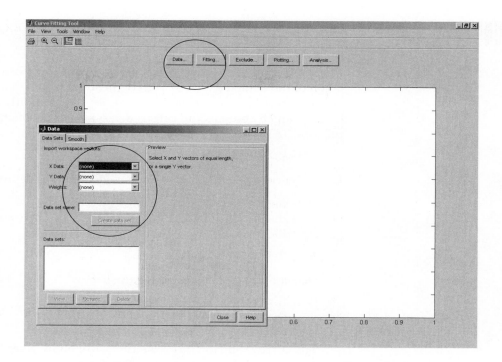

Figure 13.22

Curve-fitting tool window.

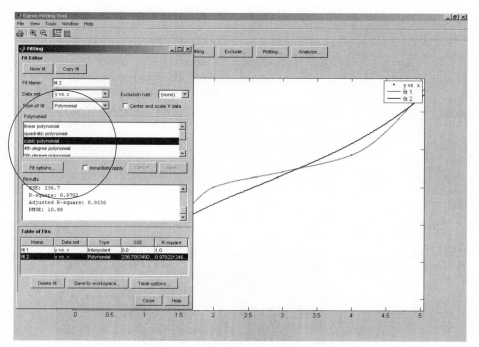

type from the `Type of fit` list. You can experiment with fitting choices to find the best one for your graph. We chose an interpolated scheme that forces the plot through all the points, and a third-order polynomial. The results are shown in Figure 13.22.

EXAMPLE 13.5

POPULATION

The population of the earth is expanding rapidly (see Figure 13.23), as is the population of the United States. MATLAB® includes a built-in data file, called census, that contains U.S. census data since 1790. The data file contains two variables: cdate, which contains the census dates; and pop, which lists the population in millions. To load the file into your workspace, type

```
load census
```

Use the curve-fitting toolbox to find an equation that represents the data.

1. State the Problem
   Find an equation that represents the population growth of the United States.
2. Describe the Input and Output

   Input   Table of population data

   Output   Equation representing the data

3. Develop a Hand Example
   Plot the data by hand.
4. Develop a MATLAB® Solution
   The curve-fitting toolbox is an interactive utility, activated by typing

```
cftool
```

which opens the curve-fitting window. You must have the curve-fitting toolbox installed in your copy of MATLAB® for this example to work. Select the data button and choose cdate as the *x*-value and pop as the *y*-value. After closing the data window, select the fitting button.

Since we have always heard that population is growing exponentially, experiment with the exponential-fit options. We also tried the polynomial option and chose a third-order (cubic) polynomial. Both approaches produced a good fit, but the polynomial was actually the best. We sent the curve-fitting window graph to a figure window and added titles and labels (see Figure 13.24).

From the data in the fitting window, we saw that the sum of the squares of the errors (SSE) was larger for the exponential fit, but that both approaches gave *R*-values greater than 0.99. (An *R*-value of 1 indicates a perfect fit.)

Figure 13.23
The earth's population
is expanding.

Figure 13.24
U.S. census data modeled with an exponential fit and a third-order polynomial.

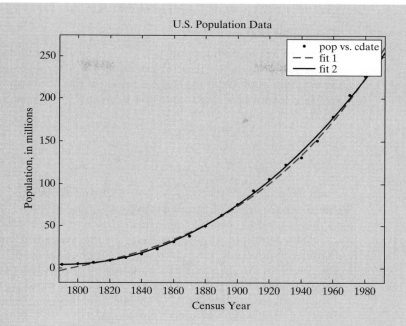

The results for the polynomial were as follows:

```
Linear model Poly3:
    f(x) = p1*x^3 + p2*x^2 + p3*x + p4
    where x is normalized by mean 1890 and std 62.05
Coefficients (with 95% confidence bounds):
    p1 = 0.921 (-0.9743, 2.816)
    p2 = 25.18 (23.57, 26.79)
    p3 = 73.86 (70.33, 77.39)
    p4 = 61.74 (59.69, 63.8)
Goodness of fit:
  SSE: 149.8
  R-square: 0.9988
  Adjusted R-square: 0.9986
  RMSE: 2.968
```

We normalized the x-values used in the equation for a better fit by subtracting the mean and dividing by the standard deviation:

```
x = (cdate-mean(cdate))/std(cdate);
```

5. Test the Solution
   Compare the fits by eye; they both appear to model the data adequately. It is important to remember that just because a solution models the data well, it is rarely appropriate to extend the solution past the measured data.

13.4 DIFFERENCES AND NUMERICAL DIFFERENTIATION

13.4.1 The Diff Function

The derivative of the function $y = f(x)$ is a measure of how y changes with x. If you can define an equation that relates x and y, you can use the functions contained in the symbolic toolbox to find an equation for the derivative. However, if all you have are data, you can approximate the derivative by dividing the change in y by the change in x:

KEY IDEA

The diff function is used both with symbolic expressions, where it finds the derivative, and with numeric arrays

$$\frac{dy}{dx} = \frac{\Delta y}{\Delta x} = \frac{y_2 - y_1}{x_2 - x_1}$$

If we plot the data from Section 13.1 that we've used throughout the chapter, this approximation of the derivative corresponds to the slope of each of the line segments used to connect the data, as shown in Figure 13.25.

If, for example, these data describe the measured temperature of a reaction chamber at different points in time, the slopes denote the cooling rate during each time segment. MATLAB® has a built-in function called `diff` that will find the difference between element values in a vector and that can be used to calculate the slope of ordered pairs of data. (The `diff` function is an example of an "overloaded" function. MATLAB® contains a version of `diff` used for symbolic algebra calculations, and a version that uses discrete data points. The software decides which version is appropriate based on the input you provide.)

For example, to find the change in our **x**-values, we type

```
delta_x = diff(x)
```

which, because the **x**-values are evenly spaced, returns

```
delta_x =
     1    1    1    1    1
```

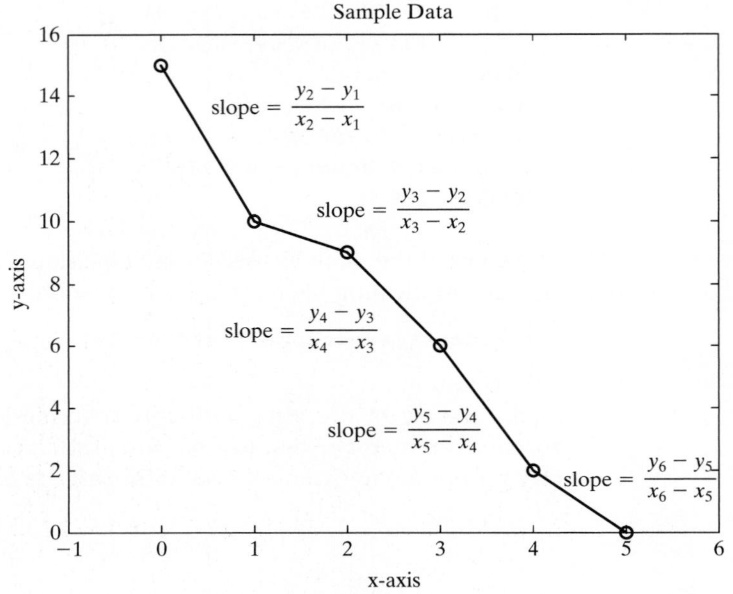

Figure 13.25

The derivative of a data set can be approximated by finding the slope of a straight line connecting each data point.

Sample Data

slope $= \dfrac{y_2 - y_1}{x_2 - x_1}$

slope $= \dfrac{y_3 - y_2}{x_3 - x_2}$

slope $= \dfrac{y_4 - y_3}{x_4 - x_3}$

slope $= \dfrac{y_5 - y_4}{x_5 - x_4}$

slope $= \dfrac{y_6 - y_5}{x_6 - x_5}$

Similarly, the difference in y-values is

```
delta_y = diff(y)
delta_y =
    -5  -1  -3  -4  -2
```

To find the slope, we just need to divide `delta_y` by `delta_x`:

```
slope = delta_y./delta_x
slope =
    -5  -1  -3  -4  -2
```

or

```
slope = diff(y)./diff(x)
slope =
    -5  -1  -3  -4  -2
```

Notice that the vector returned when you use the `diff` function is one element shorter than the input vector, because you are calculating differences. When you use the `diff` function to help you calculate slopes, you are calculating the slope between values of *x*, not at a particular value. If you want to plot these slopes against *x*, probably the best approach is to create a bar graph, since the rates of change are not continuous. The *x*-values were adjusted to the average for each line segment:

```
x = x(:,1:5)+diff(x)/2;
bar(x,slope)
```

The resulting bar graph is shown in Figure 13.26.

The `diff` function can also be used to approximate a derivative numerically if you know the relationship between *x* and *y*. For example, if

$$y = x^2$$

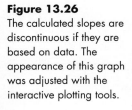

Figure 13.26
The calculated slopes are discontinuous if they are based on data. The appearance of this graph was adjusted with the interactive plotting tools.

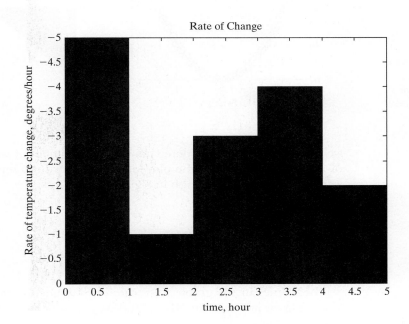

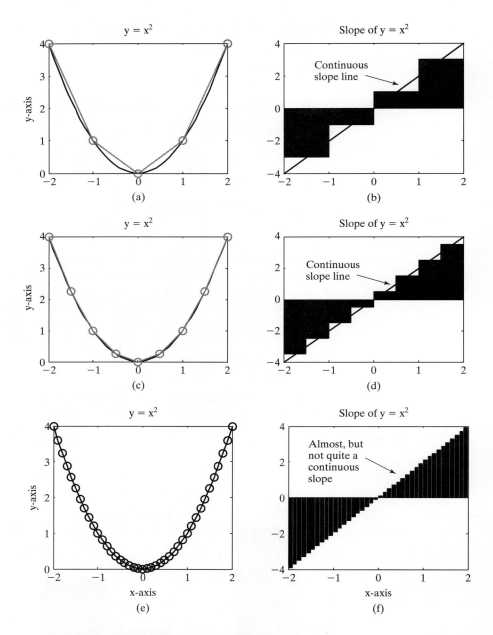

Figure 13.27
The slope of a function is approximated more accurately when more points are used to model the function.

we could create a set of ordered pairs for any number of x-values. The more values of x and y, the smoother the plot will be. Here are two sets of x and y vectors that were used to create the graph in Figure 13.27a:

```
x = -2:2
y = x.^2;
big_x = -2:0.1:2;
big_y = big_x.^2;
plot(big_x,big_y,x,y,'-o')
```

Both lines in the graph are created by connecting the specified points with straight lines; however, the big_x and big_y values are so close together that the graph looks like a continuous curve. The slope of the x–y plot was calculated with the `diff` function and plotted in Figure 13.27b:

```
slope5 = diff(y)./diff(x);
x5 = x(:,1:4)+diff(x)./2;
%These values were based on a 5-point model
bar(x5,slope5)
```

The bar graph was modified slightly with the use of the interactive plotting tools to give the representation shown in Figure 13.27b. We can get a smoother representation, though still discontinuous, by using more points:

```
x = -2:0.5:2;
y = x.^2;
plot(big_x,big_y,x,y,'-o')
slope9 = diff(y)./diff(x);
x9 = x(:,1:8)+diff(x)./2;
%These values were based on a 9-point model
bar(x9,slope9)
```

These results are shown in Figures 13.27c and 13.27d. We can use even more points:

```
plot(big_x,big_y,'-o')
slope41 = diff(big_y)./diff(big_x);
x41 = big_x(:,1:40)+diff(big_x)./2;      % 41-point model
bar(x41,slope41)
```

This code results in an almost smooth representation of the slope as a function of x, as seen in Figures 13.27e and 13.27f.

13.4.2 Forward, Backward, and Central Difference Techniques

What if you want to approximate the derivative at a point, instead of over a range, as discussed earlier? One approach is to use the slope between adjacent points as the approximation of the derivative at a single value of x.

$$\left(\frac{dy}{dx}\right)_i = \frac{y_{i+1} - y_i}{x_{i+1} - x_i}$$

We can accomplish this by using the difference function

```
dydx = diff(y)./diff(x)
```

and assigning the result as the derivative at the first point in the range. This is called a forward difference, since we are approximating the derivative by looking forward in the array to the next set of x and y values.

Take for example the sine function, whose analytical derivative is cosine. We can compare the forward difference derivative approximation to the analytical solution with the following code. First create an array of values for the independent variable, **x**, and for the dependent variable, **y**.

```
x = linspace(0,pi/2,10)
y = sin(x)
```

We know from basic calculus that the derivative of $\sin(x)$ is $\cos(x)$, which is expressed as

$$\frac{dy}{dx} = \cos(x)$$

Thus, to find the derivative analytically in MATLAB® we use the code

```
dydx_analytical=cos(x)
```

To approximate the derivative for the first nine values in the *x* array (which has a total of 10 values)

```
dydx_approx=diff(y)./diff(x)
```

It isn't possible to find an approximation for the derivative at the last point in the x array using this technique, so we use NaN (not a number) as a place holder. Notice that in order to make the code more general we've defined the last element number using the length function, which in this case returns a value of 10.

```
dydx_approx(length(x))=NaN;
```

To find the percentage error between this approximation and the analytical value we'll use the following equation:

$$\% \text{ error} = \frac{(\text{actual\_value} - \text{approximation})}{\text{actual\_value}} \times 100$$

which corresponds to the following code.

```
error_percentage = (dydx_analytical - dydx_approx)./dydx_
analytical*100;
```

Finally, to create an output table so we can evaluate the results, the following code can be used.

```
table =[x; dydx_analytical;dydx_approx;error_percentage]
disp('Forward Difference Approximation of the derivative of
sin(x)')
disp(' x dy/dx dy/dx %error')
disp(' cos(x) forward approx.')
fprintf('%8.4f\t%8.4f\t%8.4f\t%8.4f\n',table)
```

The resulting table is informative. There are significant errors in the approximation as the analytical result approaches 0, but the absolute error is fairly small.

Forward Difference Approximation of the Derivative of Sin(x)

x	dy/dx	dy/dx	%error
	cos(x)	forward approximation	(actual – est)/ actual *100
0.0000	1.0000	0.9949	0.5069
0.1745	0.9848	0.9647	2.0418
0.3491	0.9397	0.9052	3.6751
0.5236	0.8660	0.8181	5.5325
0.6981	0.7660	0.7062	7.8109
0.8727	0.6428	0.5728	10.8806
1.0472	0.5000	0.4221	15.5836
1.2217	0.3420	0.2585	24.4224
1.3963	0.1736	0.0870	49.8727
1.5708	0.0000	NaN	NaN

Figure 13.28

A comparison of the derivative approximation of sin(x), based on the number of points used.

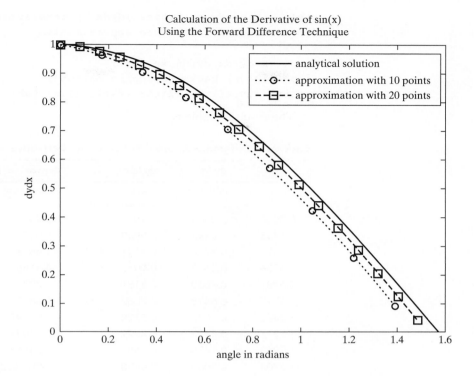

Notice that there is no approximation of the derivative for the last value of x, so (in the code) a value of NaN (not a number) was added. We repeated the calculations with 20 values and plotted the results for both 10 values and 20 values in Figure 13.28.

Clearly, we can do a better job of approximating the derivative by specifying more values of x (effectively making the points closer together).

The backwards difference is very similar. Instead of assigning the approximation of the derivative to the first value in a range, it is assigned to the last value.

$$\left(\frac{dy}{dx}\right)_i = \frac{y_i - y_{i-1}}{x_i - x_{i-1}}$$

To solve this problem in MATLAB®, we can use the diff function again. Similarly to the first example, a value of NaN was added to the dydx_approx matrix, but this time it is the first value, not the last.

```
%% Backward difference
x=linspace(0,pi/2,10);
y=sin(x);
dydx_analytical=cos(x);
dydx_approxb=diff(y)./diff(x);
dydx_approxb=[NaN,dydx_approxb];
error_percentageb = (dydx_analytical - dydx_approxb)./dydx_
analytical*100;
```

```
table =[x; dydx_analytical;dydx_approxb;error_percentageb]
disp('Backward Difference Approximation of the derivative of
sin(x)')
disp(' x dy/dx dy/dx %error')
disp(' cos(x) backward approximation')
fprintf('%8.4f\t%8.4f\t%8.4f\t%8.4f\n',table)
```

The resulting table is

Backward Difference Approximation of the Derivative of Sin(x)

x.	dy/dx	dy/dx	%error
	cos(x)	backward approximation	(actual – est)/actual *100
0.0000	1.0000	NaN	NaN
0.1745	0.9848	0.9949	−1.0279
0.3491	0.9397	0.9647	−2.6613
0.5236	0.8660	0.9052	−4.5186
0.6981	0.7660	0.8181	−6.7970
0.8727	0.6428	0.7062	−9.8667
1.0472	0.5000	0.5728	−14.5697
1.2217	0.3420	0.4221	−23.4085
1.3963	0.1736	0.2585	−48.8588
1.5708	0.0000	0.0870	−142155539756746180.0000

The absolute value of the error resulting from a forward difference technique versus a backward difference technique is very similar. (The large error for the final table entry in the backward difference table is due to the division by 0.) We can get closer by using a central difference technique, that looks both forward and backward, and therefore is centered on the actual point of interest. The approximation is therefore

$$\left(\frac{dy}{dx}\right)_i = \frac{y_{i+1} - y_{i-1}}{x_{i+1} - x_{i-1}}$$

One downside of using this technique is that it won't work for either the first or last value in the array.

MATLAB® includes a function, `gradient`, which approximates the derivative using a forward difference technique for the first point in an array, the backward difference for the last point in an array, and a centered difference for the remainder of the points. It requires two inputs, the **y** and **x** array

```
g = gradient(y,x)
```

and returns the derivative approximation. If you don't enter an x array, the program assumes the points are evenly spaced with a step size of 1. The results for all three approaches are shown in Figure 13.29.

The `gradient` function can also be used to approximate partial derivatives when used with two-dimensional arrays. Refer to the MATLAB® documentation for examples.

Figure 13.29
A superior approximation of the derivative is obtained using the centered difference approach, implemented in the `gradient` function.

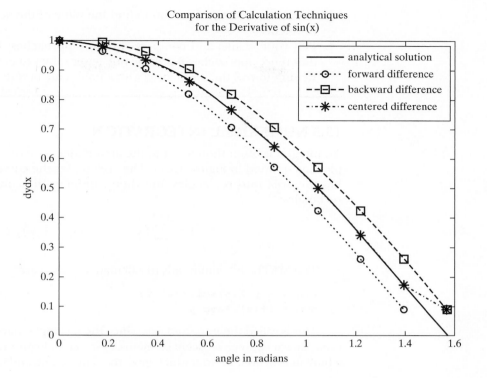

Comparison of Calculation Techniques
for the Derivative of sin(x)

PRACTICE EXERCISES 13.4

1. Consider the following equation:

$$y = x^3 + 2x^2 - x + 3$$

   Define an **x** vector from -5 to $+5$, and use it together with the `diff` function to approximate the derivative of y with respect to x, using the forward difference approach found analytically, the derivative is

$$\frac{dy}{dx} = y' = 3x^2 + 4x - 1$$

   Evaluate this function, using your previously defined **x** vector. How do your results differ?

2. Repeat Exercise 1 for the following functions and their derivatives:

Function	Derivative
$y = \sin(x)$	$\dfrac{dy}{dx} = \cos(x)$
$y = x^5 - 1$	$\dfrac{dy}{dx} = 5x^4$
$y = 5xe^x$	$\dfrac{dy}{dx} = 5e^x + 5xe^x$

3. Use the `gradient` function to find the value of the derivatives in the previous problems.
4. Plot your results and compare the two approaches. Recall that the forward difference approach will provide one fewer values than the length of the *x* array. Be sure to pad the result array with a final value of `NaN` to make plotting easier.

13.5 NUMERICAL INTEGRATION

An integral is often thought of as the area under a curve. Consider again our sample data, plotted in Figure 13.30. The area under the curve can be found by dividing the area into rectangles and then summing the contributions from all the rectangles:

$$A = \sum_{i=1}^{n-1} (x_{i+1} - x_i)(y_{i+1} + y_i)/2$$

The MATLAB® commands to calculate this area are

```
avg_y = y(1:5)+diff(y)/2;
sum(diff(x).*avg_y)
```

This is called the trapezoid rule, since the rectangles have the same area as a trapezoid drawn between adjacent elements, as shown in Figure 13.31. MATLAB® includes a built-in function, `trapz`, which gives the same result, and which uses the syntax

```
trapz(x,y)
```

We can approximate the area under a curve defined by a function instead of data by creating a set of ordered *x*–*y* pairs. Better approximations are found as we

Figure 13.30
The area under a curve can be approximated with the trapezoid rule.

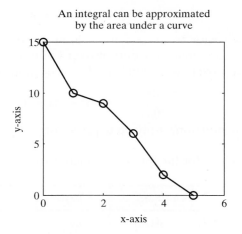

An integral can be approximated by the area under a curve

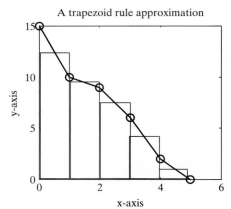

A trapezoid rule approximation

Figure 13.31
The area of a trapezoid can be modeled with a rectangle.

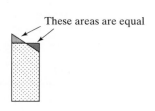

These areas are equal

Figure 13.32
The integral of a function can be estimated with the trapezoid rule.

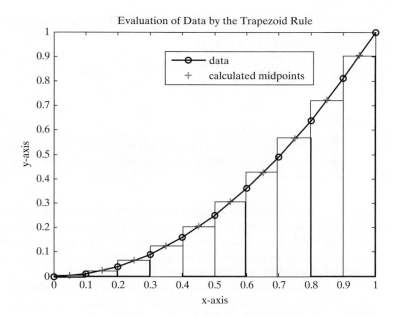

increase the number of elements in our x and y vectors. For example, to find the area under the function

$$y = f(x) = x^2$$

from 0 to 1, we would define a vector of 11 x-values and calculate the corresponding y-values:

```
x = 0:0.1:1;
y = x.^2;
```

The calculated values are plotted in Figure 13.32 and are used to find the area under the curve:

QUADRATURE
A technique for estimating the area under a curve by using rectangles

```
trapz(x,y)
```

This result gives us an approximation of the area under the function:

```
ans =
   0.3350
```

The preceding answer corresponds to an approximation of the integral from $x = 0$ to $x = 1$, or

$$\int_0^1 x^2 \, dx$$

KEY IDEA
Use **trapz** for ordered pairs of data. Use **quad** or **quadl** for functions

MATLAB® includes two built-in functions, quad and quadl, which will calculate the integral of a function without requiring the user to specify how the rectangles shown in Figure 13.32 are defined. The two functions differ in the numerical technique used. Functions with singularities may be solved with one approach or the other, depending on the situation. The quad function uses adaptive Simpson quadrature:

```
quad('x.^2',0,1)
ans =
   0.3333
```

The `quadl` function uses adaptive Lobatto quadrature:

```
quadl('x.^2',0,1)
ans =
   0.3333
```

HINT

The `quadl` function ends with the letter "l," not the number "1." It may be hard to tell the difference, depending on the font you are using.

Both functions require the user to enter a function in the first field. This function can be called out explicitly as a character string, as shown, or can be defined in an M-file or as an anonymous function. The last two fields in the function define the limits of integration, in this case from 0 to 1. Both techniques aim at returning results within an error of 1×10^{-6}.

Here's another example, using a function handle and an anonymous function, instead of defining the function inside single quotes. First we'll define an anonymous function for a third-order polynomial.

```
fun_handle = @(x) -x.^3+20*x.^2 -5
```

Now let's plot the function, to see how it behaves. The easiest approach is to use `fplot`, since it also accepts a function handle:

```
fplot(fun_handle,[-5,25])
```

The resulting plot is shown in Figure 13.33a. The integral of this function

$$\int_{-5}^{25} -x^3 + 20x^2 - 5$$

is the area under the curve, shown in Figure 13.33b.

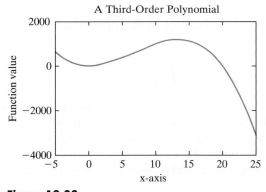
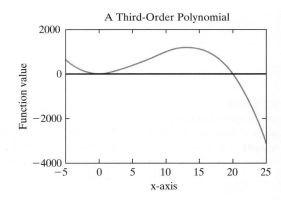

Figure 13.33
The integral of a function between two points can be thought of as the area under the curve. These graphs were created using `fplot` with a function handle representing a third-order polynomial.

Finally, to evaluate the integral we'll use the `quad` function, with the function handle as input:

```
quad(fun_handle,0,25)
ans =
  6.3854e+003
```

You can find out more about how these techniques work by consulting a numerical methods textbook, such as John H. Mathews and Kurtis D. Fink, *Numerical Methods Using MATLAB*, 4th ed. (Upper Saddle River, NJ: Pearson, 2004).

PRACTICE EXERCISES 13.5

1. Consider the following equation:

$$y = x^3 + 2x^2 - x + 3$$

   (a) Use the `trapz` function to estimate the integral of y with respect to x, evaluated from -1 to 1. Use 11 values of x, and calculate the corresponding values of y as input to the trapz function.
   (b) Use the `quad` and `quadl` functions to find the integral of y with respect to x, evaluated from -1 to 1.
   (c) Compare your results with the values found by using the symbolic toolbox function `int` and the following analytical solution (remember that the `quad` and `quadl` functions take input expressed with array operators such as `.*` or `.^`, but that the `int` function takes a symbolic representation that does not use these operators):

$$\int_a^b (x^3 + 2x^2 - x + 3)\ dx =$$

$$\left.\left(\frac{x^4}{4} + \frac{2x^3}{3} - \frac{x^2}{2} + 3x\right)\right|_a^b =$$

$$\frac{1}{4}(b^4 - a^4) + \frac{2}{3}(b^3 - a^3) - \frac{1}{2}(b^2 - a^2) + 3(b - a)$$

2. Repeat Exercise 1 for the following functions:

Function	Integral
$y = \sin(x)$	$\int_a^b \sin(x)\ dx = \cos(x)\big|_a^b = \cos(b) - \cos(a)$
$y = x^5 - 1$	$\int_a^b (x^5 - 1)dx = \left.\left(\frac{x^6}{6} - x\right)\right|_a^b = \left(\frac{b^6 - a^6}{6} - (b - a)\right)$
$y = 5x*e^x$	$\int_a^b (5e^x)dx = (-5e^x + 5xe^x)\big|_a^b =$ $(-5(e^b - e^a) + 5(be^b - ae^a))$

EXAMPLE 13.6

CALCULATING MOVING BOUNDARY WORK

In this example we'll use MATLAB®'s numeric integration techniques—both the quad function and the quadl function—to find the work produced in a piston–cylinder device by solving the equation

$$W = \int P dV$$

based on the assumption that

$$PV = nRT$$

where

P = pressure, kPa,
V = volume, m³,
n = number of moles, kmol,
R = universal gas constant, 8.314 kPa m³/kmol K, and
T = temperature, K.

We also assume that (1) the piston contains 1 mol of gas at 300 K and (2) the temperature stays constant during the process.

1. State the Problem
   Find the work produced by the piston–cylinder device shown in Figure 13.34.

2. Describe the Input and Output

   Input
   T = 300 K
   n = 1 kmol
   R = 8.314 kJ/kmol K

   $\left. \begin{array}{l} V_1 = 1\,\text{m}^3 \\ V_2 = 5\,\text{m}^3 \end{array} \right\}$ limits of integration

   Output   Work done by the piston–cylinder device

3. Develop a Hand Example
   Solving the ideal gas law

   $$PV = nRT$$

Figure 13.34
A piston–cylinder device.

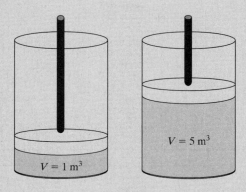

$V = 1\,\text{m}^3$

$V = 5\,\text{m}^3$

or

$$P = nRT/V$$

for P and performing the integration gives

$$W = \int \frac{nRT}{V} dV = nRT \int \frac{dV}{V} = nRT \ln\left(\frac{V_2}{V_1}\right)$$

Substituting in the values, we find that

$$W = 1 \text{ kmol} \times 8.314 \text{ kJ/kmol K} \times 300 \text{ K} \times \ln\left(\frac{V_2}{V_1}\right)$$

Since the integration limits are $V_2 = 5 \text{ m}^3$ and $V_1 = 1 \text{ m}^3$, the work becomes

$$W = 4014 \text{ kJ}$$

Because the work is positive, it is produced by (and not on) the system.

4. Develop a MATLAB® Solution

```
%Example 13.6
%Calculating boundary work, using MATLAB®'s quadrature
%functions
clear, clc

%Define constants
n = 1;        % number of moles of gas
R = 8.314;    % universal gas constant
T = 300;      % Temperature, in K

%Define an anonymous function for P
P = @(V) n*R*T./V;

% Use quad to evaluate the integral
quad(P,1,5)
%Use quadl to evaluate the integral
quadl(P,1,5)
```

which returns the following results in the command window

```
ans =
   4.0143e+003
ans =
   4.0143e+003
```

Notice that in this solution we defined an anonymous function for P, and used the function handle as input to the numerical integration functions . We could just as easily have defined the function by using a character string inside the quad and quadl functions. However, in that case we would have had to replace the variables with numerical values:

```
quad('1*8.314*300./V',1,5)
ans =
   4.0143e+003
```

The function could also have been defined in an M-file.

5. Test the Solution

We compare the results with our hand solution. The results are the same. It also helps to obtain a solution from the symbolic toolbox. Why do we need both kinds of MATLAB® solution? Because some problems cannot be solved with MATLAB®'s symbolic tools, and others (those with singularities) are ill suited to a numerical approach.

13.6 SOLVING DIFFERENTIAL EQUATIONS NUMERICALLY

MATLAB® includes a number of functions that solve ordinary differential equations of the form

$$\frac{dy}{dt} = f(t, y)$$

numerically. In order to solve higher-order differential equations (and systems of differential equations) they must be reformulated into a system of first-order expressions. This section outlines the major features of the ordinary differential equation solver functions. For more information, consult the `help` feature.

Not every differential equation can be solved by the same technique, so MATLAB® includes a wide variety of differential equation solvers (Table 13.6). However, all of these solvers have the same format. This makes it easy to try different techniques by just changing the function name.

Each solver requires the following three inputs as a minimum:

KEY IDEA
MATLAB® includes a large family of differential equation solvers

- A function handle to a function that describes the first-order differential equation or system of differential equations in terms of t and y
- The time span of interest
- An initial condition for each equation in the system

The solvers all return an array of t- and y-values:

```
[t,y] = odesolver(function_handle,[initial_time, final_time],
    [initial_cond_array])
```

If you don't specify the resulting arrays `[t,y]`, the functions create a plot of the results.

13.6.1 Function Handle Input

As we've discussed before, a function handle is a "nickname" for a function. It can refer to either a standard MATLAB® function, stored as an M-file, or an anonymous MATLAB® function. Recall that the differential equations we're discussing are of the form

$$\frac{dy}{dt} = f(t, y)$$

so the function handle is equivalent to dy/dt.

Here's an example of an anonymous function for a single simple differential equation:

dydt = @(t,y) 2*t corresponds to $\dfrac{dy}{dt} = 2t$

Although this particular function doesn't use a value of y in the result ($2t$), y still needs to be part of the input.

If you want to specify a system of equations, it is probably easier to define a function M-file. The output of the function must be a column vector of first-derivative values, as in

```
function dy=twofuns(t,y)
dy(1) = y(2);
dy(2) = -y(1);
dy=[dy(1); dy(2)];
```

Table 13.6 MATLAB®'s Differential Equation Solvers

Ordinary Differential Equation Solver Function	Type of Problems Likely to be Solved with This Technique	Numerical Solution Method	Comments
ode45	nonstiff differential equations	Runge–Kutta	Best choice for a first-guess technique if you do not know much about the function.
			Uses an explicit Runge–Kutta (4, 5) formula called the Dormand–Prince pair.
ode23	nonstiff differential equations	Runge–Kutta	This technique uses an explicit Runge–Kutta (2, 3) pair of Bogacki and Shampine. If the function is "mildly stiff," this maybe a better approach than ode45.
ode113	nonstiff differential equations	Adams	Unlike ode45 and ode23, which are single-step solvers, this technique is a multistep solver.
ode15s	stiff differential equation and differential algebraic equations	NDFs (BDFs)	Uses numerical differentiation formulas (NDFs) or backward differentiation formulas (BDFs). It is difficult to predict which technique will work best on a stiff differential equation.
ode23s	stiff differential equations	Rosenbrock	Modified second-order Rosenbock formulation.
ode23t	moderately stiff differential equations and differential algebraic equations	trapezoid rule	Useful if you need a solution without numerical damping.
ode23tb	stiff differential equations	TR–BDF2	This solver uses an implicit Runge–Kutta formula with the trapezoid rule (TR) and a second-order backward differentiation formula (BDF2).
ode15i	fully implicit differential equations	BDF	This solver uses a backward difference formula (BDF) to solve implicit differential equations of the form $f(y, y', t) = 0$.

This function represents the system

$$\frac{dy}{dt} = x$$

$$\frac{dx}{dt} = -y$$

which could also be expressed in a more compact notation as

$$y_1' = y_2$$
$$y_2' = -y_1$$

where the prime indicates the derivative with respect to time, and the functions with respect to time are y_1, y_2, and so on. In this notation, the second derivative is equal to y'' and the third derivative is y''':

$$y' = \frac{dy}{dt}, \quad y'' = \frac{d^2y}{dt^2}, \quad y''' = \frac{d^3y}{dt^3}$$

13.6.2 Solving the Problem

Both the time span of interest and the initial conditions for each equation are entered as vectors into the solver equations, along with the function handle. To demonstrate, let's solve the equation

$$\frac{dy}{dt} = 2t$$

We created an anonymous function for this ordinary differential equation in the previous section and called it `dydt`. We'll evaluate y from -1 to 1 and specify the initial condition as

$$y(-1) = 1$$

If you don't know how your equation or system of equations behaves, your first try should be `ode45`:

```
[t,y] = ode45(dydt, [-1,1],1)
```

This command returns an array of t-values and a corresponding array of y-values. You can either plot these yourself or allow the solver function to plot them if you don't specify the output array:

```
ode45(dydt, [-1,1],1)
```

The results are shown in Figure 13.35 and are consistent with the analytical solution, which is

$$y = t^2$$

Note that the first derivative of this function is $2t$ and that $y = 1$ when $t = -1$.

When the input function or system of functions is stored in an M-file, the syntax is slightly different. The handle for an existing M-file is defined as `@m_file_name`. To solve the system of equations described in `twofun` (from the previous section) we use the command

```
ode45(@twofun, [-1,1], [1,1])
```

Figure 13.35
This figure was generated automatically by the ode45 function. The title and labels were added in the usual way.

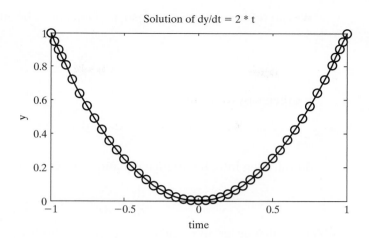

We could also assign the M-file a function handle to the M-file such as

```
some_fun = @twofun
```

and use it as input to the differential equation solver

```
ode45(some_fun,[-1,1],[1,1])
```

The time span of interest is from −1 to 1, and the initial conditions are both 1. Notice that there is one initial condition for each equation in the system. The results are shown in Figure 13.36.

13.6.3 Solving Higher-Order Differential Equations

The ode series of functions (such as ode45 or ode23) is used to solve either a single first-order differential equation, or a system of first-order differential equations. But what if you need to solve a higher-order problem? Fortunately a higher-order differential equation can be expressed as a series of equations by making some simple substitutions. Consider the following equation:

$$\frac{d^2y}{dt^2} + \frac{dy}{dt} = y + t$$

Figure 13.36
This system of equations was solved with ode45. The title, labels, and legend were added in the usual way.

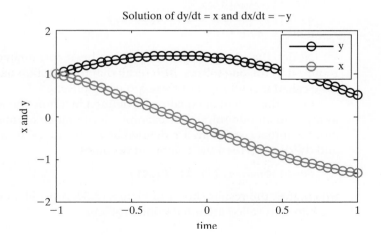

We can reformulate it into a system of equations by introducing a new variable, z. Let

$$z = \frac{dy}{dt}$$

It's then easy to see that

$$\frac{dz}{dt} = \frac{d^2y}{dt^2}$$

Substituting into the original equation we get

$$\frac{dz}{dt} + \frac{dy}{dt} = y + t,$$

which is a first-order differential equation. Effectively we've replaced

$$\frac{d^2y}{dt^2} + \frac{dy}{dt} = y + t$$

with the following two equations, which have been rearranged to solve for the first derivative of our two dependent variables, y and z

$$\frac{dy}{dt} = z$$

and

$$\frac{dz}{dt} = y + t - \frac{dy}{dt}$$

Now all we need to do is create an M-file function to use in one of the ode solvers. The function should have two inputs, which are typically called t and y. The variable t is the independent variable, and the variable y is an array of dependent variables. In this example $y(1)$ corresponds to the y used in the hand formulation, and $y(2)$ corresponds to z. The function containing the system of equations should look like this:

```
function dydt = twoeq(t,y)
dydt(1) = y(2);
dydt(2) = y(1) + t - dydt(1);
dydt = dydt'
```

Notice that the function output has been formulated as a column vector, as required by the ode solvers. Also recall that the function name is arbitrary. We could have called it anything, but twoeq is descriptive.

Once the system of equations is defined in a function M-file it is available to use as input to an ode solver. For example, if the range of time is defined as -1 to $+1$ and the initial conditions are defined as $y = 0$ and $z = 0$ (which is the same as $y = 0$ and $dy/dt = 0$), then the command becomes

```
ode45(@twoeq, [-1,1], [0,0])
```

which gives the results shown in Figure 13.37. A problem where the starting values are known is called an initial value problem.

Figure 13.37
A higher-order differential equation is solved by creating a system of equations that represents the same information. A second-order ODE requires two equations, resulting in two lines represented in the graphical output, one for y, and one for dy/dt.

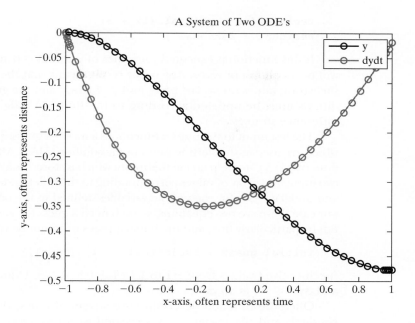

13.6.4 Boundary Value Problems

Reconsider the function from the previous section, which describes a system of two ordinary differential equations. What would happen if we didn't know the initial value of dy/dt, but instead knew the value of y at both $t = -1$ and $t = 1$? This is called a boundary value problem, and can be solved using the `bvp4c` function. Similarly to the ode solvers, the `bvp4c` function requires three inputs:

- A function handle to the system of ode's to be solved.
- A function handle to a function that solves for the residual values of the function.
- A set of guesses for the initial conditions.

The first function handle is exactly the same as we used for the ode solver set of functions. It should contain the equations for the derivatives of interest and the results must be a column vector.

To solve the problem a guess is made for the initial value of all the derivatives, then the program checks to see how it did by comparing the calculated boundary values with the actual values. For example, if:

$$\text{at } t = -1, \quad y = 0 \quad \text{and}$$

$$\text{at } t = 1, \quad y = 3$$

the program would solve the system of equations based upon an initial guess of dy/dt, and would then check to see how close the result is at $t = +1$ (i.e., it would check to see if $y = 3$). This is accomplished using a boundary condition function where the equations are arranged so that if the correct boundary condition is calculated, the function values are zero. In the case of our example,

```
function residual=bc(y_initial, y_final)
residual(1) = y_initial(1) + 0;
```

```
residual(2) = y_final (1) - 3;
residual = residual';
```

If this function is executed for values of y_initial = 0 and y_final = 3, the result will be a column of zeros. Any other result means that the program has calculated the wrong values for `y_initial` and `y_final`, and the guesses for the initial conditions must be updated according to the function's algorithm, which is a finite difference strategy.

The last input to the `bvp4c` function is a mesh of guesses for the problem solution, which are used as the starting point in the solution. MATLAB® provides a helper function, `bvpinit`, to help create this mesh, which is stored as a structure array. It requires two inputs; an array of values corresponding to the independent variable (in this case t) and initial guesses for each of the variables defined in the ode system of equations. In our case there are two equations, so we'll need a guess for y and dy/dt. The mesh need not be particularly fine, and the initial guesses need not be very good. For example:

```
initial_guess = bvpinit(-1:.5:1, [0, -1])
```

specifies five **t** values from -1 to 1 ($-1, -0.5, 0, 0.5, 1$) and initial guesses of $y = 0$ and $dydt = -1$ at all values of **t**.

Once the function describing the system of ode's, the function defining the residuals and the initial guesses created with `bvpinit` have been created, the `bvp4c` function can be executed.

```
bvp4c(@twoeq, @bc, initial_guess)
```

which returns

```
ans =
        x: [1x9 double]
        y: [2x9 double]
       yp: [2x9 double]
   solver: 'bvp4c'
```

The result is a structure array, where **x** is the value of the independent variable (denoted as t in this problem) and where an array of **y** values corresponds to the solutions to the system of ode's. In this case, y and dy/dt.

To access the array of **x** values simply use the structure syntax, `ans.x`. If we had chosen to assign a name such as `solution` to our result instead of defaulting to `ans`, the structure would be called `solution`, and the **x** values would be stored in `solution.x`. The values of most interest are the **y** values, which can also be accessed using structure syntax, such as `solution.y`. To plot the results in a manner similar to that displayed by the odesolvers use the code

```
plot(ans.x,ans.y, '-o')
```

or, if the results were named `solution`

```
plot(solution.x, solution.y, '-o'),
```

which gives the results shown in Figure 13.38. The annotations (titles, legends, etc.) were added in the usual way.

13.6.5 Partial Differential Equations

MATLAB® also includes a limited partial differential equation solver, `pdepe`. For more information, consult the MATLAB® help function.

Figure 13.38
A boundary value problem
solved using `bvp4c`.

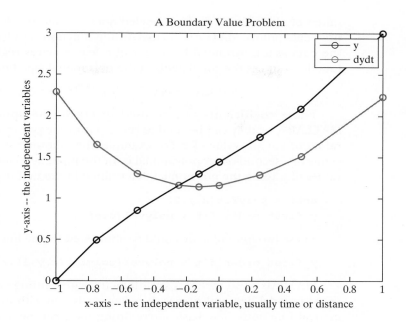

SUMMARY

Tables of data are useful for summarizing technical information. However, if you need a value that is not included in the table, you must approximate that value by using some sort of interpolation technique. MATLAB® includes such a technique, called `interp1`. This function requires three inputs: a set of x-values, a corresponding set of y-values, and a set of x-values for which you would like to *estimate* y-values. The function defaults to a linear interpolation technique, which assumes that you can approximate these intermediate y-values as a linear function of x that is,

$$y = f(x) = ax + b$$

A different linear function is found for each set of two data points, ensuring that the line approximating the data always passes through the tabulated points.

The `interp1` function can also model the data by using higher-order approximations, the most common of which is the cubic spline. The approximation technique is specified as a character string in a fourth optional field of the `interp1` function. If it's not specified, the function defaults to linear interpolation. An example of the syntax is

```
new_y = interp1(tabulated_x, tabulated_y, new_x, 'spline')
```

In addition to the `interp1` function, MATLAB® includes a two-dimensional interpolation function called `interp2`, a three-dimensional interpolation function called `interp3`, and a multidimensional interpolation function called `interpn`.

Curve-fitting routines are similar to interpolation techniques. However, instead of connecting data points, they look for an equation that models the data as accurately as possible. Once you have an equation, you can calculate the corresponding values of y. The curve that is modeled does not necessarily pass through the measured data points. MATLAB®'s curve-fitting function is called `polyfit` and models the data as a polynomial by means of a least-squares regression technique. The function returns the coefficients of the polynomial equation of the form

$$y = a_0 x^n + a_1 x^{n-1} + a_2 x^{n-2} + \cdots + a_{n-1} x + a_n$$

These coefficients can be used to create the appropriate expression in MATLAB®, or they can be used as the input to the `polyval` function to calculate values of y at any value of x. For example, the following statements find the coefficients of a second-order polynomial to fit the input x–y data and then calculate new values of y, using the polynomial determined in the first statement:

```
coef = polyfit(x,y,2)
y_first_order_fit = polyval(coef,x)
```

These two lines of code could be shortened to one line by nesting functions:

```
y_first_order_fit = polyval(polyfit(x,y,1),x)
```

MATLAB® also includes an interactive curve-fitting capability that allows the user to model data not only with polynomials, but with more complicated mathematical functions. The basic curve-fitting tools can be accessed from the `Tools` menu in the figure window. More extensive tools are available in the curve-fitting toolbox, which is accessed by typing

```
cftool
```

in the command window.

Numerical techniques are used widely in engineering to approximate both derivatives and integrals. Derivatives and integrals can also be found with the symbolic toolbox.

The MATLAB® `diff` function finds the difference between values in adjacent elements of a vector. By using the `diff` function with vectors of x- and y-values, we can approximate the derivative with the command

```
slope = diff(y)./diff(x)
```

The more closely spaced the x and y data are, the closer will be the approximation of the derivative.

The `gradient` function uses a forward difference approach to approximate the derivative at the first point in an array. It uses a backward difference approach for the final value in the array, and a central difference approach for the remainder of the points. In general, the central difference approach gives a more accurate approximation of the derivative than either of the other two techniques.

Integration of ordered pairs of data is accomplished using the trapezoidal rule, with the `trapz` function. This approach can also be used with functions, by creating a set of ordered pairs based on a set of x values and the corresponding y values.

Integration of functions is accomplished more directly with one of two quadrature functions: `quad` or `quadl`. These functions require the user to input both a function and its limits of integration. The function can be represented as a character string, such as

```
'x.^2-1'
```

as an anonymous function, such as

```
my_function = @(x) x.^2-1
```

or as an M-file function, such as

```
function output = my_m_file(x)
output = x.^2-1;
```

Any of the three techniques for defining the function can be used as input, along with the integration limits—for example,

```
quad('x.^2-1',1,2)
```

Both `quad` and `quadl` attempt to return an answer accurate to within 1×10^{-6}. The `quad` and `quadl` functions differ only in the technique they use to estimate the integral. The `quad` function uses an adaptive Simpson quadrature technique, and the `quadl` function uses an adaptive Lobatto quadrature technique.

MATLAB® includes a series of solver functions for first-order ordinary differential equations and systems of equations. All of the solver functions use the common format

```
[t,y] = odesolver(function_handle,[initial_time, final_time],
        [initial_cond_array])
```

A good first try is usually the `ode45` solver function, which uses a Runge–Kutta technique. Other solver functions have been formulated for stiff differential equations and implicit formulations.

The ode solver functions require that the user know the initial conditions for the problem. If, instead, boundary conditions are known at other than the starting conditions, the `bvp4` function should be used.

MATLAB® SUMMARY

The following MATLAB® summary lists and briefly describes all the commands and functions that were defined in this chapter:

Commands and Functions	
bvp4c	boundary value problem solver for ordinary differential equations
cftool	opens the curve-fitting graphical user interface
census	a built-in data set
diff	computes the differences between adjacent values in an array if the input is an array; finds the symbolic derivative if the input is a symbolic expression
gradient	finds the derivative numerically using a combination of forward, backward and central difference techniques
int	finds the symbolic integral
interp1	approximates intermediate data, using either the default linear interpolation technique or a specified higher-order approach
interp2	two-dimensional interpolation function
interp3	three-dimensional interpolation function
interpn	multidimensional interpolation function
ode45	ordinary differential equation solver
ode23	ordinary differential equation solver
ode113	ordinary differential equation solver
ode15s	ordinary differential equation solver
ode23s	ordinary differential equation solver

(continued)

Commands and Functions	
ode23t	ordinary differential equation solver
ode23tb	ordinary differential equation solver
ode15i	ordinary differential equation solver
polyfit	computes the coefficients of a least-squares polynomial
polyval	evaluates a polynomial at a specified value of x
quad	computes the integral under a curve (Simpson)
quad1	computes the integral under a curve (Lobatto)
trapz	approximates the integral based on ordered pairs of data

KEY TERMS

approximation	extrapolation	linear regression
backward difference	forward difference	Lobatto quadrature
boundary value problem	graphical user interface	quadratic equation
central difference	(GUI)	quadrature
cubic equation	interpolation	Simpson quadrature
cubic spline	initial value problems	trapezoidal rule
derivative	least squares	
differentiation	linear interpolation	

PROBLEMS

Interpolation

13.1 Consider a gas in a piston–cylinder device in which the temperature is held constant. As the volume of the device was changed, the pressure was measured. The volume and pressure values are reported in the following table:

Volume, m^3	Pressure, kPa, when I = 300 K
1	2494
2	1247
3	831
4	623
5	499
6	416

(a) Use linear interpolation to estimate the pressure when the volume is 3.8 m^3.

(b) Use cubic spline interpolation to estimate the pressure when the volume is 3.8 m^3.

(c) Use linear interpolation to estimate the volume if the pressure is measured to be 1000 kPa.

(d) Use cubic spline interpolation to estimate the volume if the pressure is measured to be 1000 kPa.

13.2 Using the data from Problem 13.1 and linear interpolation to create an expanded volume–pressure table with volume measurements every 0.2 m^3. Plot the calculated values on the same graph with the measured data. Show the measured data with circles and no line and the calculated values with a solid line.

13.3 Repeat Problem 13.2, using cubic spline interpolation.

13.4 The experiment described in Problem 13.1 was repeated at a higher temperature and the data recorded in the following table:

Volume, m^3	Pressure, kPa, at 300 K	Pressure, kPa, at 500 K
1	2494	4157
2	1247	2078
3	831	1386
4	623	1039
5	499	831
6	416	693

Use these data to answer the following questions:

(a) Approximate the pressure when the volume is 5.2 m^3 for both temperatures (300 K and 500 K). (*Hint*: Make a pressure array that contains both sets of data; your volume array will need to be 6×1, and your pressure array will need to be 6×2.) Use linear interpolation for your calculations.

(b) Repeat your calculations, using cubic spline interpolation.

13.5 Use the data in Problem 13.4 to solve the following problems:

(a) Create a new column of pressure values at $T = 400$ K, using linear interpolation.

(b) Create an expanded volume–pressure table with volume measurements every 0.2 m^3, with columns corresponding to $T = 300$ K, $T = 400$ K, and $T = 500$ K.

13.6 Use the `interp2` function and the data from Problem 13.4 to approximate a pressure value when the volume is 5.2 m^3 and the temperature is 425 K.

Curve Fitting

13.7 Fit the data from Problem 13.1 with first-, second-, third-, and fourth-order polynomials, using the `polyfit` function:

- Plot your results on the same graph.
- Plot the actual data as a circle with no line.
- Calculate the values to plot from your polynomial regression results at intervals of 0.2 m^3.
- Do not show the calculated values on the plot, but do connect the points with solid lines.
- Which model seems to do the best job?

13.8 The relationship between pressure and volume is not usually modeled by a polynomial. Rather, they are inversely related to each other by the ideal gas law,

$$P = \frac{nRT}{V}$$

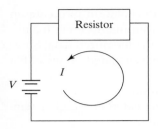

Figure P13.9
An electrical circuit.

We can plot this relationship as a straight line if we plot P on the y-axis and $1/V$ on the x-axis. The slope then becomes the value of nRT. We can use the `polyfit` function to find this slope if we input P and $1/V$ to the function:

polyfit(1./V, P,1)

(a) Assuming that the value of n is 1 mol and the value of R is 8.314 kPa/kmol K, show that the temperature used in the experiment is indeed 300 K.

(b) Create a plot with $1/V$ on the x-axis and P on the y-axis.

13.9 Resistance and current are inversely proportional to each other in electrical circuits:

$$I = \frac{V}{R}$$

Consider the following data collected from an electrical circuit to which an unknown constant voltage has been applied (Figure P13.9):

Resistance, ohms	Measured Current, amps
10	11.11
15	8.04
25	6.03
40	2.77
65	1.97
100	1.51

(a) Plot resistance (R) on the x-axis and measured current (I) on the y-axis.

(b) Create another plot with $1/R$ on the x-axis and I on the y-axis.

(c) Use `polyfit` to calculate the coefficients of the straight line shown in your plot in part (b). The slope of your line corresponds to the applied voltage.

(d) Use `polyval` to find calculated values of current (I) based on the resistors used. Plot your results in a new figure, along with the measured data.

13.10 Many physical processes can be modeled by an exponential equation. For example, chemical reaction rates depend on a reaction-rate constant that is a function of temperature and activation energy:

$$k = k_0 e^{-Q/RT}$$

In this equation,

$R =$ universal gas constant, 8.314 kJ/kmol K,

$Q =$ activation energy, in kJ/kmol,

$T =$ temperature, in K, and

$k_0 =$ constant whose units depend on characteristics of the reaction. One possibility is s^{-1}.

One approach to finding the values of k_0 and Q from experimental data is to plot the natural logarithm of k on the y-axis and $1/T$ on the x-axis. This should result in a straight line with slope $-Q/R$ and intercept $\ln(k_0)$—that is,

$$\ln(k) = \ln(k_0) - \frac{Q}{R}\left(\frac{1}{T}\right)$$

since the equation now has the form

$$y = ax + b$$

with $y = \ln(k)$, $x = 1/T$, $a = -Q/R$ and $b = \ln(k)$.
Now consider the following data:

T, K	k, s^{-1}
200	1.46×10^{-7}
400	0.0012
600	0.0244
800	0.1099
1000	0.2710

(a) Plot the data with $1/T$ on the x-axis and $\ln(k)$ on the y-axis.
(b) Use the `polyfit` function to find the slope of your graph, $-Q/R$, and the intercept, $\ln(k_0)$.
(c) Calculate the value of Q.
(d) Calculate the value of k_0.

13.11 Electrical power is often modeled as

$$P = I^2 R$$

where

$P =$ power, in watts,
$I =$ current, in amperes, and
$R =$ resistance, in ohms.

(a) Consider the following data and find the value of the resistor in the circuit by modeling the data as a second-order polynomial with the `polyfit` function:

Power, W	Current, A
50,000	100
200,000	200
450,000	300
800,000	400
1,250,000	500

(b) Plot the data and use the curve-fitting tools found in the figure window to determine the value of R by modeling the data as a second-order polynomial.

13.12 Using a polynomial to model a function can be very useful, but it is always dangerous to extrapolate beyond your data. We can demonstrate this pitfall by modeling a sine wave as a third-order polynomial.

(a) Define `x = -1:0.1:1`
(b) Calculate `y = sin(x)`

(c) Use the `polyfit` function to determine the coefficients of a third-order polynomial to model these data.

(d) Use the `polyval` function to calculate new values of **y** (`modeled_y`) based on your polynomial, for your **x** vector from −1 to 1.

(e) Plot both sets of values on the same graph. How good is the fit?

(f) Create a new **x** vector, `new_x = -4:0.1:4`.

(g) Calculate `new_y` values by finding `sin(new_x)`.

(h) Extrapolate `new_modeled_y` values by using `polyfit`, the coefficient vector you found in part (c) to model **x** and **y** between −1 and 1, and the `new_y` values.

(i) Plot both new sets of values on the same graph. How good is the fit outside of the region from −1 to 1?

Approximating Derivatives

13.13 Consider the following equation:

$$y = 12x^3 - 5x^2 + 3$$

(a) Define an **x** vector from −5 to +5, and use it together with the `diff` function to approximate the derivative of y with respect to x.

(b) Found analytically, the derivative of y with respect to x is

$$\frac{dy}{dx} = y' = 36x^2 - 10x$$

Evaluate this function, using your previously defined **x** vector. How do your results differ?

13.14 One very common use of derivatives is to determine velocities. Consider the following data, taken during a car trip from Salt Lake City to Denver:

Time, hours	Distance, miles
0	0
1	60
2	110
3	170
4	220
5	270
6	330
7	390
8	460

(a) Find the average velocity in mph during each hour of the trip.

(b) Plot these velocities on a bar graph. Edit the graph so that each bar covers 100% of the distance between entries.

13.15 Consider the following data, taken during a car trip from Salt Lake City to Los Angeles:

Time, hours	Distance, miles
0	0
1.0	75
2.2	145
2.9	225
4.0	300
5.2	380
6.0	430
6.9	510
8.0	580
8.7	635
9.7	700
10	720

(a) Find the average velocity in mph during each segment of the trip.
(b) Plot these velocities against the start time for each segment.
(c) Use the `find` command to determine whether any of the average velocities exceeded the speed limit of 75 mph.
(d) Is the overall average above the speed limit?

13.16 Consider the following data from a three-stage model rocket launch:

Time, seconds	Altitude, meters
0	0
1.00	107.37
2.00	210.00
3.00	307.63
4.00	400.00
5.00	484.60
6.00	550.00
7.00	583.97
8.00	580.00
9.00	549.53
10.00	570.00
11.00	699.18
12.00	850.00
13.00	927.51
14.00	950.00
15.00	954.51
16.00	940.00

(*continued*)

Time, seconds	Altitude, meters
17.00	910.68
18.00	930.00
19.00	1041.52
20.00	1150.00
21.00	1158.24
22.00	1100.00
23.00	1041.76
24.00	1050.00

(a) Create a plot with time on the x-axis and altitude on the y-axis.

(b) Use the `diff` function to determine the velocity during each time interval, and plot the velocity against the starting time for each interval.

(c) Use the `diff` function again to determine the acceleration for each time interval, and plot the acceleration against the starting time for each interval.

(d) Estimate the staging times (the time when a burnt-out stage is discarded and the next stage ignites) by examining the plots you've created.

Numerical Integration

13.17 Consider the following equation:

$$y = 5x^3 - 2x^2 + 3$$

Use the `quad` and `quadl` functions to find the integral with respect to x, evaluated from -1 to 1. Compare your results with the values found with the use of the symbolic toolbox function, `int`, and the following analytical solution (remember that the `quad` and `quadl` functions take input expressed with array operators such as `.*` or `.^`, but the `int` function takes a symbolic representation that does not use these operators):

$$\int_a^b (5x^3 - 2x^2 + 3)\ dx =$$

$$\left(\frac{5x^4}{4} - \frac{2x^3}{3} + 3x \right) \Bigg|_a^b =$$

$$\frac{5}{4}(b^4 - a^4) - \frac{2}{3}(b^3 - a^3) + 3(b - a)$$

13.18 The equation

$$C_P = a + bT + cT^2 + dT^3$$

is an empirical polynomial that describes the behavior of the heat capacity C_P as a function of temperature in kelvins. The change in enthalpy (a measure of energy) as a gas is heated from T_1 to T_2 is the integral of this equation with respect to T:

$$\Delta h = \int_{T_1}^{T_2} C_P\, dT$$

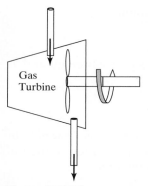

Figure P13.19
A gas turbine used to produce power.

Find the change in enthalpy of oxygen gas as it is heated from 300 to 1000 K, using the MATLAB® quadrature functions. The values of a, b, c, and d for oxygen are as follows:

$$a = 25.48$$
$$b = 1.520 \times 10^{-2}$$
$$c = -0.7155 \times 10^{-5}$$
$$d = 1.312 \times 10^{-9}$$

13.19 In some sample problems in this chapter, we explored the equations that describe moving boundary work produced by a piston–cylinder device. A similar equation describes the work produced as a gas or a liquid flows through a pump, turbine, or compressor (Figure P13.19).

In this case, there is no moving boundary, but there is shaft work, given by

$$\dot{W}_{\text{produced}} = -\int_{\text{inlet}}^{\text{outlet}} \dot{V}\,dP$$

This equation can be integrated if we can find a relationship between $\dot{V}$ and P. For ideal gases, that relationship is

$$\dot{V} = \frac{\dot{n}RT}{P}$$

If the process is isothermal, the equation for work becomes

$$\dot{W} = -\dot{n}RT \int_{\text{inlet}}^{\text{outlet}} \frac{dP}{P}$$

where

$\dot{n}$ = molar flow rate, in kmol/s
R = universal gas constant, 8.314 kJ/kmol K
T = temperature, in K
P = pressure, in kPa
$\dot{W}$ = power, in kW.

Find the power produced in an isothermal gas turbine if

$$\dot{n} = 0.1 \text{ kmol}/s$$
$$R = \text{universal gas constant, 8.314 kJ/kmol K}$$
$$T = 400 \text{ K}$$
$$P_{\text{inlet}} = 500 \text{ kPa}$$
$$P_{\text{outlet}} = 100 \text{ kPa.}$$

Differential Equations

13.20 Solve the following differential equation for values of t between 0 and 4, with the initial condition of $y = 1$ when $t = 0$,

$$\frac{dy}{dt} + \sin(t) = 1$$

(a) Analytically or using MATLAB®'s symbolic capabilities.

(b) Using the `ode45` function.

(c) Plot your results for both approaches.

13.21   Solve the following differential equation for values of t between 0 and 1, with the initial condition of $y = 0$ when $t = 0$.

$$\frac{dy}{dt} = t^2 + y$$

13.22   Blasius showed in 1908 that the solution to the incompressible flow field in a laminar boundary layer on a flat plate is given by the solution of the following third-order ordinary nonlinear differential equation

$$2\frac{d^3f}{d\eta^3} + f\frac{d^2f}{d\eta^2} = 0$$

Rewrite this equation into a system of three first-order equations, using the following substitutions:

$$h_1(\eta) = f$$

$$h_2(\eta) = \frac{df}{d\eta}$$

$$h_3(\eta) = \frac{d^2f}{d\eta^2}$$

Solve using the `ode45` function with the following initial conditions:

$$h_1(0) = 0$$

$$h_2(0) = 0$$

$$h_3(0) = 0.332$$

for $\eta = 0$ to 1

14

Advanced Graphics

Objectives

After reading this chapter, you should be able to:

- Understand how MATLAB® handles the three different types of image files
- Assign a handle to plots and adjust properties, using handle graphics

- Create an animation by either of the two MATLAB® techniques
- Adjust lighting parameters, camera locations, and transparency values
- Use visualization techniques for both scalar and vector information in three dimensions.

INTRODUCTION

Some of the basic graphs commonly used in engineering are the workhorse *x–y* plot, polar plots, and surface plots, as well as some graphing techniques more commonly used in business applications, such as pie charts, bar graphs, and histograms. MATLAB® gives us significant control over the appearance of these plots and lets us manipulate images (such as digital photographs) and create three-dimensional representations (besides surface plots) of both data and models of physical processes.

14.1 IMAGES

Let us start our exploration of some of MATLAB®'s more advanced graphics capabilities by examining how images are handled with the image and imagesc functions. Because MATLAB® is already a matrix-manipulation program, it makes sense that images are stored as matrices.

We can create a three-dimensional surface plot of the peaks function by typing

`surf(peaks)`

We can manipulate the figure we have created (Figure 14.1) by using the interactive figure-manipulation tools, so that we are looking down from the top (Figure 14.2). An easier way to accomplish the same thing is to use the pseudo color plot:

`pcolor(peaks)`

Figure 14.1
The peaks function is built into MATLAB® for use in demonstrating graphics capabilities. The title and axis labels were added in the usual way.

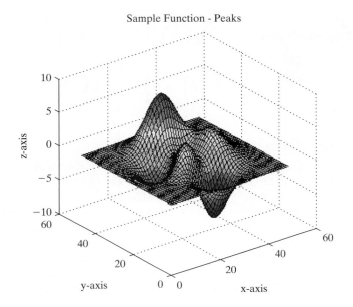

Figure 14.2
A view of the surface plot of the peaks function looking down the z-axis.

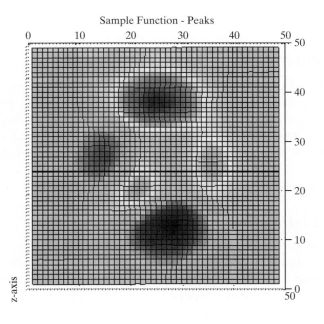

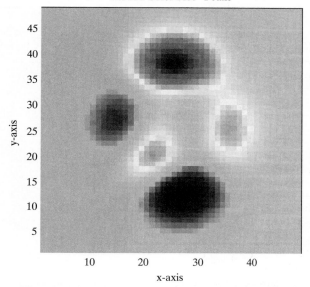

Pseudo Color Plot - Peaks

z(m,1)	z(m,2)	z(m,3)	...	z(m,n)
...	...	...	...	...
..	...	...	...	...
..	...	...	...	...
z(3,1)	z(3,2)	z(3,3)	...	z(3,n)
z(2,1)	z(2,2)	z(2,3)	...	z(2,n)
z(1,1)	z(1,2)	z(1,3)	...	z(1,n)

x-axis

Figure 14.3
A pseudo color plot (left) is the same thing as the view looking straight down at a surface plot. Pseudo color plots organize the data on the basis of the right-hand rule, starting at the (0, 0) position on the graph (right).

We can also remove the grid lines, which are plotted automatically, by specifying the shading option:

```
shading flat
```

The colors in Figures 14.1 to 14.3 correspond to the values of z. The large positive values of z are red (if you are looking at the results on the screen and not in this book, which, of course, is black and white), and the large negative values are blue. The value of z found in the first z matrix element, $z(1, 1)$, is represented in the lower left-hand corner of the graph (see Figure 14.3, right).

Although this strategy for representing data makes sense because of the coordinate system we typically use in graphing, it does not make sense for representing images such as photographs. When images are stored in matrices, we usually represent the data starting in the upper left-hand corner of the image and working across and down (Figure 14.4, left). In MATLAB®, two functions used to display images—image and imagesc—use this format. The scaled image function (imagesc) uses the entire colormap to represent the data, just like the pseudo color plot function (pcolor). The results, obtained with

```
imagesc(peaks)
```

are shown at the right in Figure 14.4.

Notice that the image is flipped in comparison to the pseudo color plot. Of course, in many graphics applications, it doesn't matter how the data are represented, as long as we understand the convention used. However, a photograph would be upside down in a vertical mirror image—clearly not an acceptable representation.

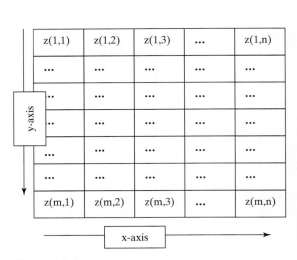

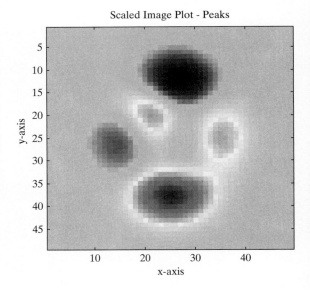

Figure 14.4
The peaks function rendered with the imagesc function. Left: images are usually represented starting in the upper left-hand corner and working across and down, the way we read a book. Right: the pcolor plot and the imagesc plot are vertical mirror images of each other.

14.1.1 Image Types

MATLAB® recognizes three different techniques for storing and representing images:

Intensity (or gray scale) images

Indexed images

RGB (or true color) images

Intensity Images

We used an intensity image to create the representation of the peaks function (Figure 14.4) with the scaled image function (imagesc). In this approach, the colors in the image are determined by a colormap. The values stored in the image matrix are scaled, and the values are correlated with a known map. (The jet colormap is the default.) This approach works well when the parameter being displayed does not correlate with an actual color. For example, the peaks function is often compared to a mountain and valley range—but what elevation is the color red? It's an arbitrary choice based partially on aesthetics, but colormaps can also be used to enhance features of interest in the image.

Consider this example: X-ray images traditionally were produced by exposing photographic film to X-ray radiation. Today many X-rays are processed as digital images and stored in a data file—no film is involved. We can manipulate that file however we want, because the intensity of X-ray radiation does not correspond to a particular color.

MATLAB® includes a sample file that is a digital X-ray photograph of a spine, suitable for display with the use of the scaled image function. First you'll need to load the file:

```
load spine
```

The loaded file includes a number of matrices (see the workspace window); the intensity matrix is named X. Thus,

```
imagesc(X)
```

produces an image whose colors are determined by the current `colormap`, which defaults to `jet`. A representation that looks more like a traditional X-ray is returned if we use the `bone` colormap:

```
colormap(bone)
```

This image is shown in Figure 14.5.

The spine file also includes a custom colormap, which happens to correspond to the `bone` colormap. This array is called `map`. Custom colormaps are not necessary to display intensity images, and

```
colormap(map)
```

results in the same image we created earlier.

Although it is convenient to think of image data as a matrix, such data are not necessarily stored that way in the standard graphics formats. MATLAB® includes a function, `imfinfo`, that will read standard graphics files and determine what type of data are contained in the file. Consider the file mimas.jpg, which was downloaded off the Internet from a NASA website (http://saturn.jpl.nasa.gov). The command

```
imfinfo('mimas.jpg')
```

Figure 14.5
Digital X-ray displayed with the use of the `imagesc` function and the `bone` colormap.

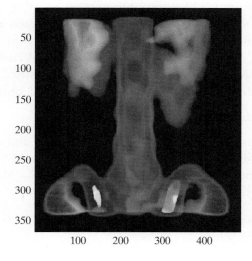

returns the following information (be sure to list the file name in single quotes—that is, as a string; also, notice that the image is `'gray scale'` —another term for an intensity image):

```
ans =
        Filename: 'mimas.jpg'
        FileModDate: '06-Aug-2005 08:52:18'
        FileSize: 23459
        Format: 'jpg'
        FormatVersion: "
        Width: 500
        Height: 525
        BitDepth: 8
        ColorType: 'gray scale'
        FormatSignature: "
        NumberOfSamples: 1
        CodingMethod: 'Huffman'
        CodingProcess: 'Sequential'
        Comment: {'Created with The GIMP'}
```

In order to create a MATLAB® matrix from this file, we use the image read function imread and assign the results to a variable name, such as X:

```
X = imread('mimas.jpg');
```

We can then plot the image with the imagesc function and gray colormap:

```
imagesc(X)
colormap(gray)
```

The results are shown in Figure 14.6a.

Indexed Image Function

When color is important, one technique for creating an image is called an *indexed image*. Instead of being a list of intensity values, the matrix is a list of colors. The image is created much like a paint-by-number painting. Each element contains a number that corresponds to a color. The colors are listed in a separate matrix called a colormap, which is an $n \times 3$ matrix that defines n different colors by identifying the red, green, and blue components of each color. A custom colormap can be created for each image, or a built-in colormap could be used.

Consider the built-in sample image of a mandrill, obtained with

```
load mandrill
```

The file includes an indexed matrix named X and a colormap named map. (Check the workspace window to confirm that these files have been loaded; the names are commonly used for images saved from a MATLAB® program.) The image function is used to display indexed images:

```
image(X)
colormap(map)
```

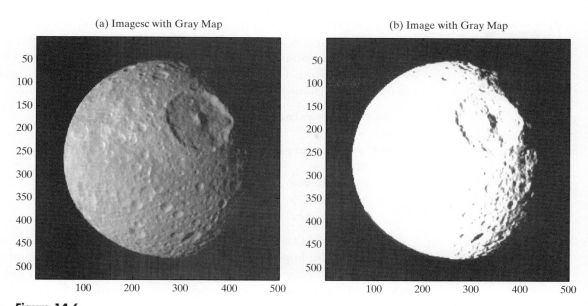

(a) Imagesc with Gray Map   (b) Image with Gray Map

Figure 14.6

(a) Image of mimas, a moon of saturn, displayed by means of the scaled image function, `imagesc`, and a `gray` colormap. (b) Image displayed with the indexed image function, `image`, and a `gray` colormap.

MATLAB® images adjust to fill the figure window, so the image may appear warped. We can force the correct aspect to be displayed by using the `axis` command:

```
axis image
```

The results are shown in Figure 14.7.

The `image` and `imagesc` functions are similar, yet they can give very different results. The image of Mimas in Figure 14.6b was produced by the `image` function instead of the more appropriate `imagesc` function. The `gray` colormap does not correspond to the colors stored in the intensity image; the result is the washed-out image and lack of contrast. It is important to recognize what kind of file you are displaying, so that you can make the optimum choice of how to represent the image.

Figure 14.7
Left: Mandrill image before the custom colormap is applied. Right: mandrill image with the custom colormap.

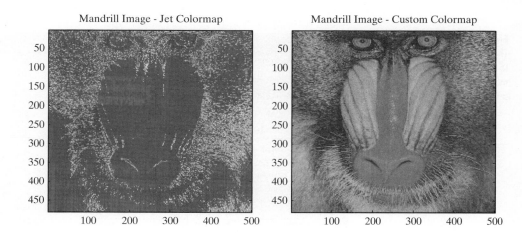

Mandrill Image - Jet Colormap   Mandrill Image - Custom Colormap

Figure 14.8
Clip art stored in the GIF file format.

Files stored in the GIF graphics format are often stored as indexed images. This may not be apparent when you use the imfinfo function to determine the file parameters. For example, the image in Figure 14.8 is part of the clip art included with Microsoft Word. The image was copied into the current folder, and imfinfo was used to determine the file type:

```
imfinfo('drawing.gif')
ans =
1x4 struct array with fields:
    Filename
    FileModDate
    FileSize
    Format etc.
```

The results don't tell us much, but if you double-click on the file name in the current folder, the Import Wizard (Figure 14.9) launches and suggests that we create two matrices: cdata and colormap. The cdata matrix is an indexed image matrix, and colormap is the corresponding colormap. Actually, the suggested name colormap is rather strange, because if we use it, it will supersede the colormap function. You'll need to rename this matrix to something different, such as map, by clicking on the variable name in the import wizard before you actually complete the import process. After importing, you can view the image with the following commands.

```
image(cdata)
colormap(map)
axis image
axis off
```

HINT

A number of sample images are built into MATLAB® and stored as indexed images. You can access these files by typing

```
load <imagename>
```

Figure 14.9
The import wizard is used to create an indexed image matrix and colormap from a GIF file.

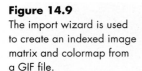

Some of the available images are

```
flujet
durer
detail
mandrill
clown
spine
cape
earth
gatlin
```

Each of these image files creates a matrix of index values called X and a color-map called map. For example, to see the image of the earth, type

```
load earth
image(X)
colormap(map)
```

You'll also need to adjust the aspect ratio of the display and remove the axis with the commands

```
axis image
axis off
```

True Color (RGB) Images

The third technique for storing image data is in a three-dimensional matrix, $m \times n \times 3$. Recall that a three-dimensional matrix consists of rows, columns, and pages. True color image files consist of three pages, one for each color intensity, red, green, or blue, as shown in Figure 14.10.

RGB
The primary colors of light are red, green, and blue

Consider a file called airplanes.jpg. You can copy this or a similar file (a colored .jpg image) into your current folder to experiment with true color images. We can use the imfinfo function to determine how the airplanes file stores the image:

```
imfinfo('airplanes.jpg')
ans =
Filename: 'airplanes.jpg'
FileModDate: '12-Sep-2005 17:51:48'
FileSize: 206397
Format: 'jpg'
```

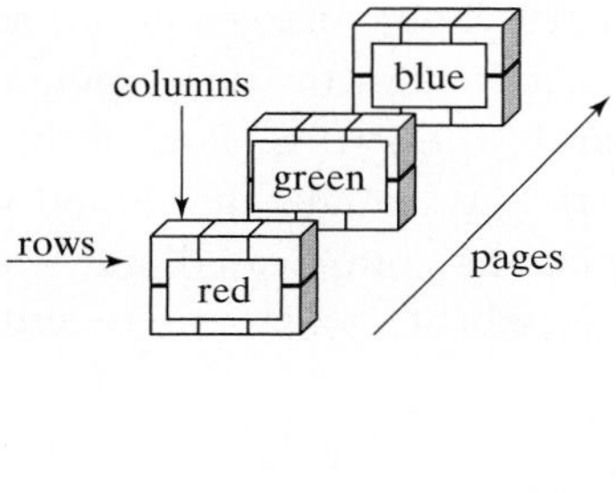

Figure 14.10
True color images use a multidimensional array to represent the color of each element.

Figure 14.11
True color image of airplanes. All of the color information is stored in a three-dimensional matrix. (Picture used with permission of Dr. G. Jimmy Chen, Salt Lake Community College, Department of Computer Science.)

```
FormatVersion: "
Width: 1800
Height: 1200
BitDepth: 24
ColorType: 'truecolor'
FormatSignature: "
NumberOfSamples: 3
CodingMethod: 'Huffman'
CodingProcess: 'Sequential'
Comment: {}
```

Notice that the color type is `truecolor` and that the number of samples is 3, indicating a page for each color intensity.

We can load the image with the `imread` function and display it with the `image` function:

```
X = imread('airplanes.jpg');
image(X)
axis image
axis off
```

Notice in the workspace window that X is a $1200 \times 1800 \times 3$ matrix—one page for each color. We don't need to load a colormap, because the color-intensity information is included in the matrix (Figure 14.11).

EXAMPLE 14.1

MANDELBROT AND JULIA SETS

Benoit Mandelbrot (Figure 14.12) is largely responsible for the current interest in fractal geometry. His work built upon concepts developed by the French mathematician Gaston Julia in his 1919 paper *Mémoire sur l'iteration des fonctions rationelles*. Advances in Julia's work had to wait for the development of the computer and computer graphics in particular. In the 1970s, Mandelbrot, then at IBM, revisited and expanded upon Julia's work and actually developed some of the first computer graphics programs to display the complicated and beautiful fractal patterns that today bear his name. Mandelbrot's work was recently described in a song by

Figure 14.12
Benoit Mandelbrot.

Jonathan Coulton. You can listen to it at http://www.jonathancoulton.com/song-details/Mandelbrot%20Set.

The Mandelbrot image is created by considering each point in the complex plane, $x + yi$. We set $z(0) = x + yi$ and then iterate according to the following strategy:

$$z(0) = x + yi$$
$$z(1) = z(0)^2 + z(0)$$
$$z(2) = z(1)^2 + z(0)$$
$$z(3) = z(2)^2 + z(0)$$
$$z(n) = z(n-1)^2 + z(0)$$

The series seems either to converge or to head off toward infinity. The Mandelbrot set is composed of the points that converge. The beautiful pictures you have probably seen were created by counting how many iterations were necessary for the z-value at a particular point to exceed some threshold value, often the square root of 5. We assume, though we can't prove, that if that threshold is reached, the series will continue to diverge and eventually approach infinity.

1. State the Problem
   Write a MATLAB® program to display the Mandelbrot set.
2. Describe the Input and Output

   Input   We know that the Mandelbrot set lies somewhere in the complex plane and that

$$-1.5 \le x \le 1.0$$
$$-1.5 \le y \le 1.5$$

   We also know that we can describe each point in the complex plane as
$$z = x + yi$$

3. Develop a Hand Example
   Let's work the first few iterations for a point we hope converges, such as $(x = -0.5, y = 0)$:

$$z(0) = -0.5 + 0i$$
$$z(1) = z(0)^2 + z(0) = (-0.5)^2 - 0.5 = 0.25 - 0.5 = -0.25$$
$$z(2) = z(1)^2 + z(0) = (-0.25)^2 - 0.5 = 0.0625 - 0.5 = -0.4375$$
$$z(3) = z(2)^2 + z(0) = (-0.4375)^2 - 0.5 = 0.1914 - 0.5 = -0.3086$$
$$z(4) = z(3)^2 + z(0) = (-0.3086)^2 + 0.5 = 0.0952 - 0.5 = -0.4048$$

   It appears that this sequence is converging to a value around (As an exercise, you could create a MATLAB® program to calculate the first 20 terms of the series and plot them.)
4. Develop a MATLAB® Solution

```
%Example 14.1 Mandelbrot Image
clear, clc
```

(continued)

```
iterations = 80;
grid_size = 500;
[x,y] = meshgrid(linspace(-1.5,1.0,grid_size),linspace
  (-1.5,1.5,grid_size));
c = x+i*y;
z = zeros(size(x));          % set the initial matrix to 0
map = zeros(size(x));        % create a map of all grid
                             % points equal to 0

for k = 1:iterations
z = z.^2 +c;
a = find(abs(z)>sqrt(5));    %Determine which elements have
                             %exceeded sqrt(5)

map(a) = k;
end
figure(1)
image(map)                   %Create an image
colormap(jet)
```

The image produced is shown in Figure 14.13.

5. Test the Solution

We know that all the elements in the solid colored region of the image (dark blue—if you are looking at the image on a computer screen) will be below the square root of 5. An alternative way to examine the results is to create an image based on those values instead of the number of iterations needed to exceed the threshold. We'll need to multiply each value by a common multiple in order to achieve any color variation. (Otherwise the values are too close to each other.) The MATLAB® code is as follows:

```
figure(2)
multiplier = 100;
map = abs(z)*multiplier;
image(map)
```

Figure 14.13
Mandelbrot image. The figure was created by determining how many iterations were required for the calculated element values to exceed the square root of 5.

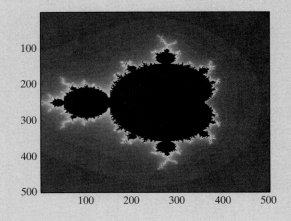

Figure 14.14
An image based on the Mandelbrot set, showing how the members of the set vary. The really interesting structure is at the boundary of the set.

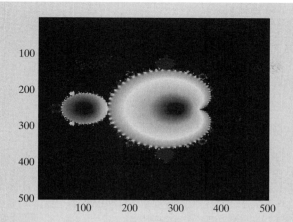

The results are shown in Figure 14.14.

Now that we've created an image of the entire Mandelbrot set, it would be interesting to look more closely at some of the structures at the boundary. By adding the following lines of code to the program, we can repeatedly zoom in on any point in the image:

```
cont = 1;
while(cont==1)
figure(1)
disp('Now let's zoom in')
disp('Move the cursor to the upper left-hand corner of the
   area you want to expand')
[y1,x1] = ginput(1);
disp('Move to the lower right-hand corner of the area you
   want to expand')
[y2,x2] = ginput(1);
xx1 = x(round(x1),round(y1));
yy1 = y(round(x1),round(y1));
xx2 = x(round(x2),round(y2));
yy2 = y(round(x2),round(y2));
%%
[x,y] = meshgrid(linspace(xx1,xx2,grid_size),linspace(yy1,
   yy2,grid_size));
c = x+i*y;
z = zeros(size(x));
map = zeros(size(x));
for k = 1:iterations
  z = z.^2 +c;
  a = find(abs(z)>sqrt(5));
  map(a) = k;
end
image(map)
colormap(jet)
```

(continued)

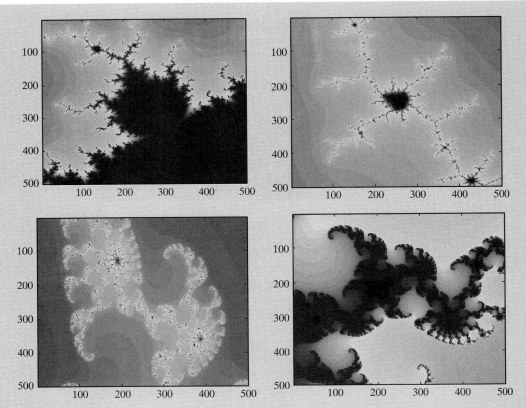

Figure 14.15
Images created by zooming in on the Mandelbrot set from a MATLAB® program.

```
again = menu('Do you want to zoom in again? ','Yes','No');
switch again
    case 1
        cont = 1;
    case 2
        cont = 0;
end
end
```

Figure 14.15 shows some of the images created by recalculating with smaller and smaller areas.

You can experiment with using both the image function and the imagesc function and observe how the pictures differ. Try some different colormaps as well.

14.1.2 Reading and Writing Image Files

We introduced functions for reading image files as we explored the three techniques for storing image information. MATLAB® also includes functions to write user-created images in any of a variety of formats. In this section, we'll explore these reading and writing functions in more detail.

Reading Image Information

Probably the easiest way to read image information into MATLAB® is to take advantage of the interactive Import Wizard. In the current folder window, simply double-click the file name of the image to be imported. MATLAB® will suggest appropriate variable names and will make the matrices available to preview in the edit window (Figure 14.9).

The problem with interactively importing any data is that you can't include the instructions in a MATLAB® program—for that, we need to use one of the import functions. For most of the standard image formats, such as .jpg or .tif, the `imread` function described in the preceding section is the appropriate technique. On the other hand, if the file is a .mat or a .dat file, the easiest way to import the data is to use the `load` function:

```
load <filename>
```

For .mat files, you don't even need to include the .mat extension. However, you will need to include the extension for a .dat file:

```
load <filename.dat>
```

This is the technique we used to load the built-in image files described earlier. For example,

```
load cape
```

imports the image matrix and colormap into the current folder, and the commands

```
image(X)
colormap(map)
axis image
axis off
```

can then be used to create the picture, shown in Figure 14.16.

Storing Image Information

You can save an image you've created in MATLAB® the same way you save any figure. Select

```
File ⟶ Save As . . .
```

Figure 14.16
Image created by loading a built-in file.

and choose the file type and the location where you'd like to save the image. For example, to save the image of the Mandelbrot set created in Example 14.1 and shown in Figure 14.13, you might want to specify an enhanced metafile (.emf), as shown in Figure 14.17.

You could also save the file by using the imwrite function. This function accepts a number of different inputs, depending on the type of data you would like to store.

For example, if you have an intensity array (gray scale) or a true-color array (RGB), the imwrite function expects input of the form

```
imwrite(arrayname,'filename.format')
```

where

    arrayname is the name of the MATLAB® array in which the data are stored,
    filename is the name you want to use to store the data, and
    format is the file extension, such as jpg or tif.

Thus, to store an RGB image in a .jpg file named flowers, the command would be

```
imwrite(X,'flowers.jpg')
```

(Consult the help files for a list of graphics formats supported by MATLAB®.)

If you have an indexed image (an image with a custom colormap), you'll need to store both the array and the colormap:

```
imwrite(arrayname, colormap_name,'filename.format')
```

In the case of the Mandelbrot set, we would need to save the array and the colormap used to select the colors in the image:

```
imwrite(map,jet,'my_mandelbrot.jpg')
```

Figure 14.17
This image of a Mandelbrot set is being saved as an enhanced metafile.

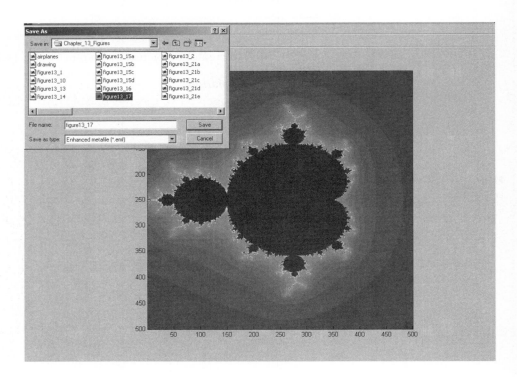

Handle
A nickname

14.2 HANDLE GRAPHICS

A *handle* is a nickname given to an object in MATLAB®. A complete description of the graphics system used in MATLAB® is complicated and beyond the scope of this text. (For more details, refer to the MATLAB® `help` tutorial.) However, we'll give a brief introduction to handle graphics and then illustrate some of its uses.

MATLAB® uses a hierarchical system for creating graphs (Figure 14.18). The basic plotting object is the figure. The figure can contain a number of different objects, including a set of axes. Think of the axes as being layered on top of the figure window. The axes also can contain a number of different objects, including a plot such as the one shown in Figure 14.19. Again, think of the plot being layered on top of the axes.

When you use a `plot` function, either from the command window or from an M-file program, MATLAB® automatically creates a figure and an appropriate axis, and then draws the graph on the axis. MATLAB® uses default values for many of the plotted object's properties. For example, the first line drawn is always blue, unless the user specifically changes it.

Figure 14.18
MATLAB® uses a hierarchical system for organizing plotting information, as shown in this representation from Matlab®'s help menu.

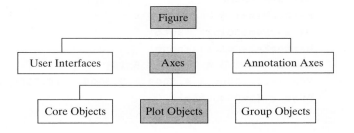

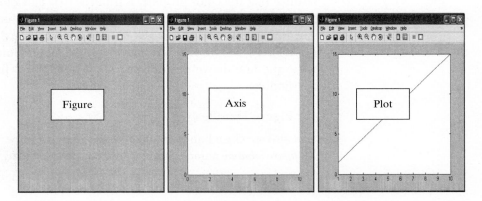

Figure 14.19
Anatomy of a graph. Left: Figure windows are used for lots of things, including graphical user interfaces and plots. In order to create a plot you need a figure window. Center: Before you can draw a graph in this figure window, you'll need a set of axes to draw on. Right: Once you know where the axes are and what the axis properties are (such as the spacing), you can draw the graph.

14.2.1 Plot Handles

Assigning a plot a name (called a handle) allows us to easily ask MATLAB® to list the plotted object's properties. For example, let's create the simple plot shown in Figure 14.19 and assign a handle to it:

```
x = 1:10;
y = x.*1.5;
h = plot(x,y)
```

The variable h is the handle for the plot. (We could have chosen any variable name.) Now we can use the get function to ask MATLAB® for the plot properties:

```
get(h)
```

The function returns a whole list of properties representing the line that was drawn in the axes, which were positioned in the figure window:

```
Color: [0 0 1]
EraseMode: 'normal'
LineStyle: '-'
LineWidth: 0.5000
Marker: 'none'
MarkerSize: 6
MarkerEdgeColor: 'auto'
MarkerFaceColor: 'none'
XData: [1 2 3 4 5 6 7 8 9 10]
YData: [1.5000 3 4.5000 6 7.5000 9 10.5000 12 13.5000 15]
ZData: [1x0 double]
                    .
                    .
                    .
```

Notice that the color property is listed as [0 0 1]. Colors are described as intensities of each of the primary colors of light: red, green, and blue. The array [0 0 1] tells us that there is no red, no green, and 100% blue. If you are looking at this graph in MATLAB®, you should notice that the plotted line is blue. The plot handle refers to the line drawn on the axis, which is different from the axis or from the figure window.

14.2.2 Figure Handles

We can also specify a handle name for the figure window. Since we drew this graph in the *figure window* named figure 1, the command would be

```
f_handle = figure(1)
```

Using the get command returns similar results:

```
get(f_handle)
Alphamap = [ (1 by 64) double array]
BackingStore = on
CloseRequestFcn = closereq
```

```
Color = [0.8 0.8 0.8]
Colormap = [ (64 by 3) double array]
CurrentAxes = [150.026]
CurrentCharacter =
CurrentObject = []
CurrentPoint = [240 245]
DockControls = on
DoubleBuffer = on
FileName = [ (1 by 96) char array]
                              .
                              .
                              .
```

Notice that the properties are different for a figure window compared to the plotted line. For example, notice the color (which is the window background color) is [0.8, 0.8, 0.8], which specifies equal intensities of red, green, and blue—which results in a light gray background. You can change the background color using

```
set(f_handle,'Color',[0.4,0.4,0.4])
```

which results in a darker gray background.

If we haven't specified a handle name, we can ask MATLAB® to determine the current figure with the gcf (get current figure) command,

```
get(gcf)
```

which returns the figure properties. Thus, using gcf and the set command we could have changed the background color with the following command.

```
set(gcf,'Color',[0.4,0.4,0.4])
```

14.2.3 Axis Handles

Just as we can assign a handle to the figure window and the plot itself, we can assign a handle to the axis by means of the gca (get current axis) function:

```
h_axis = gca;
```

Using this handle with the get command allows us to view the axis properties:

```
get(h_axis)
ActivePositionProperty = outerposition
ALim = [0.1 10]
ALimMode = auto
AmbientLightColor = [1 1 1]
Box = off
CameraPosition = [-1625.28 -2179.06 34.641]
CameraPositionMode = auto
CameraTarget = [201 201 0]
                              .
                              .
                              .
```

14.2.4 Annotation Axes

Besides the three components described in earlier sections, another transparent layer is added to the plot. This layer is used to insert annotation objects, such as lines, legends, and text boxes into the figure.

14.2.5 Using Handles to Manipulate Graphics

So what can we do with all this information? We can use the set function to change the object's properties. The set function requires the object handle in the first input field and then alternating strings specifying a property name, followed by a new value. For example,

```
set(h,'color','red')
```

tells MATLAB® to go to the plot we named h (not the figure, but the actual drawing of the line) and change the color to red. If we want to change some of the figure properties, we can do it the same way, using either the figure handle name or the gcf function. For example, to change the name of figure 1, use the command

```
set(f_handle,'name', 'My Graph')
```

or

```
set(gcf,'name', 'My Graph')
```

You can accomplish the same thing interactively by selecting View from the figure menu bar, and choosing the property editor:

```
View → Property Editor
```

You can access all the properties if you choose property inspector from the property editor pop-up window (Figure 14.20). Exploring the property inspector window is a great way to find out which properties are available for each graphics object.

Figure 14.20
Interactive property editing.

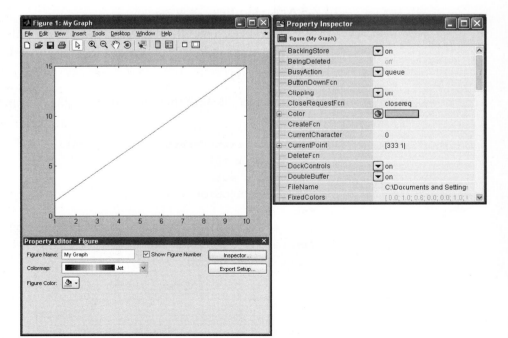

14.3 ANIMATION

There are two techniques for creating an animation in MATLAB®:

- Redrawing and erasing
- Creating a movie

We use handle graphics in each case to create the animation.

14.3.1 Redrawing and Erasing

To create an animation by redrawing and erasing, you should first create a plot and then adjust the properties of the graph each time through a loop. Consider the following example: We can define a set of parabolas with the following equation:

$$y = kx^2 - 2$$

Each value of k defines a different parabola. We could represent the data with a three-dimensional plot; however, another approach would be to create an animation in which we draw a series of graphs, each with a different value of k. The code to create that animation is:

```
clear,clc,clf
x = -10:0.01:10;      % Define the x-values
k = -1;               % Set an initial value of k
y = k*x.^2-2;         % Calculate the first set of y-values
h = plot(x,y);        % Create the figure and assign
                      % a handle to the graph
grid on
%set(h,'EraseMode','xor')  % The animation runs faster if
                           % you activate this line
axis([-10,10,-100,100])    % Specify the axes
while k<1                   % Start a loop
    k = k + 0.01;           % Increment k
    y = k*x.^2-2;           % Recalculate y
    set(h,'XData',x,'YData',y) % Reassign the x and y
                               % values used in the graph
    drawnow                 % Redraw the graph now - don't wait
                            % until the program finishes running
end
```

In this example, we used handle graphics to redraw just the graph each time through the loop, instead of creating a new figure window each time. Also, we used the XData and YData objects from the plot. These objects assign the data points to be plotted. Using the set function allows us to specify new x- and y-values and to create a different graph every time the drawnow function is called. A selection of the frames created by the program and used in the animation is shown in Figure 14.21.

In the program, notice the line

```
%set(h,'EraseMode','xor')
```

Figure 14.21
Animation works by
redrawing the graph
multiple times.

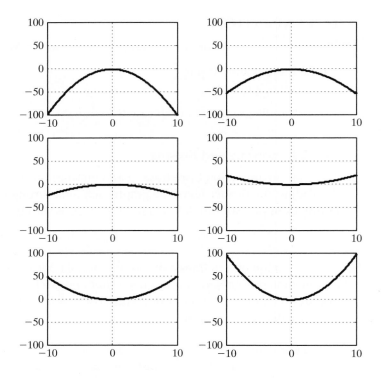

If you activate this line by removing the comment operator (%), the program does
not erase the entire graph each time the graph is redrawn. Only pixels that change
color are changed. This makes the animation run faster—a characteristic that is important when the plot is more complicated than the simple parabola used in this example.

Refer to the help tutorial for a sample animation modeling Brownian motion.

14.3.2 Movies

Animating the motion of a line is not computationally intensive, and it's easy to get
nice, smooth movement. Consider this code that produces a more complicated surface plot animation:

```
clear,clc
x = 0:pi/100:4*pi;
y = x;
[X,Y] = meshgrid(x,y);
z = 3*sin(X)+ cos(Y);
h = surf(z);
axis tight
set(gca,'nextplot','replacechildren');
%Tells the program to replace the surface each time,
%but not the axis
shading interp
colormap(jet)
for k = 0:pi/100:2*pi
    z = (sin(X) + cos(Y)).*sin(k);
```

```
    set(h,'Zdata',z)
    drawnow
end
```

A sample frame from this animation is shown in Figure 14.22.

If you have a fast computer, the animation may still be smooth. However, on a slower computer, you may see jerky motion and pauses while the program creates each new plot. To avoid this problem, you can create a program that captures each "frame" and then, once all the calculations are done, plays the frames as a movie.

```
clear,clc
x = 0:pi/100:4*pi;
y = x;
[X,Y] = meshgrid(x,y);
z = 3*sin(X)+ cos(Y);
h = surf(z);
axis tight
set(gca,'nextplot','replacechildren');
shading interp
colormap(jet)
m = 1;
for k = 0:pi/100:2*pi
    z = (sin(X) + cos(Y)).*sin(k);
    set(h,'Zdata',z)
    M(m) = getframe;    %Creates and saves each frame
                        %of the movie
m = m+1;
end
    movie(M,2)          %Plays the movie twice
```

KEY IDEA

Movies record an animation for later playback

When you run this program, you will actually see the movie three times: once as it is created, and the two times specified in the movie function. (In earlier versions of MATLAB® 7. the movie would have played one additional time as the animation was loaded.) One advantage of this approach is that you can play the movie again without redoing the calculations, since the information is stored (in our example) in the array named M. Notice in the workspace window (Figure 14.23) that M is a moderately large structure array (~90 MB).

Figure 14.22

The animation of this figure moves up and down and looks like waves in a pond.

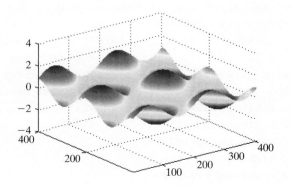

Figure 14.23
Movies are saved in a structure array, such as the M array shown in this figure.

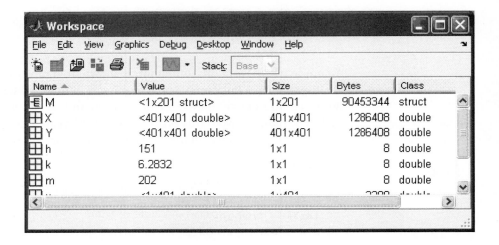

EXAMPLE 14.2

A MANDELBROT MOVIE

The calculations required to create a Mandelbrot image require significant computer resources and can take several minutes. If we want to zoom in on a point in a Mandelbrot image, a logical choice is to do the calculations and create a movie, which we can view later. In this example, we start with the MATLAB® M-file program first described in Example 14.1 and create a 100-frame movie.

1. **State the Problem**
   Create a movie by zooming in on a Mandelbrot set.
2. **Describe the Input and Output**

   Input   The complete Mandelbrot image described in Example 14.1

   Output   A 100-frame movie

3. **Develop a Hand Example**
   A hand example doesn't make sense for this problem, but what we can do is create a program with a small number of iterations and elements to test our solution and then use it to create a more detailed sequence that is more computationally intensive. Here is the first program:

```
%Example 14.2 Mandelbrot Image
%  The first part of this program is the same as Example 14.1
clear, clc
iterations = 20;        % Limit the number of iterations in
                        % this first pass
grid_size = 50;         % Use a small grid to make the
                        % program run faster
X = linspace(-1.5,1.0,grid_size);
Y = linspace(-1.5,1.5,grid_size);
[x,y] = meshgrid(X,Y);
c = x+i*y;
z = zeros(size(x));
map = zeros(size(x));
```

```
for k = 1:iterations
  z = z.^2 +c;
  a = find(abs(z)>sqrt(5));
  map(a) = k;
end
figure(1)
h = imagesc(map)

%% New code section

N(1) = getframe;                %Get the first frame of the movie
disp('Now let's zoom in')
disp('Move the cursor to a point where you"d like to zoom')
[y1,x1] = ginput(1)             %Select the point to zoom in on
xx1 = x(round(x1),round(y1))
yy1 = y(round(x1),round(y1))
%%
for k = 2:100 %Calculate and display the new images
   k            %Send the iteration number to the command window
[x,y] = meshgrid(linspace(xx1-1/1.1^k,xx1+1/1.1^k,grid_size),
   ...linspace(yy1-1/1.1^k,yy1+1/1.1^k,grid_size));
c = x+i*y;
z = zeros(size(x));
map = zeros(size(x));
for j = 1:iterations
  z = z.^2 +c;
  a = find(abs(z)>sqrt(5));
  map(a) = j;
end
set(h,'CData',map)          % Retrieve the image data from the
                            % variable map
colormap(jet)
N(k) = getframe;            % Capture the current frame
end
movie(N,2)                  % Play the movie twice
```

This version of the program runs quickly and returns low-resolution images (Figure 14.24) which demonstrate that the program works.

4. Develop a MATLAB® Solution

The final version of the program is created by changing just two lines of code:

```
iterations = 80;    % Increase the number of iterations
grid_size = 500;    % Use a large grid to see more detail
```

This "full-up" version of the program took approximately 2 minutes to run on a 3.0-GHz AMD dual- core processor with 2.0 GB of RAM. Selected frames are shown in Figure 14.25. Of course, the time it takes on your computer will be more or less, depending on your system resources. One cycle of the movie created by the program plays in about 10 seconds.

(*continued*)

Figure 14.24
Low-resolution
Mandelbrot image used
to test the animation
program.

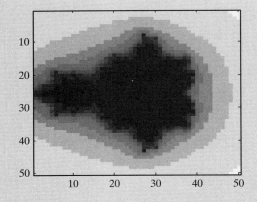

Figure 14.25
This series of
Mandelbrot images is a
selection of the frames
captured to create a
movie with the program
in this example. Each
movie will be different,
since it zooms in on a
different point of the
image.

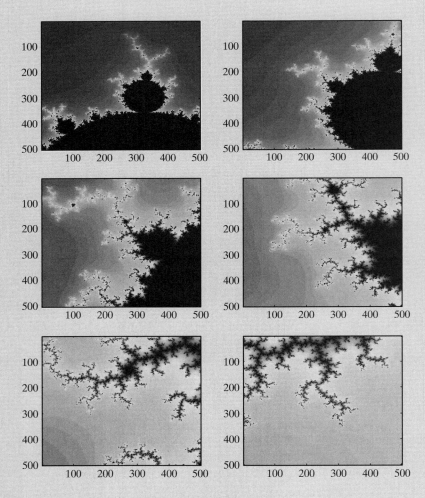

5. Test the Solution

Try the program several times, and observe the images created when you zoom
in to different portions of the Mandelbrot set. You can experiment with increasing the number of iterations used to create the image and with the colormap.

14.4 OTHER VISUALIZATION TECHNIQUES

14.4.1 Transparency

When we render surfaces in MATLAB®, we use an opaque coloring scheme. This approach is great for many surfaces, but can obscure details in others. Take, for example, this series of commands that creates two spheres, one inside the other:

```
clear,clc,clf          % Clear the command window and current
                       % figure window
n = 20;                % Define the surface of a sphere,
                       % using spherical coordinates
Theta = linspace(-pi,pi,n);
Phi = linspace(-pi/2,pi/2,n);
[theta,phi] = meshgrid(Theta,Phi);
X = cos(phi).*cos(theta);      % Translate into the xyz
                               % coordinate system
Y = cos(phi).*sin(theta);
Z = sin(phi);
surf(X,Y,Z)       %Create a surface plot of a sphere of radius 1
axis square
axis([-2,2,-2,2,-2,2])         %Specify the axis size
hold on
pause                          %Pause the program
surf(2*X,2*Y,2*Z)             %Add a second sphere of radius 2
pause                          %Pause the program
alpha(0.5)                    %Set the transparency level
```

The interior sphere is hidden by the outer sphere until we issue the transparency command,

```
alpha(0.5)
```

which sets the transparency level. A value of 1 corresponds to opaque and 0 to completely transparent. The results are shown in Figure 14.26. Transparency can be added to surfaces, images, and patch objects.

The command `alpha(0.5)` sets the transparency for all objects plotted on the axis. We can use handle graphics to specify the transparency for specific graphical objects. For example, first clear the figure window, but NOT the workspace window.

```
clf
```

Figure 14.26
Adding transparency to a surface plot makes it possible to see hidden details.

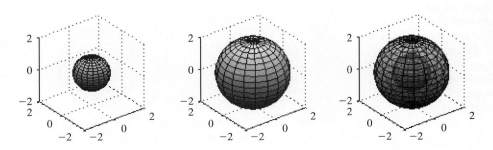

Assign a handle to each of the surface plots

```
h1 = surf(X,Y,Z);
hold on
h2=surf(2*X,  2*Y,2*Z);
```

To change the transparency of the outer sphere

```
set(h2,'Facealpha',0.3)
```

14.4.2 Hidden Lines

When mesh plots are created, any part of the surface that is obscured is not drawn. Usually, this makes the plot easier to interpret. The two spheres shown in Figure 14.27 were created with the use of the X-, Y-, and Z-coordinates calculated in the preceding section. Here are the MATLAB® commands:

```
figure(3)
subplot(1,2,1)
mesh(X,Y,Z)
axis square
subplot(1,2,2)
mesh(X,Y,Z)
axis square
hidden off
```

The default value for the hidden command is on, which results in mesh plots in which the obscured lines are automatically hidden, as shown at the left in Figure 14.27. Issuing the hidden off command gives the results at the right in Figure 14.27.

14.4.3 Lighting

MATLABV includes extensive techniques for manipulating the lighting used to represent surface plots. The position of the virtual light can be changed and even be manipulated during animations. The figure toolbar includes icons that allow you to adjust the lighting interactively, so that you can get just the effect you want. However, most graphs really need the lighting only turned on or off, which is accomplished with the camlight function. (The default is off.) Figure 14.28 shows the results achieved when the camlight is turned onto a simple sphere. The code to use is

```
Sphere
camlight
```

Figure 14.27
Left: Mesh plots do not show mesh lines that would be obscured by a solid figure. Right: The hidden off command forces the program to draw the hidden lines.

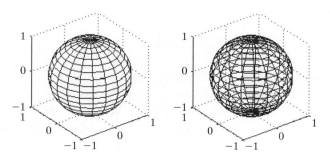

Figure 14.28
(a) The default lighting is diffuse. (b) When the `camlight` command is issued, a spotlight is modeled, located at the camera position.

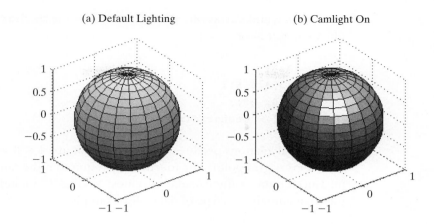

(a) Default Lighting   (b) Camlight On

The default position for the camlight is up and to the right of the "camera." The choices include the following:

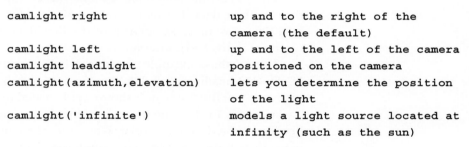

`camlight right`	up and to the right of the camera (the default)
`camlight left`	up and to the left of the camera
`camlight headlight`	positioned on the camera
`camlight(azimuth,elevation)`	lets you determine the position of the light
`camlight('infinite')`	models a light source located at infinity (such as the sun)

14.5 INTRODUCTION TO VOLUME VISUALIZATION

MATLAB® includes a number of visualization techniques that allow us to analyze data collected in three dimensions, such as wind speeds measured at a number of locations and elevations. It also lets us visualize the results of calculations performed with three variables, such as $y = f(x, y, z)$. These visualization techniques fall into two categories:

- Volume visualization of scalar data (where the data collected or calculated is a single value at each point such as temperature).
- Volume visualization of vector data (where the data collected or calculated is a vector, such as velocity).

14.5.1 Volume Visualization of Scalar Data

In order to work with scalar data in three dimensions, we need four three-dimensional arrays:

- X data, a three-dimensional array containing the *x*-coordinate of each grid point.
- Y data, a three-dimensional array containing the *y*-coordinate of each grid point.
- Z data, a three-dimensional array containing the *z*-coordinate of each grid point.
- Scalar values associated with each grid point—for example, a temperature or pressure.

The x, y, and z arrays are often created with the meshgrid function. For example, we might have

```
x = 1:3;
y = [2,4,6,8];
z = [10, 20];
[X,Y,Z] = meshgrid(x,y,z);
```

The calculations produce three arrays that are $4 \times 3 \times 2$ and define the location of every grid point. The fourth array required is the same size and contains the measured data or the calculated values. MATLAB® includes several built-in data files that contain this type of data—for example,

- MRI data (stored in a file called MRI)
- Flow field data (calculated from an M-file)

The help function contains numerous examples of visualization approaches that use these data. The plots shown in Figure 14.29 are a contour slice of the MRI data and an isosurface of the flow data, both created by following the examples in the help tutorial.

To find these examples, go to the help menu table of contents. Under the MATLAB® heading, find 3-D Visualization and then Volume Visualization techniques. When the two figures shown were created in MATLAB® 7.5 for this book, it was necessary to clear the figure (clf) each time before rendering the images—a detail not noted in the tutorial. When the clf command was not used, the plots behaved as if the hold on command were activated. This is an idiosyncrasy that may be corrected in later versions.

14.5.2 Volume Visualization of Vector Data

In order to display vector data, you need six three-dimensional arrays:

- Three arrays to define the x, y, and z locations of each grid point.
- Three arrays to define the vector data u, v, and w.

Figure 14.29
MATLAB® includes visualization techniques used with three-dimensional data. Left: Contour slice of MRI data, using the sample data file Included with MATLAB®. Right: Isosurface of flow data, using the sample M-File included with MATLAB®.

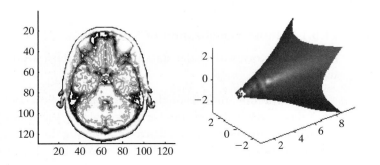

A sample set of vector volume data, called `wind`, is included in MATLAB® as a data file. The command

```
load wind
```

sends six three-dimensional arrays to the workspace. Visualizing this type of data can be accomplished with a number of different techniques, such as

- cone plots
- streamlines
- curl plots

Velocity
A speed plus directional information

Alternatively, the vector data can be processed into scalar data, and the techniques used in the previous section can be used. For example, velocities are not just speeds; they are speeds plus directional information. Thus, velocities are vector data, with components (called u, v, and w, respectively) in the x, y, and z directions. We could convert velocities to speed by using the formula

```
speed = sqrt(u.^2 + v.^2 + w.^2)
```

The speed data could be represented as one or more contour slices or as isosurfaces (among other techniques). The left-hand image of Figure 14.30 is the `contourslice` plot of the speed at the eighth-elevation (z) data set, produced by

```
contourslice(x,y,z,speed,[ ],[ ], 8)
```

and the right-hand image is a set of contour slices. The graph was interactively adjusted so that you could see all four slices.

```
contourslice(x,y,z,speed,[ ],[ ],[1, 5, 10, 15])
```

A cone plot of the same data is probably more revealing. Follow the example used in the coneplot function description in the help tutorial to create the cone plot shown in Figure 14.31.

Figure 14.30
Contour slices of the wind-speed data included with the MATLAB® program.

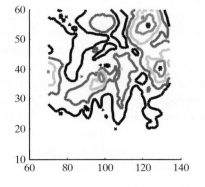
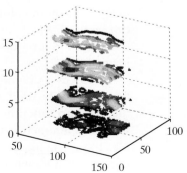

Figure 14.31
Cone plot of the wind-velocity data included with the MATLAB® program.

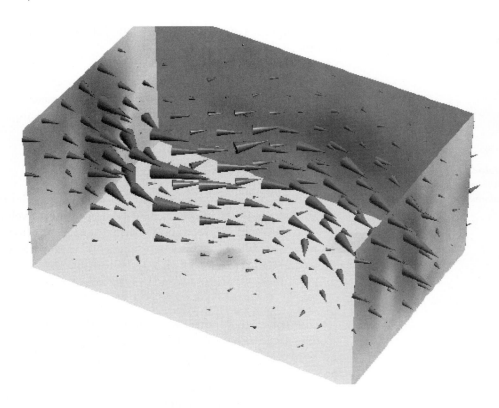

SUMMARY

MATLAB® recognizes three different techniques for storing and representing images:

Intensity (or gray scale) images
Indexed images
RGB (or true color) images

The imagesc function is used to display intensity images that are some-times called gray scale images. Indexed images are displayed with the image function and require a colormap to determine the appropriate coloring of the image. A custom colormap can be created for each image, or a built-in colormap can be used. RGB (true color) images are also displayed with the image func-tion but do not require a colormap, since the color information is included in the image file.

If you don't know what kind of image data you are dealing with, the imfinfo function can be used to analyze the file. Once you know what kind of file you have, the imread function can load an image file into MATLAB®, or you can use the software's interactive data controls. The load command can load a .dat or a .mat file. To save an image in one of the standard image formats, use the imwrite func-tion or the interactive data controls. You can also save the image data as .dat or .mat files, using the save command.

A handle is a nickname given to an object in MATLAB®. The graphics displayed by MATLAB® include several different objects, all of which can be given a handle. The fundamental graphics object is the figure. Layered on top of the figure is the axis object, and layered on top of that is the actual plot object. Each of these objects includes properties that can be determined with the `get` function or changed with the `set` function. If you don't know the appropriate handle name, the function `gcf` (get current figure) returns the current figure handle and `gca` (get current axis) returns the current axis handle. The `set` function is used to change the properties of a MATLAB® object. For example, to change the color of a plot (the line you drew) named h, use

```
set(h,'color','red')
```

Animation in MATLAB® is handled with one of two techniques: redrawing and erasing, or creating a movie. Usually, redrawing and erasing is easier for animations which represent data that can be quickly computed and are not visually complicated. For tasks that take significant computing power, it is generally easier to capture individual frames and then combine them into a movie to be viewed at a later time.

Complex surfaces are often difficult to visualize, especially since there may be surfaces underneath other surfaces. It is possible to render these hidden surfaces with a specified transparency, which allows us to see the obscured details. This is accomplished with the `alpha` function. The input to this function can vary between 0 and 1, ranging from completely transparent to opaque.

To make surfaces easier to interpret, by default hidden lines are not drawn. The `hidden off` command forces the program to draw these lines.

Although MATLAB® includes an extensive lighting-manipulation capability, it is usually sufficient to turn the direct-lighting function on or off. By default, the lighting is diffuse, but it can be changed to direct with the `camlight` function.

Volume-visualization techniques allow us to display three-dimensional data a number of different ways. Volume data fall into two categories: scalar and vector data. Scalar data involve properties such as temperature or pressure, and vector data include properties such as velocities or forces. The MATLAB® `help` function contains numerous examples of visualization techniques.

MATLAB® SUMMARY

The following MATLAB® summary lists and briefly describes all of the special characters, commands, and functions that were defined in this chapter.

Commands and Functions	
alpha	sets the transparency of the current plot object
axis	controls the properties of the figure axis
bone	colormap that makes an image look like an X-ray
cape	sample MATLAB® image file of a cape
camlight	turns the camera light on

(continued)

Commands and Functions	
clown	sample MATLAB® image file of a clown
colormap	defines which colormap should be used by graphing functions
coneplot	creates a plot with markers indicating the direction of input vectors
contourslice	creates a contour plot from a slice of data
detail	sample MATLAB® image file of a section of a Dürer wood carving
drawnow	forces MATLAB® to draw a plot immediately
durer	sample MATLAB® image file of a Dürer wood carving
earth	sample MATLAB® image file of the earth
flujet	sample MATLAB® image file showing fluid behavior
gatlin	sample MATLAB® image file of a photograph
gca	get current axis handle
gcf	get current figure handle
get	returns the properties of a specified object
getframe	gets the current figure and saves it as a movie frame in a structure array
gray	colormap used for gray scale images
hidden off	forces MATLAB® to display obscured grid lines
image	creates a two-dimensional image
imagesc	creates a two-dimensional image by scaling the data
imfinfo	reads a standard graphics file and determines what type of data it contains
imread	reads a graphics file
imwrite	writes a graphics file
isosurface	creates surface-connecting volume data, all of the same magnitude
mandrill	sample MATLAB® image file of a mandrill
movie	plays a movie stored as a MATLAB® structure array
mri	sample MRI data set
pcolor	pseudo color plot (similar to a contour plot)
peaks	creates a sample plot
set	establishes the properties assigned to a specified object
shading	determines the shading technique used in surface plots and pseudo color plots
spine	sample MATLAB® image file of a spine X-ray
wind	sample MATLAB® data file of wind-velocity information

KEY TERMS

handle	object	vector data
image plot	RGB (true color)	volume visualization
indexed image	scalar data	
intensity image	surface plot	

PROBLEMS

14.1 On the Internet, find an example of an intensity image, an indexed image, and an RGB image. Import these images into MATLAB®, and display them as MATLAB® figures.

14.2 A quadratic Julia set has the form:

$$z(n + 1) = z(n)^2 + c$$

The special case where $c = -0.123 + 0.745i$ is called Douday's rabbit fractal. Follow Example 14.1, and create an image using this value of c. For the Mandelbrot image, we started with all z-values equal to 0. You'll need to start with $z = x + yi$. Let both x and y vary from -1.5 to 1.5.

14.3 A quadratic Julia set has the form

$$z(n + 1) = z(n)^2 + c$$

The special case where $c = -0.391 - 0.587i$ is called the Siegel disk fractal. Follow Example 14.1 and create an image using this value of c. For the Mandelbrot image, we started with all z-values equal to 0. You'll need to start with $z = x + yi$. Let both x and y vary from -1.5 to 1.5.

14.4 A quadratic Julia set has the form

$$z(n + 1) = z(n)^2 + c$$

The special case where $c = -0.75$ is called the San Marco fractal. Follow Example 14.1 and create an image using this value of c. For the Mandelbrot image, we started with all z-values equal to 0. You'll need to start with $z = x + yi$. Let both x and y vary from -1.5 to 1.5.

14.5 Create a plot of the function

$$y = \sin(x) \qquad \text{for } x \text{ from } -2\pi \text{ to } +2\pi$$

Assign the plot a handle, and use the set function to change the following properties (if you aren't sure what the object name is for a given property, use the get function to see a list of available property names):

(a) Line color from blue to green
(b) Line style to dashed
(c) Line width to 2

14.6 Assign a handle to the figure created in Problem 14.5, and use the set function to change the following properties (if you aren't sure what the object name is for a given property, use the get function to see a list of available property names):

(a) Figure background color to red
(b) Figure name to "A Sine Function"

14.7 Assign a handle to the axes created in Problem 14.5, and use the set function to change the following properties (if you aren't sure what the object name is for a given property, use the get function to see a list of available property names):

(a) Background color to blue
(b) x-axis scale to log

14.8 Repeat the three previous problems, changing the properties by means of the interactive property inspector. Experiment with other properties and observe the results on your graphs.

14.9 Create an animation of the function

$$y = \sin(x - a) \quad \text{for} \quad \begin{array}{l} x \text{ ranging from } -2\pi \text{ to } +2\pi \\ a \text{ ranging from } 0 \text{ to } 8\pi \end{array}$$

- Use a step size for x that results in a smooth graph.
- Let a be the animation variable. (Draw a new picture for each value of a.)
- Use a step size for a that creates a smooth animation. A smaller step size will make the animation seem to move more slowly.

14.10 Create a movie of the function described in the preceding problem.

14.11 Create an animation of the following:
Let x vary from -2π to $+2\pi$
Let $y = \sin(x)$
Let $z = \sin(x - a)\cos(y - a)$
Let a be the animation variable.
Remember that you'll need to mesh x and y to create two-dimensional matrices; use the resulting arrays to find z.

14.12 Create a movie of the function described in the preceding problem.

14.13 Create a program that allows you to zoom in on the "rabbit fractal" described in Problem 14.2, and create a movie of the results (see Example 14.2).

14.14 Use a surface plot to plot the peaks function. Issue the hold on command and plot a sphere that encases the entire plot. Adjust the transparency so that you can see the detail in the interior of the sphere.

14.15 Plot the peaks function and then issue the camlight command. Experiment with placing the camlight in different locations, and observe the effect on your plot.

14.16 Create a stacked contour plot of the MRI data, showing the first, eighth, and twelfth layer of the data.

14.17 An MRI visualization example is shown in the help tutorial. Copy and paste the commands into an M-file and run the example. Be sure to add the clf command before drawing each new plot.

15

Creating Graphical User Interfaces

Objectives

After reading this chapter, you should be able to:

- Understand how to use the GUIDE layout editor

- Understand how to modify function callbacks
- Be able to create graphical user interfaces

INTRODUCTION

Most computer programs in use today make use of a graphical user interface (GUI) and in fact MATLAB®'s desktop environment is a graphical user interface. Any time you can click an icon to execute an action, you are using a GUI (pronounced "gooey"). Creating your own GUI's is easy in MATLAB®, especially if you use the GUIDE interface, but it does require that you understand some programming basics—all of which you have been introduced to in *MATLAB® for Engineers*. Before starting this section it would be wise to review the concepts of:

- Structure arrays
- Subfunctions
- Handle graphics

The m-file created by the GUIDE program uses a structure array to pass information between sections of the program; each of these sections is a subfunction, and components of the GUI are stored as properties of a graphics object, using handle graphics.

Generally, the first step in creating a GUI should be to carefully plan what the GUI should do and how it should look. A little planning will help you avoid a lot of frustration. However, in this chapter, we'll develop GUIs piecewise, so that we can

KEY IDEA

GUIDE makes creating GUI's easy

focus on how the program works. Be sure to try these commands out as you read through this chapter.

15.1 A SIMPLE GUI WITH ONE USER INTERACTION

15.1.1 Creating the Layout

To get started, select the guide icon from the toolbar, as shown in Figure 15.1, or type guide at the command line. The GUIDE Quick Start window will open, as shown in Figure 15.2. To start a new project, simply select the Blank GUI template, located in the list on the left-hand side of the window.

Once you select Blank GUI, a new figure window—called the GUIDE layout editor—will open, which should look similar to the one shown in Figure 15.3. You can resize it to a shape that is comfortable to work with by selecting the lower left-hand corner of the grid. If you'd like a GUI that is bigger than the figure window, just resize the figure window first.

To create a layout of buttons, textboxes, and graphics windows, use the icons on the left-hand side of the window in the "component palette." The default display for these icons is compact, but not particularly informative for new users. To change the palette of tools to a list of the item names select

File → Preferences → GUIDE

then check "Show names in component palette,"

KEY IDEA

The component palette lists the available choices for use in the layout editor

as shown in Figure 15.4. This results in a more "user friendly" list of the available options (Figure 15.5).

Let's get started with a very simple GUI that allows us to enter the number of sides on a polygon, and which then plots the polygon in polar coordinates. We'll

Figure 15.1
Select the GUIDE icon from the MATLAB® toolbar, or type guide at the command line to start the program.

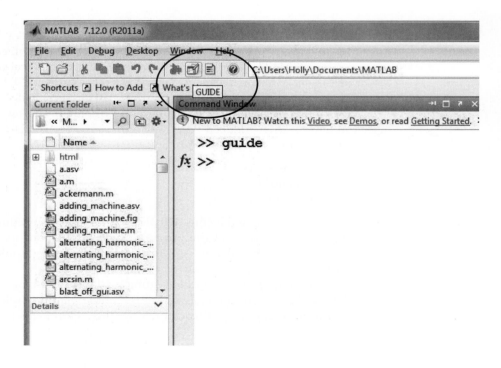

Figure 15.2
Use the GUIDE Quick Start window to get started building a graphical user interface. Select Blank GUI to start a new project.

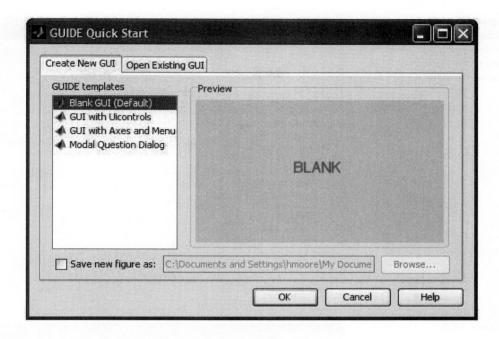

Figure 15.3
The GUIDE layout editor is used to design your GUI.

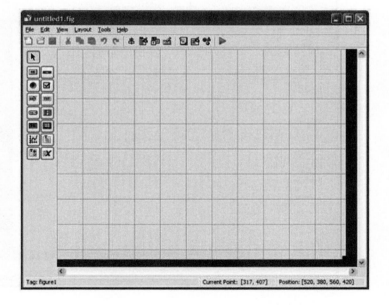

need three components in the GUI: axes, a static text box, and an edit textbox. You can pick them up from the component palette and arrange them as shown in Figure 15.6.

To modify these design elements once you have them arranged to your liking, use the Property Inspector. First, select the static text window, right click, and select the Property Inspector (Figures 15.7 and 15.8). You can also access the Property Inspector from the menu bar by selecting

View → Property Inspector

Figure 15.4
Change the component palette display to a list of item names in the preference window.

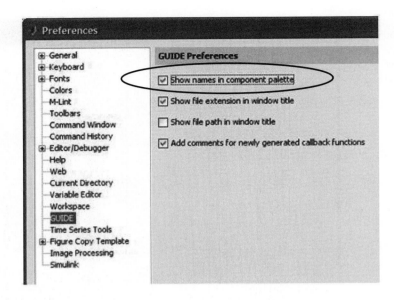

Figure 15.5
The component palette in the GUIDE layout editor can be reconfigured to show the possible actions in more detail than is possible with a simple icon.

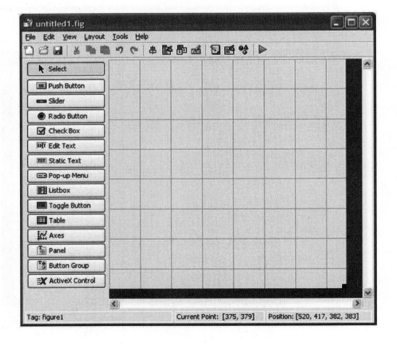

The Property Inspector lists a wide range of properties for the selected object in the GUIDE window. You can change the font of the message displayed, change the color of the text box etc. The most important property for us is the String Property. Change it from

Static Text

to

Enter the number of sides

Figure 15.6
The icons from the component palette are used to position and resize the design elements in the GUIDE window.

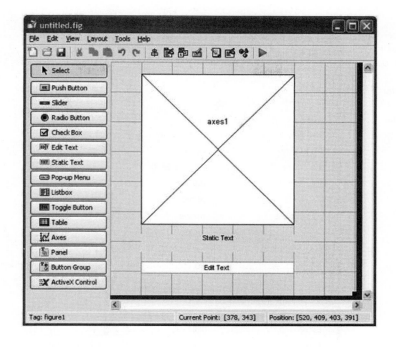

Figure 15.7
To access the property inspector, select an object from the GUIDE window, right click, and select the property inspector. You may access the same content from the menu bar by selecting View → Property Inspector.

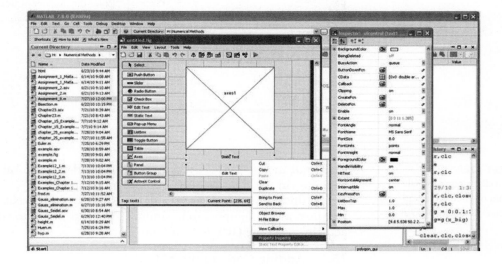

Use the same process to modify the properties of the "edit text" box. For our purposes simply delete the default text.

Now you can save and run the GUIDE window by selecting the Save and Run icon from the window toolbar (the green triangular button). You'll be prompted to enter a project name, such as polygon_gui.fig. When the file runs notice that the name of the GUIDE window changes, and an m-file is created with the appropriate code to create a figure window with which the user can interact. The m-file is

Figure 15.8
Property Inspector for the Static Textbox allows you to change properties, such as the message in the box (string property), the color of the background (Backgroundcolor), or the font size (FontSize).

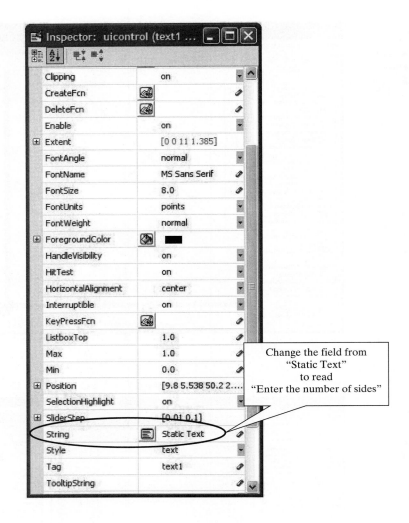

displayed in the MATLAB® edit window, and has the same name as the figure window—in this case polygon_gui.m (Figure 15.9).

At this point all we have is a figure window with an axis, a message in the static text box, and an empty input window. The next step is to add code to the m-file to actually make the GUI do something.

KEY IDEA
GUIDE creates an m-file, that is modified to add functionality to the GUI

15.1.2 Adding Code to the M-File

KEY IDEA
The GUI m-file is composed of multiple subfunctions

Just opening up the m-file and trying to interpret the code is confusing. The m-file is organized as a function, with multiple subfunctions. Some of the subfunctions create the graphics in the polygon_gui.fig window, but others are reserved for adding the code that will cause an action when a user interacts with the GUI. To see a list of the functions in the polygon_gui.m file, select the Show Functions icon on the toolbar (Figure 15.10). The only functions a user should modify are labeled as:

- gui_name_OpeningFcn
- graphics_object_name_Callback

Figure 15.9
Once the GUIDE window is activated an m-file is created along with a figure window through which the user will interact with the program.

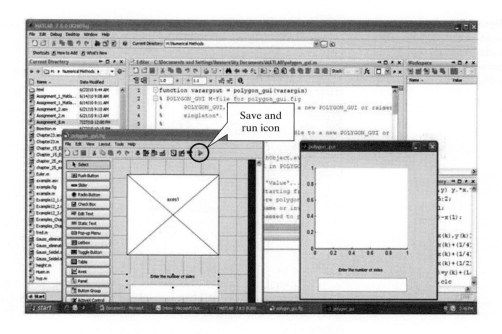

Figure 15.10
Selecting the Show Functions icon opens a list of all the subfunctions in the file. Navigate to a section of code by selecting the subfunction name from the list.

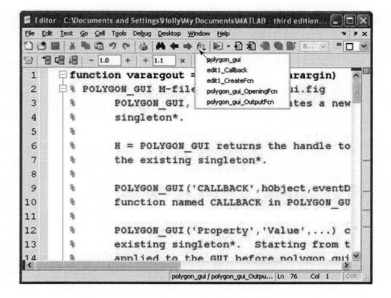

In the polygon_gui file this corresponds to:

- polygon_gui_OpeningFcn
- edit1_Callback

Callbacks
In more complicated graphical user interfaces, there will be a Callback function for each of the graphics objects on the layout, which allow the user to interact with the GUI. Clicking on the function of interest will take you to the corresponding section of code.

Figure 15.11

Right-click on the edit textbox to locate the corresponding m-file subfunction.

An alternative approach to finding the appropriate subfunction to modify is to use the layout editor. Right click on the graphics object (in this case the edit textbox), select View Callbacks, then select Callback (Figure 15.11). This will move the cursor in the m-file to the edit1_Callback subfunction, shown here.

```
function edit1_Callback(hObject, eventdata, handles)
% hObject handle to edit1 (see GCBO)
% eventdata reserved - to be defined in a future version of MATLAB®
% handles structure with handles and user data (see GUIDATA)

% Hints: get(hObject,'String') returns contents of edit1 as text
% str2double(get(hObject,'String')) returns contents of edit1 as a
  double
```

Notice that most of the code is composed of comments. The first line identifies the subfunction as edit1_Callback, with three inputs. The first, hObject, is a graphics handle that links the subfunction to the corresponding edit textbox. The eventdata argument is a placeholder that the Mathworks has included for use in later versions of the software. Finally, the handles argument is a structure array that is used to pass information between subfunctions. All callback subfunctions will have a similar structure.

Specific to a callback linked to an edit textbox are the hints listed as comments. Information typed into the textbox is interpreted using handle graphics. Recall how we modified the textbox so that it was blank by deleting the contents of the string property in the Property Inspector. When a user types in a textbox, the contents are stored as the string property. To retrieve the information and use it in our m-file, we need to "get" it using the get function.

```
get(hObject, 'String')
```

This instructs MATLAB® to retrieve the string property from the graphics object that was passed to the function as hObject—in this case the edit1 textbox. Information in the string property is stored as a character array, so if we want to use it as a numeric value it is necessary to change the array type to double. This can be accomplished either with the str2num function or the str2double function. With this in mind, add the following code to the edit1_callback subfunction.

KEY IDEA

Structure arrays are used to pass information between functions

KEY IDEA

Numbers entered as a string property are stored as character arrays, and must be converted to a numeric format before they can be used.

Figure 15.12
(a) Opening appearance of the GUI, (b) appearance once content is added to the edit textbox.

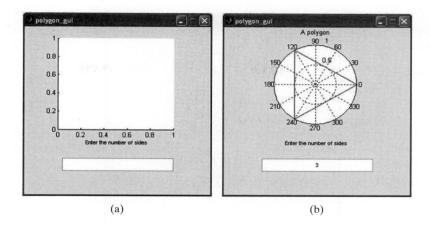

(a)   (b)

```
sides = str2double(get(hObject,'String'))
```

Now we can add additional code to draw the polygon using the polar plotting function and to annotate the graph.

```
theta = 0:2*pi/sides:2*pi;
r = ones(1,length(theta));
polar(theta,r)
title('A polygon')
```

To run your graphical user interface, select the Save and Run icon from the m-file window or from the Guide layout editor. A figure window appears, similar to Figure 15.12a. To run the GUI, type a value into the edit window, such as 3 and hit enter. This causes the edit1_callback function to execute and draw a polygon using the polar plot function (Figure 15.12b).

The opening function is the only other subfunction to be modified in this file. It executes when the GUI first runs, and can be used to control how the figure window appears before the user starts adding data. Notice that the opening version of polygon_gui displays a rectangular axis. In order to display an axis system consistent with a polar plot, we can modify the polygon_gui_OpenFcn, by adding code to create a blank polar plot.

```
function polygon_gui_OpeningFcn(hObject, eventdata, handles, varargin)
% This function has no output args, see OutputFcn.
% hObject handle to figure
% eventdata reserved - to be defined in a future version of MATLAB®
% handles structure with handles and user data (see GUIDATA)
% varargin command line arguments to polygon_gui (see VARARGIN)
polar(0,1)
title('A polygon')
% Choose default command line output for polygon_gui
handles.output = hObject; % Not necessary for this example
```

Now, when `polygon_gui.m` is executed the original figure window includes the polar plot axis system (Figure 15.13).

Figure 15.13
To modify the opening
appearance of the GUI,
add code to the
OpeningFcn subfunction.

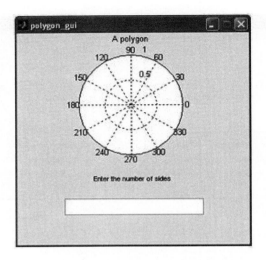

15.2 A GRAPHICAL USER INTERFACE WITH MULTIPLE USER INTERACTIONS—READY_AIM_FIRE

It's easy to create a more complicated GUI with more places for the user to enter data and with a variety of actions. Consider a GUI that plots the trajectory of a projectile launched from a cannon. The trajectory depends on both the launch angle, θ, and the initial velocity, V_0 of the projectile. The equations representing the horizontal and the vertical distances traveled are as follows:

$$h = tV_0\cos(\theta)$$

$$v = tV_0\sin(\theta) - 1/2gt^2$$

where

t is the time in seconds

V_0 is the initial velocity in m/s

θ is the launch angle in radians

g is the acceleration due to gravity, 9.81 m/s^2

To create a GUI that plots the trajectory, we'll need the following components in the layout:

Axes	for the graph
Edit textbox	for the angle input
Edit textbox	for the initial velocity input
Push button	to "fire" the canon
Static textbox	to label the angle textbox
Static textbox	to label the velocity textbox
Panel	to group the textboxes together (not necessary, but nice)

By selecting the appropriate items from the component palette, it is easy to create the layout shown in Figure 15.14. The contents of the static textboxes and the edit textboxes were modified using the property editor string property, as was the push button. Both types of textboxes were dragged onto a panel. The panel name was changed, not in the string property, but in the title property.

Figure 15.14
The GUIDE layout editor makes it easy to create more complicated GUI's. This layout represents a basic plotting program for a projectile trajectory program.

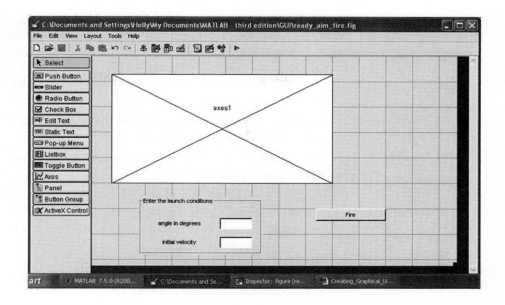

Once a GUI has multiple components, it becomes tricky to find the corresponding callbacks in the m-file based on the default names. For example, the two edit textboxes shown in Figure 15.14 default to edit1 and edit2—names that aren't very descriptive. To change the name from the default, use the tag property, which can be accessed from the property editor. For example, Figure 15.15 shows the property editor for the edit textbox corresponding to the launch angle. The tag has been changed from edit1 to launch_angle. Similarly, the tag for the initial velocity edit textbox was changed from edit2 to launch_velocity, and the tag for the push button was changed to fire_pushbutton. The contents of the layout editor were then saved and named ready_aim_fire, by selecting the Save and Run button. Recall that two files are created, a fig file containing the GUI and an m-file containing the code.

Adding code to this GUI program is not quite as straightforward as the first example. We'll need to read in the data entered into the edit textboxes in the callback functions, give the data a name, and then pass it on to the fire_pushbutton callback function to create the plot. Here are the steps to take.

First find the launch_angle_Callback subfunction, either by selecting the Show Functions icon in the m-file toolbar or by right clicking the launch angle edit textbox and navigating to the launch_angle callback. Add the following code:

```
handles.theta=str2double(get(hObject,'String'));
guidata(hObject, handles);
```

Figure 15.15
Changing the tag property in the Property Inspector changes the name of the callback functions associated with the object, making it easier for the programmer to navigate to the associated m-file.

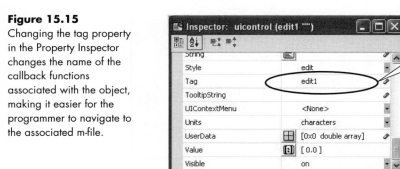

In order to pass information to other functions, we need to save the information from the edit textbox into the handles structure array. We'll store this particular information in the theta portion of the structure. Then, we need to update the rest of the program so that other functions can use the information.

Similarly the `launch_velocity` callback is modified by adding the following code:

```
handles.vel=str2double(get(hObject,'String'));
guidata(hObject, handles);
```

The graph is actually drawn when the fire push button is selected, so that's where the plotting code must go.

```
time=0:0.001:100;
h=time*handles.vel*cosd(handles.theta);
v=time*handles.vel*sind(handles.theta)-1/2*9.81*time.^2;
pos=find(v>=0);
horizontal=h(pos);
vertical=v(pos);
comet(horizontal,vertical);
```

Notice that an array called `time` was created with a small step size. This becomes important in the plotting step. Then the horizontal and the vertical distances traveled were calculated. The vertical distance will become negative, which doesn't make any physical sense, so the `find` function was used to find all the index numbers in the `v` array that are positive. Two new variables, `horizontal` and `vertical`, were defined using that information, and then plotted using the `comet` function. The `comet` function draws out the trajectory of the projectile. You can change the apparent speed by manipulating how many points are plotted—which was done by controlling the number of time values.

To run the program, select the Save and Run icon, which will open the GUI. The result of one set of input values is shown in Figure 15.16.

Figure 15.16
This GUI accepts multiple inputs, which are then used when the "Fire" push button is selected.

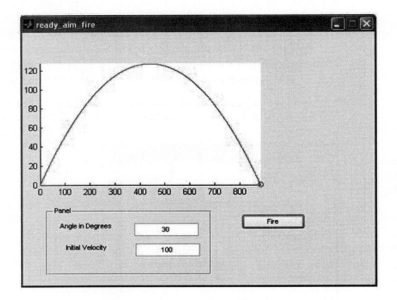

15.3 AN IMPROVED `READY_AIM_FIRE` PROGRAM

After you've run the `Ready_Aim_Fire` program a number of times, you will probably want to make some modifications. For example, each time the GUI runs, the plot resizes to completely fill the window. It makes it hard to tell what the result is of changing each of the parameters. We can modify the opening function to create an axis that never changes to alleviate this problem. While we are at it we'll also add a target, so that we can practice firing our "cannon" with a particular goal in mind.

Navigate to the opening function and add the following code:

```
plot(275,0,'s','Markersize',10,'MarkerFaceColor','r')
text(275,50,'target')
axis([0,1000,0,500])
hold on
```

The first line creates a plot of a single point, at $x = 275$ and $y = 0$. The data is shown as a square, and the size and color are adjusted so that it is easy to see. The second line adds a label to the target. The **axis** function forces the plot to cover x-axis values from 0 to 1000, and y-axis values from 0 to 500. Finally the `hold on` command forces additional plots to draw on the same graph, without erasing any of the existing lines. Figure 15.17a shows the opening screen, and Figure 15.7b shows the screen after three attempts to adjust the input parameters and hit the target.

One problem with this version of the `ready_aim_fire` GUI is that you have to completely close it to start over and clear the screen. We can remedy this by adding an additional push button to reset the plot. You'll need to:

- Return to the GUIDE layout editor and add an additional push button.
- Use the Property Inspector string property to label the push button "Reset."
- Use the Property Inspector tag property to change the name of the push button and its associated functions to "`reset_pushbutton`."
- Use the Save and Run icon to save your changes and to update the `ready_aim_fire` m-file.
- Navigate to the `reset_pushbutton_Callback` subfunction and add the appropriate code.

```
hold off
plot(275,0,'s','Markersize',10,'MarkerFaceColor','r')
text(275,50,'target')
axis([0,1000,0,500])
hold on
```

Figure 15.17
(a) The "`ready_aim_fire`" GUI opening screen, (b) the "`ready_aim_fire`" GUI after three "shots."

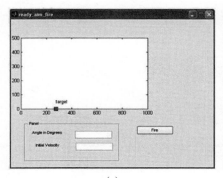
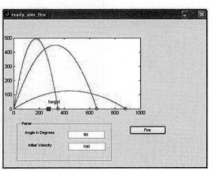

(a)   (b)

This code simply turns off the hold function and then repeats the instructions from the opening function. While modifying the code, we can also add a title and axis labels to both the opening function and the `reset_pushbutton` callback.

```
title('Projectile Trajectory')
xlabel('Horizontal Distance, m')
ylabel('Vertical Distance, m')
```

HINT

When you start to modify an existing program, close the GUI figure window (not the GUIDE layout editor window). Once you are done making changes in the m-file, select the Save and Run icon from the m-file editor tool bar. This will reinitialize the GUI figure window. If you just leave the GUI open, all the changes may not be incorporated.

15.4 A MUCH BETTER READY_AIM_FIRE PROGRAM

By now you probably want to be able to control the target position, and perhaps display an explosion if you hit the target. Let's start with moving the target, by adding a slider bar to the GUI in the GUIDE layout editor. To make the GUI neater, you'll need to move the other controls to the side, as shown in Figure 15.18. Also add a static textbox to label the slider. From the slider property inspector, change the value of the Max property to 1000 to correspond with the scale on our graph. Also change the value of the Value property to 275, so that the slider starts off at the original target position (Figure 15.19).

- Navigate to the slider callback, and notice that the "Hints" suggest how to retrieve the location of the slider. You won't need to retrieve Max and Min.

Figure 15.18
The revised layout for the "Ready_Aim_Fire" GUI.

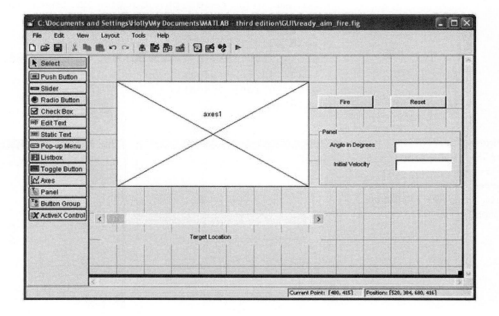

Figure 15.19
The "Slider" Property inspector. The Max property and the Value property have been adjusted.

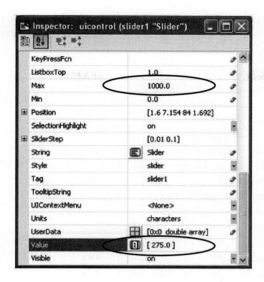

- Use the location of the slider bar to plot the target.

```
handles.location = get(hObject,'Value')
hold off
plot(handles.location,0,'s','Markersize',10,'Markerfacecolor', 'r')
axis([0,1000,0,1000])
title('Trajectory')
xlabel('Horizontal Distance')
ylabel('Vertical Distance')
text(handles.location-25,50,'Target')
hold on
guidata(hObject, handles);
```

Notice that the location of the slider is stored as part of the handles structure array. In the final line of the listed code, the handles structure for the entire program is updated, so that the `handles.location` value can be used by other functions. For example, if we don't make anymore changes, every time the reset button is pushed the target will move back to the starting location. It probably makes more sense that it should remain at the same location as the slider. Modifying the `reset_pushbutton` callback accomplishes this goal.

```
hold off
plot(handles.location,0,'s','Markersize',10,'MarkerFaceColor','r')
text(handles.location,50,'target')
axis([0,1000,0,500])
title('Projectile Trajectory')
xlabel('Horizontal Distance, m')
ylabel('Vertical Distance, m')
hold on
```

Just for fun, we'd like to show an explosion in the plot window if we select a trajectory that hits the target. The code should be added to the `fire_pushbutton` callback.

```
time=0:0.001:100;
h=time*handles.vel*cosd(handles.theta);
v=time*handles.vel*sind(handles.theta)-1/2*9.81*time.^2;
pos=find(v>=0);
horizontal=h(pos);
vertical=v(pos);
comet(horizontal,vertical);
land=pos(end);
goal=handles.location;
if (h(land)<goal+50 && h(land)>goal-50)
x=linspace(goal-100, goal+100, 5);
y=[0,80,100,80,0];                            %Code to create
z=linspace(goal-200,goal+200,9);              the "Explosion"
w=[0,40,90,120,130,120,90,40,0];
plot(x,y,'*r',z,w,'*r')
text(goal,400,'Kaboom!')
end
```

The explosion is simply a number of stars plotted at the points defined by the **x**, **y**, **z**, and **w** arrays. Notice that the `fire_pushbutton` callback uses the `handles.location` parameter, which is created in the slider callback. If the slider is never moved, this parameter is never created. This means that the attempt to create the explosion will fail, unless `handles.location` is defined in the opening function

```
handles.location = 275;
```

Figure 15.20 shows the result when a user finally hits a target.

One last refinement to the GUI is to add a textbox that congratulates you when you win. To do that, we need to add another static textbox in the GUIDE layout editor, as shown in Figure 15.21.

Figure 15.20
The "ready_aim_fire" GUI displays a new image once the target is hit.

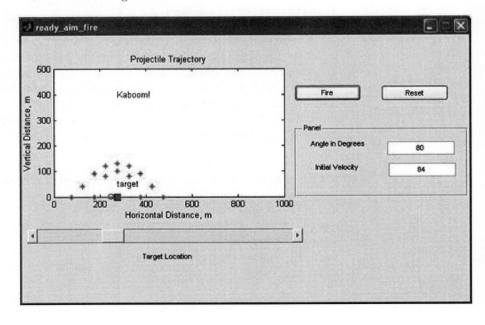

Figure 15.21

A static textbox is used to create a space for a message from MATLAB®.

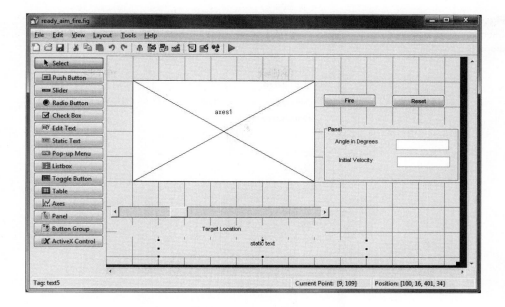

Using the property inspector, we'll need to change the string property to a blank. We'll also need to check for the tag property value, and change it to something meaningful, such as textout. Don't forget to save your changes in the GUIDE layout editor (Figure 15.22).

When you run the GUI, the opening value in the textbox will be blank. To change it to a message when the user's shot hits the target, add the following code to the `if` statement inside the `Fire_pushbutton_Callback`.

```
set(handles.textout,'string', 'You Win !','fontsize',16)
```

Notice that in addition to specifying the message, the font size has been adjusted from the default. You could have also made the adjustment from the property inspector.

The only thing left to do is make sure that when the reset button is pressed, the text box returns to a blank. This is accomplished in the `Reset_pushbutton_ Callback` with the following code:

```
set(handles.textout,'string', ' ')
```

The final version of the GUI is shown in Figure 15.23, once the user has fired the cannon and destroyed the target.

Appendix D lists the final contents of the m-file, including the following functions, which were modified to create the `ready_aim_fire` GUI:

- ready_aim_fire_OpeningFcn
- fire_pushbutton_Callback
- reset_pushbutton_Callback
- launch_angle_Callback
- launch_velocity_Callback
- slider_Callback

Figure 15.22
Change properties from the property inspector.

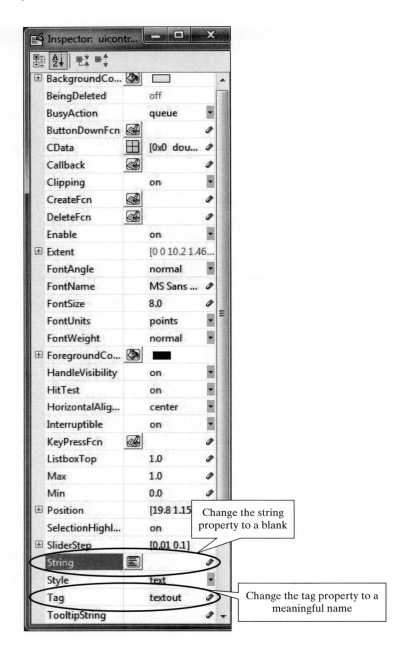

15.5 BUILT-IN GUI TEMPLATES

So far we have been working with the Blank GUI template within GUIDE. However, MATLAB® has included three other example GUI's, which you can use as a starting point for new projects, or just as examples to help you understand how to design your own GUI's. They include

- GUI with UIcontrols
- GUI with Axes and Menu
- Modal Question Dialog

Figure 15.23
The final `ready_aim_fire` GUI.

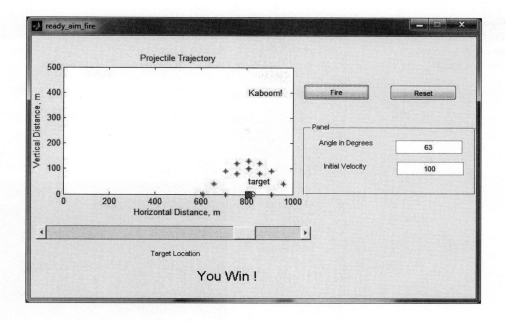

15.5.1 GUI with UIcontrols

From the GUIDE Quick Start window (Figure 15.24), select the GUI with UIcontrols template. A preview is shown in the Quick Start window to help you determine which of the built-in templates is appropriate for your needs.

The GUI with UIcontrols (user input controls) is a completely functional GUI, which performs and displays a mass calculation using either English or Metric (SI) units. The layout editing window is shown in Figure 15.25.

To see the corresponding m-file, select the Save and Run icon. This generates the appropriate MATLAB® code, which is displayed in the MATLAB® editor, and the GUI figure window shown in Figure 15.26.

Figure 15.24
The Quick Start GUIDE menu includes three example templates.

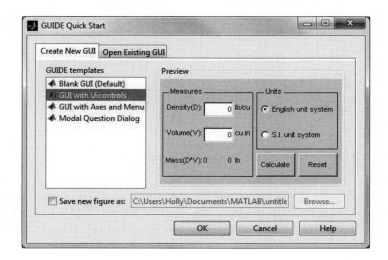

Figure 15.25
The GUIDE UIcontrols template contains a mass calculation GUI.

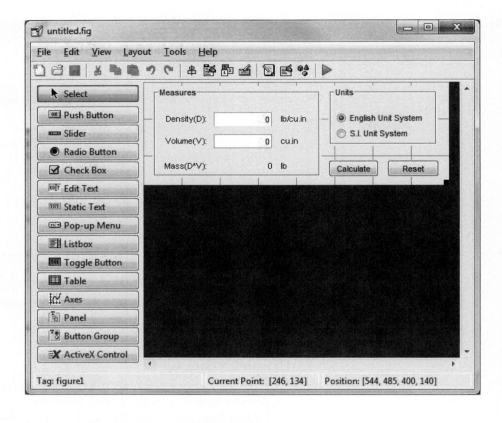

Figure 15.26
MATLAB® includes several example GUI's, which can be used as the starting point for new projects.

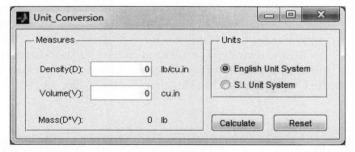

This GUI is composed of the following:

- A panel, that contains
  - Two edit textboxes
  - Seven static textboxes
- A button group that contains
  - Two radio buttons
  - Two push buttons

The only graphics objects that are new to us in this GUI are the button group and the radio buttons. When radio buttons are added to a button group, only one radio button can be active at a time. If the radio buttons had instead been added to a panel, they could all be active, all be inactive, or could be any combination of settings.

15.5.2 GUI with Axes and Menu

The GUI with Axes and Menu template illustrates how to use a popup menu (also called a dropdown menu) (see Figure 15.27). MATLAB® also includes a video demonstration that includes the use of several graphics objects, such as the popup menu, pushbuttons, and axes, which can be accessed from the help feature, and is listed under demos.

Figure 15.27
The GUI with Axes and Menu template.

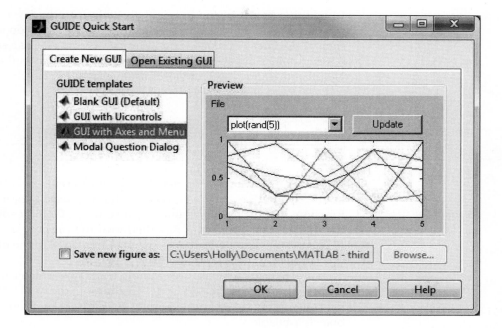

Figure 15.28
The controlsuite GUI includes examples of all the graphics objects available for use in GUIDE.

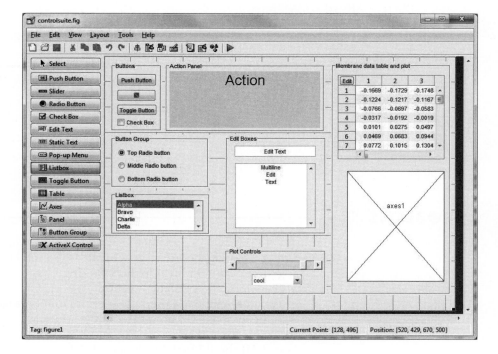

15.5.3 Modal Question Box

A modal question is one which requires a response from the user before continuing. For example, when you save a word processing document and ask the computer to overwrite an existing file, most programs ask you if you really want to do this. The modal question template demonstrates how to accomplish this in a GUI.

15.5.4 Other Examples

In addition to the example templates built into the layout editor, the MATLAB® help feature includes numerous examples that focus on single graphics objects, such as the check box or toggle buttons. It also includes a single GUI that includes all 15 graphics objects available from GUIDE. To access these resources, go to the Help feature and search on controlsuite (Figure 15.28).

SUMMARY

The GUIDE layout editor makes it easy to create graphical user interfaces in MATLAB®. It does, however, require that you have a basic understanding of subfunctions, handle graphics, and structure arrays. Graphics objects are positioned on the editor, their properties modified with the property inspector, and a function m-file created automatically. Instructions are added to the m-file in order to activate the various graphical components.

GUIDE also includes three sample templates, which can be used as the starting point for more complicated GUI's. In addition, the MATLAB® help feature offers a demonstration video and examples of GUI's showcasing each of the graphical objects available.

KEY TERMS

function callback	GUIDE	subfunctions
graphical objects	property inspector	
GUI	structure array	

PROBLEMS

15.1   Using GUIDE, create a graphical user interface to add two numbers together. It should include the following:
- Title, located in a static text box
- Two edit textboxes, used to enter the numbers to be added
- Static textboxes to hold the + and = symbols
- A static textbox to display the result

Your GUI should look like Figure P15.1.

Figure P15.1
A graphical user interface
used to add two numbers
together (a) before data is
added and (b) after a
calculation.

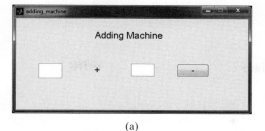

(a)

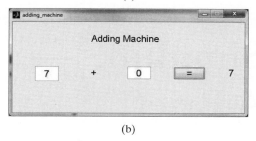

(b)

15.2 Create a GUI similar to the one in the previous problem. It should accept two numbers as input, but should allow the user to choose from the following operations by selecting a radio button.
- Addition
- Subtraction
- Multiplication
- Division

15.3 Create a GUI to simulate a cash register. It should accept the cost of an item and then display the running total. It should also display the total number of items purchased. Finally, it should accept the amount of money tendered by the user and display the change that should be returned to the customer.

15.4 Create a GUI that replicates the behavior of a simple four function calculator.

15.5 Create a GUI that accepts the name of an x, y, and z array as input. (The arrays should have been previously calculated in MATLAB®.) It should then allow the user to choose from the following graphing options:
- Surface plot (surf)
- Mesh plot (mesh)
- Contour plot (contour)

and display the graph on a set of axes in the GUI.

15.6 Forces are often represented as vectors, defined by a magnitude, and the angle from the horizontal at which the force is applied. To add them together they are placed head to tail. The resultant force is the vector drawn from the starting point to the ending point. For example consider the forces shown in Figure P15.6, and the resultant shown when they are added together.

Create a GUI that accepts both the magnitude and angle from the horizontal of three forces, then plots them end to end on a set of axes. It should also draw in the resultant, report the magnitude of the resultant and the angle from horizontal.

15.7 Repeat the previous problem in three dimensions.

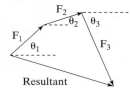

Figure P15.6
To add forces together, they
are placed head to tail.
The resultant is the vector
drawn from the starting
point to the ending point.

16

Simulink® — A Brief Introduction

Objectives

After reading this chapter you should be able to:

- Understand how Simulink® uses blocks to represent common mathematical processes

- Create and run a simple Simulink® Model
- Import Simulink® results into MATLAB®

INTRODUCTION

Simulink® is an interactive, graphics-based program that allows you to solve problems by creating *models* using a set of built-in "blocks." It is part of the MATLAB® software suite, and requires MATLAB® to run. Simulink® is included with the student edition of the software, but is not part of the standard installation of the professional edition; this means that it may or may not be included on your version of MATLAB®. LabView, produced by National Instruments, is Simulink®'s biggest competitor.

16.1 APPLICATIONS

Simulink® is designed to provide a convenient method for analyzing *dynamic systems*, i.e., systems that change with time. In particular, it found early acceptance in the signal processing community, and is reminiscent of the approach used to program *analog computers*. In fact, one way to think of Simulink® is as a virtual analog computer. Analog computers required the user to make actual physical connections between electrical components that acted as adders, multipliers, integrators, etc. Output from the computer was viewed on an oscilloscope. This is reflected in both the names of the blocks used in Simulink®, and in the icons used to represent various operations.

One shouldn't jump to the conclusion that Simulink® is only useful for analyzing electrical systems. Similar mathematical equations describe the behavior of dynamic mechanical systems, reactive chemical systems, and dynamic fluid systems. In fact, it is common to introduce students to the behavior of electricity through analogy with pipe flow problems.

Simulink®'s strength is its ability to model dynamic systems—which are modeled mathematically as *differential equations*. Usually these systems change with time, but the independent variable could also be location. Differential equations can be solved numerically in MATLAB® by making use of functions such as ode45, which utilizes Runge–Kutta techniques. They can also be solved analytically using the symbolic algebra toolbox, which utilizes the MuPad engine. Simulink® uses similar methods, but they are transparent to the user. Instead of programming equations directly, a visual model is created by collecting appropriate Simulink® blocks and connecting them together, using a graphical user interface.

16.2 GETTING STARTED

To start Simulink®, open MATLAB® and type

    simulink®

into the command window. (Or select the Simulink® icon from the Shortcut toolbar as shown in Figure 16.1).

The Simulink® Library Browser opens, showing the available libraries of blocks used to create a Simulink® model (Figure 16.2). The browser is the location where you'll select blocks and drag them into the model workspace. Spend a few minutes exploring the browser. To view the blocks available in each library, either select the library from the left-hand pane or double click on the icons in the right-hand pane. In particular, take a look at the Commonly Used Blocks library—the Source and Sink libraries and the Math Operations library.

Simulink®'s strength is in solving complex dynamic systems, but before we try to work on a complex system, it would be better to build some very simple static

Figure 16.1
Access Simulink® either from the command window, or by selecting the icon from the shortcut toolbar.

Figure 16.2
The Simulink® Library
Browser contains numerous
blocks that are used to
create a Simulink® model.

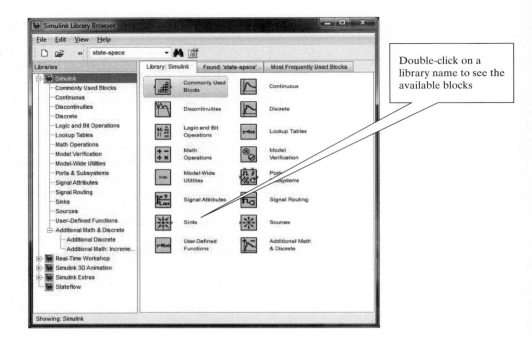

Figure 16.2
The Simulink® Library
Browser contains numerous
blocks that are used to
create a Simulink® model.

models to demonstrate the problem-solving process. To create a new model, select
File → New → Model from the browser window. The model window opens on top
of the library browser (Figure 16.3). For convenience, resize the library browser
window and the model window so that you can see both on the computer screen.
You'll also want to keep the MATLAB® desktop open, but resize it so that it also
fits on your computer screen without overlapping the other windows. See, for
example, Figure 16.4.

Our first model will simply add two numbers. From either the Source library or
the Commonly Used Blocks library, click and drag the constant block into the
model window. Repeat the process, so that you have two copies of the constant
block in the model, as shown in Figure 16.5.

Now drag the sum block into the model. It is found both in the Commonly
Used Blocks library and the Math Operations library. Notice that the sum block has
two "ports." You can draw connections between the constants and the sum block by

Figure 16.3
The model window is the
workspace where
Simulink® models are
created and executed.

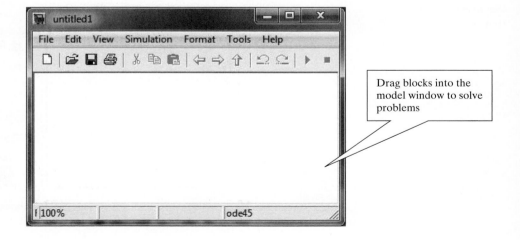

Figure 16.4

Simulink® uses multiple windows. Arrange them on your computer desktop so that you can easily drag blocks from the Simulink® Library Browser to the model window.

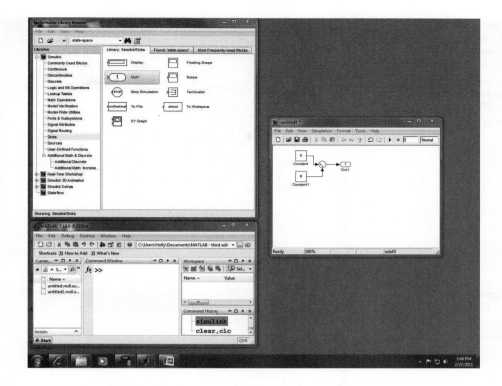

Figure 16.5

Two copies of the constant block were added to the model.

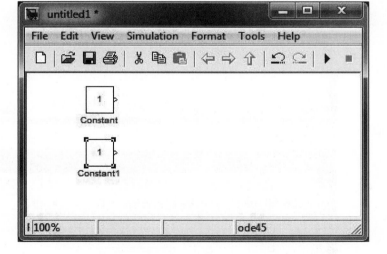

clicking and dragging between the ports, as shown in Figure 16.6. You should notice that the cursor changes to a cross-hair as you connect the ports. The model we've created thus far just adds 1 + 1, and doesn't display the answer. We'll need to modify the constant blocks to specify a value different from the default, which in this case is 1. Double click on each constant block, and change the "constant value" field, for example, to 5 in the top block and 6 in the bottom block.

To add a display option, look in the sink library. For this case, the display block is all we need, so drag it to the model and connect it to the output port of the sum block. The last thing we need to do before running the model is to adjust the simulation time, from the box on the menu bar (see Figure 16.7). Since nothing in this

Figure 16.6
The constants are connected to the sum block. Change the values in the constant blocks by double clicking and modifying the "constant value" field.

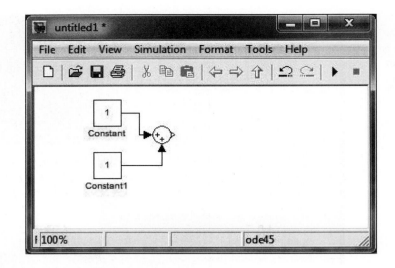

Figure 16.7
(a) The completed model.
(b) Results are shown in the display block.

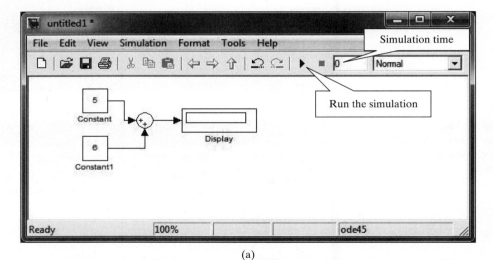

(a)

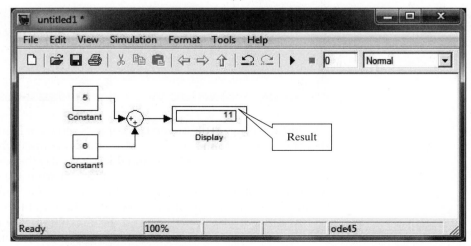

(b)

Figure 16.8

The sum block can be used to perform subtraction operations, as well as for adding more than two input values.

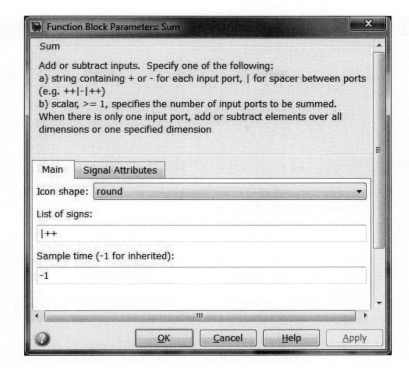

calculation will change with time, we can change the value to zero. Run the simulation by selecting the run button on the toolbar (the black triangle) or by selecting Simulation → Start from the menu bar.

Save this model in the usual way, by selecting File → Save and adding an appropriate name. The files are stored with the extension, .mdl.

As the sum block serves both the addition and subtraction functions, you could use this same model to perform subtraction operations. Double click the sum block in the model and the block parameter window opens, as shown in Figure 16.8.

The block description is located near the top of the window, and provides information on how to use the block—in this case the sum block. This description includes instructions to change the block into a "subtraction block" by changing the input from |++ to |+−. We could also adjust the block to add three inputs by changing the list of signs field to the number 3. Adjust your model and run it several more times as you explore the possibilities for the sum block.

HINT

Simulink® includes a "subtraction" block, but if you open its block parameter window you'll notice the block title is "sum."

The previous example was trivial. A slightly more complex model, with results that change with time, is described in Example 16.1.

EXAMPLE 16.1

RANDOM NUMBERS

As we saw in Example 3.5, random numbers can be used to simulate the noise we hear on the radio as static. Although we could solve a similar problem in MATLAB®, let's use Simulink®. In this case, instead of a music file use a sine wave as the input to which we want to add the noise, using the following equation:

$$y = 5*\sin(2t) + \textit{noise}$$

The noise should be the result of a uniform random number generator, with a range of 0 to 1.

1. **State the Problem**
   Create a Simulink® model of the equation

   $$y = 5*\sin(2t) + \textit{noise}$$

   where the noise is based on a random number.

2. **Describe the Input and Output**

   Input Use Simulink®'s built-in sine wave generator to provide the sine wave. Use Simulink®'s built-in random number generator to simulate the noise.

   Output View the results using the Simulink® Scope block.

3. **Develop a Hand Example**
   In this case, since we are well versed in MATLAB®, a MATLAB® solution will substitute for a hand example.

```
t=0:0.1:10;
noise = rand(size(t));
y=5*sin(2*t)+noise;
plot(t,y)
title('A sine wave with noise added')
xlabel('time,s'), ylabel ('function value')
```

   which results in the plot shown in Figure 16.9.

4. **Develop a Simulink® Solution**
   Simulink® includes blocks for creating both sine waves, and for uniform random number generators. You can find both in the Source Library. You'll also

Figure 16.9
Adding noise to a sine wave can be accomplished using MATLAB®, as well as Simulink®.

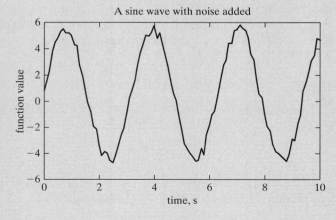

need to include an add block. Finally, add a scope (the name comes from the word "oscilloscope") to view the plotted result. Your model should resemble the one shown in Figure 16.10. Notice that the time field in the upper right corner of the model is set to 10 seconds, and that two additional scopes were added so that we can observe the behavior of the sine wave generator, the random number generator, and the combined output.

The model specifies only a sine wave, not the entire sine portion of the expression, 5*sin(2t). Open the Sine Wave block by double clicking on the icon inside the model. The Source Block Parameters window opens (as shown in Figure 16.11), allowing us to specify the amplitude, the frequency and additional parameters as needed. By changing the amplitude to 5 and the frequency to 2, the block now represents the first term in our equation.

Similarly, the random number generator parameter window can be modified to specify a minimum value of 0 and a maximum value of 1. Run the model by selecting the black start simulation triangle, or by selecting Simulation → Start. To view the output, double click on each of the scopes. Scale the images by selecting the binocular icon as shown in Figure 16.12, which shows the results of the combined inputs.

5. Test the Solution

Compare the results to those found with the MATLAB® solution. We could also revise the model, so that the results are sent to MATLAB® by replacing the scope for the combined output with the simout block, as shown in Figure 16.13. The simout block is found in the sinks library. Before running the model, you'll need to modify the block parameters (double click on the block to open the window). Change the Save format from Structure to Array. Re-execute the model, and observe that two new arrays have appeared in the MATLAB® workspace window, `simout` and `tout`, both of which are 101x1 double precision arrays. The values in the arrays can now be used for plotting, or in other calculations.

Figure 16.10
Simulink® model to add noise to a sine wave.

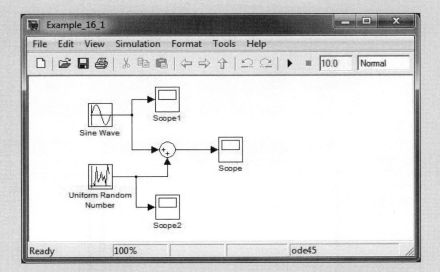

(continued)

Figure 16.11
(a) The Sine Wave parameter window.
(b) The Uniform Random Number parameter window. The Source Block Parameter window for each Simulink® block allows the user to modify the default values of the input parameters. Access the parameter window by double clicking on the block in the model window.

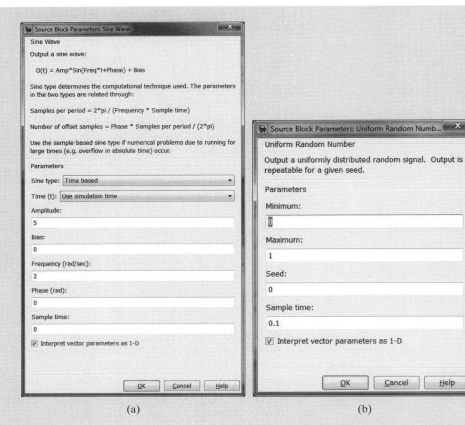

(a)

(b)

Figure 16.12
The scope output from the three oscilloscopes specified in the Simulink® model.

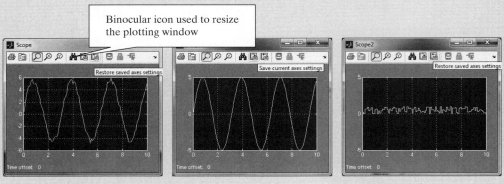

Binocular icon used to resize the plotting window

Figure 16.13
The simout block sends simulation results to the MATLAB® workspace, where they can be used in other calculations as needed.

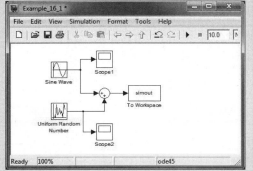

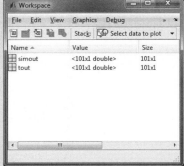

16.3 SOLVING DIFFERENTIAL EQUATIONS WITH SIMULINK®

Thus far the problems we've solved by creating models in Simulink® could have been solved more readily in MATLAB®. Where Simulink® really excels is in solving differential equations. In general, a differential equation includes a dependent variable, an independent variable, and the derivative of the dependent variable with respect to the independent variable. For example,

$$\frac{dy}{dt} = t^2 + y$$

is a differential equation. In this case y is the dependent variable, t is the independent variable, and dy/dt is the derivative with respect to t. In function notation,

$$\frac{dy}{dt} = f(t, y)$$

To find y, we could integrate

$$y = \int \frac{dy}{dt} dt = \int f(t, y) dt$$

This equation has an infinite number of solutions, unless the initial value of y is defined. For this problem we'll set $y(0) = 0$.

To solve this problem in Simulink®, create a model by dragging the appropriate blocks onto the model window, and connecting them as shown in Figure 16.14.

The blocks include the following:

- A clock, to generate times (Source library)
- A math function block, modified in the parameter window to square the block input (Math Operations library)
- A sum block (Commonly Used Blocks library)
- An integrator block (Continuous library)
- A scope block (Sink library)

Adjust the integrator block in the parameter window so that the initial condition is 0. The scope output, after running the model, is shown in Figure 16.15. (You may need to click on the binocular icon to see the entire plot in the scope screen.)

Figure 16.14
Simulink® model to solve the differential equation $\frac{dy}{dt} = t^2 + y$.

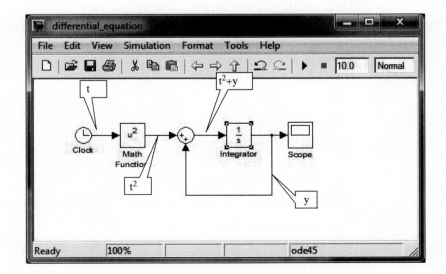

Figure 16.15
A plot of the solution to the ordinary differential equation, $\mathbf{d}y/\mathbf{d}t = t^2 + y$, with $y(0) = 0$. (a) Plot created with Simulink®. (b) Plot created in MATLAB® using symbolic algebra.

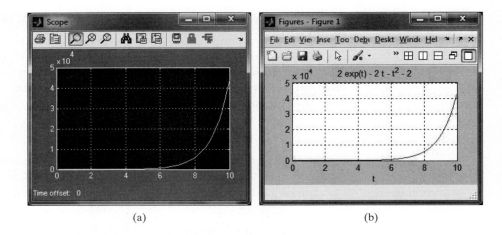

(a) (b)

An alternative approach to this problem might be to use MATLAB®'s symbolic algebra capability to solve the same problem, as discussed in an earlier chapter. Because this is a simple differential equation, the `dsolve` function can be used.

```
y = dsolve('Dy = t^2 + y','y(0) = 0')
ezplot(y,[0,10])
```

The solution to the differential equation is shown analytically in the command window as

```
y =
2*exp(t) - 2*t - t^2 - 2
```

and the plot is shown in Figure 16.15b.

EXAMPLE 16.2

VELOCITY OF A FALLING OBJECT

Consider an object, falling toward the ground. A widely reported equation describing the resulting velocity is the differential equation:

$$\frac{\mathbf{d}v}{\mathbf{d}t} = g - \frac{c}{m}v^2$$

where

g is the acceleration due to gravity

v is the velocity

m is the mass

c is the second-order drag coefficient

Solve this equation by finding velocity as a function of time, for the first 15 seconds.

1. **State the Problem**
   Use Simulink® to find the time versus velocity behavior of a falling object.

2. Describe the Input and Output

   Input $g = 9.81 \text{ m/s}^2$
                $m = 70 \text{ kg}$
                $c = 0.3 \text{ kg/m}$
                $v(0) = 0 \text{ m/s}$

   Output   Plot of velocity versus time from 0 to 15 seconds

3. Develop a Hand Example
   Given that the initial velocity is 0, we would expect that the velocity would rapidly increase—but would eventually level off and reach a terminal value. We would expect a plot much like the sketch shown in Figure 16.16.

4. Develop a Simulink® Solution
   The Simulink® model is shown in Figure 16.17 along with the resulting plot displayed on the scope. It is composed of

   - Three constant blocks
   - Both a divide and product block
   - An add block
   - An integrator block
   - Math function block, set to square the output of the integrator block

   As you build the model, you will notice that some of the blocks are reversed from their standard orientation. You can accomplish this by placing the block into the model, right clicking the icon, and selecting Format from the drop-down menu. There are a number of choices that allow the user to select a convenient block orientation. In particular, notice that the math function block has been flipped to accommodate the data flow leaving the integrator block. Also notice that the time block has been set to 15 seconds.

   If a simout block is used to replace the scope block, the output data is sent to MATLAB®, where it could be used in other programs, or plotted in the usual manner.

5. Test the Solution
   Because we are well versed in MATLAB®, we could also solve the problem using MATLAB® and the tools found in the symbolic algebra toolbox.

   ```
   clear,clc
   y = dsolve('Dv = g-c/m*v^2','v(0) = 0')
   y = subs(y,{'g','c','m'},{9.81,0.30,70})
   ezplot(y,[0,15])
   title('A falling object'), xlabel('time,s')
     ylabel('velocity, m/s')
   ```

   The resulting plot is shown in Figure 16.18, and corresponds well to the scope output from the Simulink® model.

Figure 16.16
Projected behavior of the velocity versus time curve, for a falling object.

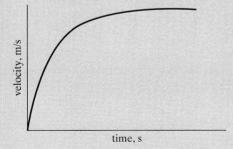

time, s

(*continued*)

Figure 16.17
Simulink® model to solve the falling object problem.

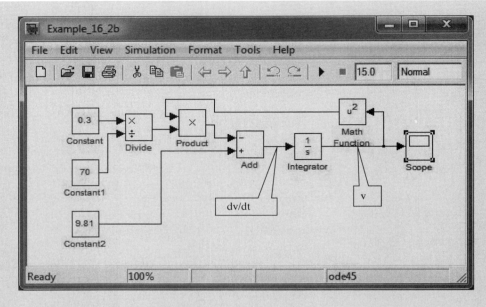

Figure 16.18
The velocity plot for a falling object. (a) Plot created using Simulink®. (b) Plot created using MATLAB®'s symbolic algebra tools. Notice that in both cases the velocity levels-off around 47 m/s.

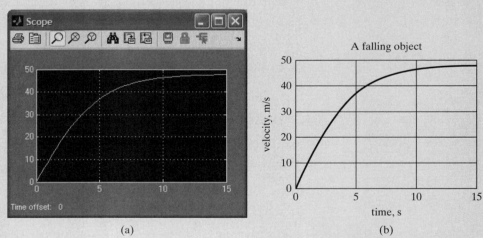

(a)   (b)

EXAMPLE 16.3

POSITION OF A FALLING OBJECT

In the previous example, we solved the following differential equation for velocity as a function of time.

$$\frac{dv}{dt} = g - \frac{c}{m}v^2$$

However, velocity can also be described as a derivative; it is the rate of change of position with time.

$$v = \frac{dx}{dt}$$

We could reformulate the velocity equation in terms of position as

$$\frac{d^2x}{dt^2} = g - \frac{c}{m}\left(\frac{dx}{dt}\right)^2$$

Use Simulink® to create a plot showing how far the object has fallen, as a function of time.

1. State the Problem
   Solve the second-order differential equation

$$\frac{d^2x}{dt^2} = g - \frac{c}{m}\left(\frac{dx}{dt}\right)^2$$

   for x as a function of t.

2. Describe the Input and Output

   Input

   $g = 9.81$ m/s^2
   $m = 70$ kg
   $c = 0.3$ kg/m
   $v(0) = 0$
   $x(0) = 0$
   $t = 0{-}15$ seconds

   Output   Create a plot, showing how x changes with time, using the Simulink® Scope. Also send the results to MATLAB®.

3. Develop a Hand Example
   As the object falls, it eventually reaches a terminal velocity, as shown in the previous example problem. At that point, x should be increasing at a steady rate. A sketch of the expected behavior is shown in Figure 16.19.

4. Develop a Simulink® Solution
   The model created in the previous example can be expanded by adding an integration block, and by splitting the output into feeds leading to both the scope and the simout block (see Figure 16.20). Be sure to adjust the simout block to report the data as an array.
   The plot created in the scope is shown in Figure 16.21a.

5. Test the Solution
   Once again we could use MATLAB® to solve this second-order differential equation using the symbolic algebra toolbox.

```
x = dsolve('D2x = g-c/m*Dx^2','x(0) = 0','Dx(0) = 0')
x = subs(x,{'g','c','m'},{9.81,0.30,70})
ezplot(x,[0,15])
title('A falling object'), xlabel('time,s'), ylabel('position, m')
```

Figure 16.19
Expected position of an object reaching terminal velocity.

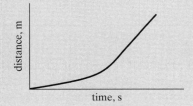

distance, m

time, s

(*continued*)

Figure 16.20
Simulink® model to solve the second-order differential equation, $\dfrac{d^2x}{dt^2} = g - \dfrac{c}{m}\left(\dfrac{dx}{dt}\right)^2$.

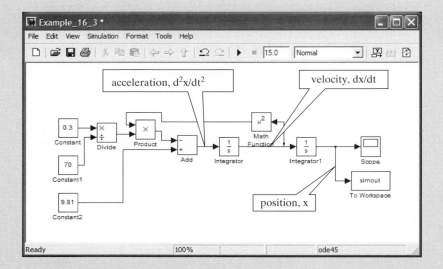

Figure 16.21
The position of a falling object. (a) Results from the Simulink® Scope. (b) Results from MATLAB® using a symbolic algebra solution.

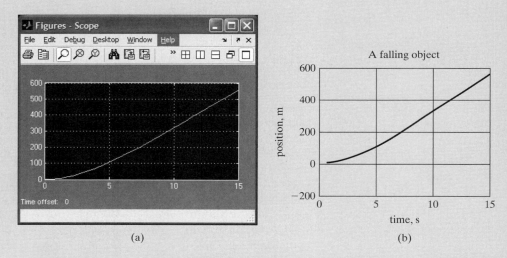

The resulting plot (Figure 16.21b) matches the scope output, and thus verifies our calculations.

By using both Simulink® and a symbolic algebra approach, we can develop confidence in the Simulink® solutions. Not all problems can be solved symbolically, so having both approaches available is important. This example was inspired by Steven Chapra's use of a skydiver to illustrate techniques to solve differential equations, in 'Numerical Methods for Engineers', McGraw-Hill, 2010.

SUMMARY

Simulink® is part of the MATLAB® family of programs. It uses a graphical user interface to facilitate the development of models that represent real systems. Simulink® is especially useful for modeling dynamic systems—those that can be mathematically described as differential equations.

Simulink® relies on a large library of blocks, which can be combined to solve a wide variety of problems. Its visual approach offers an alternative to building m-file programs using the numerical techniques described in earlier chapters. However, these same techniques (for example, ode45) are used by Simulink® when its models are executed.

The MATLAB® help function includes an extensive tutorial on using Simulink®, including many examples.

Command and Function	
simulink®	opens the Simulink® library browser

KEY TERMS

dynamic systems	differential equations	block
analog computers	model	

PROBLEMS

16.1 The sinc function is often used in electrical engineering applications. It is defined as

$$\text{sinc}(x) = \frac{\sin(x)}{x}$$

Use Simulink® to model the behavior of the sinc function, from -20 to 20 seconds. Display your results using Simulink®'s scope block. To adjust the simulation time, in the model window menu bar select Simulation → Configuration Parameters.

16.2 The equation of a circle can be represented parametrically as

$$x = \sin(t)$$
$$y = \cos(t)$$

where t varies from 0 to 2*pi. Create a Simulink® model to parametrically graph a circle using the xy graph block found in the sink library. To model cosine, you will need to modify the Sin block.

16.3. The multiplexer block (Mux) accepts multiple inputs that can then be sent to a scope block to create a graph with multiple signal plots. Use two sine blocks to create a signal representing the $\sin(t)$ and the $\cos(t)$. Combine the signals with the Mux block (found in the Commonly Used Blocks library), and plot the results from 0 to 20 seconds, using a Scope block.

16.4 The derivative block finds the derivative (rate of change) of the incoming signal. Create a Simulink® model that finds the derivative of

$$y = \frac{1}{t}$$

and which plots both y and dy/dt in the scope window, for times from 0 to 10 seconds. You'll need a Clock (time) block, the Math Function block, the Derivative block, and a Mux block, in addition to the Scope block.

Applications

16.5 The change in internal energy (kJ/kmol) of an ideal gas over a given temperature range can be represented by the equation:

$$\Delta u = \int_{T_1}^{T_2} (a - R + bT + cT^2 + dT^3)\,\mathrm{d}T$$

where T is the temperature in kelvin.
For nitrogen, the constant values are:

$$a = 28.90$$
$$b = -0.1571 \times 10^{-2}$$
$$c = 0.8081 \times 10^{-5}$$
$$d = -2.873 \times 10^{-9}$$
$$R = 8.31447 \text{ kJ/kmol K}$$

Use Simulink® to plot the value of the change in internal energy (Δu) between 0 K and a temperature of 1000 K. (Use the time block to simulate the values of T.)
Data Source: B.G. Kyle, *Chemical and Process Thermodynamics* (Englewood Cliffs, NJ: Prentice Hall, 1984).

16.6 Newton's law of cooling tells us that the rate at which an object cools is proportional to the difference in temperature between the object and the surroundings (Figure P16.6). In other words,

$$\frac{\mathrm{d}T}{\mathrm{d}t} = k(T - T_{\text{surroundings}})$$

where k is a proportionality constant. If for a cup of hot coffee, the surroundings temperature is 70°F, the constant is 0.5 min^{-1} and the initial temperature is 110°F, plot the temperature of the object as a function of time for 10 minutes.

16.7 The rate of a chemical reaction is related to the concentration of the reactants. For example, a first-order reaction would have the following relationship between the rate of change of the reactant and the concentration of said reactant:

$$\frac{\mathrm{d}[A]}{\mathrm{d}t} = -k*[A]$$

A slightly more complicated reaction might be dependant upon the square of the reactant concentration:

$$\frac{\mathrm{d}[A]}{\mathrm{d}t} = -k*[A]^2$$

Model the change in concentration, $[A]$, with time using Simulink® for both the first- and second-order reaction problems. Assume $k = 0.1$ min^{-1} for the first-order reaction and $k = 0.1$ 1/mol min for the second-order reaction. The initial concentration, $[A]$, is 5 mol/l. Display the results using a Simulink® Scope block. (Choose an appropriate length of time for the simulation, based upon your intermediate results.)

16.8 Blasius showed in 1908 that the solution to the incompressible flow field in a laminar boundary layer on a flat plate is given by the solution of the following third-order ordinary nonlinear differential equation.

$$2\frac{\mathrm{d}^3 f}{\mathrm{d}\eta^3} + f\frac{\mathrm{d}^2 f}{\mathrm{d}\eta^2} = 0$$

Figure P16.6
A cup of hot coffee cools according to Newton's law of cooling.

To solve this system for f, first solve for the highest order derivative.

$$\frac{\mathbf{d}^3 f}{\mathbf{d}\eta^3} = -0.5 f \frac{\mathbf{d}^2 f}{\mathbf{d}\eta^2}$$

Now use Simulink® to create a model. You'll need three integration blocks plus a multiplier and a gain block (the gain block multiplies by a constant), in addition to a scope block to view the output. The initial conditions are:

$$\frac{\mathbf{d}^2 f(0)}{\mathbf{d}t^2} = 0.332$$

$$\frac{\mathbf{d} f(0)}{\mathbf{d}t} = 0$$

$$f(0) = 0$$

16.9 If a projectile such as a bullet or a rocket is fired vertically, the only force acting on it is the force due to gravity. A force balance yields the equation:

$$\frac{\mathbf{d}^2 x}{\mathbf{d}t^2} = -g \left(\frac{R^2}{(R + x)^2} \right)$$

where
x is the vertical distance measured from the surface of the earth in meters
R is the radius of the earth, 6.4×10^6 m
g is the acceleration due to gravity, 9.81 m/s²

Model this equation using Simulink®. Display a graph of the projectile height, x, as a function of time. Assume that the initial height is 0, and the initial velocity is 100 m/s. ($dx/dt = 100$ at time = 0.)

16.10 The motion of a pendulum (Figure P16.10) can be modeled with an ordinary second-order differential equation as:

$$\frac{\mathbf{d}^2 \theta}{\mathbf{d}t^2} = -\frac{g}{L} \sin(\theta)$$

where
θ is the vertical angle
g is the acceleration due to gravity, 9.81 m/s²
L is the length of the pendulum, 2 m

Model the behavior of the pendulum (i.e., the angle as a function of time) with Simulink®. Assume the initial angle, θ, is 30° ($\pi/6$ radian) and that the initial angular velocity is 0 ($d\theta/dt = 0$).

16.11 Consider the simple RC series circuit shown in Figure P16.11.
At time zero the switch is turned on, allowing to current to flow. Assuming that constant voltage is applied, the response of the circuit can be described by the differential equation:

$$R \frac{\mathbf{d}i}{\mathbf{d}t} + \frac{i}{C} = 0$$

which can be rearranged to

$$\frac{\mathbf{d}i}{\mathbf{d}t} = -\frac{1}{R * C} * i$$

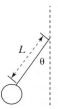

Figure P16.10
The Motion of a pendulum is described by a second-order differential equation.

Use Simulink® to model the system response, assuming that $R = 100,000\ \Omega$ and $C = 1 \times 10^{-6}$ F. Calculate the initial current value from Ohm's law

$$V = iR$$

with a constant voltage value of 5 V applied to the system.

Figure P16.11
A simple RC series circuit.

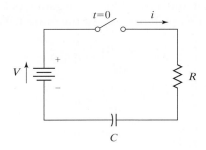

16.12 The current, i, flowing through the circuit shown in Figure P16.12, can be described by a second-order differential equation:

$$L\frac{\mathbf{d}^2 i}{\mathbf{d}t^2} + R\frac{\mathbf{d}i}{\mathbf{d}t} + \frac{1}{C}i = 0$$

which can be rearranged to give

$$\frac{\mathbf{d}^2 i}{\mathbf{d}t^2} = -\frac{R}{L}\frac{\mathbf{d}i}{\mathbf{d}t} - \frac{1}{L * C}i$$

Figure P16.12
A simple RCL circuit can be described by a second-order differential equation.

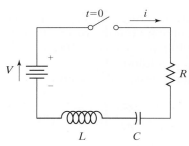

The behavior of this system depends upon the relative values of L, C, and R (the inductance, capacitance, and the resistance). When

$$R^2 > \frac{4L}{C} \quad \text{the system is "over-damped," when}$$

$$R^2 < \frac{4L}{C} \quad \text{the system is "under-damped." And when}$$

$$R^2 = \frac{4L}{C} \quad \text{the system is "critically damped."}$$

Use Simulink® to model the system response, assuming that $R = 100,000\ \Omega$ and $C = 1 \times 10^{-6}$ F. Select values of L to meet each of the damping conditions described above. Calculate the initial current value from Ohm's law

$$V = iR$$

with a constant voltage value of 5 V applied to the system.

A Special Characters, Commands, and Functions

The tables presented in this appendix are grouped according to category, which roughly parallels the chapter organization.

Special Characters	Matrix Definition	Chapter
[ ]	forms matrices	Chapter 2
( )	used in statements to group operations;	Chapter 2
	used with a matrix name to identify specific elements	
,	separates subscripts or matrix elements	Chapter 2
;	separates rows in a matrix definition;	Chapter 2
	suppresses output when used in commands	
:	used to generate matrices;	Chapter 2
	indicates all rows or all columns	

Special Characters	Operators Used in MATLAB® Calculations (Scalar and Array)	Chapter
=	assignment operator: assigns a value to a memory location; not the same as an equality	Chapter 2
%	indicates a comment in an M-file	Chapter 2
%%	creates a cell, used to organize code	Chapter 2
+	scalar and array addition	Chapter 2
-	scalar and array subtraction	Chapter 2
*	scalar multiplication and multiplication in matrix algebra	Chapter 2
.*	array multiplication (dot multiply or dot star)	Chapter 2
/	scalar division and division in matrix algebra	Chapter 2
./	array division (dot divide or dot slash)	Chapter 2
^	scalar exponentiation and matrix exponentiation in matrix algebra	Chapter 2
.^	array exponentiation (dot power or dot carat)	Chapter 2
...	ellipsis: continued on the next line	Chapter 4
[]	empty matrix	Chapter 4

Commands	Formatting	Chapter
format +	sets format to plus and minus signs only	Chapter 2
format compact	sets format to compact form	Chapter 2
format long	sets format to 14 decimal places	Chapter 2
format long e	sets format to 14 exponential places	Chapter 2
format long eng	sets format to engineering notation with 14 decimal places	Chapter 2
format long g	allows MATLAB® to select the best format (either fixed point or floating point), using 14 decimal digits	Chapter 2
format loose	sets format back to default, noncompact form	Chapter 2
format short	sets format back to default, 4 decimal places	Chapter 2
format short e	sets format to 4 exponential places	Chapter 2
format short eng	sets format to engineering notation with 4 decimal places	Chapter 2
format short g	allows MATLAB® to select the best format (either fixed point or floating point), using 4 decimal digits	Chapter 2
format rat	sets format to rational (fractional) display	Chapter 2

Commands	Basic Workspace Commands	Chapter
ans	default variable name for results of MATLAB® calculations	Chapter 2
clc	clears command screen	Chapter 2
clear	clears workspace	Chapter 2
diary	saves both commands issued in the workspace and the results to a file	Chapter 2
exit	terminates MATLAB®	Chapter 2
help	invokes help utility	Chapter 2
load	loads matrices from a file	Chapter 2
quit	terminates MATLAB®	Chapter 2
save	saves variables in a file	Chapter 2
who	lists variables in memory	Chapter 2
whos	lists variables and their sizes	Chapter 2
help	opens the help function	Chapter 3
helpwin	opens the windowed help function	Chapter 3
clock	returns the time	Chapter 3
date	returns the date	Chapter 3
intmax	returns the largest possible integer number used in MATLAB®	Chapter 3
intmin	returns the smallest possible integer number used in MATLAB®	Chapter 3
realmax	returns the largest possible floating-point number used in MATLAB®	Chapter 3
realmin	returns the smallest possible floating-point number used in MATLAB®	Chapter 3
ascii	indicates that data should be saved in a standard ASCII format	Chapter 2
pause	pauses the execution of a program until any key is hit	Chapter 5

Special Functions	Functions with Special Meaning That Do Not Require an Input	Chapter
pi	numeric approximation of the value of π	Chapter 2
eps	smallest difference recognized	Chapter 3
I	imaginary number	Chapter 3
Inf	Infinity	Chapter 3
j	imaginary number	Chapter 3
NaN	not a number	Chapter 3

Functions	Elementary Math	Chapter
abs	computes the absolute value of a real number or the magnitude of a complex number	Chapter 3
erf	calculates the error function	Chapter 3
exp	computes the value of e^x	Chapter 3
factor	finds the prime factors	Chapter 3
factorial	calculates the factorial	Chapter 3
gcd	finds the greatest common denominator	Chapter 3
isprime	determines whether a value is prime	Chapter 3
isreal	determines whether a value is real or complex	Chapter 3
lcn	finds the least common denominator	Chapter 3
log	computes the natural logarithm, or log base $e(\log_e)$	Chapter 3
log10	computes the common logarithm, or log base $10(\log_{10})$	Chapter 3
log2	computes the log base $2(\log_2)$	Chapter 3
nthroot	finds the real nth root of the input matrix	Chapter 3
primes	finds the prime numbers less than the input value	Chapter 3
prod	multiplies the values in an array	Chapter 3
rats	converts the input to a rational representation (i.e., a fraction)	Chapter 3
rem	calculates the remainder in a division problem	Chapter 3
sign	determines the sign (positive or negative)	Chapter 3
sqrt	calculates the square root of a number	Chapter 3
sum	sums the values in an array	Chapter 3

Functions	Trigonometry	Chapter
asin	computes the inverse sine (arcsine)	Chapter 3
asind	computes the inverse sine and reports the result in degrees	Chapter 3
cos	computes the cosine	Chapter 3
sin	computes the sine, using radians as input	Chapter 3
sind	computes the sine, using angles in degrees as input	Chapter 3
sinh	computes the hyperbolic sine	Chapter 3
tan	computes the tangent, using radians as input	Chapter 3

MATLAB® includes all of the trigonometric functions; only those specifically discussed in the text are included here.

Functions	Complex Numbers	Chapter
abs	computes the absolute value of a real number or the magnitude of a complex number	Chapter 3
angle	computes the angle when complex numbers are represented with polar coordinates	Chapter 3
complex	creates a complex number	Chapter 3
conj	creates the complex conjugate of a complex number	Chapter 3
imag	extracts the imaginary component of a complex number	Chapter 3
isreal	determines whether a value is real or complex	Chapter 3
real	extracts the real component of a complex number	Chapter 3

Functions	Rounding	Chapter
ceil	rounds to the nearest integer toward positive infinity	Chapter 3
fix	rounds to the nearest integer toward zero	Chapter 3
floor	rounds to the nearest integer toward minus infinity	Chapter 3
round	rounds to the nearest integer	Chapter 3

Functions	Data Analysis	Chapter
cumprod	computes the cumulative product of the values in an array	Chapter 3
cumsum	computes the cumulative sum of the values in an array	Chapter 3
length	determines the largest dimension of an array	Chapter 3
max	finds the maximum value in an array and determines which element stores the maximum value	Chapter 3
mean	computes the average of the elements in an array	Chapter 3
median	finds the median of the elements in an array	Chapter 3
min	finds the minimum value in an array and determines which element stores the minimum value	Chapter 3
mode	finds the most common number in an array	Chapter3
nchoosek	finds the number of possible combinations when a subgroup of k values is chosen from a group of n values	Chapter 3
numel	determines the total number of elements in an array	Chapter 3
size	determines the number of rows and columns in an array	Chapter 3
sort	sorts the elements of a vector	Chapter 3
sortrows	sorts the rows of a vector on the basis of the values in the first column	Chapter 3
prod	multiplies the values in an array	Chapter 3
sum	sums the values in an array	Chapter 3
std	determines the standard deviation	Chapter 3
var	computes the variance	Chapter 3

Functions	Random Numbers	Chapter
rand	calculates evenly distributed random numbers	Chapter 3
randn	calculates normally distributed (Gaussian) random numbers	Chapter 3

Functions	Matrix Formulation, Manipulation, and Analysis	Chapter
meshgrid	maps vectors into a two-dimensional array	Chapters 4 and 5
diag	extracts the diagonal from a matrix	Chapter 4
fliplr	flips a matrix into its mirror image from left to right	Chapter 4
flipud	flips a matrix vertically	Chapter 4
linspace	linearly spaced vector function	Chapter 2
logspace	logarithmically spaced vector function	Chapter 2
cross	computes the cross product	Chapter 9
det	computes the determinant of a matrix	Chapter 9
dot	computes the dot product	Chapter 9
inv	computes the inverse of a matrix	Chapter 9
rref	uses the reduced row echelon format scheme for solving a series of linear equations	Chapter 9

Functions	Two-Dimensional Plots	Chapter
bar	generates a bar graph	Chapter 5
barh	generates a horizontal bar graph	Chapter 5
contour	generates a contour map of a three-dimensional surface	Chapter 5
comet	draws an x–y plot in a pseudo animation sequence	Chapter 5
fplot	creates an x–y plot on the basis of a function	Chapter 5
hist	generates a histogram	Chapter 5
loglog	generates an x–y plot with both axes scaled logarithmically	Chapter 5
pcolor	creates a pseudo color plot similar to a contour map	Chapter 5
pie	generates a pie chart	Chapter 5
plot	creates an x–y plot	Chapter 5
plotyy	creates a plot with two y-axes	Chapter 5
polar	creates a polar plot	Chapter 5
semilogx	generates an x–y plot with the x-axis scaled logarithmically	Chapter 5
semilogy	generates an x–y plot with the y-axis scaled logarithmically	Chapter 5

Functions	Three-Dimensional Plots	Chapter
bar3	generates a three-dimensional bar graph	Chapter 5
bar3h	generates a horizontal three-dimensional bar graph	Chapter 5
comet3	draws a three-dimensional line plot in a pseudo animation sequence	Chapter 5
mesh	generates a mesh plot of a surface	Chapter 5
peaks	creates a sample three-dimensional matrix used to demonstrate graphing functions	Chapter 5
pie3	generates a three-dimensional pie chart	Chapter 5
plot3	generates a three-dimensional line plot	Chapter 5
sphere	sample function used to demonstrate graphing	Chapter 5
surf	generates a surface plot	Chapter 5
surfc	generates a combination surface and contour plot	Chapter 5

Special Characters	Control of Plot Appearance	Chapter
Indicator	**Line Type**	
-	Solid	Chapter 5
:	dotted	Chapter 5
-.	Dash-dot	Chapter 5
--	dashed	Chapter 5
Indicator	**Point Type**	
.	point	Chapter 5
o	circle	Chapter 5
x	x-mark	Chapter 5
+	Plus	Chapter 5
*	Star	Chapter 5
s	square	Chapter 5
d	diamond	Chapter 5
v	triangle down	Chapter 5
^	triangle up	Chapter 5
<	triangle left	Chapter 5
>	triangle right	Chapter 5
p	pentagram	Chapter 5
h	hexagram	Chapter 5
Indicator	**Color**	
b	blue	Chapter 5
g	green	Chapter 5
r	red	Chapter 5
c	cyan	Chapter 5
m	Magenta	Chapter 5
y	Yellow	Chapter 5
k	Black	Chapter 5

Functions	Figure Control and Annotation	Chapter
axis	freezes the current axis scaling for subsequent plots or specifies the axis dimensions	Chapter 5
axis equal	forces the same scale spacing for each axis	Chapter 5
colormap	color scheme used in surface plots	Chapter 5
figure	opens a new figure window	Chapter 5
gtext	Similar to text. The box is placed at a location determined interactively by the user by clicking in the figure window	Chapter 5
grid	adds a grid to the current plot only	Chapter 5
grid off	turns the grid off	Chapter 5
grid on	adds a grid to the current and all subsequent graphs in the current figure	Chapter 5
hold off	instructs MATLAB® to erase figure contents before adding new information	Chapter 5
hold on	instructs MATLAB® not to erase figure contents before adding new information	Chapter 5
legend	adds a legend to a graph	Chapter 5
shading flat	shades a surface plot with one color per grid section	Chapter 5
shading interp	shades a surface plot by interpolation	Chapter 5
subplot	divides the graphics window up into sections available for plotting	Chapter 5
text	adds a text box to a graph	Chapter 5
title	adds a title to a plot	Chapter 5
xlabel	adds a label to the x-axis	Chapter 5
ylabel	adds a label to the y-axis	Chapter 5
zlabel	adds a label to the z-axis	Chapter 5

Functions	Figure Color Schemes	Chapter
autumn	optional colormap used in surface plots	Chapter 5
bone	optional colormap used in surface plots	Chapter 5
colorcube	optional colormap used in surface plots	Chapter 5
cool	optional colormap used in surface plots	Chapter 5
copper	optional colormap used in surface plots	Chapter 5
flag	optional colormap used in surface plots	Chapter 5
hot	optional colormap used in surface plots	Chapter 5
hsv	optional colormap used in surface plots	Chapter 5
jet	default colormap used in surface plots	Chapter 5
pink	optional colormap used in surface plots	Chapter 5
prism	optional colormap used in surface plots	Chapter 5
spring	optional colormap used in surface plots	Chapter 5
summer	optional colormap used in surface plots	Chapter 5
white	optional colormap used in surface plots	Chapter 5
winter	optional colormap used in surface plots	Chapter 5

Functions and Special Characters	Function Creation and Use	Chapter
addpath	adds a directory to the MATLAB® search path	Chapter 6
function	identifies an M-file as a function	Chapter 6
nargin	determines the number of input arguments in a function	Chapter 6
nargout	determines the number of output arguments from a function	Chapter 6
pathtool	opens the interactive path tool	Chapter 6
varargin	indicates that a variable number of arguments may be input to a function	Chapter 6
@	identifies a function handle, such as any of those used with anonymous functions	Chapter 6
%	comment	Chapter 6
matlabFunction	converts a symbolic expression into a MATLAB® funciton	Chapter 13

Special Characters	Format Control	Chapter
'	begins and ends a string	Chapter 7
%	placeholder used in the **fprintf** command	Chapter 7
%f	fixed-point, or decimal, notation	Chapter 7
%d	decimal notation	Chapter 7
%e	exponential notation	Chapter 7
%g	either fixed-point or exponential notation	Chapter 7
%s	string notation	Chapter 7
%%	cell divider	Chapter 7
\n	linefeed	Chapter 7
\r	carriage return (similar to linefeed)	Chapter 7
\t	tab	Chapter 7
\b	backspace	Chapter 7

Functions	Input/Output (I/O) Control	Chapter
disp	displays a string or a matrix in the command window	Chapter 7
fprintf	creates formatted output which can be sent to the command window or to a file	Chapter 7
ginput	allows the user to pick values from a graph	Chapter 7
input	allow the user to enter values	Chapter 7
pause	pauses the program	Chapter 7
sprintf	similar to **fprintf** creates formatted output which is assigned to a variable name and stored as a character array	Chapter 7
uiimport	launches the Import Wizard	Chapter 7
wavread	reads wave files	Chapter 7
xlsimport	imports Excel data files	Chapter 7
xlswrite	exports data as an Excel file	Chapter 7
load	loads matrices from a file	Chapter 2
save	saves variables in a file	Chapter 2
celldisp	displays the contents of a cell array	Chapter 11
imfinfo	reads a standard graphics file and determines what type of data it contains	Chapter 14
imread	reads a graphics file	Chapter 14
mwrite	writes a graphics file	Chapter 14

Functions	Comparison Operators	Chapter
<	less than	Chapter 8
<=	less than or equal to	Chapter 8
>	greater than	Chapter 8
>=	greater than or equal to	Chapter 8
==	equal to	Chapter 8
~=	not equal to	Chapter 8

Special Characters	Logical Operators	Chapter
&	and	Chapter 8
|	or	Chapter 8
~	not	Chapter 8
xor	exclusive or	Chapter 8

Functions	Control Structures	Chapter
break	causes the execution of a loop to be terminated	Chapter 9
case	sorts responses	Chapter 8
continue	terminates the current pass through a loop, but proceeds to the next pass	Chapter 9
else	defines the path if the result of an **if** statement is false	Chapter 8
elseif	defines the path if the result of an **if** statement is false, and specifies a new logical test	Chapter 8
end	identifies the end of a control structure	Chapter 8
for	generates a loop structure	Chapter 9
if	checks a condition resulting in either true or false	Chapter 8
menu	creates a menu to use as an input vehicle	Chapter 8
otherwise	part of the case selection structure	Chapter 8
switch	part of the case selection structure	Chapter 8
while	generates a loop structure	Chapter 9

Functions	Logical Functions	Chapter
all	checks to see if a criterion is met by all the elements in an array	Chapter 8
any	checks to see if a criterion is met by any of the elements in an array	Chapter 8
find	determines which elements in a matrix meet the input criterion	Chapter 8
isprime	determines whether a value is prime	Chapter 3
isreal	determines whether a value is real or complex	Chapter 3

Functions	Timing	Chapter
clock	determines the current time on the CPU clock	Chapter 9
etime	finds elapsed time	Chapter 9
tic	starts a timing sequence	Chapter 9
toc	stops a timing sequence	Chapter 9
date	returns the date	Chapter 3

Functions	Special Matrices	Chapter
eye	generates an identity matrix	Chapter 10
magic	creates a "magic" matrix	Chapter 10
ones	creates a matrix containing all ones	Chapter 10
pascal	creates a Pascal matrix	Chapter 10
zeros	creates a matrix containing all zeros	Chapter 10
gallery	contains example matrices	Chapter 10

Special Characters	Data Types	Chapter
{ }	cell array constructor	Chapters 11 and 12
"	string data (character information)	Chapters 11 and 12
abc	character array	Chapter 11
⊞	numeric array	Chapter 11
▯	symbolic array	Chapter 11
✓	logical array	Chapter 11
�${\\\\}$	sparse array	Chapter 11
{}	cell array	Chapter 11
⊟	structure array	Chapter 11

Functions	Data Type Manipulation	Chapter
celldisp	displays the contents of a cell array	Chapter 11
char	creates a padded character array	Chapter 11
double	changes an array to a double-precision array	Chapter 11
int16	16-bit signed integer	Chapter 11
int32	32-bit signed integer	Chapter 11
int64	64-bit signed integer	Chapter 11
int8	8-bit signed integer	Chapter 11
num2str	converts a numeric array to a character array	Chapter 11
single	changes an array to a single-precision array	Chapter 11
sparse	converts a full-format matrix to a sparse-format matrix	Chapter 11
str2num	converts a character array to a numeric array	Chapter 11
uint16	16-bit unsigned integer	Chapter 11
uint32	32-bit unsigned integer	Chapter 11
uint64	64-bit unsigned integer	Chapter 11
uint8	8-bit unsigned integer	Chapter 11

Functions	Manipulation of Symbolic Expressions	Chapter
collect	collects like terms	Chapter 12
diff	finds the symbolic derivative of a symbolic expression	Chapter 12
dsolve	differential equation solver	Chapter 12
expand	expands an expression or equation	Chapter 12
factor	factors an expression or equation	Chapter 12
int	finds the symbolic integral of a symbolic expression	Chapter 12
matlabFunction	converts a symbolic expression into an anonymous MATLAB® function	Chapter 12
mupad	opens the MuPad workbook	Chapter 12
numden	extracts the numerator and denominator from an expression or an equation	Chapter 12
simple	tries and reports all the simplification functions, and selects the shortest answer	Chapter 12
simplify	simplifies using Mupad's built-in simplification rules	Chapter 12
solve	solves a symbolic expression or equation	Chapter 12
subs	substitutes into a symbolic expression or equation	Chapter 12
sym	creates a symbolic variable, expression, or equation	Chapter 12
syms	creates symbolic variables	Chapter 12

Functions	Symbolic Plotting	Chapter
ezcontour	creates a contour plot	Chapter 12
ezcontourf	creates a filled contour plot	Chapter 12
ezmesh	creates a mesh plot from a symbolic expression	Chapter 12
ezmeshc	plots both a mesh and contour plot created from a symbolic expression	Chapter 12
ezplot	creates an x–y plot of a symbolic expression	Chapter 12
ezplot3	creates a three-dimensional line plot	Chapter 12
ezpolar	creates a plot in polar coordinates	Chapter 12
ezsurf	creates a surface plot from a symbolic expression	Chapter 12
ezsurfc	plots both a mesh and contour plot created from a symbolic expression	Chapter 12

Functions	Numerical Techniques	Chapter
bvp4c	boundary value problem solver for ordinary differential equations	Chapter 13
cftool	opens the curve-fitting graphical user interface	Chapter 13
diff	computes the differences between adjacent values in an array if the input is an array; finds the symbolic derivative if the input is a symbolic expression	Chapter 13
fminbnd	a function that accepts a function handle or function definition as input and numerically finds the function minimum between two bounds – known as a "function-function"	Chapter 6
fzero	a function that accepts a function handle or function definition as input and finds the zero point nearest a specified value – known as a "function-function"	Chapter 6
gradient	finds the derivative numerically using a combination of forward, backward, and central difference techniques	Chapter 13
interp1	Approximates intermediate data, using either the default linear interpolation technique or a specified higher order approach	Chapter 13
interp2	two-dimensional interpolation function	Chapter 13
interp3	three-dimensional interpolation function	Chapter 13
interpn	multidimensional interpolation function	Chapter 13
ode45	ordinary differential equation solver	Chapter 13
ode23	ordinary differential equation solver	Chapter 13
ode113	ordinary differential equation solver	Chapter 13
ode15s	ordinary differential equation solver	Chapter 13
ode23s	ordinary differential equation solver	Chapter 13
ode23t	ordinary differential equation solver	Chapter 13
ode23tb	ordinary differential equation solver	Chapter 13
ode15i	ordinary differential equation solver	Chapter 13
polyfit	computes the coefficients of a least-squares polynomial	Chapter 13
polyval	evaluates a polynomial at a specified value of x	Chapter 13
quad	computes the integral under a curve (Simpson)	Chapter 13
quad1	computes the integral under a curve (Lobatto)	Chapter 13

Functions	Sample Data Sets and Images	Chapter
cape	sample MATLAB® image file of a cape	Chapter 14
clown	sample MATLAB® image file of a clown	Chapter 14
detail	sample MATLAB® image file of a section of a Dürer wood carving	Chapter 14
durer	sample MATLAB® image file of a Dürer wood carving	Chapter 14
earth	sample MATLAB® image file of the earth	Chapter 14
flujet	sample MATLAB® image file showing fluid behavior	Chapter 14
gatlin	sample MATLAB® image file of a photograph	Chapter 14
mandrill	sample MATLAB® image file of a mandrill	Chapter 14
mri	sample MRI data set	Chapter 14
peaks	creates a sample plot	Chapter 14
seamount	sample MATLAB® data file of a seamount	Chapter 5
spine	sample MATLAB® image file of a spine X-ray	Chapter 14
wind	sample MATLAB® data file of wind velocity information	Chapter 14
sphere	sample function used to demonstrate graphing	Chapter 5
census	a built-in data set used to demonstrate numerical techniques	Chapter 13
handel	a built-in data set used to demonstrate the sound function	Chapter 3

Functions	Advanced Visualization	Chapter
alpha	sets the transparency of the current plot object	Chapter 14
camlight	turns the camera light on	Chapter 14
coneplot	creates a plot with markers indicating the direction of input vectors	Chapter 14
contourslice	creates a contour plot from a slice of data	Chapter 14
drawnow	forces MATLAB® to draw a plot immediately	Chapter 14
gca	gets current axis handle	Chapter 14
gcf	gets current figure handle	Chapter 14
get	returns the properties of a specified object	Chapter 14
getframe	gets the current figure and saves it as a movie frame in a structure array	Chapter 14
image	creates a two-dimensional image	Chapter 14
imagesc	creates a two-dimensional image by scaling the data	Chapter 14
imfinfo	reads a standard graphics file and determines what type of data it contains	Chapter 14
imread	reads a graphics file	Chapter 14
imwrite	writes a graphics file	Chapter 14
isosurface	creates surface connecting volume data of the same magnitude	Chapter 14
movie	plays a movie stored as a MATLAB® structure array	Chapter 14
set	establishes the properties assigned to a specified object	Chapter 14
shading	determines the shading technique used in surface plots and pseudo color plots	Chapter 14

B Scaling Techniques

Plotting data using different scaling techniques is a useful way to try to determine how y-values change with x. This approach is illustrated in the following sections.

LINEAR RELATIONSHIPS

If x and y are related by a linear relationship, a standard x–y plot will be a straight line. Thus, for

$$y = ax + b$$

an x–y plot is a straight line with slope a and y-intercept b.

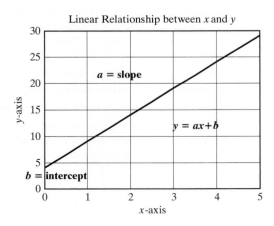

LOGARITHMIC RELATIONSHIP

If x and y are related logarithmically

$$y = a \log_{10}(x) + b$$

a standard plot on an evenly spaced grid is curved. However, a plot scaled evenly on the y-axis but logarithmically on the x-axis is a straight line of slope a. The y-intercept doesn't exist, since $\log_{10}(0)$ is undefined. However when $x = 1$, the value of $\log_{10}(1)$ is zero and y is equal to b.

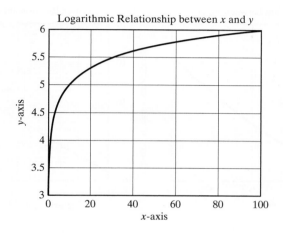

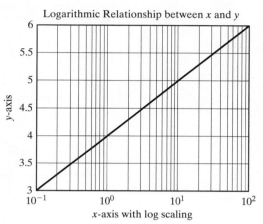

EXPONENTIAL RELATIONSHIP

When x and y are related by an exponential relationship such as

$$y = b * a^x$$

a plot of $\log_{10}(y)$ versus x gives a straight line because

$$\log_{10}(y) = \log_{10}(a) * x + \log_{10}(b)$$

In this case, the slope of the plot is $\log_{10}(a)$.

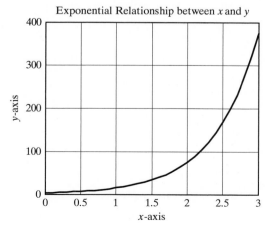

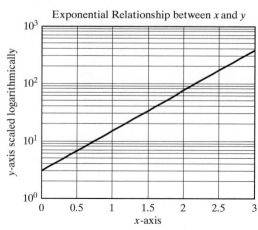

POWER RELATIONSHIP

Finally, if x and y are related by a power relationship such as

$$y = bx^a$$

a plot scaled logarithmically on both axes produces a straight line with a slope of a. When x is equal to 1, the $\log_{10}(1)$ is zero, and the value of $\log_{10}(y)$ is $\log_{10}(b)$.

$$\log_{10}(y) = a * \log_{10}(x) + \log_{10}(b)$$

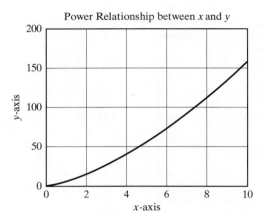

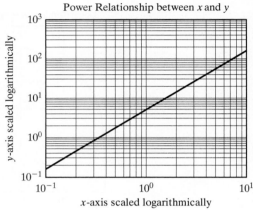

C

The Ready_Aim_ Fire GUI

```
function varargout = ready_aim_fire(varargin)
% READY_AIM_FIRE M-file for ready_aim_fire.fig
% READY_AIM_FIRE, by itself, creates a new READY_AIM_FIRE or raises the existing
% singleton*.
%
% H = READY_AIM_FIRE returns the handle to a new READY_AIM_FIRE or the handle to
% the existing singleton*.
%
% READY_AIM_FIRE('CALLBACK',hObject,eventData,handles,...) calls the local
% function named CALLBACK in READY_AIM_FIRE.M with the given input arguments.
%
% READY_AIM_FIRE('Property','Value',...) creates a new READY_AIM_FIRE or raises the
% existing singleton*. Starting from the left, property value pairs are
% applied to the GUI before ready_aim_fire_OpeningFcn gets called. An
% unrecognized property name or invalid value makes property application
% stop. All inputs are passed to ready_aim_fire_OpeningFcn via varargin.
%
% *See GUI Options on GUIDE's Tools menu. Choose "GUI allows only one
% instance to run (singleton)".
%
```

```
% See also: GUIDE, GUIDATA, GUIHANDLES

% Edit the above text to modify the response to help ready_aim_fire

% Last Modified by GUIDE v2.5 29-Aug-2010 17:17:24

% Begin initialization code - DO NOT EDIT
gui_Singleton = 1;
gui_State = struct('gui_Name', mfilename, ...
                'gui_Singleton', gui_Singleton, ...
                'gui_OpeningFcn', @ready_aim_fire_OpeningFcn, ...
                'gui_OutputFcn', @ready_aim_fire_OutputFcn, ...
                'gui_LayoutFcn', [] , ...
                'gui_Callback', []);
if nargin && ischar(varargin{1})
   gui_State.gui_Callback = str2func(varargin{1});
end


if nargout
    [varargout{1:nargout}] = gui_mainfcn(gui_State, varargin{:});
else
   gui_mainfcn(gui_State, varargin{:});
end
%  End initialization code - DO NOT EDIT


% --- Executes just before ready_aim_fire is made visible.
function ready_aim_fire_OpeningFcn(hObject, eventdata, handles, varargin)
% This function has no output args, see OutputFcn.
% hObject handle to figure
% eventdata reserved - to be defined in a future version of MATLAB®
% handles  structure with handles and user data (see GUIDATA)
% varargin command line arguments to ready_aim_fire (see VARARGIN)
plot(275,0,'s','Markersize',10,'MarkerFaceColor','r')
text(275,50,'target')
axis([0,1000,0,500])
title('Projectile Trajectory')
xlabel('Horizontal Distance, m')
ylabel('Vertical Distance, m')
hold on
handles.location=275;
% Choose default command line output for ready_aim_fire
handles.output = hObject;

% Update handles structure
guidata(hObject, handles);

% UIWAIT makes ready_aim_fire wait for user response (see UIRESUME)
% uiwait(handles.figure1);
```

```
% --- Outputs from this function are returned to the command line.
function varargout = ready_aim_fire_OutputFcn(hObject, eventdata, handles)
% varargout cell array for returning output args (see VARARGOUT);
% hObject handle to figure
% eventdata reserved - to be defined in a future version of MATLAB®
% handles structure with handles and user data (see GUIDATA)

% Get default command line output from handles structure
varargout{1} = handles.output;


% --- Executes on button press in Fire_pushbutton.
function Fire_pushbutton_Callback(hObject, eventdata, handles)
% hObject handle to Fire_pushbutton (see GCBO)
% eventdata reserved - to be defined in a future version of MATLAB®
% handles  structure with handles and user data (see GUIDATA)
time=0:0.001:100;
h=time*handles.vel*cosd(handles.theta);
v=time*handles.vel*sind(handles.theta)-1/2*9.81*time.^2;
pos=find(v>=0);
horizontal=h(pos);
vertical=v(pos);
comet(horizontal,vertical);
land=pos(end);
goal=handles.location;
if (h(land)<goal+50 && h(land)>goal-50) % Code to create the "Explosion"
  x=linspace(goal-100, goal+100, 5);
  y=[0,80,100,80,0];
  z=linspace(goal-200,goal+200,9);
  w=[0,40,90,120,130,120,90,40,0];
  plot(x,y,'*r',z,w,'*r')
  text(goal,400,'Kaboom!')
  set(handles.textout,'string', 'You Win !','fontsize',16)
end




function launch_angle_Callback(hObject, eventdata, handles)
% hObject handle to launch_angle (see GCBO)
% eventdata reserved - to be defined in a future version of MATLAB®
% handles structure with handles and user data (see GUIDATA)

% Hints: get(hObject,'String') returns contents of launch_angle as text
%   str2double(get(hObject,'String')) returns contents of launch_angle as a double
handles.theta=str2double(get(hObject,'String'));
guidata(hObject, handles);

% --- Executes during object creation, after setting all properties.
function launch_angle_CreateFcn(hObject, eventdata, handles)
% hObject handle to launch_angle (see GCBO)
```

```
% eventdata reserved - to be defined in a future version of MATLAB®
% handles empty - handles not created until after all CreateFcns called

% Hint: edit controls usually have a white background on Windows.
% See ISPC and COMPUTER.
if ispc && isequal(get(hObject,'BackgroundColor'), get(0,'defau ltUicontrolBackgroundColor'))
  set(hObject,'BackgroundColor','white');
end



function launch_velocity_Callback(hObject, eventdata, handles)
% hObject handle to launch_velocity (see GCBO)
% eventdata reserved - to be defined in a future version of MATLAB®
% handles structure with handles and user data (see GUIDATA)

% Hints: get(hObject,'String') returns contents of launch_velocity as text
% str2double(get(hObject,'String')) returns contents of launch_velocity as a double
handles.vel=str2double(get(hObject,'String'));
guidata(hObject, handles);

% --- Executes during object creation, after setting all properties.
function launch_velocity_CreateFcn(hObject, eventdata, handles)
% hObject  handle to launch_velocity (see GCBO)
% eventdata reserved - to be defined in a future version of MATLAB®
% handles  empty - handles not created until after all CreateFcns called

% Hint: edit controls usually have a white background on Windows.
% See ISPC and COMPUTER.
if ispc && isequal(get(hObject,'BackgroundColor'), get(0,'defaultUicontrolBackgroundColor'))
  set(hObject,'BackgroundColor','white');
end

% --- Executes on button press in Reset_pushbutton.
function Reset_pushbutton_Callback(hObject, eventdata, handles)
% hObject  handle to Reset_pushbutton (see GCBO)
% eventdata reserved - to be defined in a future version of MATLAB®
% handles  structure with handles and user data (see GUIDATA)
hold off
plot(handles.location,0,'s','Markersize',10,'MarkerFaceColor','r')
text(handles.location,50,'target')
axis([0,1000,0,500])
title('Projectile Trajectory')
xlabel('Horizontal Distance, m')
ylabel('Vertical Distance, m')
hold on
set(handles.textout,'string', ")

% --- Executes on slider movement.
function slider1_Callback(hObject, eventdata, handles)
```

```
% hObject  handle to slider1 (see GCBO)
% eventdata reserved - to be defined in a future version of MATLAB®
% handles structure with handles and user data (see GUIDATA)

% Hints: get(hObject,'Value') returns position of slider
% get(hObject,'Min') and get(hObject,'Max') to determine range of slider
handles.location = get(hObject,'Value')
hold off
plot(handles.location,0,'s','Markersize',10,'Markerfacecolor','r')
axis([0,1000,0,1000])
title('Trajectory')
xlabel('Horizontal Distance')
ylabel('Vertical Distance')
text(handles.location-25,50,'Target')
hold on
guidata(hObject, handles);

% --- Executes during object creation, after setting all properties.
function slider1_CreateFcn(hObject, eventdata, handles)
% hObject  handle to slider1 (see GCBO)
% eventdata reserved - to be defined in a future version of MATLAB®
% handles  empty - handles not created until after all CreateFcns called

% Hint: slider controls usually have a light gray background.
if isequal(get(hObject,'BackgroundColor'), get(0,'defaultUicontrolBackgroundColor'))
  set(hObject,'BackgroundColor',[.9 .9 .9]);
end
```

Index